Klaus Buchholz and
John Collins

Concepts in Biotechnology

Related Titles

Nicolaou, K. C., Montagnon, T.

Molecules that Changed the World

385 pages

2008

Hardcover

ISBN: 978-3-527-30983-2

Sneader, W.

Drug Discovery

A History

472 pages

Softcover

ISBN: 978-0-471-89980-8

Ho, R. J. Y., Gibaldi, M.

Biotechnology and Biopharmaceuticals

Transforming Proteins and Genes into Drugs

576 pages

2003

Softcover

ISBN: 978-0-471-20690-3

Greenberg, A.

The Art of Chemistry

Myths, Medicines, and Materials

384 pages

2003

Hardcover

ISBN: 978-0-471-07180-8

Sapienza, A. M., Stork, D.

Leading Biotechnology Alliances

Right from the Start

216 pages

2001

Hardcover

ISBN: 978-0-471-18248-1

Klaus Buchholz and John Collins

Concepts in Biotechnology

History, Science and Business

WILEY-VCH

WILEY-VCH Verlag GmbH & Co. KGaA

The Authors

Prof. em. Dr. Klaus Buchholz
Technical University Braunschschweig
Institute for Chemical Engineering
Hans-Sommer-Str. 10
38106 Braunschweig
Germany

Prof. em. Dr. John Collins
Technical University Braunschschweig
Life Sciences Faculty
c/o Helmholtz Centre for Infection Research – HZI
AG Directed Evolution
Inhoffenstr. 7
38124 Braunschweig
Germany

All books published by **Wiley-VCH** are carefully produced. Nevertheless, authors, editors, and publisher do not warrant the information contained in these books, including this book, to be free of errors. Readers are advised to keep in mind that statements, data, illustrations, procedural details or other items may inadvertently be inaccurate.

Library of Congress Card No.: applied for

British Library Cataloguing-in-Publication Data
A catalogue record for this book is available from the British Library.

Bibliographic information published by the Deutsche Nationalbibliothek
The Deutsche Nationalbibliothek lists this publication in the Deutsche Nationalbibliografie; detailed bibliographic data are available on the Internet at http://dnb.d-nb.de.

© 2010 WILEY-VCH Verlag & Co. KGaA, Boschstr. 12, 69469 Weinheim, Germany

Cover Illustration:
Production fermenters with kind permission by Roche Penzberg, Germany
Cover Adam Design, Weinheim
Typesetting Thomson Digital, Noida, India
Printing and Binding betz-druck GmbH, Darmstadt

Printed in the Federal Republic of Germany
Printed on acid-free paper

ISBN: 978-3-527-31766-0

For
Diana and Marie-Christiane

Contents

Concepts in Biotechnology: History, Science and Business. Klaus Buchholz and John Collins
Copyright © 2010 WILEY-VCH Verlag GmbH & Co. KGaA, Weinheim
ISBN: 978-3-527-31766-0

Preface

Over the last century the development of *Biotechnology* (BT) has followed fascinating pathways to influence ever more aspects of our lives and to provide significant contributions to the improvement of the quality of life. BT flourished in parallel with biological sciences as a result of insights into the molecular details of genetics and the control of biochemical reactions. Following a long-standing tradition, this knowledge was translated by commercial application for human benefit. It enabled biological pathways to be manipulated and even created for the purpose of manufacturing products and developing processes and services on an industrial scale. Historically, controlled fermentation was used to provide efficient storage for food thus enabling a population to survive periods of cold or drought. By the end of the last century biotechnology had developed into a science and engineering discipline in its own right and is considered to be a field of industrial activity with major economic relevance. The applications of BT extend beyond historical tradition, ranging from production of chemicals, bio-fuels and pharmaceuticals to ensuring a continued supply of clean water.

This book reviews the progress of biotechnology over time and highlights the seminal events in this field. It gives an introduction to the main developments, the principles or concepts, and key researchers involved in pioneering work and in conclusion, attempts to extrapolate to further advances expected in the near future. In view of the extensive range of biotechnological activities it was necessary to concentrate on essentials, illustrated with selected examples, as opposed to using an encyclopedic approach. This book is intended to guide the reader through the diverse fields of activity in BT and encourage further reading in the form of books, specialised reviews and original literature as provided in the reference sections. It is envisaged that the readership of this book will include students of biology, biotechnology and biochemical engineering, in addition to scientists and engineers already engaged in or proposing to work in the fields of BT and related disciplines. It may also serve as a broad introduction to BT for other readers who are interested in an overview of the subject, ranging from historical aspects to the latest developments which are largely a result of the accelerated research in molecular biology and bioinformatics that has taken place over the last 20 years.

Concepts in Biotechnology: History, Science and Business. Klaus Buchholz and John Collins
Copyright © 2010 WILEY-VCH Verlag GmbH & Co. KGaA, Weinheim
ISBN: 978-3-527-31766-0

The *historical aspects* of BT are discussed in the opening chapters which highlight the role of inquisitiveness and the thirst for knowledge and understanding of natural processes. This involves a discussion of reputation-building, the interplay of economics and business as well as the role of and dependence on theories. We trace the developments in chemistry and physics that became a prerequisite for the study of the chemical nature of the components involved in biological processes such as brewing, wine and bread making. Heated discussions centring on both the vitalist and chemical theories resulted not only in the emergence of theories and paradigms but also in their reversal. The close interaction of scientists, craftsmen and industry together with significant stimulus, promoted continued research.

Pasteur and Koch established the science of microbiology. A few decades later Buchner finally refuted the last metaphysical hypothesis that processes in living cells required a 'vis vitalis', a vital factor and following this biochemistry emerged as a new speciality. Biotechnological engineering was based on more precise control of the microbial fermentations involved in food processing including large-scale processes for the manufacture of beer, wine, cheese, bread etc. together with the use of sterile starting materials. This led to the subsequent production of fuels and chemical components for polymers and explosives particularly during war time, and the manufacture of antibiotics and vaccines. This in turn stimulated detailed studies on the manufacture of products from microbial fermentations. By the mid-twentieth century, biotechnology had become an accepted speciality.

Basic research in *biochemistry, molecular biology* and *genetics* dramatically broadened the field of life sciences and at the same time unified them by the study of genes and their relatedness throughout the evolutionary process. In Part 2 we discuss the development of this fruitful interplay and describe how it broadened the scope of accessible products and services, at the same time making production cheaper, safer, more reproducible and more reliable. Rapid acceleration of gene and protein analysis caused an explosion of data which led to the emergence of *bioinformatics*. This opened up new avenues for medical analysis that was orientated more towards preventive measures rather than corrective intervention. This is a continuing trend which is substantiated by the prediction that during the next few years affordable analysis of the complete genetic potential of an individual will be available within hours. New areas of research have evolved such as *systems biology* in which living systems can be successfully modelled as networks of ever-increasing complexity. As the volume of information increases and modelling improves so does the probability that insight into potential targets for pharmaceuticals can be better translated into developing successful medicines. To foster such aims, centres for translational medicine are being founded in many cities where medical schools and hospitals participate in close interactions with basic research institutes.

The understanding of the fundamental *programming of animal cells* in the developing embryo and in particular the discovery of a small number of proteins capable of guiding stem cell differentiation and even the reprogramming of already differentiated cells, has opened up perspectives for a completely new and very exciting branch of biotechnology in the area of tissue and organ synthesis for *regenerative medicine*. In combination with advances in fertility medicine this has also led to the cloning of animals and the production of *transgenic animals*. One aspect of this technology is the use of tissue cloning to produce human tissue cultures as models for inherited disease.

In Part 3 we discuss engineering and applied topics. *Biochemical and bioprocess engineering* constitute the basis for translating scientific innovation and development into industrial processes. They represent an interdisciplinary field based on molecular biology, biochemistry and engineering disciplines. As a result of the progress in molecular biology, new tools known as the 'omics' were developed: genomics, proteomics and metabolomics, to mention only the most common. *Biosystems engineering* or systems biotechnology, integrates the approaches and the extensive volume of data derived from these specialities and from bioreaction engineering in a 'holistic' approach, using bioinformatics tools.

Industrial biotechnology, with its historical roots, continues in diverse industrial fields of activity including food and feed and commodities such as enzymes for use in detergents, bio-fuel and energy production, polymer manufacture and the development and production of many drug constituents, as well as providing services, for example in waste treatment and other processes related to environmental protection.

The approval in 1982 of recombinant human insulin produced in *E. coli* and developed by Genentech in cooperation with Eli Lilly in the late 1970s, was an historical landmark. By 2006, some 165 biopharmaceuticals had been approved in the EU and/or the USA for human use. This illustrates the emergence and rise of recombinant technologies which constitute the basis of *pharmaceutical biotechnology*. Today, approximately one in four of all genuinely new drugs currently entering the market is a biopharmaceutical and in 2008 over 400 biopharmaceuticals were in various stages of clinical evaluation. These include hormones, soluble hormone receptors (as hormone antagonists), blood factors, thrombolytics, interferons, monoclonal antibodies, vaccines and therapeutic enzymes. Selected aspects of engineering and production processes together with information relating to their use are discussed in this chapter. Data on industrial development, products, companies and economics are also presented.

The potential of transgenic *plant biotechnology* is to create crops that produce higher yields and are able to grow on less fertile land in order to feed the growing world population. Crops should be resistant to pests and require less chemical treatment, notably with insecticides, fungicides, herbicides and fertilizers, and exhibit low environmental impact. The majority of agricultural scientists are convinced that such crops can be delivered by the exploitation of molecular breeding strategies. Food production has risen considerably over the decades in terms of a 'Green Revolution', most notably in developing countries, but the increase in per capita food supply has been small. Hence research in recombinant food production is considered to be a necessary part of the strategies to ensure adequate nutrition. Nevertheless, debates over the risks of the technology have evoked conflicts and created a critical, even negative publicity, particularly in Western Europe.

BT offers in general a sustainable method of production, based mostly on renewable resources with minimal or no waste and by-products that can be recycled or reused, for example as feed components. There are manifold interactions with political, social, economic and environmental issues. Laws, regulations and ethical concerns pertinent to biotechnology are important topics of discussion although there are dramatic differences in legislation between countries. Current efforts are centred around establishing common global regulations including the removal of unfair unilateral advantages and support for health care and economies in developing countries. The regulatory influences which affect how science is carried out and technology is applied are addressed in each chapter. In addition to the underlying scientific concepts, further information is presented in each chapter on the use of products, along with data on industrial activities and production.

The increase in computing power due to the invention and continued development of microchips via nanotechnology has pioneered and driven a revolution in communication during the last three decades. At least one computer, television and mobile telephone have found their place in essentially every home. Biotechnology has also undergone a corresponding development, although perhaps not so immediately identifiable at the level of consumer goods in the shops. There is, however, hardly an area of human activity which has not been affected by the recent biotechnological revolution. We hope that after completing our book, the readers will feel that they have a better understanding of how and why this revolution took place, its roots and its further potential to improve so many aspects of our lives.

Acknowledgements

We are most grateful for comments on the manuscript from a number of friends and colleagues as well others who agreed to have their photographs taken for inclusion in the book. Their comments contributed to the readability of the text, led to the avoidance of certain errors and extended the knowledge base. We, the authors take full responsibility for any remaining mistakes.

In particular thanks are due to Anthony (Tony) C. R. Samson, Karl Simpson, Raimo Franke, Heidi Lloyd-Price, and Erik Pollmann, Ulrich Behrendt, Sonja Berensmeier and Volker Kasche for significant information and relevant advice.

Further valuable information and assistance was contributed by Robert Bud, Arnold Demain, Albert J. Driesel, Reinhard Hehl, Dietmar Hempel, Gerhard Höfle, Hans-Joachim Jördening, Peter Rapp, Jürgen Seibel and Hermann Stegemann. We would also like to thank Frank Weinreich for his wise suggestions during the final stages of converting our manuscript into a book. JC is grateful to the Helmholtz Centre for Infection Research, Braunschweig (HZI; formerly the GBF) for funding the transport costs to international meetings.

This book is dedicated to our wives Diana Buchholz and Marie-Christiane Collins without whose support we could not have completed this project.

Abbreviations and Glossary

Acre	$4046 \, m^2$
ADM	Archer Daniels Midland (starch producing and converting company, USA)
ADP	adenosine diphosphate
7-ACA	7-Aminocephalosporanic acid
7-ADCA	7-Aminodesoxycephalosporanic acid
6-APA	6-Aminopenicillanic acid
AIChE	American Institute of Chemical Engineers
AMP	adenosine monophosphate
Array CGH	Array *Comparative Genome Hybridization*, for example for comparing (malignant) biopsy material with DNA from normal tissue
ATP	adenosine triphosphate
BAC libraries (BACs)	bacterial artificial chromosome libraries
BHK	baby hamster kidney (cells)
BMP	bone morphogenetic protein
BMS	Bristol Meyers Squibb (USA)
bn	billion
BOD	Biological oxygen demand (of waster water)
BP	Before present
BPTI	Bovine pancreatic trypsin inhibitor
BT	Biotechnology
Bt	*Bacillus thuringiensis*
C&EN	Chem. Eng. News
CCD	computational cell dynamics
cDNA	copy DNA, reverse transcribed from mRNA
CDR	complementarity-determining region of an antibody
CEPH	Centre d'études des polymorphisms humains, Paris, France (The Centre for the Study of Human Polymorphisms)
CFD	computational fluid dynamics
CFTR	Cystic fibrosis transmembrane conductance regulator
cGMP	current Good Manufacturing Practice

Concepts in Biotechnology: History, Science and Business. Klaus Buchholz and John Collins
Copyright © 2010 WILEY-VCH Verlag GmbH & Co. KGaA, Weinheim
ISBN: 978-3-527-31766-0

CHO	Chinese hamster ovary (cells)
CIP	clean in place
CMV	Cytomegalie virus
CNV	copy number variation
CP	capsid or coat protein (of virus)
CSF	Colony stimulating factor
Cultivars	cultivated plant varieties
2D	two dimensional
DARPins	Designed Ankyrin Repeat Proteins
2DE	two dimensional electrophoresis
2DE IEF/SDS-PAGE	two dimensional electrophoresis combined with IEF and SDS-PAGE
DGT	direct gene transfer (including particle bombardmet)
DHA	docosahexanoic acid
dm	dry matter
2D-PAGE	two-dimensional gel electrophoresis
2DE IEF/SDS-PAGE	two dimensional electrophoresis method
DH	dehydrogenase
DOE US	Department of Energy, United States of America
DPN^+	diphosphonucleotide (is identical with NAD^+)
DPNH	hydrogenated diphosphonucleotide (is identical with NADH)
dt/ha	decitonnes (0,1 t) per hectare
€	EURO, 1.40 $ (Oct. 2010, mean)
EBIT	earnings before interest and taxes
E. coli	*Escherichia coli*
EF	environmental factor
EI	environmental index
ELISA	enzyme linked immunosorbent assay
EMEA	European authority for approval of pharmaceuticals
EP	epothilone
EPA	Environmental Protection Agency (USA)
EPA	eicosapentanoic acid
Epitope	specific region on a protein recognized by an antibody
EPC	European patent convention (5 October 1973)
EPO	European patent office or Erythropoietin
ER	endoplasmatic reticulum
ESC or ES	embryonic stem cells
ESI-MS	electrospray-ionisation mass spectrometry
ESI-TOF MS/MS	electrospray-time of flight-mass spectrometry
EST	expressed sequence tags; short DNA fragments obtained by random sequencing of clones from cDNA libraries
EU	European Union
FAO	Food and Agriculture Organization (USA)
FBA	flux balance analysis

FDA	Food and Drug Administration (USA)
FDP	fructose-1,6-diphosphate
Fluxome	flux distribution of the central metabolic pathways
Ft.	feet (30.5 cm)
Gal	gallon (3,78 L)
GC-MS	coupled gas chromatography-mass spectrometry
GM	genetically modified;
GMO	genetically modified organism
GRAS	generally recognized as safe
GMP	Good Manufacturing Practice
GPCRs	G protein coupled receptors
GSK	GlaxoSmithKline
ha	hectar, 10 000 m2
hGH	human growth hormone
HIV	Human immunodeficiency virus
hl	hectoliter (100 l)
HR	hypersensitive response
HTS	high throughput screening
IEF	isoelectric focusing
IFN	interferon
IgG	immune globulin G
IL	interleukin
In.	inch (2.54 cm)
i.v.	intravenous
JACS	Journal Am. Chem. Soc.
J&J	Johnson & Johnson
LC-MS	liquid chromatography-mass spectrometry
LD (LOD score)	linkage disequilibrium in population genetics
LRR	leucine-rich-repeat proteins, for example ankyrin
mAB	monoclonal antibody
MALDI-TOF-MS	Matrix-Assisted-Laser-Desorption/Ionization – Time-Of-Flight-Mass-Spectrometry
MDR	multi drug resistant
MFA	metabolic flux analysis
MI	mass Index
miRNAs	micro RNAs
mn	million
Mtoe	million tons oil equivalents
Mw	molecular weight, molar mass
m-Arrays	micro-arrays
NAD	Nicotinamide-adenine-dinucleotide
NADH	hydrogenated NAD
NBF	new BT firm
NCE	new chemical entity
NGOs	non governmental organizations

NIH	National Institutes of Health (USA)
NK cells	natural killer cells
NMR	nuclear magnetic resonance
NRRL	Northern Regional Research Laboratory (USA)
NSO	mouse myeloma derived mammalian cells
ON	oligonucleotides
ORF	open reading frame
OS	oligosaccharides
OTA	Office of Technology Assessment (USA)
PAGE	polyacrylamide gel electrophoresis
PAT	process analytical technology
PDO	1,3-propanediol
PDR	pathogen-derived resistance
PEG	polyethylene glycol
PEGylation	attachment of polyethylene glycol
PET	positron emission tomography
PHB	polyhydroxybutyrate (a polyester)
pI	Ionic strength (logarithmic scale)
Plastids	Intracellular organelles, e.g., chloroplasts that have their own double stranded DNA
pO_2	oxygen partial pressure
Pound	453 g
PR	pathogenesis related
PR	plant disease resistance
PS	iPS and piPS Pluripotent stem cells, induced pluripotent stem cells, protein-induced pluripotent stem cells
PSTI	Human pancreatic secretory trypsin inhibitor
QTL	quantitative trait locus
QM	quality management
R	resistance (genes)
rasiRNAs	repeat-associated small interfering RNA.s
R&D	research and development
rDNA	recombinant DNA
rDNA technologies	recombinant DNA technologies
rh	recombinant human
RNAi	interfering RNA, RNA interference
rPC	real time PCR
rRNA	ribosomal RNA
$	US $, corresponding to 0,71 € (Oct. 2010, mean)
SAGE	serial analysis of gene expression
SDA	stearidonic acid
SDS-PAGE	sodium dodecyl sulfate polyacrylamide gel electrophoresis
SEC	size exclusion chromatography
SIP	sterilization in place
siRNA	small interfering RNA

shRNA	short hairpin RNA
SNP	single nucleotide polymorphism
STR	stirred tank reactor
SUB	single use bioreactor
t/a	tonnes per year
TM	trade mark
TNF	Tumor necrosis factor
tPA	Tissue plaminogen activator
Translation capacity	the number of times a transcript is translated.
USDA	US Department of Agriculture
US$	US dollar (see $)
YAC libraries.	yeast artificial chromosome. libraries

Part One
History

Concepts in Biotechnology: History, Science and Business. Klaus Buchholz and John Collins
Copyright © 2010 WILEY-VCH Verlag GmbH & Co. KGaA, Weinheim
ISBN: 978-3-527-31766-0

1
Introduction

Historical events in early biotechnology comprise fascinating discoveries, such as yeast and bacteria as living matter being responsible for the fermentation of beer and wine. The art of Biotechnology emerged from agriculture and animal husbandry in ancient times through the empirical use of plants and animals which could be used as food or dyes, particularly where they had been preserved by natural processes and fermentations. Improvements were mostly handed on by word of mouth, and groups that maintained these improvements had a better chance of survival during periods of famine and drought. In such processes alcohol was produced and provided highly acceptable drinks such as beer and wine, or acids were formed which acted as the preservative agent in the storable food produced.

In this book we follow how the study of the chemical nature of the components involved in biotechnology first became possible subsequent to the development of chemistry and physics. Serious controversies about the theories both vitalist and chemical, resulted in the reversal of theories and paradigms; significant interaction with and stimulus from the arts and industries prompted the continuing research and progress. Last but not least, it was accepted that the products produced by living organisms should not be treated differently from inorganic materials. Pasteur's work led to the abandonment of the idea which had been an anathema to exact scientific enquiry in the life sciences, namely 'spontaneous generation'. He established the science of microbiology by developing pure monoculture in sterile medium, and together with the work of Robert Koch the experimental criteria required to show that a pathogenic organism is the causative agent for a disease were also recognised. Several decades later Buchner disproved the hypothesis that processes in living cells required a metaphysical 'vis vitalis' in addition to what was necessary to understand general chemistry. Enzymes were shown to be the chemical basis of bioconversions. Biochemistry emerged as a new speciality prompting dynamic research in enzymatic and metabolic reactions. However, the structure of proteins was not established until more than 40 years later.

The requirements for antibiotics and vaccines to combat disease, and chemical components required for explosives particularly in war time, stimulated exact studies in producing products from microbial fermentations. By the mid-twentieth century,

Concepts in Biotechnology: History, Science and Business. Klaus Buchholz and John Collins
Copyright © 2010 WILEY-VCH Verlag GmbH & Co. KGaA, Weinheim
ISBN: 978-3-527-31766-0

Biotechnology was becoming an accepted speciality with courses being established in the life sciences departments of several Universities.

Basic research in Biochemistry and Molecular Biology dramatically widened the field of life sciences and at the same time unified them considerably by the study of genes and their relatedness throughout the evolutionary process. The scope of accessible products and services expanded significantly. Economic input accelerated research and development, by encouraging and financing the development of new methods, tools, machines, and robots. The discipline of 'New Biotechnology', one of the lead sciences which resulted from an intimate association between business and science, is still the subject of critical public appraisal in many Western countries due to a particular lack of confidence in the notion that improved quality of life will prevail in spite of commercial interest.

The study of the history of science, and specifically that of Biotechnology, should go beyond documenting and recording events and should ideally contribute to an understanding of the motives and mechanisms governing the dynamics of sciences; the role of inquisitiveness in gaining new knowledge and insights into nature, of reputation-building, of the interplay between economics and business; the role of and dependance on theories, the role of analytical and experimental, as well mathematical methods, and more recently computation and robotics.

In the first part of this book the emphasis is on scientific discoveries and results, theories, technical development, the creation of leading paradigms, and on their decay. Industrial problems and political issues, and their influence on and correlation with developments in science are also addressed.

Although biotechnology has historical roots, it continues to influence diverse industrial fields of activity, including food and feed and commodities, for example polymer manufacture, providing services such as environmental protection, biofuel and energy including biofuel cells, and the development and production of many of the most effective drugs.

2
The Early Period to 1850

2.1
Introduction

Fermentation has been of great practical and economic relevance as an art and handicraft for thousands of years, yet, in the absence of analytical tools, there was initially no understanding of the changes in constituents that took place. There were serious controversies regarding the vitalistic or chemical nature of these processes. The vitalists believed that mysterious events and forces would be involved such as spontaneous generation of life and a specific vital force, while the chemical school, with Liebig at its head, was in contrast, convinced that only chemical decay processes took place, denying that any living organism was involved in fermentation. Nevertheless most of the phenomena that are relevant to the understanding of the role of microorganisms were noted. It was only during the following period, from 1855 onwards, that Pasteur proposed a theory of fermentation which discredited the hypothesis of spontaneous generation as well as that of Liebig and his school.

'Natural' processes would have been identified and adopted by human populations as soon as food hoarding became of interest. Water-tight vessels, such as pots and animal skins would have aided these processes as compared to drying, salting and smoking. This would have allowed the spontaneous discovery of processes for making beer, wines, yoghurt, and sauerkraut from fruit juices, milk, and vegetables. 'Natural' fermentations would have been discovered spontaneously but have only been documented in more recent history, although such practices presumably predate writing by many thousands of years.

Barley, which is the basic raw material for beer preparation but not for bread making, was the first cereal to be cultured about 12 500 years BP (= before present), and was grown 6000 years before bread became a staple food; the first document on food preparation was written by the Sumerians 6000 years ago and describes the technique of brewing. A new theory by Reichholf [1] claims that mankind formed settlements after the discovery of fermentation and used the alcohol produced by this process for indoctrination into a cult or for purposes of worship

Concepts in Biotechnology: History, Science and Business. Klaus Buchholz and John Collins
Copyright © 2010 WILEY-VCH Verlag GmbH & Co. KGaA, Weinheim
ISBN: 978-3-527-31766-0

Figure 2.1 Olympus, Nectar Time (Dionysos: god's thunder – the miracle of wine formation) [5].

[1, p. 265–269, 259–264].[1] Thus beer and wine manufacture form the roots of Biotechnology practices which were developed in ancient times. The vine is assumed to originate from the Black and Caspian seas, and to have been cultivated in India, Egypt and Israel during this early period. In Greek mythology gods such as Dionysos or Bacchus, granted the availability of wine (Figure 2.1) and the birthplace of Dionysos is believed to have been in the Indian mountain Nysa (Hindukusch) [2, p. 441], [3, p. 591–595], [4, Vol. 12, p. 1, 2].

Written documentation on beer and wine manufacture which form the roots of Biotechnology can be traced back in ancient history: about 3500 BC brewers in Mesopotamia manufactured beer following established recipes [6]. In 3960 BP King Osiris of Egypt is assumed to have introduced the production of beer from malted

---

1) The author presents several reasonable arguments in favour of this theory, among them: collecting grain for adequate food production would have been too laborious as compared to the established process of hunting and eating meat (p. 259–264). However, the author admits that no final conclusive arguments can be given in support of his theory as there is a dearth of information relating to this early period of human history [1, p. 287–300].

cereals [7, p. 1001]. In Asia, fermentation of alcoholic beverages has been documented since 4000 BP, and fermentation starters[2] are estimated to have been produced about 6000 BP by the daughter of the legendary king of Woo, known as the Goddess of rice-wine in Chinese culture [8, p. 38, 39, 45]. Soya fermentation was established in China around 3500 BP. Around about 2400 BP, Homer described in *The Iliad* the coagulation of milk using the juice produced from figs which contains proteases. Pozol, a non-alcoholic fermented beverage, dates back to the Maya culture in Yucatan, Mexico [9]. There is a mythical report that Quetzalcoatl, a Toltec king of the tenth century, was seduced by demons to drink wine with his servants and his sister so that they became drunk and addicted to desire and pleasure; later Quetzalcoatl set fire to himself as an act of repentance and was resurrected as a king on another planet [10].

Tacitus reports that the Germans have a history of beer-making [11, p. 299]. The famous German law of 1516 on brewing has its origins in Bavaria. The medieval tradition of *brewing* can be traced back through the literature, such as the first books by the 'Doctor beider Rechte' Johannes Faust, who wrote five books on the '*divine and noble art of brewing*' [12, Vol. 2, p. 409, 410].

Thus in the absence of detailed knowledge of the process, fermentation became a rational method of utilizing living systems. The fermentation of tobacco and tea were also established in ancient times. These fermentations were presumably initially adopted as fortuitous processes. In fact all of them utilize living systems, however, the early users of fermentation had no understanding of neither the origins of the essential organisms involved nor their identities nor the way in which they predominated over possible contaminants. A most significant step was the description of tiny 'animalcules' in drops of liquids, which *Leeuwenhook* observed with his microscope (about 1680, the year in which he became a member of the Royal Society in London). This, however, was not seen in the context of, or correlated to fermentation. Stahl in his 1697 book *Zymotechnika Fundamentalis* (the Greek 'zyme', meaning *yeast*) explored the nature of fermentation as an important industrial process, whereby zymotechnica was used as a descriptor for the scientific study of such processes [13].

Various descriptions were associated with 'bad air' in marshy districts. In 1776 A. Volta observed the formation of 'combustible air' (methane, 'hidrogenium carbonatrum', as analysed by Lavoisier in 1787) from sediments and marshy places in lake Lago Maggiore in Italy. He noted that 'This air burns with a beautiful blue flame...' [14]. The first enzymatic reactions, then considered to be fermentative in nature, were observed by the end of the eighteenth century. Thus the liquefaction of meat by the gastric juices was noted by Spallanzani as early as 1783 [15], the enzymatic hydrolysis of tannin was described by Scheele in 1786 [16], and Irvine detected starch hydrolysis in the aqueous extract of germinating barley in 1785 [17, p. 5]. Such processes or reactions were distinguished from simple 'inorganic' reactions on the basis that the reaction could be stopped by heat denaturation of the 'organic' components.

In addition to providing pleasure, beer and wine manufacture have become economically important because for several thousand years dating back to the early economy of Mesopotamia and Egypt, it has been a major source of tax revenue. The manufacture of alcoholic beverages developed into major industrial activities during

---

2) Starting a fermentation by transferring a small quantity of an old ferment to a new substrate.

the nineteenth century, and further fermentation processes were embraced enthusiastically in order to widen the horizon for new business opportunities, the expression of which was the foundation of numerous research institutes in several European countries during the nineteenth century.

2.2
Experimental Scientific Findings

2.2.1
Alcoholic Fermentation

Due to its great practical relevance alcoholic fermentation was the focus of technical as well as scientific interest. The essential results are reported here and also include the early findings on acetic acid fermentation. In the second half of the eighteenth century Spallanzani had undertaken microscopic investigations on microbial growth. He showed that microorganisms do not arise spontaneously [18, p. 14], [19, p. 112]. A key event was the development of the scientific foundation of modern chemistry (as opposed to alchemy) by Lavoisier [20]. He established quantitative correlations and mass balances, based on exact experimentation supporting the idea that specific compounds contained specific compositions and proportions of atoms (see Figure 2.2

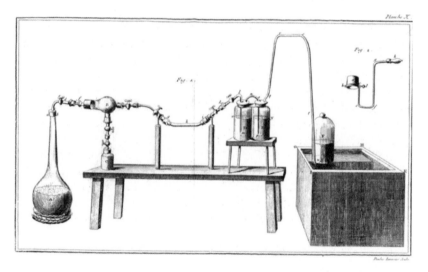

Figure 2.2 Lavoisier's apparatus for the investigation of fermentation. In the first vessel A the material to be fermented, for example sugar, and beer yeast is added to water of a weight determined exactly; the foam formed during fermentation is collected in the following two vessels B and C; the glass tube h holds a salt, for example nitrate, or potassium acetate; this is followed by two bottles (D, E) containing an alkaline solution which absorbs carbon dioxide, and only air is collected in the last bottle F. This device allowed determination 'with high precision' of the weight of the substances undergoing fermentation and formed by the reaction. [20, p. 139, 140, Planche X].

for the experimental set up). Lavoisier simply accepted the existence of the ferment and restricted his work to the chemistry of the reaction which occurred during alcoholic fermentation, showing that the only products were ethanol and carbon dioxide. In his famous book of 1793 he gave a phenomenological description of fermentation. Concluding on the products he stated that '...when the fermentation is complete the solution does not contain any more sugar....'. 'So I can say: juice = carbon dioxide + alkool' [20, p. 141]. In 1810 Gay-Lussac gave the quantitative correlation, with glucose being converted into two moles of ethanol and two of carbon dioxide [17, p.5].

From the end of the eighteenth century efforts to find a solution to the fermentation problem bear witness to the attempt to approach and explain this phenomenon either as the result of the activity of living organisms or as purely interactions of chemical compounds. However from the mid-1830s evidence began to accumulate to indicate that biological aspects formed the basis of fermentation [21, p. 24].

Important findings based on well-designed experiments were published by Schwann and Cagniard-Latour in 1837 and 1838. They showed independently that yeast is a microorganism, an 'organized' body, and that alcoholic fermentation is linked to living yeast. In 1838 Cagniard-Latour reported that 'In the year VIII (1799–1800) the class of physical and mathematical sciences of the Institute (Institut de France) had proposed for the subject (of fermentation) a prize (for solving) the following question: 'What are the characters in vegetable and animal matter....' 'I have undertaken a series of investigations but proceeding otherwise than had been done. That is by studying the phenomena of this activity by the aid of a microscope. ... This attempt ... was useful since it has supplied several new observations with the following principal results: 1. That the yeast of beer (this ferment of which one makes so much use and which for this reason was suitable for examination in a particular manner) is a mass of little globular bodies able to reproduce themselves, consequently organized, and not a substance simply organic or chemical, as one supposed. 2. That these bodies appear to belong to the vegetable kingdom and to regenerate themselves into two different ways. 3. That they seem to act on a solution of sugar only as long as they are living. From which one can conclude that it is very probably by some effect of their vegetable nature that they disengage carbonic acid from this solution and convert it into spirituous liquor.... I will add that the question formerly proposed by the Institut appears now to be solved... I have communicated (the results) to the Philomatic Society during the years 1835 and 1836.' [22].

Schwann [23] first reported his experiments concerning spontaneous generation to the Annual Assembly of the Society of German naturalists and Physicians, held in Jena in September 1836. He demonstrated that provided the air was heated neither mould nor infusoria appeared in an infusion of meat and that the organic material did not decompose and become putrid. Schwann perceived that these experiments did not support those of the proponents of spontaneous generation. They could be explained on the basis that air normally contained germs (Keime). Schwann concluded that alcoholic fermentation was promoted by proliferating yeast organisms which he classified as sugar fungi, derived from 'Zuckerpilz' or sugar fungus

[21, p. 26]. From experiments with known poisons he drew the conclusion that 'a plant was probably the organism to be expected.'...

'With microscopic examination of beer yeast there appear the known granules which form the ferment; most of them hanging together in rows. ...Frequently also one sees ...small granule seated sideways as the foundation of a new row that is without doubt a plant'. 'Besides sugar a nitrogen-containing body is necessary. One must therefore picture the vinous fermentation as the decomposition which is so brought about that the sugar fungus draws from the sugar and a nitrogen-containing body the substances necessary for its own nutrition and growth. Whereby the elements of these bodies not entering the fungus (probably amongst several other substances) combine preferentially to alcohol'.

'Beer yeast is made up almost entirely of these fungi. ... These grow visibly under the microscope, so that already after $^1/_2$ to1 hour one can observe the increase in volume of a very small granule which sits on a larger one.... It is highly probable that the latter by its development causes the phenomena of fermentation'. [23].

Also independently, Kützing [24] performed microscopic investigations on yeast and 'mother of vinegar' ('Essigmutter', in acetic acid fermentation). He confirmed the thesis of living organisms both with respect to yeast and the 'vegetabile' organisms (vegetabilische Organismen) active in acetic acid fermentation ('Essigmutter') and also recognized nucleation in yeasts. Turpin [25] refers to Cagniard-Latour, dealing with 'organisation, vegetation, reproduction and growth of yeast,... repeating his observations carefully'. The aim of these recent publications is to '...elucidate this mysterious process from a microscopic-physiological viewpoint'. The author's investigations were carried out in a large brewery and took into account the procedures of inoculation and growth of yeast. He had no doubts concerning the 'vegetable organized existence' that circumscribes the nature of living entities ('vegetabile organisierte Existenz'). He described his procedure in extensive detail, including taking samples from the technical vessel at various times. The observations included inoculation by spores or reproducing bodies ('Sporen oder reproduktiven Körpern'), and growth of spheres. Turpin concluded that fermentation is a purely physiological process, 'beginning and ending with the existence of infusoria plants or animals...' (living species).

A fermentation process that was different from alcoholic fermentation was described by Gay-Lussac and Pelouze [26]. It was used to isolate, purify and characterize lactic acid (later called lactic acid fermentation). Mixed fermentations were described by Gaultier de Claubry [27] and by Schill [28]. Schill refers to a considerable number of earlier studies on milk fermentation from 1754 onwards. Fermentation occurred both with and without the addition of ferment (in some cases yeast, in others cheese).

In his book on chemical technology Knapp [11] described the current technology for beer and wine fermentation and presented a detailed description of growing yeast including figures showing the increase in the number of yeast cells, similar to 'primitive plant cells', over time and several generations. He also noted some details concerning the cell wall and an internal, protein-like substance. He concluded that yeast is not a non-living precipitate but an organized being ('Wesen') of the 'lowest type', an initial, early state ('Anfangsstufe') of plants [11, p. 277]. Knapp came to the conclusion that no one of the '...hypotheses (Ansichten)... is up to now accepted as unequivocal truth' [11, p. 271]. A detailed review of the work of the scientists mentioned is given by Barnett [29].

2.2.2
'Unformed, Unorganized Ferments'

Unformed or unorganized ferments had characteristics that were obviously different from those of yeast: it was definitely not living matter; the substances were water soluble, they could be precipitated and thus isolated. They are enzymes in today's terms, as first proposed by Kühne [30]. Interestingly several isolated ferments, such as diastase (amylases and glucamylases in today's terminology), could be characterized to a considerable extent. Their nature however remained unknown and even obscure.

Payen and Persoz [31] investigated in detail and with high precision, the action of extracts of germinating barley known as diastase, on starch and formulated some basic principles of enzyme action (see also [17, p. 5]):

- the active principle can be isolated by precipitation and thus purified
- it is water soluble
- small amounts of the preparation were able to liquefy large amounts of starch
- the material was thermolabile, that is it loses its (catalytic) potential when boiled.

Payen and Persoz [31] also described in detail the isolation and purification of diastase from germinating barley by the procedures of maceration, pressing, filtration and repeated precipitation with alcohol. The precipitated ferments could not be crystallized, remained amorphous and chemically undefined. The fact that no definite chemical composition could be established was in contrast to the fact that the products which were sugars, as obtained by Guerin-Varry [32] who established the chemical composition [$C_{12}H_{28}O_{14}$] (although not fully correct), could be crystallized, for example glucose.

The first industrial processes that used enzymes were established from the 1830s onwards in France based on Payen's work (see Section 2.3).[3]

In 1830 Robiquet and Boutron observed that an extract from bitter almonds which was named 'emulsin' by Liebig and Wöhler in 1837, hydrolysed a glycoside, amygdaline [17, p. 5]. Schwann [34] precipitated the active principle from mucus 'as an individual substance' and named it pepsin; he also characterized its action in precipitating casein [21, p. 22]. 'By all these reactions the digestive principle is characterized as an individual substance, to which I have given the name pepsin' [34].

---

3) In this context, and with respect to Pasteur's later work on this topic, a remarkable analytical tool was reported by Biot [33] which could be used to discriminate between sucrose and glucose ('Traubenzucker', obtained from dextrin) by the optical rotation of polarized light.

2.3
Application

Handicraft and art became a source for, and a subject of scientific investigation. 'To Stahl . . . science was the basis of technology, . . . providing key ideas, . . . the basis . . . of that important German industry of Gärungskunst – the art of brewing'. 'His concern was with its chemical interpretation'. Bud considers Stahl's 'Zymotechnica Fundamentalis' (published in 1697) 'to be the founding text of biotechnology' [13].

The application of fermentation as well as the use of unformed ferments (the enzyme diastase) proceeded pragmatically during the early nineteenth century despite the fact that no commonly accepted theory had been established. Several books on chemical technology gave detailed and advanced information on the procedures of beer, wine, bread and acetic acid production and although based on ample technical experience, the discussion of the theoretical approaches to the fermentation phenomena relied in part on mysterious and contradictory background information. Nevertheless both handicraft and industrial fermentation represented important and successful production processes.

It is important to note that inoculation was used for yeast fermentation in industrial processes. Thus Lampadius [35] describes the utilization of yeast from the previous year's fermentation for the current industrial production of wine in Saxony, and Knapp [11] reported the common practice of inoculation in brewing.

Work on technological issues was equally important and even dominated over that on the basic aspects from the beginning of the nineteenth century. This is apparent from the large amount of text that was devoted to fermentation processes in the books of the time on technology and chemical engineering ([36, 380 ff., based on [37], [38, 75 ff.], [11, 249 ff.], [39, 74 ff.], [2, 195 ff.]). From these books the close relationship of technology, industry and scientific research was also obvious. Knapp [11, p. 367] stated that 'no fermentation has for industry, and notably for agriculture, such an importance, or weight, as alcoholic fermentation because the production of all alcoholic beverages, of wine, beer, . . .has this fermentation as a basis'. He considered scientific and practical and industrial interests to be equally important.

In his book on technology Poppe [38, p. 387] introduces the chapter on beer with enthusiastic characterization, as a marvellous, refreshing, caloric and healthy wine-ous drink. Knapp [11, p. 333–349] mentions that brewing was carried out in Germany at the level of a handicraft, in vessels (Bottichen) of 1000–2000 l in volume, whereas in the UK it was carried out on an industrial scale in large factories with major investment of capital; the fermenters were up to 240 000 l in volume (Figure 2.3). In the many details of the process, he states that the normal process proceeds by inoculation with yeast and procedures which did not employ inoculation were in danger of failure. Furthermore the breweries always produced a surplus of yeast which was then supplied to the baking and distilling industries. Thus the growth of yeast became an obvious, even economically important fact, making Liebig's theory obsolete.

It is estimated that in Germany in 1840 about 22.7 million hectolitres of alcohol were produced. Wagner enumerates 42 different beer specialities, including

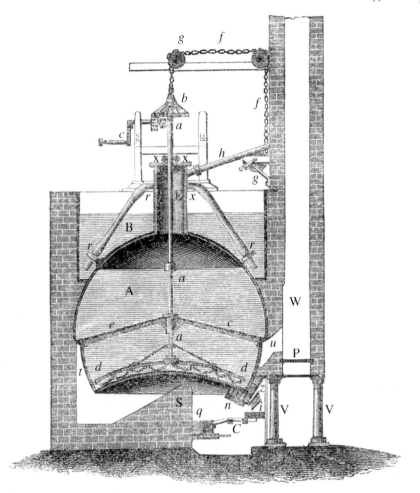

Figure 2.3 A brewing vessel as used in the UK and Belgium. The substrate is preheated in part B of the vessel where water vapour is introduced via the tubes rr; fermentation takes place subsequently in part A (the next substrate being preheated in B); part A is equipped with a stirrer dd (held by aa) which is equipped with chains that agitate the sediments on the bottom [11, p. 332].

Bavarian and English products with their different characteristics, such as alcohol content. Thus beer, and more particularly fermentation products including wine, acetic and lactic acids, became most relevant contributors to the national economies.

An acetic acid fermentation process known as 'fast acetic acid manufacture' ('Schnellessigfabrikation') was developed by Schützenbach in 1823. It worked, remarkably, with active acetic acid bacteria (of course not recognized at that time) immobilized on beechwood chips. The process was carried out in vessels made of wood, some 1–2 m wide and 2–4 m high, and was aerated to oxidize the substrate alcohol. First acetic acid was introduced to wet the wood chips, and alcohol was then continuously added and oxidized. The process took only 3 days in contrast to the classical fermentation that required several weeks. Since the bacteria were immobilized on the wood chips, a period of slow growth phase was not needed so that the oxidation of the alcohol took place immediately (Figure 2.4) [40, p. 514], [2, Vol. 2, p. 480–498].

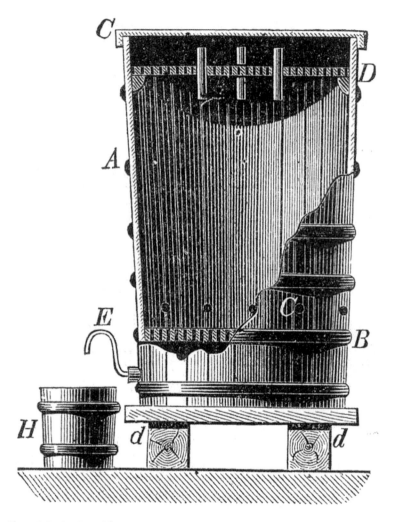

Figure 2.4 Acetic acid fermentation using immobilized bacteria. The vessel was equipped with sieve plates in positions D and B. Space A was filled with beechwood chips (on which the bacteria were immobilized). A 6–10% alcohol solution was added from the top to a solution containing 20% acetic acid and beer (containing nutrients). Air for oxidation was introduced through holes in a position above B, the temperature was maintained at 20–25 °C. The product containing 4–10% acetic acid was continuously removed via position E [40, p. 514].

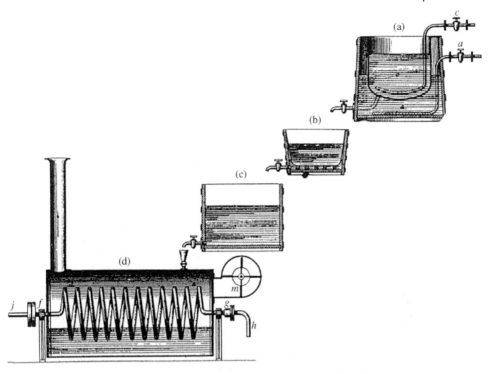

Figure 2.5 Process for dextrin manufacture showing the reaction vessel (a), filter (b), reservoir (c) and concentration unit (d) [2, Table XXV, in Vol. 1].

In his description of the chemical technology Wagner [39, p. 365/6], as did Poppe [38], enumerates three different fermentations: alcoholic, with yeast as the ferment, lactic and acetic acid fermentation, and putrefaction. The nature of the ferments is still not considered to be clear, either as an organized (living) body ('Wesen') such as yeast or a proteic substance ('Körper') undergoing decay. Alcoholic fermentation ('geistige Gährung') is the basis of both wine and beer production and baking [39].

The first industrial processes using an unformed ferment (enzyme), diastase, were established from the 1830s onwards in France and were based on Payen's work (Figure 2.5). Payen and Persoz [31, p. 74–78] had presented details for the production of dextrin on a large scale (500 kg) using 6–10% diastase. The product, dextrin, was used in bakeries, as well as for the production of beer and wines from fruits [11, 39].

2.3.1
Education

As a result of the economic relevance of fermentation, education in this field was established from the time it originated in the nineteenth century. The schools for commerce in the German speaking countries began to combine knowledge obtained from the arts with scientific knowledge. In Germany, the first agricultural high school was founded in 1806 and in 1810 became part of the new Berlin University. A further 20 agricultural schools were subsequently established up until 1858 in German speaking countries, amongst them schools in Vienna and Braunschweig.[4] In France, Boussigault founded a private agricultural research laboratory in 1835 and in London Lawes and Gilbert founded their laboratory in 1842. The brewers of Bohemia asked the director of the schools of engineering in Prague to establish an institute to train experts for their industry. As a consequence, in 1818 the pharmacist J. J. Steinmann offered the first course in fermentation chemistry, presumably the first worldwide. Balling, his successor, introduced a science-based but nevertheless practically orientated course in brewing [6, p. 21–24].

2.4
Theoretical Approaches

Beginning in the time of the renaissance during the sixteenth century, handicraft and art became a source of science. The transition from handicraft and art to science has been analysed in detail by Böhme *et al.* [42], who refer for example to Powers' experimental philosophy of 1633 (p. 7), and by Mittelstrass [43, p. 167–179] in the context of 'La Nuova Scienza' with reference to eminent artists, scientists and engineers such as Brunelleschi, Leonardo da Vinci and Tartaglia. The academies founded subsequently, the Royal Society in London in 1662 and the Académie des Sciences in Paris in 1666, accepted and introduced observation, experimentation and measurement as the basis of scientific work. This also led to the accumulation of an empirical knowledge base that was described in the form of laws [42, p. 13–128, p. 136–139]. The traditional processes used in arts and handicraft that were developed in 'workshops' ('Werkstätten') together with technical phenomena including baking and brewing, also became sources of empirical knowledge [42, p. 188–190]. With respect to biotechnology, Bud [13] considers Stahl's 'Zymotechnica Fundamentalis' (published in 1697) 'to be the founding text'. 'To Stahl … science was the basis of technology, … providing key ideas, … the basis … of that important German industry of Gärungskunst – the art of brewing'. 'His concern was with its chemical interpretation'. [13]. However, when Lavoisier later laid the foundations of modern chemistry, Stahl's theories were considered to be incorrect.

Major progress in the understanding of fermentation was made by Schwann, Cagniard-Latour and others in the 1830s and had its source and origin of interest in practical fermentation not in the observation of natural processes. The investigations of Cagniard-Latour as well as those of Turpin (see Section 2.2.1), were

---

4) In the Vienna Polytechnic Institute a chair for 'special technical Chemistry' was introduced which offered courses on brewing, wine-making, distilling and the production of vinegar [41]. At the school in Braunschweig, the 'Colleguim Carolinum' had also offered courses on agriculture and chemistry including fermentation since 1835. Otto, the professor of technical chemistry and pharmaceuticals, published a book [36] describing the products made by fermentation generally.

undertaken in a brewery thus making it clear that the *origin of the scientific problem was rooted in an industrial process*. Extensive microscopic investigations in breweries demonstrated that microorganisms were the source of fermentation. However, Schwann and Cagniard-Latour, were not able to establish a commonly accepted theory, even if the experimental basis was broad and scientifically strong. This was in contrast to Pasteur who was able to establish his theory during the 1860s; – the question is why? One strong counter-current which delayed arrival at the correct conclusion (the vitalist view) was the opposition by the chemical school, first and foremost, led by Liebig and Berzelius.

2.4.1
The School of Vitalism

Early theories: Schwann, Cagniard-Latour: living and growing cells

An exciting debate was in progress which referred to the questions of whether

- fermentation was due to living organisms and a vital factor? – the vitalist view,
- or was it a purely chemical process? – the viewpoint of the chemical, notably Liebig's, school.

Furthermore, the vitalists discussed the question of whether fermentation was a

- spontaneous phenomenon, due to spontaneous generation of living organisms, or
- if an agent, the addition of a ferment (inoculation) was necessary to initiate it, as had been shown experimentally by Schwann;
- a living force, the *vis vitalis*, governs the activity and reactions observed in fermentation and surpasses the pure chemical forces such as affinity.

As to the origin of ferments, *Gay-Lussac* had postulated a '*generatio spontanea*', a spontaneous generation, in 1810, the hypothesis that a continuing – rather myste-rious – chain of events should be the cause of spontaneous generation (of organisms) ([44, p. 130], summary by [45]). Typical controversial points of view may be illustrated by the subsequent citations: Some vital factor, '*le principe vital*', was considered to be an important principle in the chemical processes associated with the synthesis of materials isolated from living matter: 'All simple bodies in nature are subject to the action of two powers, of which one, that of attraction, tends to unite the molecules of bodies one with another, while the other, produced by caloric, forces them apart. . . A certain number of these simple bodies in nature are subject to a third force, to that caused by the vital factor, which changes, modifies and surpasses the two others, and whose limits are not yet understood' [46]. Similarly rather mysterious concepts were summarized in an early book on technology. 'Fermentation is seen as a – at a time and under circumstances spontaneous – occurring mighty movement in a liquid of different compounds . . ., which is due to the fact that several compounds act in harmony with each other, others in opposition to each other, so that the first attract, the latter reject each other' [38, p. 229]. A vital factor may be considered as an incoherence, or inconsistency in theories of the new science of chemistry, based on Lavoisier's [20] and others' experimental and empirical foundations as well as the theory of affinities by Berthollet [47, p. 2] which states that every substance reacts according to its affinity and quantity. Non-physical phenomena were not considered.

Schwann [23] and Cagniard-Latour [22] published important basic findings based on well-designed experiments. Both presented a theory of fermentation in terms of the vitalist approach essentially corresponding to that of Pasteur which would be put forward about two decades later. The fact that fermentation depends on inoculation was shown by conclusive experiments undertaken by Schwann [23]. The growth of yeast had been definitively observed by Schwann, Cagniard-Latour and Turpin. Inoculation with yeast in brewing was common practice (Section 2.3). 'Now, that these fungi are the cause of fermentation, follows, first, from the constancy of their occurrence during the process; secondly, from the cessation of fermentation under any influence by which they are known to be destroyed, . . ., a phenomenon which is met with only in living organisms.' (citation from [44, p. 139]). These findings were not accepted by the leading chemists of the time, Berzelius, Liebig and Wöhler (see below).

The myth of 'living force' continued to be discussed and Kützing [24] confirmed that it created organic mass (which was a living body in his terminology) from inorganic substances which corresponded to the hypothesis of spontaneous generation. He developed and advanced his theory on the vital force ('organisierende Lebenskraft'), based on numerous observations on yeast and acetic acid bacteria ('Essigmutter') which he had published since 1834, and on the 'generatio spontanea' ('Urbildung organischer Materie, Gebilde') that had been proposed by Gay-Lussac. Kützing concluded from his findings, that '. . . two forces, the organizing living force (Lebenskraft) and the chemical affinity (chemische Verwandschaft) are in operation'. Quevenne [48] also believed that fermentation resulted from the 'secret of living force'.[5] In his book on 'chemical technology' ([11], p. 271) which was of a high scientific and technical standard, Knapp supported the concepts of the leading chemists (see below) in stating that ferments are not living bodies, but he nevertheless underlined the fact that inoculation and growth of yeast was common practice in breweries.

The vitalist theory created a strong empirical foundation broadly based on observations from technical processes and scientific investigations together with the experimental observations of several scientists working independently. This provided convincing arguments which constituted the vitalist theory of fermentation [29].

However, in today's terms there were several irrational arguments in the vitalist theoretical concept:

---

5) Quevenue (1838) observed that 'yeast spheres sometimes seem to carry a second little sphere that seems to originate from the first. I however could not be convinced, that this would be a form of reproduction. . .'.

- the '*vis vitalis*', a mysterious vital force, was assumed to be essential, responsible for, and dominant in the reactions observed during fermentation; this argument was disproved by Buchner's work in 1897;
- the '*generatio spontanea*', spontaneous generation (Urzeugung) of living (micro-) organisms, occurred in fermentations where the starting materials were solutions of chemical compounds (not living matter, in contrast to the common practice of inoculation).[6] Pasteur's work of around 1860 disproved this second thesis.

It was much later that it was proposed anew that life (at least once) had been generated spontaneously and autonomously and this inevitably became seen as an emergent property of inorganic matter (see [49]). In 1929 Haldane assumed that this property had been established 3.5 billion years ago in the 'primordial soup' of organic molecules that had been formed by the energetic reactions between water and the components of the prebiotic atmosphere ([50], see also [51]). Later similar assumptions led to the Urey-Miller experiment in 1953 and subsequent investigations [52]. Spontaneous generation was considered to be a singular event or a series of a few rare events. The mechanistic basis and details still cannot be traced in detail and remain unknown, but are assumed to be based on known chemical principles.[7]

Unformed ferments (enzymes) had been identified and characterized and even used in industrial processes, but their chemical nature remained obscure as they could not be crystallized.[8] Progress resulting from studies on enzymes stagnated from 1840 onwards for several decades, almost certainly due to the two opposing theories put forward by Liebig (decomposition hypothesis) and Pasteur (*vis vitalis*), a situation which discouraged scientific work on ferments.

---

6) Remarkably, neither Kützing nor Quevenne saw any contradiction; inoculation was not seen as contradictory to spontaneous generation but as an additional argument in favour of Kützing's theory – both inoculation by spores or cells of yeast led to growth of the organism and the reactions involved: formation of alcohol or lactic acid, respectively. Spontaneous formation of living bodies in solutions containing chemicals such as sugar and nitrogen-containing compounds might well occur. However this latter view may also be seen as naïve, inasmuch as such highly complex systems as living bodies whose constitution, reactions and mode of reproduction were not understood, should be generated or created from simple chemical compounds in flasks within hours or days.

7) It is interesting that the concept discussed is not far removed from that of Anaximenes

(588–524 B.C) and his mentor, Anaximander, whose ideas predicted evolution and who thought that air was the element that imparted life, motion and thought, and supposed there was a primordial terrestrial slime, a mixture of earth and water, from which the sun's heat formed plants, animals and human beings directly.

8) Knapp argued: 'Even if the so called diastase can only be a mixture of substances, and even if one does not know anything on its chemical composition, and one even is in doubt if it contains nitrogen or if it is free of nitrogen, - one nevertheless has given this totally hypothetical body its rights of a citizen in science and literature. It thus is only a type of symbol and must be taken as such *titles* that (as some *shares*) circulate as a *promise* on a fact to be raised in future'. [11, p. 303].

2.4.2
The Chemical School

The conclusion drawn from the investigations of Cagniard-Latour, Schwann, Kützing and others that yeast was the organism accountable for alcoholic fermentation was severely criticized by Berzelius. He saw in this hypothesis the detrimental expression of a philosophical influence on science (*Naturphilosophie*) which was thought to have been discredited a long time ago [21, p. 29]. Berzelius [53, 54] developed a new concept for reactions in chemistry which included fermentation and also the action of unformed ferments such as diastase: he published a chapter on 'Some ideas on a power active during the formation of organic compounds, not realized before'. 'This is a new force for developing chemical activity belonging both to inorganic and to organic nature'.... 'I will therefore name it the catalytic force of bodies and breakdown caused by it catalysis'.... 'The catalytic force appears to consist intrinsically in this: that bodies through their mere presence, and not through their affinity, may awaken affinities slumbering at this temperature'. [53, p. 237]. Berzelius even speculated on a vision of synthetic processes in nature which remarkably did not become a topic in scientific investigations and discussions for decades to come: 'We have well-grounded reason to conjecture... that in the living plants and animals thousands of catalytic processes go on between the tissues and the fluids, and produce the amount of dissimilar chemical syntheses...'. As a theoretical background, Berzelius referred to the affinities which were dealt with by Berthollet, by stating that every substance reacts according to its affinity and quantity, and a power which drives or directs chemical reactions according to the affinities of the components. He thus defined catalysts as materials that promoted chemical reactions which did not occur in the absence of those materials[9] [54, p. 19–24]. Interestingly, Berzelius also raised several open questions concerning such phenomena as selectivity, which he considered to be valid research topics [17, p. 6].

Liebig argued against both the concepts of living bodies being active in fermentation processes and a catalytic power, and advanced his 'mechanical theory of ferments' that dates back to Stahl: a body undergoing decomposition transfers its disturbed equilibrium onto other metastable substances. The ferment transfers this reactivity to the substrate via molecular collision or molecular oscillations. The changes which organic atoms undergo by external influence insofar as a new order of its elements is established, is generally understood as metamorphosis [56, p.131, 132]. 'There is therefore no specific body, no substance or matter which

---

9) Different catalytic substances in reactions were recognized by Berzelius, for example acids in the decomposition of starch to glucose (described by Kirchhoff) or the fact that alkali or silver and platinum remained unaltered and did not contribute to the formation of any of the new compounds. He also included the transformation of sugar into carbon dioxide and alcohol during fermentation in the presence of an insoluble body which was known as ferment, and diastase, referring to Payen and Persoz (see before) within these phenomena [54, p. 19–22, 467, 468].

Referring to organized, or living matter, Berzelius used the term 'organic chemistry' ('The part of physiology which describes the composition of living bodies, and the chemical processes which occur in them, is termed organic chemistry') [55, p. 1].

brings about breakdown, but these are only carriers of an activity which extends beyond the sphere of the decomposing body' [58]. 'From the preceding it follows, that during the fermentation of pure sugar with ferment both undergo a decay in parallel, in the consequence of which they disappear totally. . ..The ferment thus is a body undergoing decay, putrefaction (Zersetzung, Fäulnis)...'[10] [56, p. 150–152]. It is difficult to identify Liebig's point of view on catalysis and fermentation. Not surprisingly, it is vague, even obscure, self-contradictory and incorrect in the essential points, for example the statement that the ferments vanish completely (in striking contrast to the experimental evidence presented by Schwann and Cagniard-Latour and the fact that breweries produced and sold surplus yeast). It was Liebig and his friend Wöhler who published an anonymous skit, a bad satirical article which mocked the microscopical findings they had rejected, in an attempt to rebut the views of their adversaries who were promoting the vitalist position. They described the microscopic investigation of little beasts which had developed from eggs and had organs such as teeth, stomach, anus, and so on, working in the fermentation process and resembling a vessel for distillation [29, 57]. The rise and fall of Liebig's metabolic theories is discussed in more detail by Florkin [58, p. 145–162].

An important event provided further criticisms of the vitalist position: Wöhler had synthesized urea in 1828 and for the first time showed that products of living organisms could be made chemically; he thus rejected the paradigm that organic compounds could only have their origin in reactions 'mediated' by organic material or living tissues/cells [55, 59].

2.4.3
History of Science Approaches: Preparadigmatic Research; Fermentation – An Epistemic Thing

In order to understand progress and dynamics in science and technology we refer to theories developed under the heading history and sociology of science and technology. These approaches in part follow the established concepts in the history of science but do not give a comprehensive account of such theories. Few explanations can be given here and more questions will remain open.[11]

The academic situation in fermentation research, a topic that dominated the chemical and biological sciences, was extremely controversial with regard to the theoretical approaches. Nevertheless, all the books on chemical technology that dealt with fermentation processes in detail were of great importance in scientific, technical and economic terms. However, a discipline of fermentation sciences could not be

---

10) Liebig himself calculated a precise and correct mass balance for sugar and its products and concluded that one additional molecule (called atom by him) of water was also found in the products. But he erroneously concluded that yeast or the ferment would disappear (possibly because he did not add any nitrogen or mineral sources which are required by yeast during fermentation). He also dealt with the action of emulsin, an enzyme that hydrolysed amygdalin, and that of diastase (amylases and amyloglucosidases) on starch [56, p. 165].

11) A considerable number of theories have been put forward in recent decades but then abandoned, probably as a result of changing fashions.

established under these conditions, as would otherwise have been expected with respect to the importance of and the level of activity devoted to the problem. The model advanced by van den Daele *et al.* [60] comprises three phases at the macroscopic level: the explorative or preparadigmatic, the paradigmatic, and the postparadigmatic research phases.[12] The type of research discussed here corresponds to the explorative, preparadigmatic phase which used trial and error methods, in contrast to theory-guided research, as it progressed in the subsequent period with Pasteur's work (see Section 3.4.2). In a similar approach Foucault [61, p. 254–260] described phases of practical discourse and procedures preceding the formation of disciplines. Aspects observed in such phases comprise descriptions, correlations of phenomena, rules, coherent propositions, and (a series of) theoretical approaches which constitute the preliminary forms of understanding the reality of elements that constitute a scientific discursus and which may subsequently lead to the formation of disciplines.

Referring to another model, an inside view, fermentation tended to become an epistemic thing. It represented a fascinating topic to the scientists involved, a challenging problem which required vivid interest. Analogies to epistemic things to use Rheinberger's term [62], may be seen in the early research on fermentation during the first half of the eighteenth century which takes into account the considerable research activity which took place most notably during the 1830s. Epistemic things may be characterized as a field of certain, yet unidentified objects, experimental techniques and implicit knowledge – a field of problems of phenomena, reactions, and structures which were not understood unequivocally, nor explained. As technical things they tended to give a structure to a whole new experimental field; they may have provided stabilized or reproducible experimental conditions with technical objects that frame epistemic things and integrate them into scientific fields. Methods may fail or remain insufficient and the complexity of the field increases with the amount of data and results that were accumulated but with the perspective of restructuring the field of interest. 'To put it as a paradox: epistemic things represent that what remains unknown. They may be characterized by a list of activities and properties'. The field also constituted a hybrid assembly or

---

12) The three phases of the model advanced by van den Daele *et al.* [62] are as follows:

1) Preparadigmatic, explorative research: no unifying theory guides work, trial and error methods being applied. Typical of a pre-paradigmatic phase are continuing debates on methods, problems and solutions; a tendency to form different schools which compete and hold controversial views with respect to theories and concepts.

2) Paradigmatic research: guided by a comprehensive theoretical framework (paradigma); oriented towards understanding and explanation of natural (and technical) phenomena. Dynamics of science follow theory development. Theory-guided investigations of phenomena have been highly regarded in sciences and continue to be so. In Pasteur's research theory orientation coincided with investigations of practical and technical problems and their solution based on fundamental concepts and insight.

3) Postparadigmatic research: understanding and explaining the phenomena that is advanced in the field. Research expands into more differentiated and detailed problems and phenomena following established approaches. The field is more open for external, practical or technical purposes and goals.

configuration of social, research and instrumental, as well as technical entities [62, p. 25].

The characteristics of the field of fermentation, relating research to epistemic things, may be summarized as follows: clear experimental facts were known including analytical results concerning products, a precise mass balance, similarities to living bodies including several analogous phenomena; however doubt remained and no unambiguous facts showing reproduction (of living entities) were accumulated. Adequate methods of working on the topic to give unambiguous results were not available; several different approaches such as microscopy, product analysis and quantitative investigations including mass balances, were applied. Ferments as living entities were observed under the microscope but they could not be analysed by chemical methods as they were not accessible to direct analysis or established analytical methods such as chemical analysis of ferments. Thus there was a field of problems relating to phenomena, reactions and structures which were not understood unequivocally nor explained. Results were difficult to reproduce and often contradictory in different studies. Ferments were considered to be due to living matter by most of the scientists involved in experimental research. 'Unorganized ferments' which were obviously not living matter but were responsible for the transformation of organic material were isolated. They could not be identified as substances, but were characterized only by a list of activities and properties.

Although inoculation was found to be the initiating process, a 'generation spontanea', a process of spontaneous generation of living matter, was claimed to have been observed by leading scientists; fermentation processes were believed by many, including Pasteur during his later investigations, not simply to follow the laws of chemistry but to require a 'vis vitalis', a vital force. The field was considered by many scientists to be a difficult subject with the mysteries or secrets of vital force and spontaneous generation still unresolved. Strong opposition came from the chemical school, notably put forward by Liebig and his scholars. Thus several groups which used different techniques established a field of common interest although their approaches were conflicting or even opposed.

In contrast to the study published by Rheinberger [62], no paradigmatic guidance in fermentation research was achieved until Pasteur's theory was shown to be effective. The dynamics of the topic were of both scientific and of eminent practical interest. Thus a trait significantly different to the topic analysed by Rheinberger formed the major economic interest that stimulated and in part governed research into fermentation throughout the nineteenth century – nevertheless it is legitimate to use the term 'epistemic things' in cases, when new scientific topics for research emerge from practical problems (J. Rheinberger, personal communication, Internat. Conf. Berlin, July 2008).

On a macroscopic level, this phase of fermentation research can be described as phase 1 of the model advanced by van den Daele et al. [60]: explorative research, as there was no unifying theory guiding the work, trial and error methods were being applied, and there was still no consensus on methods, problems and solutions.

References

1 Reichholf, J.H. (2008) Warum die Menschen sesshaft wurden, in *Das größte Rätsel unserer Geschichte*, S. Fischer, Frankfurt.

2 Payen, A. (1874) *Handbuch der Technischen Chemie. Nach A. Payens Chimie industrielle*, vol. II (frei bearbeitet von F. Stohmann and C. Engler), E. Schweizerbart'sche Verlagsbuchhandlung, Stuttgart.

3 (1995) *Brockhaus' Konversations-Lexikon*, Wein, Vol. 16, Berlin, p. 591–595.

4 Ullmann, F. (1923) Wein, *Enzyklopädie der technischen Chemie*, **12**, 1–63.

5 Searle, R. (1986) *Something in the Cellar*, Souvenir Press, London.

6 Bud, R. (1993) *The Uses of Life, A History of Biotechnology*, Cambridge University Press, German translation: Wie wir das Leben nutzbar machten. Vieweg, Braunschweig, 1995.

7 (1894) *Brockhaus' Konversations-Lexikon*, Bier, Vol. 2, Berlin, p. 992–1002.

8 Lee, C.-H. (2001) *Fermentation Technology in Korea*, Korea University Press, Seoul.

9 Olivares-Illana, V. et al. (2002) Characterization of a cell-associated inulosucrase from a novel source: A *Leuconostoc citreum* strain isolated from *Pozol*, a fermented corn beverage of Mayan origin. *Journal of Industrial Microbiology and Biotechnology*, 28, 112–117.

10 Nicholson, I. (1967) *Mexikanische Mythologie*, E. Vollmer Verlag, Wiesbaden, Germany.

11 Knapp, F. (1847) *Lehrbuch der chemischen Technologie*, vol. 2, F. Vieweg und Sohn, Braunschweig.

12 Ullmann, F. (1915) Bier, *Enzyklopädie der technischen Chemie*, 2, 409–410.

13 Bud, R. (1992) The zymotechnic roots of biotechnology. *The British Journal for the History of Science*, 25, 127–144.

14 Wolfe, R.S. (1993) An historical overview of methanogenesis, in *Methanogenesis* (ed. J.G. Ferry), Chapman and Hall, New York.

15 Sumner, J.B. and Somers, G.F. (1953) *Chemistry and Methods of Enzymes*, Academic Press, New York, p. XIII–XVI.

16 Tauber, H. (1949) *The Chemistry and Technology of Enzymes*, Wiley, New York.

17 Hoffmann-Ostenhof, O. (1954) *Enzymologie*, Springer, Wien.

18 Birch, B. (1990) *Louis Pasteur*, Exley, Watford, Herts, UK.

19 Vallery-Radot, R. (1948) *Louis Pasteur*, Schwarzwald-Verlad Freudenstadt, Germany, Original: La vie de Pasteur, Flammarion, Paris, 1900.

20 Lavoisier, M. (1793) *Traité Élémentaire de Chimie. Tome premier*, 2nd edn, Cuchet, Paris.

21 Teich, M. and Needham, D.M. (1992) *A Documentary History of Biochemistry 1770–1940*, Associated University Press, Cranbury, NJ.

22 Cagniard-Latour, C. (1838) Mémoire sur la fermentation vineuse. *Annales de Chimie*, **68**, 206, (Memoir on vinous fermentation – *cited from Teich*).

23 Schwann, Th. (1837) Vorläufige Mittheilung, betreffend Versuche über die Weingärung und Fäulnis. *Annalen der Physik*, **11** (2), 184.

24 Kützing, F. (1837) Microscopische Untersuchungen über die Hefe und Essigmutter, nebst mehreren anderen dazu gehörigen vegetabilischen Gebilden. (Microscopic researches on yeast and mother of vinegar, besides several other relevant plant formations.). *Journal für Praktische Chemie*, **11**, 385–409.

25 Turpin, E. (1839) Ueber die Ursache und Wirkung der geistigen und sauren Gährung. *Annalen der Pharmazie*, **29**, 93–100, (translation from Comptes rendus seconde semestre No. 8, author mentioned in the text).

26 Gay-Lussac, J. and Pelouze, J. (1833) Ueber die Milchsäure. *Annalen der Pharmazie*, **7**, 40–47.

27 de Claubry F Gaultier (1836) Ueber Stärkemehlgewinnung ohne Fäulnis. *Annalen der Pharmazie*, **20**, 194–196, (translated from Jour. des connaiss. usuelles. March 1836, 124).

28 Schill, A.F. (1839) Ueber den Milchbranntwein. *Annalen der Pharmazie*, **31**, 152–168.

29 Barnett, J.A. (1998) A history of research on yeasts 1: work by chemists and biologists 1789–1850. *Yeast*, **14**, 1439–1451.

30 Kühne, W. (1877) Ueber das Verhalten verschiedener organisierter und sog. Ungeformter Fermente Verhandlungen des naturhistorisch-medicinischen Vereins, 1, 190.

31 Payen, A. and Persoz, J.F. (1833) Mémoire sur la diastase, les principaux produits de ses réactions, et leurs applications aux arts industriels (Memoir on diastase, the principal products of its reactions, and their applications to the industrial arts). *Annales de Chimie et de Physique*, 2me. *Série*, **53**, 73–92, Translated and reprinted in: *Ann. Pharm.*, 1834, 12, 295–299.

32 Guerin-Varry (1836) Wirkung der Diastase auf Kartoffelstärkemehl. *Annalen der Pharmazie*, **17**, 261–269.

33 Biot, J. B. (1833) Über eine optische Eigenschaft, mittels deren man unmittelbar erkennen kann, ob irgend ein vegetabilischer Saft Rohrzucker oder Traubenzucker geben kann. *Annales de Pharmacie*, **7**, 257–260.

34 Schwann, Th. (1836) Ueber das Wesen des Verdauungsprocesses. *Annalen der Pharmazie*, **20**, 28–34.

35 Lampadius, W.A. (1834) Die Benutzung der Hefen von der Zubereitung der Stärkzuckerweine zu einer neuen Weinbereitung. *Journal für Praktische Chemie*, **2**, 299–301.

36 Otto, J. (1838) *Lehrbuch der rationellen Praxis der landwirtschaftlichen*, Gewerbe, Vieweg, Braunschweig, Reprint VCH, Weinheim, 1987.

37 Graham, T. (1838) *Elements of Chemistry*, (including the application of science in the arts), London.

38 Poppe, J.H.M.v. (1842) *Volks-Gewerbslehre Oder Allgemeine und Besondere Technologie*, Carl Hoffmann, Stuttgart, (fifth edition, the first three were before 1837).

39 Wagner, R. (1857) *Die chemische Technologie*, O. Wiegand, Leipzig.

40 Ost, H. (1900) *Lehrbuch der Chemischen Technologie*, Gebrüder Jännecke, Hannover.

41 Roehr, M. (2000) History of biotechnology in Austria. *Advances in Biochemical Engineering*, **69**, 125–149.

42 Böhme, G., van den Daele, W., and Krohn, W. (1977) *Experimentelle Philosophie*, Suhrkamp, Frankfurt.

43 Mittelstrass, J. (1970) Neuzeit und Aufklärung, in *Studien zur Entstehung der neuzeitlichen Wissenschaft und Philosophie*, De Gruyter, Berlin.

44 Florkin, M. (1972) A history of biochemistry, in *Comprehensive Biochemistry*, vol. 30, p. 129–144 (eds M. Florkin and E.H. Stotz), Elsevier.

45 Anonymous (1862) Ueber die Gährung und die sogenannte generatio aequivoca (summary article). *Journal für Praktische Chemie*, **85**, 465–472.

46 Béral, 1815, cited in: Roberts, S.M., Turner, N.J., Willets, A.J., and Turner, M.K. (1995) *Biocatalysis*, Cambridge University Press, Cambridge, p. 1.

47 Berthollet, C.L. (1803) *Essai de Statique Chimique*, a) 1ère partie, b) 2ème partie F. Diderot, Paris.

48 Quevenne, T.A. (1838) Mikroskopische und chemische Untersuchungen der Hefe, nebst Versuchen über die Weingährung. *Journal für Praktische Chemie*, **14**, 328–349 and 458-L 478.

49 Kauffman, S.A. (1996) *At home in the Universe*, Oxford University Press.

50 Bernal, J. (1967) *The Origin of Life*, Weidenfeld and Nicolson, London, p. 199–234.

51 Oparin, Aleksandr Ivanovic (1924) Proiskhozhozhdenie zhizny, Moscow, in *The Origin of Life*, (Translated by Ann Synge in J. Bernal(1967)) Weidenfeld and Nicolson, London, p. 199–234.

52 Johnson, Adam P., Cleaves, H. James, Dworkin, Jason P., Glavin, Daniel P., Lazcano, Antonio, and Bada, Jeffrey L. (2008) The miller volcanic spark discharge experiment. *Science*, **322**, 404.

53 Berzelius, J. (1836) Einige Ideen über eine bei der Bildung orgnischer Verbindungen in der lebenden Natur wirksame, aber bisher nicht bemerkte kraft. *Jahres-Berichte*, **15**, 237.

54 Berzelius, J. (1837) *Lehrbuch der Chemie*, vol. 6 (translated by F. Woehler) Arnoldische Buchhandlung, Dresden und Leipzig.

55 Fruton, J.S. and Simmonds, S. (1953) *General Biochemistry*, Wiley, New York, a) 1–14, b) 17–19, c) 199–205.

56 Liebig, J. (1839) Ueber die Erscheinungen der Gährung, Fäulnis und Verwesung und ihre Ursachen. *Journal für Praktische Chemie*, **18**, 129–165.

57 Anonymous (1836) Das enträthselte geheimnis der geistigen gährung. *Annalen der Pharmazie*, **29**, 100–104.

58 Florkin, M. (1972) The rise and fall of Liebig's metabolic theories, in *A history of Biochemistry* (ed. Florkin, M.) Comprehensive Biochemistry, Vol. 30.

59 Wöhler, F. (1828) Über künstliche Bildung des Harnstoffs *Pogg. Ann. für Physik und Chemie*, **12**, 253.

60 van den Daele, W., Krohn, W., and Weingart, P. (1979) Die politische Steuerung der wissenschaftlichen Entwicklung, in *Geplante Forschung* (eds W. van den Daele, W. Krohn, and P. Weingart), Suhrkamp Verlag, Frankfurt, p. 11–63.

61 Foucault, M. (1981) *Archäologie des Wissens, Suhrkamp, Frankfurt; Original: L'Archéologie du savoir*, Gallimard, Paris.

62 Rheinberger, H.-J. (2001) *Experimentalsysteme und epistemische Dinge*, Wallstein –Verlag, Göttingen, also: H-J Rheinberger, Toward a History of Epistemic Things, Stanford University Press, Stanford, 1997.

3
The Period from 1850 to 1890

The Logic of Discovery and Science Meets the Logic of Technology, Pasteur

3.1
Introduction

Although the application of fermentation processes was well established, there were still problems with the manufacture and quality of the most important products, alcohol, beer and wine. All the available books on chemical technology dealt only with the details of the fermentation processes because these were of the greatest importance in terms of technology, science and economics. In contrast, the academic stance on fermentation research which was a dominant topic in the chemical and biological sciences remained controversial with regard to the theoretical approaches (Section 2.4). The results produced by Schwann, Cagniard-Latour and others solved the mystery and most of the questions surrounding fermentation in principle and from the current viewpoint the action of microorganisms, notably yeast, and the role of inoculation had also been established. However, doubt and strong opposition to his theory persisted until the mid-1950s.

It was only with Pasteur's work that the debate was settled in favour of the role of living microorganisms in fermentation. Starting from hypotheses based on empirical results provided by sophisticated experiments and ingenious theoretical conclusions, Pasteur solved the problem and laid the foundations of microbiology as a scientific discipline.

3.2
Experimental Findings

3.2.1
Fermentation – Definite Results by Pasteur

'In the late 1850s a strong counter current in favour of the biological explanation of fermentation made itself felt. It arose from the work of Louis Pasteur (1822–1895), a

Concepts in Biotechnology: History, Science and Business. Klaus Buchholz and John Collins
Copyright © 2010 WILEY-VCH Verlag GmbH & Co. KGaA, Weinheim
ISBN: 978-3-527-31766-0

Figure 3.1 Portrait of Pasteur by Adalbert Edelfelt (1885).

remarkable fabric woven from intellectual, experimental and social threads' [1, p. 35] (Figure 3.1). Pasteur's outstanding accomplishments, experimental results, discoveries and theoretical and practical progress have been analysed and documented in several biographies and analytical and science research papers, all of which are different in scope and character and even controversial in their analysis. The most important work with respect to biotechnology will be summarized here with reference to some of these biographies and to the papers published by Pasteur.

The first basic question which Pasteur answered definitively was that of the origin and character of fermentation: Was it brought about by living microorganisms or chemical phenomena as Liebig, Berzelius and their school believed? The second key question was that of the 'generatio spontanea', the spontaneous generation of life. Early in his investigations on fermentation Pasteur was engaged in several industrial problems that suggested specific questions. He solved these with sophisticated, elaborated scientific methods which had practical consequences and led to improvements in the technical processes – an ingenious combination of scientific and technical progress with mutual interaction.

An overview of the sequence of events which led to the establishment of microbiology as a scientific discipline has been given in [2, p. 13, 14] and is shown in Table 3.1. Pasteur began his academic career with research on the phenomenon of optical rotation of polarized light by certain organic compounds formed in

Table 3.1 The origins of microbiology [2, p. 13, 14], [6, p. 62].

The origins of microbiology	1789 Lavoisier	Identification and first quantification of products of alcohol fermentation
	1815 Gay-Lussac	Quantitative analysis of products of alcohol fermentation
	1837/38 Schwann and Cagniard-Latour	Experimental demonstration of living yeast as an agent in alcoholic fermentation
	1856–77 Pasteur	Investigations on fermentation: Beginning of studies on fermentation (1856) Memoire on lactic acid fermentation (1857) Investigations into alcohol fermentation (1858) Studies on spontaneous generation (1859–62) Demonstration of the germ theory at the Sorbonne (1864) Studies on butyric acid fermentation and detection of anaerobic fermentation (1861) Studies on wine fermentation, invention of Pasteurization (1864) Studies on beer fermentation (1871) Detection of facultative anaerobic fermentation of yeast. Studies on silk (1865–70) Sterilization technique (collaboration with Chamberland, 1877)
	1850 Rayer and Davaine	Detection of the origin of anthrax and the role of microorganisms in diseases
	1865 Lister	Praxis of antisepsis in chirurgical operations
	1876 Koch	Work on the bacterium causing anthrax
	1877–86 Pasteur	Beginning of investigations into anthrax (1877)
	1877 Tyndall	Discovery of spores
	1880 Pasteur	Investigations on chicken cholera and discovery of immunization (1879)
	1880 Winogradsky	Soil microorganisms: the bacterial nature of nitrification
	1880 Pasteur Koch	Studies on rabies
	1881 Pasteur	Vaccination against anthrax First vaccination of a person against rabies
The origins of biochemistry	1868 Miescher	Finding of nucleic acids
	1888 Waldeyer	Proposal of the term 'chromosome'
	1897 Buchner	Alcohol formation from sugar in the absence of yeast cells but using enzyme(s) from yeast extract
	1900 De Vries, Tchermak and Correns	Rediscovery of Mendel's heredity laws

fermentations: first, tartaric acid, then amyl alcohol and lactic acid. Pasteur early recognized, from a report of the Academy of Sciences of 1844, the optical activity of salts of tartaric acid (by-products of wine manufacture), and he began, in1848 with studies on this phenomenon which fascinated him: two otherwise chemically identical organic substances showed different optical behavior with respect to the rotation of polarized light. Pasteur supposed molecular asymmetry to be the cause of this, as also reflected in the crystals of two isomers which behaved like mirror images. Pasteur also recognized the plant origin of the different forms (isomers) of tartaric acid and associated the phenomenon with life processes. He also found that one of the isomers was consumed during fermentation, whereas the other was not, which stimulated his interest in investigations of fermentation.

He noted that there seemed to be a high degree of correlation between optical activity and organic (living) processes and expressed this correlation in the following form 'if a substance exhibits optical activity, then it most likely has an animal or plant origin' [3, 4, p. 143–277, 5].

Pasteur had been a lecturer in chemistry at Strasbourg University since 1849 and soon became Professor of Chemistry and Dean of the newly established Faculty of Sciences at Lille. He was expected to apply scientific knowledge to the local chemical and brewing industries. In his inaugural address as Dean in December 1854, Pasteur strongly supported this goal and reiterated his support later in Paris. The first technical questions and request for assistance with fermentation work were put to Pasteur by Mr Bigo, a French manufacturer of alcohol located in Lille who asked for his advice in 1856 [7, p. 92–96]. Bigo and others who owned factories which produced alcohol using beet as the raw material, had suffered considerable losses when the juice turned sour. Pasteur visited Bigot's factory on a nearly daily basis and took samples of the fermentation broth which he investigated in his laboratory. In series of experiments and after numerous microscopical observations he observed yeast buds in normal fermentation runs, but rods that he soon identified as lactic acid 'yeast', when the fermentation 'ran sour' (due to the formation of acetic or lactic acid). Thus from his microscopic observations, Pasteur was convinced that the fermentation was due to living microscopic entities and that different species were responsible for different products [5, p. 92, 93], [6, p. 25–29], [7, p. 92–97] [8, 9].

In his paper on lactic acid fermentation Pasteur [8] presented the essentials of fermentation processes. The introduction gave a description of the conversion of sugar into lactic acid and the remark that this phenomenon was very difficult to understand. He went on to note that the facts seemed to be in favour of Liebig's theory and that the latter was supported by several papers by Fremy, Boutron and Bethelot. Pasteur then clearly pointed out that his investigations gave a basically different result, since in all cases where sugar is transformed into lactic acid a special ferment, a 'lactic acid yeast' can be found. He presented the means with which to isolate it in a pure culture. Pasteur's brief discussion of lactic acid fermentation introduced (1) the biological concept of fermentation as the result of the activity of living microorganisms; (2) discussion of the practice of inoculation for starting a reliable fermentation which was also common practice in beer fermentation; (3) the notion of specificity, according to which each fermentation could be traced to a specific microbe (probably taking into account his findings from the alcohol factory in 1856) (see also [10–12]); (4) the essential experimental factor, namely that the fermentation medium must provide the nutrients for the microbes; (5) specific chemical features characterized by the main products of fermentation and the by-products. Furthermore, a large quantity of sugar could be transformed by very few of these 'lactic acid yeast' ([5, p. 90], [8, 9]). When considering Pasteur's very clear statement on a ferment being the agent of sugar transformation it should be noted that the prior high level investigations of Schwann and Cagniard-Latour were already known to the public and of course to Pasteur, including their results on yeast fermentation and their theory of microorganisms being the source of alcohol fermentation [7, p. 97], [6, p. 28]. In his investigations on alcohol fermentation Pasteur [9] communicated further basic findings: beer yeast decomposes in pure aqueous sugar solution, a finding that would be in agreement with Liebig's theory on fermentation, but in the presence of proteinaceous matter the yeast regenerates; thus yeast requires a nitrogen source in order to survive.

Pasteur made a strong and radical attack against the chemical school (see Section 2.4.2) of which Liebig was the head. In his classical memoir of 1860, he provided a summary of his findings using previous abstracts, notes and letters on alcoholic fermentation, the discussion being focussed on the dispute on fermentation. It 'inflicted on the chemical theory a series of blows' [5, 13]. As Pasteur noted, from his findings it became obvious that Liebig had failed to undertake precise experimental investigations prior to the formulation of his theory. In contrast to Liebig's view yeast consumes a minor quantity of the available sugar as a substrate for its growth. The growth of yeast as shown by the increasing mass and formation of new yeast spheres was unequivocally established experimentally. All investigations were based on an elaborate experimental protocol for analysing microbial processes [8]. Results, findings and theoretical conclusions were summarized in a review [14].[1] The decomposition of sugar to alcohol and carbon dioxide correlated with the life cycle of yeast and its physiological functions and was thus not purely a chemical act as Berthelot insisted. Other types of fermentation were identified and found to be due to the presence of different microorganisms and included butyric and lactic acid and a slimy fermentation; it was these reactions that were responsible for the losses in alcoholic fermentation (Figures 3.2 and 3.3). Beer yeast can survive in the presence or absence of oxygen ('vie sans air'); in the absence of oxygen it assimilates oxygen bound to the sugar and in the presence of oxygen the yeast grows rapidly but does not act as a ferment [14, 15].

Pasteur's memoir of 1860 and the papers of 1862 'marked a watershed in the debate over biological vs. chemical explanations of fermentation'. Even if it did not end the debate it is 'easy enough to declare Pasteur the victor in his long and rancorous dispute with Liebig ... Pasteur's work on fermentation was immensely fruitful, both scientifically and practically. ...it led Pasteur into the study of spontaneous generation' [5], a battlefield of outstanding scientists for more than five

---

1) Review in a German journal, a translation and summary from several original papers in *Comptes Rendue*, including a paper in preparation.

Une des levûres des fruits acides au début de *la* fermentation
de leurs moûts naturels

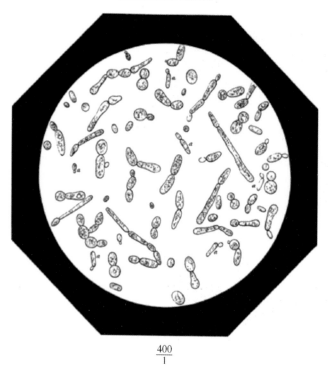

$$\frac{400}{1}$$

Figure 3.2 Natural yeast of fruit at the beginning of fermentation (Une des levûres des fruits acides au début de la fermentation de leurs mouts naturels) ([16], Pl.X, p. 164).

decades. But *technical problems* also continued to raise questions for leading scientists, notably Béchamp and Pasteur (see Section 3.3).

One of the mysteries of fermentation remained highly controversial, the hypothesis of a '*generatio spontanea*', *spontaneous generation*. Pasteur [14] addressed this basic and controversial question efficiently. He referred to Schwann and others whose 'serious work' he repeated and confirmed, with significant experimental modifications (see also [5, p. 115]). In addition to highly precise experiments using various methods, Pasteur undertook something of a show in 1860 with expeditions to high altitude mountains, most spectacularly to the Alps and the glacier Mer de Glace, to demonstrate the existence of germ free air[2] (these expeditions were replicated by his colleagues, Pouchet and coworkers who favoured the hypothesis of spontaneous

2) In Excursions starting from the plains, going up to the alps near Chamonix with nearly pure air free of microorganims, Pasteur took with him sealed sterile bottles containing nutrient solution that he opened at different height in the mountains in order to demonstrate the effect of germs in the air which became increasingly pure – germ free – with increasing altitude, and thus growth by inoculation from the air became less frequent, until no growth was observed on the glacier Mer de Glace.

PRINCIPAUX FERMENTS DE MALADIE
DU MÔUT ET DE LA BIÈRE

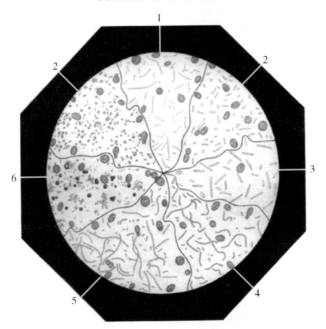

Figure 3.3 Main ferments of the diseases of mash and beer (Principaux ferments de maladie du môut et de la bière) ([16], Pl. I, p. 6).

generation but who met with less success). The results of these experiments were presented by Pasteur first in a lecture to the Société Chimique de Paris in 1861 and then in a famous lecture at the Sorbonne in 1864, a show for 'tout Paris', in order to convince not only the scientific community but also the general public (including some such famous names as Alexandre Dumas) of his theory [7, p. 119–121, 131–2], [5, p. 118–120]. With the current state of knowledge and his repeated experiments on this matter he claimed the downfall of spontaneous generation.[3] In his book 'Sur la Bière' Pasteur summarized his achievements in detail, both experimental theoretical and technical. Making ingenious use of his insights he solved the problems of the French brewers (see Section 3.3). He also described in minute detail his experimental tools, the flasks he used for fermentation experiments in the laboratory which he designed himself, showing his extraordinary talent for the design of laboratory apparatus (Figure 3.4). He also discussed possible sources of errors that should be avoided in order to prevent infection [16, p. 27, 29, 54, 55, 81].

3) Pasteur undertook a critical review notably of the theory of spontaneous generation including an extended discussion of errors that are obvious only on the basis of his highly sophisticated experiments and his profound concept of microbial growth, inoculation, pathways and sterility. He referred to findings by Gay-Lussac published in 1810, Trécul and Fremy, who in a session of the Academy in 1872 continued to defend the theory of spontaneous generation ([16], Chapter 3, [52]).

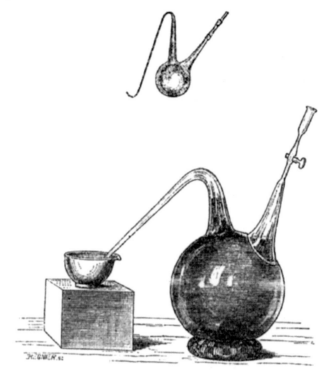

Figure 3.4 Flasks used for fermentation experiments [16, Figure 4, p. 29; Figure 59, p. 232]. Above: Figure 4: Flask conceived for sterile fermentation experiments (the tube at the right allows water vapour to escape during boiling for sterilization of the substrate solution, pure yeast is added as an inoculum and the tube is subsequently closed; the capillary at the left serves for pressure equilibration without allowing organisms or spores to enter the tube. Below: Figure 59: Flask conceived for anaerobic fermentation experiments (a small amount of yeast suspension was introduced from the right into the substrate solution that has been previously boiled; the carbon dioxide formed can escape through the left capillary and a beaker containing mercury).

From 1865 onward, fermentation continued to be a current topic. However, there were no dramatic developments or basic new findings until the work of E. Buchner in the 1890s which led to a turn in the theoretical concept of fermentation (see Section 4.4.1). The publication rate was rather low during the 1870s and 1880s (see [17]). There were several publications by other authors concerning fermentation but they did not have the same impact as those written by Pasteur. Berthellot [18] observed the fermentation of substrates other than sugar including glycerol, mannitol and sorbitol, in which not only was alcohol and carbon dioxide formed but also hydrogen (clearly anaerobic fermentations). Several papers on fermentation were published in the *Bulletin* between 1864 and 1870 most notably by Béchamp [19–23] but also by others.

3.2.2
'Unformed, Soluble Ferments' – Enzymatic Reactions

Several new active substances from different sources (e.g. flowers and fruits, pancreas) were discovered by Béchamp [21, 22], Dobell [24], Berthelot [25, 26] and Marckwort and Hüfner [27], including invertase, 'nefrozymase' (a substance analogous to diastase which was isolated from urine), lipase activities, fibrinolytic activity, emulsin (degradation of amygdalin) and so on (see also [28]). Hüfner [29] used a new procedure in order to extract several soluble ferments from the pancreas. Schwarzer [30] as well as Marckwort and Hüfner [27] carried out investigations on what today would be termed the kinetics of an enzymatic reaction, including the influence of concentration, time, temperature and the inactivation of the active ferment by temperature.

In trying to identify the active principle, Berthelot showed according to his own interpretation, that a peculiar substance formed by yeast can transform sugar into alcohol. This substance was nitrogen containing, could be precipitated by alcohol and was comparable to diastase. He stated that this substance was different from yeast and that there was no production of yeast cells. Thus he claimed to have observed the transformation of sugar into alcohol without the activity of any living organisms. He was convinced that the formation of alcohol from sugar was a genuinely chemical process – a hypothesis that was established only some 30 years later by Buchner. Berthelot's experiments, however, were not carried out under sterile conditions. Although it was unknown at the time, anaerobic bacteria (in tests where oxygen was excluded) must have produced small amounts of alcohol and carbon dioxide in addition to hydrogen, acetic and butyric acid [18].

By the 1870s studies had established the existence of two types of ferments. They became known as unformed (unorganized) and formed (organized) ferments (the latter referred to living bodies, such as yeast). The terminology reflected the fundamentally different but prevailing views of the two groups. The German physiologist, Willy Kühne [31] referred to the pepsin type of ferments as 'enzymes'. This particular term was chosen to indicate that the activity was due to a constituent to be found 'in yeast' (*en zyme*, in Greek). Only about a dozen 'unformed ferments' (enzymes) had been identified by 1880 since there was little research activity during this period [17].

3.3
Practical Application, Technical Progress and Institutional Development

3.3.1
Products and Progress in Processes

Several fermentation products became an important part of the overall economy in European, North American and Asian countries, thus the fermentation industry was growing fast and included the manufacture of beer and wine, industrial alcohol, yeast extract, acetic and lactic acid, soy sauce and sake. Alcoholic fermentation is based on the conversion of sugar, either sucrose or glucose and fructose, to ethyl alcohol and carbon dioxide and a range of by-products, many of which have characteristic flavours and tastes which clearly affect the quality of the product.

Beer manufacture represented one of the most important economic activities. Thus in Germany 6–8 m hl (million hectolitre, 1 hl corresponding to 0.1 m^3) of beer were produced at the beginning of the nineteenth century, a figure that had grown to 22.7 m hl by around 1840, 36 m hl in1873 and 50 m hl in 1890 [32, p. 533]. Large breweries were essentially industrial companies, the largest in Munich produced 233 500 hl and the equivalent in Vienna produced 450 000 hl per year [33, Vol. 2, p. 1000–02]. The economic importance of beer manufacture derived from the tax on beer (Biersteuer) as it has been common practice for centuries for governments to levy such a tax which makes up an important part of their income [33, Vol. 2, p. 991, 992]. The consumption of beer per capita (per inhabitant and year) varies significantly with nation. The maximum consumption was recorded among the Belgians who consumed 168 l per capita, followed by Britain and Germany where the per capita consumption was 97 l in 1885 [32, Vol. 2, p. 532–534]. By 1890, this figure had increased to 106 l in the German Tax Union, but was 229 l in Bavaria, 22 l in France, 36 l in the Netherlands, and had also increased in Britain and Belgium to 127 l and 184 l respectively. The alcohol content of the beers produced varied significantly ranging from 3.7% in Weissbier to 4.7% in Hofbräu Bock both manufactured in Munich, to 4.9% in Ale and 5.3% in Porter [33, Vol. 2, p. 1000].

The production process (described in all technology textbooks of the nineteenth century) comprised malt formation using five operational steps: (1) mashing (Einmaischen) of cereals (barley in general, and exclusively according to German law) in water, (2) germination (Keimen), (3) kiln drying (Darren), (4) the separation of the germs (Keime), (5) grinding (shooting, Schroten). Brewing comprised four operational steps: (1) mashing (Maischen) and solubilization of the extract (starch conversion to glucose), (2) cooking and addition of hops (Hopfen), (3) cooling, (4) fermentation. The analytical procedures for this important commercial product were well developed [34, Vol. 2, p. 404, 415].

Technical problems continued to raise questions among the leading scientists. The investigations by Béchamp and Pasteur's highly accurate work and meticulous scientific investigations on beer fermentation, led to the solution of the most urgent problems of the French brewers by providing the art of brewing with improved technology, resulting in good quality ('the health') beer. Pasteur's book *Études sur la Bière* (1876) [16] gave a thorough experimental, theoretical and scientific account of his investigations, results, theoretical consequences, technical solutions and his suggestions which together led to the majority of the technical problems being solved. Further investigations by Béchamp and Pasteur were concerned with practical questions and problems relating to wine fermentation, most notably the souring of wine which occurs when acetic acid is formed from alcohol, a phenomenon that had been frequently observed when wine, beer or cider was exposed to air. Béchamp [19, 20] observed the presence of spherical or long filamentous rod-like fungi when the wine was exposed to air and this resulted in the formation of acetic acid. Pasteur undertook similar investigations with the aid of a microscope which showed that microorganisms ('microscopic vegetations') are the cause of the diseases of wine; he also produced diagrams of the lactic acid ferment and butyric acid 'infusoria' (Figures 3.2 and 3.3). He showed [35] that this is due exclusively to an organized body, *Mycoderma aceti*. One type of fungus causes bitterness in wine while another is responsible for the condition known as 'turned wine' ('vins tournés'). An essential consequence of Pasteur's insight into microbial growth due to inoculation and exposure to air which contains a multitude of germs and spores, was to protect the fermentation process from air-borne infections, for example by using closed vessels for brewing (Figure 3.5) [16, p. 328], [36].

Yeast as a commercial product was mainly produced in high yield in distilleries after it had been washed, mixed with starch and pressed (pressed yeast, Presshefe), it was then sold for use in other industrial processes, for example bread manufacture [34, Vol. 2, p. 403]. In Denmark, Hansen made major progress in breeding pure yeast by working with solid culture media (e.g. agar plates, as did Koch) and isolating single cells which he could then propagate. This became the basis for pure yeast fermentation and commercial applications which was then adopted by other countries, most notably by the German brewing industry. The Berlin Institute (see below) and its first director Delbrück played a major role in adopting these techniques which led to the production of unusually pure and high quality beer by breweries in northern Germany [37, p. 22, 23].

The tradition of wine making had always been and continued to be of major economic importance in Europe. Not only was it agreeable to drink but health benefits were also attributed to its consumption when administered properly. The physiological effects of wine were described as follows 'stimulation of the nervous

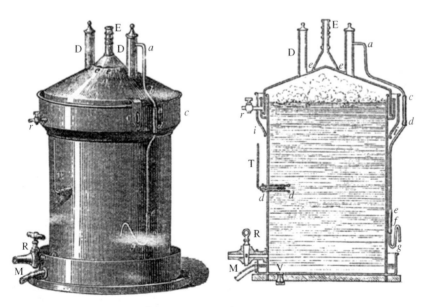

Figure 3.5 Pasteur's technical fermenter [16, p. 328].

system and blood circulation, improving or enhancing the subjective feeling and performance, for weakness and illness with fever the protein saving capacity of alcohol represents a positive effect. Wine therefore is a remarkable means for preserving the forces and improving the resistance to infections'. [38, Vol. 16, p. 591–595]. In addition to a wide variety of European wines, wines from outside Europe were also becoming popular, for example the famous Madeira wine (the addition of sugar, honey and cognac to the fermentation product was responsible for its unique flavour) and Arabian, Chinese, Japanese, American and Australian wines [39, Vol. 12, p. 2–31]. The best grapes grown in Europe for producing high quality wine were considered to be (and still are) Riesling, Sylvaner, Chablis (all white), Cabernet, Sauvignon and Syrah (red) [38, Vol. 16, p. 591–595]. The production methods were described as a whole, giving details of viticulture, mashing, pressing and fermentation, as well as analytical methods. Analytical procedures which were essential for quality assessment together with the details of the by-products of fermentation were described in a textbook by Payen which even included a chapter on the utilization of residues, a topic of major importance at the present time [34, p. 457–478]. Pasteur had identified the microorganisms found on the surfaces of the grapes at the start of fermentation and found that they included different types of yeast and also lactic and acetic acid bacteria which were the sources of the 'diseases' of wine, as a result of these findings breeding of pure wine yeast became common practice [39, Vol. 12, p. 12–14, 52–56]. Pasteur [35, 36, 40] highlighted the *practical conclusions* of his work.

The production of wine around 1890 was estimated at 120 m hl (million hectolitre) worldwide, 113 m hl in Europe, 39 m hl in France which was clearly the most productive country by far, 24.5 m hl in Italy, 24 m hl in Spain and 5 m hl in Germany. The consumption per capita in 1890 was highest in Spain at 115 l, followed by Greece at 109.5 l, Bulgaria 104 l, Portugal 95 l, Italy 95 l, France 94 l and Germany, where beer was more frequently consumed, at only 5 l [38, Vol. 16, p. 591–595].

Alcohol was produced both for human consumption and industrial use from a variety of raw materials and sources such as wine, fruit, molasses, cereals and potato [34, Vol. 2, p. 499–591]. The volume of alcohol produced was considerable and increased over time particularly for industrial purposes. In Germany the overall production of alcohol was 2.7 m hl (million hectolitres) in 1888/89 and up to 3.7 m hl in 1893/94, of which 2.2 m hl was used for human consumption during both periods. The 71 500 plants of varying size that were in operation may also indicate the economic importance of this product in Germany. In Russia over 2000 plants produced 3.6 m hl of alcohol in 1893/94 and in Austria and Hungary 118 000 distilleries (1580 of which were large) produced 2.3 m hl of alcohol. Advanced technology had been developed and applied in large factories: the process using starch as the raw material was operated at high pressure to ensure gelatinization (Aufschluss, 'Henzedämpfer'), hydrolysis was then achieved by adding diastase (malt) to stirred tank reactors, followed by fermentation with yeast that had been produced separately for 72 h; distilleries were controlled automatically [38, Vol. 15, p. 172–178].

Other fermentation products that were manufactured on an industrial scale included bread, acetic and lactic acid. Historically, bread has been a staple constituent of the human diet worldwide and was presumably made using fermentation.[4] Inoculation had become common practice and was achieved either by using a sample of the current fermentation (the sourdough) for the next fermentation or by adding so-called 'artificial yeast' which during the period from 1850 to 1890 could be easily purchased [34, Vol. 2, p. 169, 170]. Acetic acid made from wine had been used for centuries by the Jews, Greeks and Romans. It was exclusively produced by bacterial oxidation (by *Mycoderma aceti* as it was called by Pasteur) of alcohol in the presence of air, although various chemical oxidation reactions were known. A process known as 'Schnellessigfabrikation' (quick acetic acid production process) has been described in Section 2.3 [34, Vol. 2, p. 480–498]. Historically, lactic acid fermentation has also been used for the preparation of sour milk products, cheese, yoghurt and sour vegetables although the nature of the reaction was then unknown. A variety of cheeses was produced, mainly in Europe, from various sources of milk using many different procedures. 'More or less numbers of vessels, or equipment are used, and modern plants are based on scientific principles...'. An example of the production of Emmentaler is given which states that this cheese 'is the most difficult to produce; it should exhibit a taste of nut...and have evenly distributed holes of the size of peas or cherries, with 4–5 cm distance.' [41, Vol. 10, p. 211–214].

Ferments containing enzymes had been used since the 1840s (see Section 2.3) and diastase was used on a major industrial scale with a few others being introduced in the second half of the nineteenth century. The first company to carry out fermentations based on an enzyme process, the Christian Hansen's Laboratory which is still in existence today, was founded by Christian Hansen in Copenhagen (Denmark) and pioneered the use of rennet (laboratory ferment, chymosin) in cheese manufacture [41, Vol. 10, p. 863/4], [42]. Other pancreatic enzymes such as trypsin or pancreatin and pepsin which were isolated from pig or cow were used as drugs, for example as digestive aids. 'Pepsin, is a rational drug insofar ... that a weakened function of the stomach (dyspepsia) is reinforced by little doses of pepsin, and, in fact, numerous positive reports by doctors are available' [41, Vol. 12, p. 1007].

Remarkably the potential problems of waste water were recognized at that time and discussions regarding its disposal were already underway. This included recommendations to collect the waste water in dumps or special cavities or to dispose of it by using it for irrigation or by sprinkling on the soil [34, Vol. 2, p. 590].

Thus, the decades following Pasteur's work brought innovation and increasing numbers of biotechnological processes, mainly related to alcohol, acetic and lactic acid production, which were based on the knowledge that specific microorganisms produce special products in association with new technology. Progress in the type and number of food and comestible goods which could be produced by biotechnological methods must also be seen in the context of the food shortage and famine that dominated the mid-nineteenth century. Increased and rational production,

---

4) From the Bible it follows that at the time of Abraham sourdough was not known yet, but Moses had forbidden its consumption when eating the Paschal Lamb (Osterlamm) [41, Vol. 3, p. 582–585].

improved processing to give high yields and food preservation (e.g. by Pasteurization, sterilization and lactic acid fermentation) became increasingly important [43, p. 20, 21], [44].

3.3.2
Application Research, Bacteriology, and New Institutions

Scientific knowledge and new methods laid the foundations for the shift in production processes from small scale handicraft-based units to large industrial complexes employing rational process development and control, using for example pure yeast as inoculum, control of infection in fermentation, improved fermentation reactors and process and product analysis based on advanced analytical techniques, as well as the first automation protocols.[5]

Following the progress in the understanding of fermentation notably due to Pasteur's work, new fields of research or disciplines were established with titles such as Bacteriology or Mycology. There was a sudden upsurge in the number of research institutions most of which were government funded, that were established for research into beer, wine and food manufacture, hygiene, medical care and water, however research laboratories were also established for the brewing and bakery industries. Thus institutions for brewing research and education were established in Weihenstephan near Munich (1872/76), Berlin ('Institut für Gärungsgewerbe', 1874), Hohenheim (1888) (all in Germany), in Copenhagen and, in Paris, the famous 'Institut Pasteur' (1888). In the Berlin institute about 80 scientists worked on both the laboratory- and the pilot plant scale on the chemistry, biology, physiology, engineering and economics of fermentation. They addressed brewing and yeast production and alcohol and acetic acid fermentation [37, p. 20]. In Vienna the 'Institut für Gärungsphysiologie und Bakteriologie' was established at the Technical High School (1897) [45]. In Britain the 'British School of Malting and Brewing' was founded at the University of Birmingham in 1899, followed 10 years later by a Scottish school, and at about the same time by the private Carlsberg Institute in Copenhagen, Denmark. By 1863 Germany had 17 research stations and by 1877 this number had increased to 59; in the USA agricultural research institutions which eventually became the origins of well-known universities such as MIT, Cornell and Wisconsin, were funded by state decrees from 1863 onwards [43, p. 21–31], [37, p. 20].

Both the terms 'Zymotechnologie' and microbiology denominated the new research field. A. Jorgensen undertook the commercialization of dry yeast on a worldwide basis. In 1885, he founded the journal *Zymotechnisk Tidende* as well as an institute known as the 'Fermentology Laboratory'; he became highly regarded after the publication of his book on 'Microorganisms and fermentation'. One year later in Chicago, J. E. Siebel established the journal *Zymotechnic Magazine: Zeitschrift für*

---

5) Improved fermentation reactors were for example described in German patents of 1881 and 1889, for the introduction of air pipe systems and agitation in order to improve yeast growth before fermentation (G. Ueckermann, 1881, Einrichtung zum Lüften von Bierwürzen und Maischen mit kalter gereinigter Luft. Patentschrift No. 17979; C.A. Holz, 1889, Hefe-Aufzieh- und Reinigungsapparat. Patentschrift No. 48791).

Gärungsgewerbe and Food and Beverage Critic. In 1872 he founded an analytical laboratory, followed by an experimental station and brewer's school, the 'Zymotechnic College' in 1884 [43, p. 21–31], [32, Vol. 2, p. 410]. Ten journals on brewing were issued in German speaking countries at that time [33, Vol. 2, p. 1002]. Several books on mycology and/or bacteria or microorganisms were published, for example those by A. de Bary *et al.* (1884), Flügge (1886) and others in the 1880s [33, 38], Vol. 2, p. 312, Vol. 13, p. 151] in addition to a handbook entitled *Technische Mikrobiologie* published in 1897 to 1903, followed by a second edition comprising 5 volumes published from 1904 onwards [46]. These publications alone illustrate not only the great economic importance of the manufacture of beer, wine and other fermentation products but also the new role of research and education which was recognized as being essential for maintaining the quality and improving the technology required to produce fermented food.

Following Pasteur and Koch's success in identifying causative agents of disease and establishing pure cultures, pharmaceutical companies also established Bacteriology departments which produced vaccines or tested substances for their antimicrobial properties [46].[6]

Thus the *'age of Bacteriology'* began with a new paradigm, research institutions, journals and books, and a broadened industrial and economic base.

3.4
Theoretical Approaches

3.4.1
Fermentation Theory, an Epistemic Thing

Retracing the main events in fermentation theory as outlined above, we see a considerable amount of work on fermentation that had been carried out following the pioneering work of Cagniard-Latour and Schwann, both of whom were well recognized. They had shown that fermentation was due to living bodies. Yeast and also other types of ferments had been identified as essential for the most economically important fermentations, that of beer and wine. Schwann had shown that inoculation was necessary since no activity was observed under conditions known today as sterile. No spontaneous generation took place. Inoculation was also common practice in breweries. However these results were not acknowledged by the dominant chemical school of which Berzelius, Liebig and Wöhler were the leading scientists who however were not seriously engaged in experimental research on fermentation.[7]

6) In the field of medical care and hygiene the work of R. Koch, P. Ehrlich and E.v. Behring, in addition to that of Pasteur, represents the most impor-tant progress in basic innovations for methods, such as staining and cultivation of microorganisms, notably of pure cultures, and gelatin or agar plates as a basic new methodology [46].

7) During the 1850s, while accepting fermentation as underlying (most) biochemical acitvities *(vital-chemische Prozesse)*, Traube turned Schwann's approach upside down in the sense that he claimed ferments were chemical and not living entities. Thinking that they were proteins he effectively connected their ability to bring about fermentation with their chemical structure [53].

The chemical paradigm postulated a purely chemical process, and that no living entities were involved.

The situation represents that of an *epistemic thing* [47] (see Section 2.4.3). Controversial discussion even within the vitalist school continued and included such topics as the growth and creation of new yeast, different forms of ferments (formed and unformed – clearly not living), inoculation was shown to be essential and was common practice in brewing industries but spontaneous generation of new life was nevertheless assumed by several scientists who also presented experimental results in support of this notion. Fermentation processes were believed by many not simply to follow the laws of chemistry, but to require a 'vis vitalis', a vital force, fermentation being essentially linked to a 'vital act' as Pasteur had also claimed [16, p. 306]. Thus the field represented certain as yet unidentified objects, experimental techniques and implicit knowledge. Fermentation continued to represent a fascinating topic to the scientists involved, with respect to both scientific and industrial aspects. Finding a solution to the problems central to fermentation represented a strong motive for experimental work and theoretical approaches. Finally the whole field was restructured in terms of the establishment of a new scientific discipline (see below) [47, p. 8,9,31, 32].

3.4.2
Pasteur Establishes a New Paradigm

To emphasize an earlier quotation '...a strong counter current in favour of the biological explanation of fermentation ...arose from the work of Louis Pasteur, a remarkable fabric woven from intellectual, experimental and social trends'. [1, p. 35]. Problems with industrial applications certainly initiated and fuelled part of this work. It was the fermentation industries producing acetic acid and wine and most importantly the brewing industries which encountered severe problems with the quality of their products, and it was Pasteur who made the greatest contribution to overcoming these difficulties. Thus it is not surprising that his famous book in which he also summarized his theories, was entitled *Études sur la Bière* [16]. Within a short period of time Pasteur had elaborated the major elements which contributed to solving the questions and problems underlying the most controversial scientific debates. Between 1855 and 1875 he established unequivocally (a) the role of yeast in alcoholic fermentation, (b) fermentation as a physiological phenomenon (of living organisms), (c) differences between aerobic and anaerobic microbial activity [28], and in addition, the characterization of specific microorganisms by their physiology and typical products. It is essential to note that Pasteur was familiar with the work preceding his own and that of his contemporaries who worked on fermentation. He thus referred to, and in most cases discussed at length the work of the most important scientists, notably Béchamp (highlighting his errors), Cagniard-Latour, Schwann, Berthelot, and notably Liebig. Sometimes Pasteur noted how he refuted the arguments and results of opponents (in the question on spontaneous generation) with 'malicious pleasure' ('malicieux plaisir').

The roots of Pasteur's interest in fermentation have been described in detail by Gale [4, p. 151, 152], Geison [5, p. 95, 96][8] and Barnett [28]. 'As his work on optically active substances proceeded he came across amyl alcohol [after a suggestion by Biot in 1849], which is a mixture of [optically active] alcohols produced as by-products of souring milk...'. [4, p. 151, 152]. The major shift in Pasteur's research interest was announced in a now famous paper on lactic acid fermentation [5, p. 90, 95, 96]. Pasteur [8] introduced the paper by providing a clear and concise account of how he came to focus his attention on fermentation: 'I think I should indicate in a few words how I have been led to occupy myself with researches on the fermentations.' Pasteur than turned to his studies on amyl alcohol and the phenomenon of light polarization by two of its forms. 'Preconceived ideas' in this context (the fermentation origin) 'sufficed to determine me to study what influence the ferment might have in the production of these two amyl alcohols'. 'Ultimately, I hope to be able to show the connexion between the phenomena of fermentation and the molecular asymmetry characteristic of substances of organic origin' [10]. 'The constitution of substances – considered from the point of view of their molecular asymmetry or non asymmetry ... plays a considerable role in the most intimate laws of the organization of living organisms.... Such has been the motive of [my] new experiments on fermentations.' The 'preconceived idea' (idée préconcue), that is the hypothesis that living organisms are the origin of optical activity as well of fermentation processes, which guided Pasteur in his research, was termed a 'creative act' '... not an exercise in formal logic' by Gale ([4, p. 157, 158], see also [5, p. 95, 96, 133]). The paper also included central theoretical and technical precepts that characterized all of his subsequent work on fermentation. It was a rational concept based on experimental results and it soon became a paradigm.

But other questions obviously had their origin in technical problems. 'Even while he was investigating amyl alcohol, he had been employed by the big breweries of Lille as a researcher....He soon found an answer,...but more importantly, he clarified the conceptual system involved in the brewing art. That is to say, he came up with a theory ...' [4, p. 151, 152]. Pasteur identified the origin of the souring process in fermentation as the formation of acetic acid from alcohol by microorganisms other than yeast which he described as lactic acid yeast (actually lactic acid bacteria). Thus the motives for his basic research and research on technical problems in relation to specific questions and their solution were intimately linked. Pasteur himself clearly stated later: 'The idea for this research has been inspired by our problems' (with beer manufacturing) [16, Préface].

---

8) Geison's book (1995) comprises both meticulous and in depth studies of science research and although he acknowledged Pasteur's outstanding achievements, his book also contains a mix of psychological and political points and views which are not appropriate to science history and together with some malevolent remarks akin to socio-political theatre and disgust. These factors weaken the value of his book.

From the beginning Pasteur [8, 10] established a clear, rigorous concept. In his system of work in progress he first presented his hypothesis (proposition, preconceived idea) and then gave experimental approaches. The titles of some of the chapters in his book *Études sur la Bière* published in 1876 show the comprehensiveness of his work. Four of these titles will therefore be mentioned here: I. The near correlation that exists between the easy alteration of the beer or the mash that serves for its production and the procedures of its manufacture; II. Investigations on the causes of the diseases of beer and those of the mash that serves for its production; VI. Physiological theory of fermentation; VII. Novel process for the manufacture of beer. This strict concept laid the basis for his early success: 'In his first paper on alcoholic fermentation . . . (he) made a fundamental contribution to (micro-) biology, based on meticulous and ingenious experiments' [28].

The concept of *experimental systems* [47, p. 34, 148–154, 195/6] may contribute to the understanding of the outcome of the controversy between Liebig and Pasteur. Liebig neglected the experimental techniques and the evidence derived from fermentation research, and thus he did not appreciate the results adequately, nor did he develop a strong empirical basis for his own arguments. In contrast, Pasteur developed the methods which had been available since the 1830s (those of Schwann and others) to the highest precision, reproducibility and conclusiveness. Furthermore it was his analysis of the concept of specific organisms producing special products, that led Pasteur to his conclusions. In addition, it was also the new technique of sterile experimentation developed by Pasteur which opened the way to reorientation, progress, and understanding [16, p. 229, 232,233]. With this information Pasteur was thus able to convince the scientific community of the veracity of his hypotheses.

Pasteur had finally established the basic facts relating to fermentation that had been under discussion for decades:

- Fermentation is due to living organisms.
- Inoculation was necessary in order to start fermentation. No spontaneous generation ever took place. 'The finding of yeasts and their living nature, as well as the knowledge of their origin, *eliminates the mystery of the spontaneous occurrence* ('suppriment le mystère. . .') of fermentations of natural sugar juices. . .'. [16, p. 229]. These findings, and their establishment, may be considered to be a new *paradigm* guiding further research. Pasteur thus laid the foundations of a new scientific discipline, microbiology, known as bacteriology at the time ([48, Avant-propos, p. 22]), and of industrial microbiology [49]. Thus, remarkably, a single outstanding scientist clarified most of the mysteries of fermentation, and he thus solved major technical problems (Pasteur's major contributions to research into various types of disease, prevention of infection, and vaccination will not be discussed here since they are more relevant to the field of medical sciences and public health than to biotechnology).

3.4.3

**The Question of Living and Unformed Ferments (Enzymes),
and Explanatory Concepts**

Pasteur also established a *dogma* stating that the reactions, namely the chemical processes taking place during fermentation, should necessarily be linked to a vital act ('phénomène correlatif d'un acte vital').[9] '...so must one see in these fermentations reactions which cannot be explained by the ordinary laws of chemistry?... the fermentations gain a place outside the body (ensemble) of chemical and biological phenomena' [16, p. 229, 230]. This dogmatic point of view seemed to inhibit further research on enzymatic reactions within the framework of common chemical principles [17].

Amongst the theoretical concepts put forward by other scientists a remarkable paper by Hüfner [29] is worthy of mention. It discussed the question of 'unformed ferments' (in terms of enzymes), their role in living organisms, and the dogma of life being essential for fermentations. He asked 'what are the ... mighty chemical forces that ownes this cell [yeast]? Are they normal chemical forces, or are other facts playing a role ... that one calls 'organized bodies?'. He referred to Pasteur and continued: 'But one went further ... and proclaimed a dogma, that no fermentation could proceed without the presence and the growth of organized material.' With reference to investigations on 'unformed' ferments, Hüfner expressed the hope, 'that an insight into the origin and process of fermentative degradation would be possible also without the assumption of those still unknown forces ... following the same laws ... as all chemical processes ...' [29, p. 388, 391]. Remarkably a similar progressive, integrated biological and chemical view was expressed in Payen's textbook on industrial chemistry, where it was taken as probable 'that the type of activity of both classes of ferments are the same, that the (the organized) ferment of the second class produces, by its vital activity, a body of the first (non organized substance under decay) that itself originates the fermentation process either inside, or outside of the ferment.' [34, German edition, Vol. 2, p. 399].

With respect to unformed ferments, called enzymes by Kühne [31], their nature was characterized as follows: the primary feature is their role (later identified as catalytic) in reactions (diastase, invertase, etc.). Secondary features included precipitation with alcohol, loss of activity when heated, non crystallisable and not exhibiting a definite chemical composition. Their chemical nature remained

---

9) A different viewpoint was that of Traube, 'exemplified by the following words, published in 1878, which express an interpretation of biological catalysis that differed greatly from his previously cited views [53] which he had held 20 years earlier': 'The ferments are not, as Liebig assumed, substances in the process of decomposition, ... but are chemical com- pounds related to the albuminous bodies, which ... undoubtly have ... a specific chemical composition and produce changes in other compounds by means of specific chemical affinities. Schwann's hypothesis (later adopted by Pasteur) of fermentation as a consequence of the vital activity of lower organisms is inadequate ...'.

unknown until the work of Sumner, Northrup and Kunitz who succeeded in crystallizing enzymes in 1925 and in 1930–31, respectively.

An interim conclusion concerning the *epistemic thing* – fermentation, the transformation of juices into alcohol and other compounds, the chemical steps, the bodies and forces involved in this transformation – may be conceived – on a *macroscopic level* – to include three stages of its development (following the concepts of Van den Daele *et al.* [50] (see Section 2.4.3, note 11), and Kuhn [51]:

1) The period beginning with early scientific experiments and concepts, that of Stahl, Spallanzani, Lavoisier's chemical investigations in the late eighteenth century, Schwann, Cagniard-Latour and others in the 1830s, investigations starting from rather well defined substrates (grape juice, beer mould or sugar solutions) to yield well defined products, alcohol and carbon dioxide in defined proportions. However, the interpretation of the phenomena involved and the material that reacted, participated in and pushed this reaction, remained highly controversial, even contradictory. The interpretation included – the vitalist concept – living organisms, with experimental proof, but with different concepts promoted as to their origin, further the concept of catalysis of Berzelius, unclear in the catalyst of alcoholic fermentation, and, in striking contrast, that of Liebig, strictly chemical, but without sound experimental basis and speculative and ill defined in its terms and formulation. This phase may be termed *preparadigmatic* [50].

2) The *paradigmatic* phase established by Pasteur in the late 1850s was rigorously based on experimental evidence and showed that living microorganisms carried out the transformations of fermentations. Paradigms have been defined by Kuhn as 'universally recognizable scientific achievements that for a time provide model problems and solutions to a community of practitioners' [51]. Research is guided by a comprehensive theoretical frame (paradigm) orientated towards understanding and explanation of natural (and technical) phenomena.

3) The subsequent phase of research from about 1880 onwards, during which government research institutions were established and became involved in research programmes essentially orientated towards practical needs may correspond to the *postparadigmatic phase*, or phase three, of the model advanced by Van den Daele *et al.* [50].

However there was no answer to the question of how fermentation reactions occurred – Pasteur frankly left that question open, and furthermore with his dogma suggested that the reaction was strictly linked to a vital act of the microorganisms involved, a mystery that continued to endure.[10]

---

10) Remarkably Pasteur did not advance further and ask about that complex of questions. 'Now . . . in what does the chemical act of decomposing the sugar exists; and what is its precise cause? I confess I simply do not know' [13, p. 359, 360].

The hypothesis that enzymatic reactions – chemical catalysis by unformed ferments – were taking place during fermentation was not substantiated with clear, stringent experimental evidence.

Thus the next epistemic thing already began to appear on the horizon.

It was during the subsequent period, initiated by the work of E. Buchner in the 1890s, that this problem was solved, leading to the biochemical paradigm that no hidden, mysterious act was involved thus reversing Pasteur's dogma (Section 4.4.1). Thus it took more than 150 years to solve the problems of fermentation. This involved a considerable number of scientists, and many significant controversies, hypotheses, theories, and the establishment of paradigms. Furthermore the question of the nature of unformed, soluble ferments, and the structure of proteins, remained open for decades (see Section 4.2.1).

References

1 Teich, M. and Needham, D.M. (1992) *A Documentary History of Biochemistry 1770–1940*, Associated University Press, Cranbury, NJ.

2 Scriban, R. (1982) *Biotechnologie*, Technique & Documentation-Lavoisier, Paris.

3 Pasteur, L. (1856) Ueber den Amylalcohol. *Journal für praktische Chemie*, **67**, 359–362;(1855) *Comptes rendus de l'Académie des Sciences*, **XLI** (8), 269.

4 Gale, G. (1979) *Theory of Science*, Mc Graw-Hill, New York, (a) 143–168; (b) 169–277.

5 Geison, G.L. (1995) *The Private Science of Louis Pasteur*, Princeton University Press, Princeton, New Jersey.

6 Birch, B. (1990) *Louis Pasteur*, Exley Publications, Watford, Herts, UK.

7 Vallery-Radot, R. (1948) *Louis Pasteur*, Schwarzwald-Verlad Freudenstadt, Germany;(1900) Original: *La vie de Pasteur*, Flammarion, Paris.

8 Pasteur, L. (1857) Mémoire sur la fermentation appelée lactique. *Comtes Rendus hebdomadaires des séances de l'Académie des Sciences (Paris)*, **XLV** (22), 913–916;(1858) Ueber die Milchsäuregährung. *Journal für praktische Chemie*, **73**, 447–451.

9 Pasteur, L. (1857) Mémoire sur la fermentation alcoolique. *Les Comptes rendus de l'Académie des Sciences*, **45**, 1032–1036;(1858) Ueber die alkoholische Gährung. *Journal für praktische Chemie*, **73**, 451–455.

10 Pasteur, L. (1858) Mémoire sur la fermentation appelée lactique. *Annales de Chimie et de Physique*, 3e sér, **52**, 404–418.

11 Pasteur, L. (1858) Bernsteinsäure als Produkt der alkoholischen Gährung (Succinic acid as product of alcoholic fermentation). *Journal für praktische Chemie*, **73**, 456;(1858) *Les Comptes rendus de l'Académie des Sciences*, **XLVI** (4), 179.

12 Pasteur, L. (1858) Ueber die Bildung von Glycerin bei der alkoholischen Gährung (On the formation of glycerol during alcoholic fermentation). *Journal für praktische Chemie*, **73**, 506;(1858) *Les Comptes rendus de l'Académie des Sciences*, **XLVI** (18), 857.

13 Pasteur, L. (1860) Mémoire sur la fermentation alcoolique. *Annales de Chimie et de Physique*, **58**, 323–426.

14 Pasteur, L. (1862) Ueber die Gährung und die sogenannte *generatio aequivoca*. *Journal für praktische Chemie*, **85**, 465–472; *Comtes Rendus hebdomadaires des séances de l'Académie des Sciences (Paris)*, **52** (LII), 344; 1260; **48** (XLVIII), 753; 1149.

15 Pasteur, L. (1863) Nouvel exemple de fermentation déterminée par des animalcules infusoires pouvant vivre sans oxygène libre et en dehors de tout conact

avec l'air. *Bulletin de la Société Chimique de Paris*, 221–223.

16 Pasteur, L. (1876) *Études sur la Bière. Avec une Théorie Nouvelle de la Fermentation*, Gauthier-Villars, Paris.

17 Buchholz, K. and Poulson, P.B. (2000) Introduction/overview of history of applied biocatalysis, in *Applied Biocatalysis* (eds A.J.J. Straathof and P. Adlercreutz), Harwood Academic Publishers, Amsterdam, p. 1–15.

18 Berthellot, M. (1857) Ueber die geistige Gährung. *Journal für praktische Chemie*, **71**, 321–325; (1857) *Comtes Rendus hebdomadaires des séances de l'Académie des Sciences (Paris)*, **XLIV**, 702.

19 Béchamp, M.A. (1864) Ueber die Weingährung. *Journal für praktische Chemie*, **93**, 168–171; (1857) *Comtes Rendus hebdomadaires des séances de l'Académie des Sciences (Paris)* **LVII**, 112.

20 Béchamp, M.A. (1864) Sur la fermentation alcoolique. *Bulletin de la Société Chimique de Paris*, (1), 391–392.

21 Béchamp, M.A. (1865) Matière albuminoide, ferment de l'urine. *Bulletin de la Société Chimique de Paris*, (3), 218–219.

22 Béchamp, M.A. (1866) Sur les variations de la néfrozymase dans l'état physiologique et dans l'état pathologique. *Bulletin de la Société Chimique de Paris*, (1), 231–232.

23 Béchamp, M.A. (1866) Sur l'épuisement physiologique et la vitalité de la levûre de bière. *Bulletin de la Société Chimique de Paris*, (1), 396–397.

24 Dobell, M.H. (1869) Action du Pancréas sur les graisses et sur l'amidon. *Bulletin de la Société Chimique de Paris*, (2), 506.

25 Berthelot, M. (1860) Sur la fermentation glucosique du sucre de cannes. *Les Comptes rendus de l'Académie des Sciences*, **50** 980–984.

26 Berthelot, M. (1864) Remarques sur la note de M. Béchamp relative à la fermentation alcoolique. *Bulletin de la Société Chimique de Paris*, (1), 392/3.

27 Marckwort, E. and Hüfner, G. (1875) Ueber ungeformte Fermente und ihre Wirkungen. *Journal für praktische Chemie*, **11** (new series), 194–209.

28 Barnett, J.A. (2000) A history of research on yeast: Louis Pasteur and his contemporaries 1850–1880. *Yeast*, **16**, 755–771.

29 Hüfner, G. (1872) Untersuchungen über "ungeformte Fermente" und ihre Wirkungen. *Journal für Praktische Chemie*, **113**, 372–396.

30 Schwarzer, M.A. (1870) Sur la transformation de l'amidon par la diastase. *Bulletin de la Société Chimique de Paris*, (2), 400–402.

31 Kühne, W. (1877) *Verhandlungen des Naturhistorische-Medicinischen Vereins*, vol. 1, p. 190, "Ueber das Verhalten verschiedener organisierter und sog Ungeformter Fermente" (On the behaviour of different organized and so-called unformed ferments).

32 Ullmann, F. (1915) *Enzyklopädie der technischen Chemie*, vol. 2, Bier, Urban & Schwarzenberg, Berlin, p. 408–535.

33 Brockhaus (1894) *Brockhaus' Konversations-Lexikon*, Berlin, vol. 2, Bier.

34 Payen, A. (1874) *Handbuch der Technische Chemie. Nach A. Payen's Chimie industrielle, II. Band* (eds F. Stohmann and C. Engler), E. Schweizerbartsche Verlagsbuchhandlung, Stuttgart.

35 Pasteur, L. (1865) Recherches sur la fermentation acétique. *Bulletin de la Société Chimique de Paris*, (1), 306–308.

36 Pasteur, L. (1873) Improvement in the Manufacture of Beer and Yeast, US patent 141 072.

37 Dellweg, H. (1974) Die Geschichte der Fermentation, in: *100 Jahre Institut für Gärungsgewerbe und Biotechnologie zu Berlin 1874–1974*, Institut für Gärungsgewerbe und Biotechnologie, Berlin.

38 Brockhaus (1895) *Brockhaus' Konversationslexikon*, Berlin vols. 15–16. Spiritusfabrikation Wein.

39 Ullmann, F. (1923) *Enzyklopädie der technischen Chemie*, vol. 12 Urban & Schwarzenberg, Berlin, p. 1–63.

40 Pasteur, L. (1866) *Etudes sur le Vin*, 2nd edn, 1873, Savy, Paris.

41 Brockhaus (1894) vol. 5, p. 255, Diastase; vol. 6, p. 679, (1894) Fermente; vol. 12, p. 845, Pankreatin; vol. 12, p. 1007, Pepsin, Brockhaus, Konversationslexikon, Berlin.

42 Poulson, P.B. and Buchholz, K. (2003) History of enzymology with emphasis on food production, in *Handbook of Food Enzymology* (eds J.R. Whitaker, A.G.J. Voragen, and D.W.S. Wong), M. Dekker, New York, p. 11–20.

43 Bud, R. (1993) *The Uses of Life, A History of Biotechnology*, Cambridge University Press; (1995) *German translation: Wie wir das Leben nutzbar machten*, Vieweg, Braunschweig.

44 Pasteur, L. (1866) Sur l'emploi de la chaleur pour conserver les vins. *Bulletin de la Société Chimique de Paris*, (1), 468.

45 Röhr, M. (2001) Ein Jahrhundert Biotechnologie und Mikrobiologie an der Universität Wien, personal communication; CD Rom.

46 Metz, H. (1997) Nutzung der Biotechnologie in der chemischen Industrie, in *Technikgeschichte als Vorbild moderner Technik*, vol. 21, Schriftenreihe der Georg-Agrikola-Gesellschaft, Göttingen, p. 105–115.

47 Rheinberger, H.-J. (2001) *Experimentalsysteme und Epistemische Dinge*, Wallstein –Verlag, Göttingen, D, also:Rheinberger, H.-J. (1997) *Toward a History of Epistemic Things*, Stanford University Press, Stanford.

48 Delaunay, A. (1951) *Pasteur et al Microbiologie*, Presses Universitaires de France, Paris.

49 (1991) *Annual Review of Microbiology*, **45**, 89–106.

50 Van den Daele, W., Krohn, W., and Weingart, P. (1979) Die politische Steuerung der wissenschaftlichen Entwicklung, in *Geplante Forschung* (eds W. van den Daele, W. Krohn, and P. Weingart), Suhrkamp Verlag, Frankfurt, p. 11–63.

51 Kuhn, T.S. (1973) *Die Struktur Wissenschaftlicher Revolutionen*, Suhrkamp Verlag, Frankfurt (original: The structure of scientific revolutions).

52 Pasteur, L. (1872) Comptes rendus de l'Academie, séance du 28 octobre 1872.

53 Traube, M. (1858) Zur Theorie der Gährungs- und Verwesungserscheinungen, wie der Fermentwirkungen überhaupt. (On the theory of fermentation and decay phenomena, also of ferment activity in general). *Annual Review of Physical Chemistry*, **103**, 331–344.

4
The Period from 1890 to 1950

4.1
Introduction

Biochemistry was initiated jointly by E. Fischer who investigated enzymatic properties at an advanced level such as stereo-selectivity, and E. Buchner who showed definitively that enzymes catalyse life processes thereby ending the myth of the vital force. The biochemical paradigm thus established provided the impetus for research in biochemistry or physiological chemistry, leading to the elucidation of glycolysis which encompassed a multi enzyme-coenzyme sequence. The coenzymes performed as redox-equivalents and energy carriers. As a result of a long series of scientific experiments the citric acid cycle was established. Although Sumner and Northrup succeeded in crystallizing enzymes, the controversial debate nevertheless continued. Resolution of the protein structure was only achieved after several decades. Fleming made a key discovery: penicillin, and with the work of Florey, Chain and Heatley laid the foundations for the antibiotic era.

Meanwhile enzyme technology expanded. War requirements prompted the production of acetone, butanol and glycerol by large scale fermentation. Other new fermentation products, enzymes for food processing and washing, and citric acid, were established. Production of penicillin on a large scale was a technological breakthrough which occurred a considerable time after its discovery and was again driven by the requirements of war. The antibiotics era began with secondary metabolites being used as medicines and involved science, bioengineering and business on a new scale.

4.2
Research – Advances in the Basics of Biotechnology: Experimental Findings

4.2.1
Biochemistry

The breakthrough in the development of structural biochemistry came with the work of Emil Fischer (1852–1919) (Figure 4.1), who in the course of his scientific career,

Concepts in Biotechnology: History, Science and Business. Klaus Buchholz and John Collins
Copyright © 2010 WILEY-VCH Verlag GmbH & Co. KGaA, Weinheim
ISBN: 978-3-527-31766-0

Figure 4.1 Emil Fischer.

completely shifted the orientation of research in chemistry towards the principal organic components of living matter: sugars, fats, and proteins [1]. In 1891 Fischer established the field of *stereochemistry*, through the *configurations of glucose and isomers* [2]. From about 1894, in a series of experiments he investigated the action of different enzymes using several glycosides and oligosaccharides of known structure; the results revealed specificity as one of the key characteristics of enzymes. Fischer derived his *theory on specificity* with the famous analogy of a lock and key: 'To use a picture, I will say that enzyme and glucoside must fit like *lock and key* in order to interact chemically...' [3, p. 843]. Since the agents of the living cell (enzymes) are optically active ...'one might assume that the yeast cells with their asymmetric agents can utilize only those sugars, the geometry of which is essentially similar to the natural hexoses...'. He investigated for example the influence of the stereochemistry of α- and β-methylglycosides on the action of two different enzymes, called 'invertin' and 'emulsin', that hydrolyzed either the one or the other substrate (Figure 4.2) [4]. These were the most important observations for Fischers theory [3, p. 843]. After Pasteur had recognized the correlation between optical activity due to three-dimensional structure, and life processes, Fischer then introduced a key for understanding biological phenomena: the relevance and essential role of the three-dimensional structure of both substrates (sugars) and enzymes to catalyse biological reactions on a chemical basis in which no living systems were involved.

Of the immense progress made in biochemistry, only two major developments will be discussed shortly. They were of outstanding importance for biotechnology in

101. Emil Fischer: Einfluß der Konfiguration auf die Wirkung der Enzyme. I.

Berichte der deutschen chemischen Gesellschaft **27**, 2985 [1894].
(Vorgetragen in der Sitzung vom Verfasser.)

Das verschiedene Verhalten der stereoisomeren Hexosen gegen Hefe hat Thierfelder und mich zu der Hypothese geführt, daß die aktiven chemischen Agenten der Hefezelle nur in diejenigen Zucker eingreifen können, mit denen sie eine verwandte Konfiguration besitzen.[1])

Diese stereochemische Auffassung festzustellen

ist mir nun in unzweideutiger Weise zunächst für zwei glucosidspaltende Enzyme, das Invertin und Emulsin, gelungen.

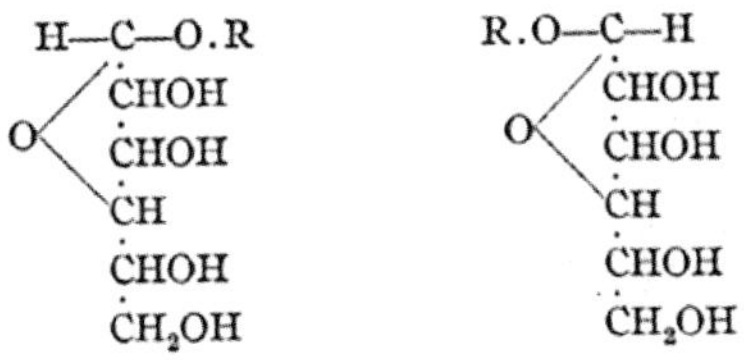

$$
\begin{array}{ll}
\text{H—C—O.R} & \text{R.O—C—H} \\
\quad\diagup\text{CHOH} & \quad\diagup\text{CHOH} \\
\text{O}\diagdown\text{CHOH} & \text{O}\diagdown\text{CHOH} \\
\quad\diagdown\text{CH} & \quad\diagdown\text{CH} \\
\quad\text{CHOH} & \quad\text{CHOH} \\
\quad\text{CH}_2\text{OH} & \quad\text{CH}_2\text{OH}
\end{array}
$$

Figure 4.2 Influence of stereochemistry on the action of different enzymes: α- and ß-methyl glycosides which are hydrolysed by either invertin, or emulsin, respectively [4].

the long term: glycolysis which represents the key steps of the conversion of glucose into products such as alcohol and lactic acid, and the elucidation of the structure of proteins and thus enzymes.

In the mid-1890s Buchner rejected the hypothesis of *vis vitalis* which still postulated hidden mysterious forces in fermentation, when he published a series of papers [5–10], which signalled a *breakthrough in fermentation and enzymology*. Buchner presented the proof that (alcoholic) fermentation did not require the presence of '...such a complex apparatus as is the yeast cell'. The agent was in fact a soluble substance, without doubt a protein body, which he called 'zymase' and which much later turned out to be the enzyme system of the whole glycolytic pathway [5]. The experiments seemed simple, but required the highest precision possible at the time.[1]) The whole experience of Pasteur's work was necessary to exclude unknown and unexpected effects, notably the concept of sterile work excluding infection by microorganisms. His key experiment was to prepare a press juice from yeast, which contained all the enzymes and coenzymes required for the transformation of glucose into alcohol and carbon dioxide and to demonstrate that no living cells were present. He was then able to show that this solution could perform the same reactions as did living yeast during fermentation.

---

1) The initiative for these experiments came from E. Buchner's brother, Hans Buchner, an immunologist, with the aim of isolating microbial protoplasmic proteins which he considered to be of primary importance to immunity. In order to stabilize the juices obtained by grinding yeast, Eduard Buchner added 40% glucose and observed the occurrence of fermentation [11].

During the period up to 1930, the discovery of cell-free fermentation stimulated the molecular approach to the study of the pathway of alcoholic fermentation, mainly research on the successive intermediates in metabolism. It encompassed studies that merged into a common trend, that were carried out in relation to alcoholic fermentation and muscle glycolysis, which were recognized as a common pattern involving two different cellular phenomena. A considerable number of researchers and their coworkers were involved [11, p. 59–68, 91]. Buchner himself, even when he initially mentioned one enzyme, did not assume a simple one step transformation of sugar into alcohol and he continued to work on cell-free fermentation investigating intermediate compounds and activities both in cell-free press juice as well as in living yeast that would convert possible intermediates including trioses.[2] Working at the Institut Pasteur, Fernbach [18] also systematically investigated metabolic intermediates with various microorganisms (see Section 4.2.2).

Glycolysis[3] was studied under different headings with increasing intensity in an effort to provide understanding of how cells provide energy for life processes and how microorganisms make products such as alcohol and organic acids. An initial concept of a metabolic pathway in alcoholic fermentation involving pyruvate and acetaldehyde as intermediates came to be generally recognized by 1930 [11, p.59]. It turned out to be a puzzle to discover the different intermediates, the coenzymes required, leading to the conversion of glucose either into pyruvate and the final products, alcohol or lactic acid, or entering the citric acid cycle to be finally oxidized to CO_2 and H_2O. A number of outstanding groups participated in these investigations. It took nearly 50 years to establish the correct scheme of reaction sequences from the early 1900s to the late 1940s. Only a short summary of key events in the many single steps of investigations is presented here, since the results did not have an immediate impact on large fermentation processes. Investigations were undertaken in different fields, anaerobic, notably alcoholic fermentation, aerobic fermentation, and muscle glycolysis, incrementally increasing understanding of these processes fundamental to life. Different hypotheses, correct as well erroneous, were established and discussed over the years [1, 11].

It was not until 1925 that it was realized that both phenomena, aerobic and anaerobic glucose metabolism, while ending in different ways, were functional variations of a common biochemical mechanism. Since then 'its unity through life has become clear' [1, 11, p. 23]. A review paper by Kluyver and Donker [21] was entitled 'The unity of biochemistry' ('Die Einheit in der Biochemie'). A series of hypotheses

---

2) Thus Buchner and coworkers [12–14], with the title 'chemical processes in alcoholic fermentation', published investigations on a number of substances, including glyceraldehyde and dihydroxyacetone that were converted by cell free preparations and living cells [11, p. 59–68] [12, 13]. They also found that a so-called co-enzyme which contained an organic phosphoric acid ester, exhibited significant influence on the reactions, referring also to results by Harden and Young on press-juice [15–17].

3) Glycolysis represents the transformation of glucose into pyruvate via intermediate steps with the subsequent formation of end products either alcohol or lactic acid, or products which enter the tricarboxylic acid cycle to be finally oxidized to CO_2 and H_2O (Figure 4.3). The history of biochemistry has been described in an excellent and extended review by Florkin in two issues of *Comprehensive Biochemistry* [19, 11, 20], and in the textbook by Fruton and Simmonds [1], which are the main sources of this short account.

was put forward. Neuberg and Wieland forwarded the theory of the chemistry of aerobic respiration. Warburg revealed the role of iron in oxidative biochemical catalysis.[4] Many results were mentioned that show that the sugar is first cleaved to give two C_3-units, with glycerol aldehyde assumed as an intermediate [21, p. 138, 143, 145].

An important observation was made by Harden and Young in 1905, the formation of a hexose diphosphate. Phosphorylated compounds were being recognized as intermediates in the path from glucose to pyruvic acid. Phosphorylation was also observed by Embden and Lacquer in the metabolism of carbohydrates by muscle where lactic acid was the metabolic product. Pyruvic acid was recognized as an intermediate in alcoholic fermentation. From the work accomplished up to 1930, the concepts which were to survive are contained in a scheme, according to which glucose is phosphorylated, decomposed to pyruvate, which, in the presence of a carboxylase yields acetaldehyde and CO_2, while the acetaldehyde is reduced to ethanol (see figure 4.3). A striking argument in favour of the unified pathway was provided when Meyerhof showed in 1924 that the same 'coenzyme', later recognized as the same complex system, was active in both fermentation and muscle glycolysis [21, p. 138, 143, 145], [1, p. 420, 421], [11, p. 48–50, 76]. The sequence of steps between the hexose phosphate and pyruvate, which was also recognized as an intermediate, remained unknown. In 1933, Embden and coworkers published an important preliminary paper on the cleavage of fructose diphosphate (FDP) into (phospho-) dihydroxyacetone and glyceraldehyde-3-phosphate, with subsequent transformation to give phosphoglyceric acid and finally pyruvic acid[5] [11, p. 91–103, 105–110]. In 1934 Meyerhof and Lohmann discovered in the dialysed extracts of muscle and yeast an enzyme they called 'zymohexase' (an aldolase) which was responsible for this reaction [11, p. 105–110].

In 1929 Lohmann reported the isolation of a new compound which could be split into AMP and pyrophosphate, which became known as adenosine triphosphate (ATP), and which promised to be of great importance. After contradictory results by von Euler on Harden's coenzyme, a compound later identified as NAD, and a polemic between Lohmann and von Euler, it became apparent that both schools were partially correct, ATP being required for the phosphorylation and NAD for the oxido-reduction. Thus a long delay separates the discovery of the 'coenzyme of fermentation' by Harden and Young in 1906 and the elucidation of its composition [11, p. 117–131]. Harden's 'coenzyme' turned out to be a coenzyme system, a shuttle for redox equivalents and energy. The end-reaction of alcoholic fermentation has a

---

4) In 1918, having observed the formation of equivalent quantities of acetaldehyde and glycerol, Neuberg had proposed a scheme of alcoholic fermentation with methylglyoxal as an intermediate and although his proposal was erroneous it was accepted until about 1930 [1, pp. 428]. A number of further hypotheses for fermentation reactions were proposed but also found to be incorrect [21, p. 138, 143, 145].

5) Cori and Cori [22] described phosphorylases of different origins that catalysed the reversible cleavage reaction. This reaction was shown to occur in extracts of muscle, heart, liver and yeast, as well as in extracts of a variety of plant tissues.

very long story of *confused concepts* about the reductive action involved in the reaction leading from acetaldehyde to ethanol. The *reverse reaction* was familiar enough, as the appearance of acetaldehyde from alcohol had been used for a long time by the vinegar industry. A large number of studies were carried out on the enzyme involved [11, p. 146–148]. Finally the studies of Meyerhof, Needham and Warburg during the late 1930s revealed the roles of the coenzymes involved (ADP, ATP, DPN^+, DPNH) in oxidation and reduction steps, and in the energy transfer in the coupled reactions involved. In 1952 Lipmann elucidated the acetyl transfer to coenzyme A in the oxidative decarboxylation of pyruvic acid and the formation of lactic acid. He thus opened a new chapter in the study of the metabolic roles of both ATP and coenzyme A [1, p. 440–442].

An essentially correct scheme of glycolysis and alcoholic fermentation was summarized by Nord [23], and since much of it grew out of the work and theories of Embden and Meyerhof, it is frequently termed the 'Embden-Meyerhof' scheme (Figure 4.3) [1, p. 349–355, 428–437, 440–442]. A scheme of glycogenolysis was presented by Parnass [24]. Nevertheless, controversial and even contradictory conclusions remained and are discussed critically by [23, p. 998–1000].

A most decisive discovery in the study of the mechanism of aerobic oxidation of carbohydrates was provided by Krebs in 1937. He showed that minced pigeon breast muscle could convert oxaloacetic acid into citric acid. This conversion of a four-carbon acid into a six-carbon acid clearly involved the addition of two carbons from some metabolite. On the basis of the facts to hand, Krebs proposed a metabolic cycle involving citric acid as an intermediate, a scheme that has been variously termed the 'citric acid cycle', 'Krebs cycle' and 'tricarboxylic acid cycle'. The great merit of Krebs in addition to his experiments, was his conception of a *cyclic scheme* of oxidations (dehydrogenations) defining the stages through which the carbon atoms of the carbohydrates pass in a cycle of acetyl transfer. This concept represented a key to the understanding of a central range of metabolic as well as industrial processes (Figure 4.4) (The citric acid cycle formulated by Krebs was at the time essentially a theory). Fully established pathways for glycolysis, anaerobic alcoholic fermentation, and oxidative carbohydrate catabolism via the citric acid cycle were finally summarized by Fruton and Simmonds [1, p. 419–465, 466–486] (see also [11, p. 276–283]). Krebs who lost his university position in 1933 left Germany together with other leading scientists because of the Nazi regime, and as a consequence core biochemical research shifted to the UK and USA.

The first *enzymatic syntheses* were shown by Croft-Hill in 1898 with the synthesis of isomaltose from glucose and by Kastle and Lovenhart in 1900 with the pancreatic-lipase catalysed synthesis of butyric acid ethyl ether [25, 26]. The *nature of enzymes* and the *structure of proteins* had remained unknown throughout the nineteenth century and took more than 40 years to be established from 1900 onwards; the topic remained controversial for decades (see also Section 4.4.1 and Chapter 9).

Remarkably, technological applications of these compounds had nevertheless been put into practice since the middle of the century and based on their action they were eventually recognized as catalysts (Section 4.3.1). Traube, as early as 1860, had assumed that enzymes were proteins [27, p. 4]. The entry of E. Fischer

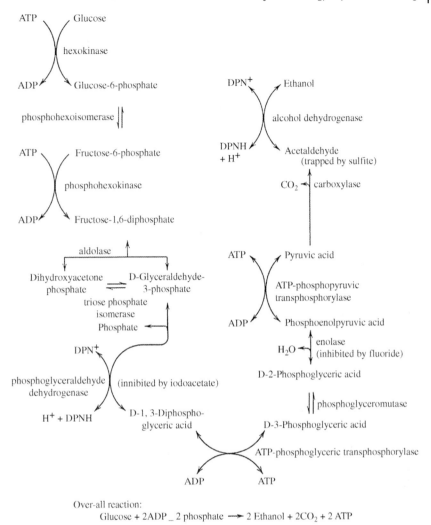

Figure 4.3 Glycolysis: reactions of the glycolytic pathway and anaerobic formation of ethanol and carbon dioxide, as summarized correctly in the early 1950s. Energy is provided via the formation of two ATPs (DPN$^+$ is NAD$^+$, DPNH is NADH in currently used nomenclature; one additional phophate is introduced to give D-1,3-diphosphoglyceric acid [1, p. 435, 436].

into the protein field changed the understanding of these compounds completely when he showed that the amino acids (he recognized about 15) present in an acid hydrolysate were protein constituents. Most important was his systematic attack on the problem of synthesizing long-chain peptides, he provided overwhelming evidence in favour of the hypothesis that peptides containing amino acids were in turn constituents of proteins [28]. Fischer was also convinced that enzymes were proteins. However, Willstätter's view of the structure of enzymes found wide acceptance.

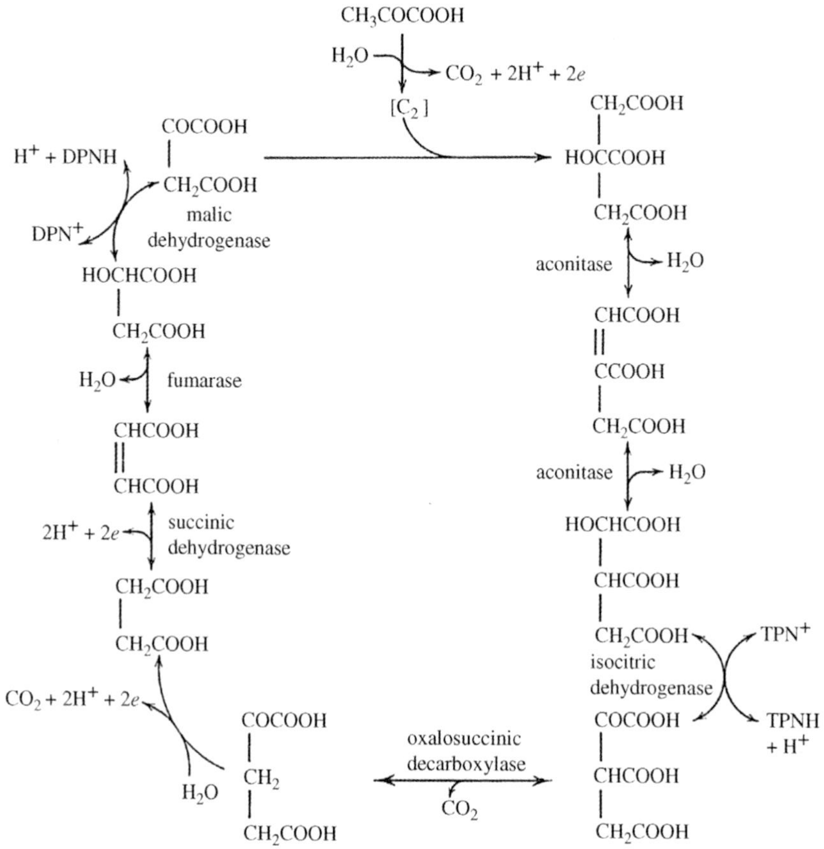

$CH_3COCOOH$

$H_2O \longrightarrow$

$\longrightarrow CO_2 + 2H^+ + 2e$

$[C_2]$

$COCOOH$

$H^+ + DPNH$

$|$

CH_2COOH

malic

DPN^+ dehydrogenase

$HOCHCOOH$

$|$

CH_2COOH

$H_2O \longrightarrow$ fumarase

$CHCOOH$

$||$

$CHCOOH$

$2H^+ + 2e \longrightarrow$ succinic
dehydrogenase

CH_2COOH

$|$

CH_2COOH

$CO_2 + 2H^+ + 2e$

$COCOOH$

$|$

H_2O CH_2

$|$

CH_2COOH

oxalosuccinic
decarboxylase

CO_2

CH_2COOH

$|$

$HOCCOOH$

$|$

CH_2COOH

aconitase $\longrightarrow H_2O$

$CHCOOH$

$||$

$CCOOH$

$|$

CH_2COOH

aconitase $\longrightarrow H_2O$

$HOCHCOOH$

$|$

$CHCOOH$

$|$

CH_2COOH TPN^+

isocitric
dehydrogenase

$COCOOH$ $TPNH$

$|$ $+ H^+$

$CHCOOH$

$|$

CH_2COOH

Figure 4.4 Citric acid cycle, as formulated by Krebs [1, p. 467–481].

According to this theory, enzymes are not proteins, but are composed of one active group responsible for specificity, and one more or less unspecific, and therefore changeable, colloidal bearer [29]. The final proof that enzymes were in fact proteins was the crystallization of urease by Sumner in 1926, and further enzymes (e.g. trypsin) by Northrup and Kunitz in 1930–31 who showed that the pure enzyme crystals were in fact proteins [26].

Amongst many other results, Rudolf Brill's 1923 publication was of special importance in the study of proteins. Based on X-ray photographs, Brill proposed two possible structures for silk protein, either a small peptide or a long polypeptide chain. 'This alternative was, of course, the *last* the chemists wanted to hear because it was the time of the controversy between small aggregates and macromolecules' [30]. The X-ray crystallography method Brill used was developed by William Lawrence Bragg in 1915 who at the age of 25 years was the youngest Nobel Prize winner ever. The method was refined for the analysis of organic molecules, largely through the work of Dorothy Hodgkin (e.g. cholesterol in 1937; penicillin in 1945; vitamin B12 in 1954). In 1934, J.D. Bernal recorded the first X-ray diffraction pattern of a globular protein crystal, pepsin. Throughout the 1930s and 1940s there were many attempts to devise model structures for protein molecules. Finally in 1951, using all the accurate X-ray analysis of the Pasadena school, Pauling, Branson and Corey proposed precise pleated sheet models for the atomic arrangement in β-folded chains and a wholly new helical model, the α-helix, for α-folded chains in protein molecules [30].[6]

4.2.2
Microbiology

The era from 1860 to about 1900 has often been called the 'golden age of microbiology' [31, p. 15], [32, p. 4]. Since the end of that era, work in microbiology or bacteriology, was continued in part following the routes established by Pasteur: collecting data, information and details about microorganisms, their characterization by microscopy, the substrates utilized and products formed. The field was extended by advances in methodology and elaboration of the theory that had been established some decades before, apart from the dogma of vitalism that had been refuted by Buchner. Nevertheless, some outstanding progress in microbiology deserves to be mentioned despite the fact that it had no immediate impact on industrial microbiology (or biotechnology) [31, p. 3–29], [32, p. 16]:

> 1889–1897: The virus concept (tobacco mosaic virus) by Beijerink and Löffler
> 1915–1920: Bacteriophages by Twort and D'Herelle
> 1929: Penicillin by Fleming
> 1930 onwards: Crystallization of TMV virus by Stanley
> Late 1930s: Systematic study by Florey, Heatley and Chain on penicillin, demonstration of its therapeutic value
> 1944: DNA as genetic material by Avery, Mcleod and McCarthy. Streptomycin discovered by Waksman and Schatz

The requirements for products produced by fermentation during both World Wars I and II provided the impetus for progress in the field and led to burgeoning research efforts in applied industrial microbiology.

Work in the institutes founded during the late nineteenth century (see Section 3.3.2) was centred round practical problems of industrial fermentation processes, mainly of beer, alcohol, yeast, acetic and lactic acid. It included characterization of microorganisms of practical relevance and providing pure cultures and fermentation conditions for industrial processes (Section 4.3) [33, p. 35–65], [34]. 'Impure' fermentations lowered the yield of alcoholic fermentations by more than 20%, since other micoorganisms diverted the sugar into other products, notably acetic and lactic acid. Hansen at the Carlsberg Laboratory in Copenhagen laid the foundations for the production of pure yeast cultures. Based on these results, Delbrück in Berlin established rules for practical work that he introduced into the yeast, brewing and

---

6) The history of protein structure research has been published in comprehensive manner by many of the pioneers involved in *The Origins of Modern Biochemistry*, edited by Srinivasan, P.R., Fruton, J.S. and Edsall, J.T., New York Academy of Sciences.

alcohol industries. Further work in Berlin concerned questions of yeast nutrition and the yeast air process ('Lufthefeverfahren'), thus improving the yield of yeast by aeration by a factor of three. This was followed by studies on the process of (semi-) continuous addition of glucose ('Zulaufverfahren') based on insight into the Pasteur-effect (glucose-repression) [34, 35].

Of considerable importance was the finding that *Acetobacter suboxydans* converted sorbitol into L-sorbose with high regioselectivity and yield. Sorbitol is easily obtained by the hydrogenation of glucose. L-sorbose can be converted to vitamin C in a straightforward but complex reaction sequence using protecting group chemistry, the so-called Reichstein process. It was used from the 1930s and continues to be used today to produce vitamin C on a large industrial scale. Another process established in the 1930s was the manufacture of L-ephedrine as a pharmaceutical using the stereoselective condensation of benzaldehyde and acetaldehyde by yeast [36, 37]. A range of enzymes, including diastase (amylolytic enzymes), proteinases and pectinases, were isolated from different organisms for commercial use, mainly from *Bacillus subtilis* and other species such as *Aspergillus oryzae* and *Aspergillus niger*. Thus Takamine began isolating bacterial amylases in the 1890s in what later became known as the Miles Laboratories, USA, (today a part of Genencor) [38, p. 396–494].

Working at the Institut Pasteur in Paris Fernbach's investigations of diverse microorganisms had a major impact on industrial microbiology or biotechnology, respectively [18]. He systematically investigated metabolic intermediates of glycolysis in *Tyrothrix tenuis* and established corresponding enzymatic activities in terms of biochemistry, taking into account the results of E. Buchner. In several papers, Fernbach identified the formation of trioses, aldehydes, acids, notably acetic, succinic and pyruvic acids (he considered the latter to be an intermediate in alcoholic fermentation) by yeast and other organisms. He undertook these investigations both from the aspect of the mechanism of fermentation and that of practical relevance for acid production [18, 39, 40]. Using extracts of the microorganisms, Fernbach identified further enzymatic activities, for example amylases, maltase, sucrase, diastases and other oxidative enzymes.

During the years 1908–1910 numerous investigations were in progress in European countries in connection with the synthesis of rubber. An alliance was formed between several organic chemists and the business community in England, initiated by Perkin [41] and A. Fernbach of the Pasteur Institute. The production of butyl alcohol was developed for the synthesis of butadiene which could be polymerized to yield synthetic rubber. This work formed the foundations of the synthetic rubber industry [42, 43].[7]

The requirements of World War I led to intense research and development work on *new processes* with the aim of producing chemicals required for explosives by fermentation (see Section 4.3.2). Fernbach had obtained patents for the production of acetone and higher alcohols [44, 45]. Weizmann, who had worked in Fernbach's

---

7) The case represents an historically interesting example where industrial interest initiated scientific and practical work long before a theoretical basis for polymerization was established some 15 years later by Staudinger.

laboratory, continued similar research and in 1915 made a new contribution using a more abundant source of raw material, namely maize. He also eventually succeeded in isolating a sporulating organism 'of the long rod type' which, under anaerobic conditions actively fermented a fairly concentrated mash and gave good and constant yields of acetone and butyl alcohol for the explosives industry (solvent for cardite). The investigations were carried out in the Biochemistry Department of Manchester University and by 1914/15 there was a small vessel holding five gallons of mash running almost continuously. These investigations included cultivation conditions and preparation of appropriate cultures, all aimed at industrial production; Gill [46] highlighted the need for scientific control of the process [43, 47].

On the other side of the North Sea in Germany, glycerol fermentation for the manufacture of explosives became a highly important issue. Neuberg had undertaken his research on glycerol formation in the context of his theory of alcoholic fermentation whereas the motives of Connstein and Lüdecke [48] were technical and economic (see Section 4.3.2).

Food and feed production was of critical importance during the war years and a protein supplement for feed was developed at the Berlin institute using the rapidly growing yeast *Torula utilis* (actually *Candida utilis*) which provided high yields. Some 10 factories were erected with capacities of up to 10 000 t of dry yeast which were mainly used for the commercial production of feed [34].

The formation of oxalic and citric acids by *Apergillus* and *Penicillium* sp. had been observed by Wehmer in the 1890s. At the time when citric acid production became a major industrial activity in about 1933, citric acid fermentation was introduced as a subject for academic studies. In 1916 Thom and Currie showed that black aspergilli were capable of producing appreciable quantities of citric acid under defined conditions of culture. Currie then undertook what came to be considered as a classic investigation of the factors controlling the production of citric acid by a selected strain of *Aspergillus niger*, including substrate and the concentration of nutrient nitrogen and the initial pH which had to be as low as 3.5 in order to suppress the formation of oxalic acid (see Section 4.3.3). Whereas citric acid was produced in surface cultures, Klyver and Perquin invented the so-called shake flask technique in 1933 for a fermentation process novel for filamentous fungi known as submerged fermentation. A vast array of hypotheses was published which attempted to explain the conversion of the straight chain form of glucose via diverse intermediates to a branched chain six-carbon acid. Finally citric acid was identified in 1949 as the principal metabolite of the tricarboxylic acid cycle [49–51].

In addition to service work for industry, the role of institutions included education. The Institute in Berlin offered various courses from 1888 (and later a curriculum for brewers), as did the 'Institut für Gärungsphysiologie und Bakteriologie' that was established at the Technical High School in Vienna in 1897 (see Section 3.3.2) [34, 52]. The Institute for Agricultural Bacteriology ('Institut für landwirtschaftliche Bakteriologie') renamed the Institute for Microbiology at the University of Göttingen (Germany) in 1935, became one of the leading institutes for applied microbiology [53]. At Stanford University, the Department of Bacteriology and Experimental Pathology offered courses in bacteriology in the early 1930s. Courses on fermentation were offered by Bernhauer at the German University in Prague in the 1930s, at the University of California, Berkeley from 1947 on, and later at the universities of Columbia and Pennsylvania. The publication of Stephenson's *Bacterial Metabolism* and Kluyver's *Chemical Activities of Micro-Organisms* served to shift the emphasis from the kinetic aspects of growth and death to more biochemical aspects. Bacterial physiology developed rapidly in the decade 1940–50, and during that time formal courses in this field were initiated in many universities. The title of a classical article published in 1930 by Smyth and Obold was 'Industrial Microbiology', and another classical text, the 'Gärungschemische Praktikum', by Bernhauer was published in 1936 [33, p. 60, 61, 104, 132, 202, 203], [54, 108].

4.2.3
The Case of Penicillin – the Beginning of the Antibiotics Era

Fleming's findings on penicillin stimulated further research. However, there was a rather long delay before Florey, Heatley and Chain entered this new field of metabolic products. In his enthusiastic review of the penicillin story, Coghill [55] (Coghill participated in the development of penicillin at NRRL) refers to Shakespeare: 'There is a tide in the affairs of men which, taken at the flood, leads on to fortune'. Perhaps the most significant advance in industrial microbiology during the first half of the century took place during World War II with the development of large-scale fermentation plants for the production of penicillin (see Section 4.3.4) [31, p. 3–29] (The penicillin story has been described in detail both enthusiastically and critically by Bud [56], and by the scientists and engineers engaged in the development of AIChE in the USA [57]). Some early observations prior to Fleming's discovery had identified the antibiotic activities of certain microorganisms, as for example Pasteur's work in 1877 [58, 59].

The breakthrough event was in 1928 when Fleming observed that a culture of a *Penicillium notatum* inhibited the growth of bacteria (Figure 4.5). A spore from that mould had settled on a Petri dish in which he had plated out a culture of *Staphylococcus*. Sometime later before discarding his plates Fleming noticed that the mould colony was surrounded by a zone of growth inhibition and lysis. To his credit, and for which he was awarded a Nobel Prize, he realized the potential of his observation. He demonstrated the production of an antibacterial substance in the culture broth and named it penicillin. He further studied a number of human pathogens and found them to be penicillin sensitive. He published the first paper in the *British Journal of Experimental Pathology* in 1929, and further results in an easily accessible journal, and then widely distributed the culture that produced the active substance.[8] Although he suggested the promising therapeutic utility of penicillin, none of the attempts to isolate the substance was successful, nor were those of Raistrick in London in 1930.[9] Fleming's discovery did not attract any further

---

8) Fleming, A, 1929, *Brit. J. Exp. Path.* **10**, 226; 1944, *Brit. Med. Bulletin*, No. 1, **2**, 4; No. 1, **2**, 5.

9) Raistrick's group characterized pernicillin with respect to its stability and published their results in 1932: Clutterbuck, P.W., Lovell, R., Raistrick, H. 1932, *Biochem. J.* **26**, 1907.

Figure 4.5 Professor Alexander Fleming, 1930 (with kind permission of The Science Museum, London/SSPL).

attention for the next decade [55, 56, 58, 59]. The problems were great and ill-defined, for example the very low concentration of penicillin in the broth (due to its role as a secondary metabolite), no analytical procedures to detect the substance and its unknown structure were available [60].

Towards the end of the 1930s, Florey, Chain and co-workers at Oxford University began to investigate penicillin in the course of their systematic study of antibacterial substances, looking for a better treatment for burns at the time of the fire-bombing of England by Germany. Chain again took up the chemical challenge of isolating penicillin, and Heatley suggested the creation of a much more concentrated solution by extracting penicillin with an organic solvent. The Oxford group certainly deserve the credit for resurrecting penicillin which at the time was described 'as unstable as an opera singer'. They developed an assay, found a method of producing penicillin in surface culture, and demonstrated the marked activity and therapeutic value of penicillin against bacterial infections in a clinical trial in 1940 on a small number of patients. Essentially no side effects were observed. Although modest in scale, the achievements of the Oxford team up to the summer of 1941 have nevertheless been deemed to be of the highest scientific calibre. Their work represented the first instance of a complete evaluation of a drug carried out by a single group of investigators [55, 56, 58, 59, 61, p. 28–30]. Initial yields and recovery however, were very discouraging and the restrictions of wartime England led them to seek assistance from authorities, laboratories and industrial companies in the USA in July 1941 (see Section 4.3.4).

The prospects, and later the success of penicillin, prompted further research on antibiotics. Waksman isolated actinomycin in 1940, streptotricin in 1942 and streptomycin in 1944 from cultures of actinomycetes. Brotzu began a search for antibiotic-producing organisms and examined a culture of *Cephalosporium* sp. He sent the isolated organisms to Oxford, where several antibiotics were isolated from the culture and one was named cephalosporin [59]. A range of derivatives, notably of penicillin, were found and produced [55]. The era of antibiotics had begun.

4.3
Technological Development, Progress and Application

4.3.1
Expanding Enzyme Technology

At the beginning of the twentieth century, enzymes which were mainly produced by fermentation began to be used by the food industries. Early patents and applications have been summarized by Neidleman [62]. J. Takamine began isolating bacterial amylases in the 1890s in the institution which later became known as Miles Laboratories (today a part of Genencor). In 1894, he obtained a patent for the production of a diastatic enzyme preparation from moulds, which he called 'Takadiastase'. The enzyme was produced in a surface culture of *Aspergillus oryzae* [38, p. 401]. In 1895, Boidin had discovered a new process for the manufacture of alcohol called the 'Amyloprocess'. This process involved cooking cereals, inoculating them with a mould which formed saccharifying enzymes and subsequent fermentation with yeast. Together with Effront who had been working on enzymes for alcohol production since 1900, Boidin founded the Société Rapidase (later a part of DSM-Gist-Brocades) in 1920 [63, p. 6]. Proteolytic enzymes have been successfully used for chill-proofing beer since 1911 in the USA [38, p. 327–340]. So urgent was the demand throughout the brewing industry in the USA for a practical solution to this problem that in 1909 and again in 1910 the then US Brewmaster's Association offered two cash awards [62]. As early as 1890, Lintner observed that wheat diastase was important in dough making. This effect was extensively studied and led to the addition of malt extract becoming common practice. In 1922 American bakers used 30 million pounds (13.5 × 10³ t) of malt extract valued at 2.5 million dollars [38, p. 419–427]. The production of pectinases began around 1930 for use in the fruit industry (Schweizerische Ferment, now part of Novo Nordisk) [64].

The use of enzymes became important in the manufacture of leather. For the preparation of hides and skins for tanning, the early tanners kept the de-haired skins in a warm suspension of dog or bird dung. In 1907, Röhm patented the application of a mixture of pancreatic extract and ammonium salts as a bating (tanning) agent, replacing the unpleasant use of dung. In the same year he founded a company, which successfully entered the market and expanded rapidly (Figure 4.6). (It later became the Rohm and Haas company in the USA). Different enzyme preparations used for bating were produced by fermentation using *Bacillus* or *Aspergillus* sp. [38, p. 485–496]. The history of the Röhm company makes it obvious that the market for a new product providing technical progress is an important factor, but that the background of scientific knowledge on the principles of enzyme action, that Röhm had learned from Buchner, is equally important as a condition, leading experiments to a technically feasible solution [65].

4.3.2
Wartime Production, First World War: Acetone, Butanol, Glycerol

An *industrial incentive* – an alliance between several organic chemists and the business community in England together with A. Fernbach of the Pasteur Institute – stimulated the development of new industrial processes since there was a shortage of rubber on the world market. In 1908 the company, Messrs. Strange and Graham, Ltd., 'directed attention to the subject'. Perkin [41] proposed an alliance comprising

1909 am 22. Juli
Verlegung der Firma nach Darmstadt.
Errichtung von Filialen in Philadelphia (durch Otto Haas) und Lyon.

Werk Darmstadt 1910

Figure 4.6 The Röhm company in 1910 [65].

an extended list of chemists and bacteriologists including A. Fernbach, and C. Weizmann of Manchester University. There was no doubt that rubber might be obtained synthetically by the polymerization of isoprene and its homologues, notably butadiene. 'The dramatic nature of the race in progress can be seen by the dates of patents' that was highlighted in a paper by Perkin and referred to several patents applied for by BASF (Badische Anilin und Soda Fabrik) and the Bayer Co. [33, p. 48–55], [41]. Access to appropriate raw materials at a reasonable price was a major problem. In the meantime, around 1910 Fernbach had discovered fermentation processes which were of 'the greatest importance', the main products of which were acetone or butanol (butyl alcohol). He had obtained patents for the fermentation process to produce acetone and higher alcohols and acids [44, 45]. Butanol was assumed to undergo chemical conversion to butadiene which could then be polymerized to yield good quality rubber. The new processes and raw materials were relatively inexpensive [18, 41].

Shortly after this initiative, it was the demands of *the First World War* which drove the technical innovation in the fermentation industries and led to the establishment of new processes which in part disappeared after the war ended. 'During the war period greatly increased demands for many essential commodities had to be satisfied . . .' [43]. Acetone and butanol fermentation became a key technology for the production of explosives since acetone which ran short in Britain under war conditions was required as a solvent.[10] The process that was established industrially was developed by Chaim Weizmann (who later became the first President of Israel) and is usually referred to as the Weizmann process [47, 66, 67] (for more details of the political background see [33]).

Weizmann working at Manchester University had learned this particular fermentation process in Fernbach's laboratory (see Section 4.2.2). He developed two important modifications; first, he used maize as a substrate as it was available in large quantities and at a low price, in contrast to potato which had been used previously, and second he identified an organism (*Clostridium acetobutylicum Weizmann*) that produced four times more acetone than those previously used and which was able to utilize maize as a substrate [68]. Weizmann recognized the potential assistance which such a new development might give to the war departments. In the spring of 1915, he brought his laboratory experiments to the notice of the Admiralty. He asked Winston Churchill, the first Lord of the Admiralty, to build a plant and in July of that year a pilot plant was erected in Nicholson's London gin distillery [33, 43, p. 57]. An experimental plant (with a 500-gallon seed tank and a 12 000-gallon fermenter, 1.9 and 45 m³, respectively) was ready for use at the Royal Navy Cordite factory by about Christmas 1915. It was due to the large scale of this plant that great difficulties were encountered (mainly with respect to the sterile operation of the whole process, including vessels, pipes, etc.). These experiences brought to the fore the necessary fundamental principles required in the design of such a plant and its manipulation, and highlighted the necessity for scientific control of the process work [43, 46].

---

10) The British Service explosive, cordite, was a gelatinized mixture of nitrocellulose and nitroglycerin with a percentage of mineral jelly; the solvent used for gelatinizing the carrier varieties of cordite was acetone [47].

At the same time, research work was being carried out under Weizmann's leadership. It was also decided by the Ministry of Munitions to adapt some of the spirit distilleries for the process: three in London and three in Scotland [47].

The large-scale manufacture of acetone by the Weizmann process attained its greatest success at the British Acetones factory, Toronto, Ltd. in Canada. The plant in Toronto profited from earlier experiences (including negative ones) and operations commenced in May 1916. Strains of *Clostridium acetobutylicum* were mainly used. A total of 14 new fermenters were constructed, holding 24 000 gallons (91 m³) of mash each (Figure 4.7). The plant which had a capacity of 3458 runs and an output of nearly 200 tons a month was operated successfully from 1916 until the armistice was signed in 1918 when operations ceased. This result 'represents the united and disinterested efforts of bacteriological, engineering, and chemical departments'. Several factories had to be closed down after the war. But, due to requirements

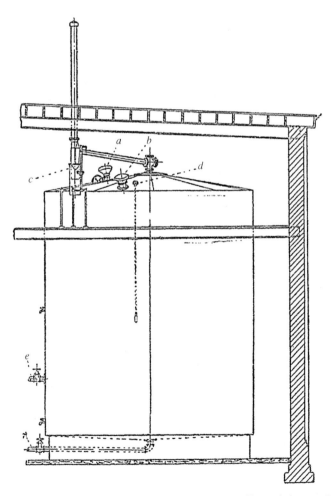

Figure 4.7 Fermenter for the production of acetone and butanol, diameter 5.5 m, height 6.1 m [43].

following the increase in the production of airplanes, new factories went on stream in the USA [43, 47, 68, 69].

In Germany, the major war requirement was for glycerol (glycerine) after the supplies of fat were dramatically curtailed as a result of the imposition of a blockade. Investigations initiated by Lüdecke with the object of obtaining glycerol on an industrial scale by means of the fermentation of sugar, assumed supreme importance after the outbreak of the war. The process was developed by the Protol Company, and 'as many as 63 factories were at first pressed into the service, although only the few largest were finally retained, the monthly output of glycerol being about 1000 tons' [70]. Connstein and Lüdecke [71] [48] began their investigations in 1914 but the German military wanted to keep the results secret. It was known from Pasteur's work that during alcoholic fermentation about 3% glycerol was formed. Neuberg had found that the yield of glycerol in fermentations using yeast was considerably enhanced by the addition of sulphite. For economic reasons molasses was used as a substrate, and the yeast could be used repeatedly after regeneration. On a manufacturing scale, a 20–25% yield was obtained with the avoidance of infections. After extraction and distillation the glycerol could be used for nitration, and thus for the manufacture of explosives. During the war work was also carried out in the USA using *Bacillus acetoethylicus*, the details of which were published by Northrop *et al.* [72]. The process could be worked continuously.

4.3.3
New Fermentation Products: Citric Acid, Fermented Food, Enzymes; Waste Water Treatment

Citric acid represented another major fermentation product from the 1920s onward. The formation of oxalic and citric acids by *Apergillus* and *Penicillium* sp. had been observed by Wehmer in the 1890s. Being aware of the practical potential, he filed patents, and attempts were made to run a plant in France which failed, however, due to contamination. Of pivotal importance was the work of Currie, a dairy chemist in the Bureau of Animal Industry in Washington who worked in cooperation with Thom, a mycologist at the Bureau of Chemistry. Currie had observed that several strains of *Apergillus niger* produced larger quantities of citric acid than oxalic acid, and that without neutralizing the fermentation fluid citric acid was produced abundantly and much more rapidly, and thus with a considerably reduced risk of contamination. Currie further elaborated optimum conditions for the production of citric acid. He left the Bureau to join Chas. Pfizer in New York, where a plant was established which went into production in 1923. Major engineering difficulties were encountered in converting the laboratory fermentation to a plant process. These centred around the problem of sterilization and handling a large number of comparatively small shallow pans of surface-growing *A. niger*. Later, in 1952, successful pilot plant fermentations were established which employed stirred and aerated tanks and produced yields in excess of 75% [51]. Within a few years, other countries were competing in the production of citric acid from citrus fruit, partly because Italy and the USA imposed high taxes on exports and imports, respectively. By 1933, this industry already accounted for 85% (Europe 5100 t and USA 3500 t) of the world's citric acid production of 10 400 t [49–51]. Further products manufactured by fermentation were gluconic acid and lactic acid (by lactic acid bacteria, produced from glucose, sugar or corn starch). The first commercial plant was established in the USA and had been in production since 1881 [49, 73, 74].

Alcohol continued to be a major product of outstanding economic importance. The volume of production was (in 1911/12, in mn (million) l) 550 in Russia, USA 355, Germany 346 (including 128 mn l of industrial alcohol), France 331, Austria-Hungary 277 and Britain 119. In Germany, potato was the dominant raw material. There were 13 400 agricultural installations and 770 industrial plants. The price of pure alcohol was 52 Mark per hl, the income of the German state was 203 mn Mark from taxes, plus an additional 5 mn from customs in 1911/12 [75, p. 636–795], [33, p. 57]. The large potential of fermentation processes to manufacture industrial products was recognized by 1930 from the many-fold intermediate fermentation products identified since Pasteur's work and following that of Buchner. This has been summarized by Fulmer [76], highlighting microorganisms as unique tools for the production of chemicals.

In the *food industries* an important number of mostly traditional fermentation products were manufactured which were of considerable commercial value. Yeast, in the form of pressed yeast, was a major industrial product that was developed during Pasteur's later years, and his results on aeration increased the yield remarkably. Further significant improvement was due to the work of Hansen on pure cultures. The education of bakers in the use of compressed yeast as a leavening agent in bread making had been undertaken at the Centennial Exposition in the USA in 1876. A large number of plants were built to ensure adequate distribution to various parts of the USA, and around 1927, it was estimated that more than half of the bread consumed in the country was made by bakers using compressed yeast. In the region of 9–11 billion pounds (4.1–5.0 mn t) of bread was produced and production was worth 694 mn dollars; the production of yeast amounted to some 230 mn pounds (104 000 t) in the USA. Yeast was also used in the brewing and the alcohol industries. In China and Japan, it was applied to the manufacture of sake [35, 77]. Other traditional products included tea, coffee and cocoa, sauerkraut, vinegar, sauces and cheese. Sauerkraut, besides its traditional role in Germany, had also attained a high degree of popularity in the USA, as had food products from the Far East which were produced using moulds and were used in condiments, such as soya sauce and Worcestershire sauce. The most popular foods not only in Europe but also in the USA were different varieties of cheese such as camembert and Roquefort which were prepared using *Penicillium camembertii* and *Penicillium roqueforti*, respectively [77].

Again it was the specific demands of *World War II* that drove technical innovation in the fermentation industries and the establishment of new processes: alcohol fermentation using wheat as a new and readily available raw material (the supply of molasses had been largely cut off from early on during the war) (the extensive programme concerning the industrial production of penicillin is discussed in Section 4.3.4). During the five years prior to 1941, an average of 123 mn gallons (465 000 m^3) of industrial alcohol was made, around 80% by fermentation [78]. In order to solve the severe problems associated with the fermentation of wheat the War Production Board in the USA set up a collaborative research programme coordinated by the Northern Regional Research Laboratory in Peoria which included some 17 companies and other institutions. The success of this programme is reflected in the peak production of alcohol in 1944 of 606 mn gallons (2.3 mn m^3, over 90% produced by fermentation) [78].

At the beginning of the twentieth century, *biological waste water treatment* became a standard procedure for the purification of municipal and industrial waste water which in many cases could be treated together. The main aim was to avoid or reduce sources of infection and the development of putrescent odour caused by the degradation of organic substances in solution, as well as fine suspended organic material, and pathogens. The first processes applied were the fixed bed (Füllkörper; discontinuous operation) and trickling filter (Tropfverfahren; continuous operation) processes. Both had to provide aeration which is essential for aerobic degradation. A process using suspended bacterial flocs and aeration was under development, and the first plant was erected in Essen (Germany) around 1928 for the treatment of 40 000 m^3 of waste water per day. The biological reaction required a sufficient supply of air which was monitored by the decrease in oxygen consumption using titration with permanganate [79, 80].

It had long been known that under anaerobic conditions, one of the products of the decomposition of organic matter by bacteria is *methane*, the major component of *biogas*. It was in 1897 that the first waste-disposal tank in Bombay was equipped with gas collectors and the gas used to drive gas engines. At about the same time, the waste-disposal tanks in Exeter, England, were partially equipped with gas collectors and the gas used for heating and lighting at the disposal works, followed some years later, by similar installations in Australia, the USA, and in Germany. In 1925, at the large waste water plant in Essen, the sludge-reduction tank was also equipped with gas collectors that were connected to the city mains to supply gas for general municipal use. In order to gain some insight into the chemical reactions which occur under anaerobic fermentation, Buswell and Neave performed experiments with pure substances and established mass balance equations which were proved to be correct [79].

4.3.4
War Requirements of the Second World War: Penicillin Production – A New Quality of Technology

Penicillin production on an industrial scale became a major and even dominating challenge from 1941 onwards. Early yields and recovery, however, were very dis-couraging, and led Florey and Heatley to seek assistance from authorities, labora-tories and industrial companies in the USA in July 1941. 'Thus began a wartime collaboration which was to involve the efforts of literally hundreds of biochemists, chemists, bacteriologists, biologists, chemical engineers, physicians, toxicologists, pharmacologists, and pathologists on both sides of the Atlantic, managed and coordinated by industrial executives, academic administrators, and government leaders' [61]. It was the 'antibiotic area', perhaps the most significant advance in industrial microbiology during the first half of the century. Many new products such as other antibiotics, hormones, vitamins and so on, subsequently became available as a result of new biotechnological processes based on this achievement.

Florey and Heatley were advised to visit the Northern Regional Research Laboratory (NRRL), Peoria, Illinois (USA), because this institution had just organized a fermentation division. Biosynthesis work began immediately on 15 July 1941. Dr Heartley would stay on to share his expertise with the Peoria team in addition to providing a culture of Fleming's *P. notatum* which had been used at Oxford. Dr Moyer, who had been assigned to the problem, began with routine microbial work on a new fermentation problem. Changing from surface to submerged culture and most notably, including corn steep liquor and lactose in the medium, were found to be key steps in improving yields. Within a short time and by the fall of 1941, the yields of penicillin obtained began to increase from 6 to 10 and then 24 Oxford units per ml, as compared to about 3 units per ml obtained by the Oxford group. These results convinced industry representatives that the industrial production of penicillin in the kg range was a possibility. Strain screening and development, including mutation procedures, proved to be another key factor for increasing yields. The US Army collected many hundreds of strains of penicillin producers from a variety of sources including soil samples from around the world which were then isolated. The best producer of all (labelled NRRL 1951), ironically, came from a mouldy cantaloupe (a type of melon) picked up in a Peoria fruit market. In addition to the research being undertaken in Peoria, research studies were also initiated at the Office of Production Research and Development (OPRD) at the Universities of Minnesota (on microbial strains), Wisconsin (on fermentation), Penn State (on recovery), the Carnegie Institute, Wisconsin and Stanford (on mutation), and at MIT (on drying and packaging). Dramatic results were obtained using an ultraviolet-induced mutant created at the University of Wisconsin which together with further improvements in the composition of the medium led to an increase in penicillin yield of several orders of magnitude to about 1500 units/ml, the success of this mutant made engineering work and technical development much more efficient [55]. A significant hindrance to this work was the considerable emphasis placed on the chemical synthesis of penicillin by the Committee on Medical Research of the Office of Scientific Research and Development. The number of the chemists working on this synthesis which was never published was estimated to be between 200 and 400[11),12)] [81].

In December 1941, the US Government became interested in this work. The US Department of Agriculture (USDA) called a meeting in New York which included representatives from the National Research Council and from four companies,

---

11) In order to illustrate the early technical perspective it must be taken into account that, based on the carbon content, the penicillin yield was about 0.00 001%. With this very low yield it was obvious to any technically trained person that a synthetic process would be more economical [81].

12) Efforts in Britain and elsewhere outside the USA have been summarized by Bud [56, pp. 47, 48] and by Pieroth [58].

Merck, Squibb, Pfizer and Lederle, an event that was considered to represent the real turning point for penicillin production [55, 61]. By 1943, the amazing curative properties of penicillin were becoming pretty well known, and there was a huge demand for the drug. The prime goal established by the government representatives was to have ample stocks on hand for the US army's invasion of Europe in the spring of 1944. That goal was met [55]. Dr A.L. Elder was appointed coordinator of the penicillin programme by the US War Production Board in 1943 [57, Introduction].

At a meeting held in October 1941 with Dr Richards, Chairman of the Committee on Medical Research in the USA, industry representatives agreed to make research teams available to work on the problem of supplying adequate quantities of penicillin. 'As a result, we began to get a trickle of a supply of penicillin during the early months of 1942', as Richards reported [61, 82]. Major emphasis was placed on the fermentation and coordination efforts of Elder at the US War Production Board as demonstrated by his letters to institutions and companies, each accompanied by a memorandum with highly precise and differentiated specifications of records to be made, even 'to disclose to all other producers all of their present information on the production and recovery of penicillin' [81].

By June 1942, enough penicillin to treat 10 patients had been produced; by February 1943 there was sufficient to treat approximately 100 patients. The early surface culture flasks contained about 0.4 l culture medium and for the treatment of 2500 patients, 100 000 flasks were necessary (Figure 4.8). This was however, the most reliable method at the time. Pfizer produced 416 000 Oxford units per month

Figure 4.8 Early surface culture production of penicillin at Pfizer using flasks [61].

Figure 4.9   Penicillin fermenters in operation at E.R. Squibb & Sons, 1946 [83].

in 1942; within two years the level of output of penicillin had been increased by about 140 000-fold. The most efficient approach was submerged or deep-tank fermentation, but there were a number of severe practical problems, the solutions of which were not obvious (Figure 4.9) [61, 82]:

- The mould had to be adapted to growth under these conditions
- The laboratory aseptic techniques had to be translated to the plant scale
- The tanks had not only to be sterilized at the outset, but they had to be maintained in this condition for days while the fermentation proceeded
- Large volumes of air had to be filtered to maintain aseptic conditions and supplied to the fermenters
- Aeration, agitation and mass transfer throughout the operation had to be established at high levels to maintain optimal growth conditions (highly efficient stirring)
- Fittings, ports, pipes, valves, and instrument probes had to be maintained free of foreign microorganisms.

Sterile fermentation processes and enzyme production had been established previously for citric acid (for example by Pfizer), but the difficulties that arose with penicillin fermentation were much more complex due to the instability of the product, the protracted period of fermentation and very severe problems of infection.

Downstream operations, the isolation of penicillin, also represented a challenge. Of the two isolation methods used initially, adsorption onto carbon and liquid–liquid extraction, the second proved to be more efficient but it was not without severe problems: appropriate manipulation of pH and speed were critical, penicillin being very unstable at the low pH required to extract it from the aqueous into the organic phase. Centrifugal extractors provided an answer. The final aqueous solution of penicillin which was highly unstable, had to be isolated to yield the water-free product. This was initially achieved by freeze drying. The very substantial quantities of product solution necessitated the use of vacuum tanks of exceedingly large capacity, a technology that was new to this type of operation (Figure 4.10). Finally crystallization proved to be the best way to isolate penicillin, for example as the potassium salt [82, 84]. A considerable number of companies were involved from the beginning of this work, Merck and Co., Inc., E. R. Squibb and Sons, and Chas. Pfizer and Co. Inc. and some 18 more companies became involved subsequently [81].

On a scale that was unprecedented in applied microbiology, later known as biotechnology, interdisciplinary teams of collaborating scientists from a variety of scientific and engineering disciplines were assembled to meet the challenge of efficient downstream processing on a time scale that was to be much shorter than that of preceding steps (Figure 4.11) [56]. A significant improvement in penicillin producing strains was one condition for success (Figure 4.12). The other was 'integrated innovation' of biology and engineering skills, with much improved understanding of aeration, agitation and mass transfer, and how they could affect

Figure 4.10 Vacuum tanks used to evaporate solvents and to concentrate penicillin [61].

CHRONOLOGICAL SCALE

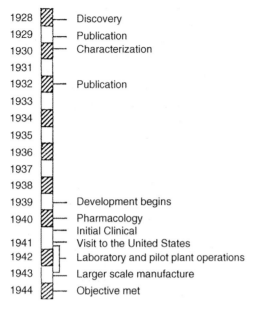

1928	Discovery
1929	Publication
1930	Characterization
1931	
1932	Publication
1933	
1934	
1935	
1936	
1937	
1938	
1939	Development begins
1940	Pharmacology
	Initial Clinical
1941	Visit to the United States
1942	Laboratory and pilot plant operations
1943	Larger scale manufacture
1944	Objective met

Figure 4.11 Chronological scale of penicillin discovery, development and production [61].

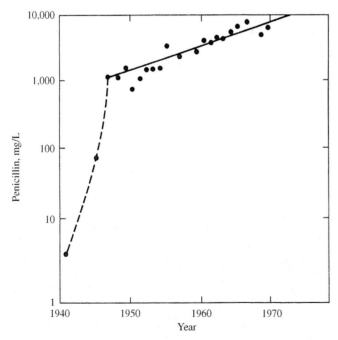

Figure 4.12 Maximum attainable penicillin yields over a 30-year period (yields on a logarithmic scale) [85].

the performance of the biological system [82]. At the *political level* 'the injection of funds, people, companies' and government interest meant a transformation in the ways of doing science' [56, p. 23–53].

The drug became a great topic of interest among the *public*. 'In autumn of 1942, penicillin became a public property and part of the history of wartime culture, resembling a Hollywood movie with an enormous cast, a major marketing campaign, big business, and the great instruments of wartime propaganda' [56, p. 54–74]. Newspapers, magazines and radio publicity stimulated the public's demand for penicillin and increased the pressure to make penicillin available, by for example such headlines as 'The Yellow Magic of Penicillin' (*Readers' Digest*, August 1943), or 'Penicillin, "New Miracle" Drug Combats Bacterial Infections (*Evening Star*, August 1, 1943) [81]' and as highlighted in the film "The Third Man" with Orson Welles (GB, 1949).

4.4
Theoretical Approaches

4.4.1
The Biochemical Paradigm, and the New Discipline of Biochemistry

In 1894, E. Fischer [3] had shown the unique, special potential of enzymes as catalysts, providing a high degree of precision in regio- and stereo control of reactions of the cell, linked to the three-dimensional structure of substrates and products. This concept obviously is far from that of the former of unorganized ferments (the term for soluble enzymes). Now an organized structure with high stereochemical precision was considered to be essential for the chemical potential to catalyse highly selective reactions.

Buchner [5, 6] finally established a new paradigm, the *biochemical paradigm* which was in strict contrast to the vitalist dogma and resolved the lengthy dispute on vitalism that still postulated hidden mysterious forces, the *vis vitalis*, in fermentation. Buchner published a series of papers that signalled a breakthrough in fermentation and enzymology ([5–10]; for details see [86]; a detailed history of the key research events has been presented by Florkin [11]): they showed that enzyme catalysis, including complex phenomena like that of alcoholic fermentation, was a chemical process not necessarily linked to the presence and action of living cells; all reactions in living matter are catalysed by enzymes which by themselves are sufficient for catalytic activity [25].[13] No hidden laws beyond those of physics and chemistry remained.

---

13) Buchner commented on the new paradigm as follows: '...the pioneering work of *Pasteur*, that he began in the midst of this century, led to the final conclusion ...: *no fermentation without organisms*.... It was the victory of the vitalist theory' [6]. This theory, or hypothesis, was superseded by *Buchner's* new results. 'As the agent of the fermentation effect a soluble substance has to be acknowledged, without doubt a protein body: it shall be called zymase' [5].

'However, with respect to *Pasteur's* theory, it is impossible that older theories can disprove new experimental results' – thus highlighting the central role of the experiment [7]. 'The proof..., by the way,... is not a 'circumstance', that is in strict contradiction to Pasteur's views... and it only needs a modification: No fermentation without zymase, that is formed by organisms' [7–10].

Table 4.1 One hundred years of scientific dispute and mechanistic–vitalistic controversy relating to fermentation and biocatalysis.

1790/1793	A.L. Lavoisier: rational, quantitative analysis of fermentation products
1810	Gay-Lussac: quantitative analysis; 'GENERATIO SPONATANEA' (spontaneous generation of organisms)
1815	Béral: 'VITALIST PRINCIPLE' ('le pricipe vital') – MYSTERIOUS 'VIS VITALIS'
1834	J.J. Berzelius: CONCEPT OF CATALYSIS, including fermentation and enzymatic reactions
1837/38	Schwann, Cagniard-Latour: evidence for LIVING MICROORGANISMS as the cause of fermentation
1839	J. Liebig: chemical, but not catalytic concept – fermentation as an ability transferred by a body under decomposition
1861	L. Pasteur: fermentation as a consequence of microbial action – unequivocal experimental basis, confirmation of the vitalist principle
1894	E. Fischer: Specificity of enzyme action – lock and key hypothesis
1897	E. Buchner: BIOCHEMICAL SOLUTION TO THE MECHANISTIC–VITALISTIC CONTROVERSY: alcoholic 'fermentation' without cells – breakthrough of the biochemical concept of enzymes active in fermentation

Thus, *biochemistry* was orientated at the chemical-molecular level, assuming enzymatic reactions of pure chemical character and structures of molecules to be the basis of the transformations observed in fermentation. What Buchner provided was more than a theory; it was the basis of a new conceptual system, that may be termed a new *paradigm*, guiding – on the macroscopic level – future research in the field [11, p. 91]. A lengthy dispute which had continued for nearly 100 years since the establishment of scientific chemistry was finally settled; many of the major scientific events are summarized in Table 4.1. The changes brought with the biochemical paradigm may not however, be interpreted in terms of Kuhn's scientific revolution – it did not reverse the experimental results obtained before by Pasteur. Buchner's work is in fact based on Pasteur's theory and experimental techniques relating to inoculation and sterility. The essentials of this theory with the exception of the concept of vis vitalis, remained valid for future microbial research and the large amount of empirical data remained valid within the body of the knowledge that was accumulating in the biological sciences. The biochemical paradigm also provided the field of technical development with a new, scientific basis for enzyme processes giving important insight into such problems as for example Röhm's search for a solution for a new bating process (see Section 4.3.1) [87].

A new *epistemic thing* emerged which informed the unknown, open questions, the driving forces of research [88]. Whereas Buchner initially supposed one single enzyme, called zymase, to be the catalyst, it was soon discovered that there was a complex multistep sequence of intermediates and different enzymes, in fact the whole glycolytic pathway which remained unidentified for another 30 years.[14] The

---

14) Buchner's own position as to whether one or several enzymes were responsible for the conversion of sugar into alcohol, changed during his own subsequent research 'with a view to the elucidation of the mechanism of the process and of determining the character of the active substance' [13, 15, 17, 89] (see also Section 4.2.1).

chemical nature of enzymes also remained unknown, and again it took more than 30 years to establish their protein nature. Obvious questions arose in the context of alcohol formation by 'zymase': what are the intermediate steps and intermediates; what is the reaction sequence (see Figure 4.3)? These questions represented *epistemic things*, in part unidentified at first, substances identified in different biochemical contexts, the identification of singular reaction steps, molecular transformations, intermediates, energetics, guided by the biochemical paradigm, as well as aiming at identifying the structure of molecules. This included proteins, and the three-dimensional arrangement of their atoms. They tended to serve two different cellular phenomena: fermentation and muscle glycolysis (see Section 4.2). In summary, at the microscopic level they made up the fascination and dynamics of research [88]. It was the molecular approach to intermediary metabolism that was stimulated and it favoured the development of a community of biochemists who shared the views expressed by E. Buchner and F.G. Hopkins, and considered that the cell represents a machinery of chemical reactions in which substances identifiable by chemical methods undergo changes which can be followed by chemical methods.

There was, however, strong opposition to the enzymatic concept of zymase, and the new theory was met with scepticism. Critical arguments were published by Neumeister (1897), Stavenhagen, Schunk (1898) and others. Buchner and Rapp [7, 9], responded to several objections, and published further quantitative data. Many biologists persisted in the opinion that zymase was nearer to 'protoplasm' than to an enzyme, a piece of surviving protoplasm that was alive, in contrast to 'dead ferments'. The myth of protoplasm thus was still not overcome. It was commonly believed that it had no chemical structure and that its components did not exert their chemical properties in the sense of laboratory chemistry [11]. Buchner answered the criticisms with what was to become the credo of the new community of biochemists: 'The presence of such mysterious agents in press-juice, these pieces of living plasma, must be demonstrated before the simpler enzyme theory, which agrees with all the facts, is discarded' (Buchner, as cited by [11, p. 33). The scientific discussions remained controversial, for example in the Société Chimique de Paris. There Maumené still argued using an 'obscure' status of fermentations [90]. A number of reports soon came out stating that the authors had been unable to repeat Buchner's experiments (which required, as mentioned, the highest precision possible at the time). The reception was uniformly negative in the circles of brewing technologists, including Max Delbrück and Stavenhagen who reported that press juice was inactive, and who refused to recognize zymase on the grounds that it 'was in complete contradiction to Pasteur's theory' [11, p. 32, 33]. Buchner's clear objection was that 'it is impossible that older theories can disprove new experimental results' [7–10]. Most active support came surprisingly from the disciples of Pasteur.[15] Opposition disappeared with time-even though some scientists never change their minds, they eventually die (M. Planck) [19, p. 187, 188]. Kohler's analysis (see [11]) takes into

---

15) Emile Duclaux, director of the Pasteur Institute, hailed the discovery with enthusiasm. Roux similarly stated in a lecture of November 1998, that the science of ferments and that of microbes depends on the study of chemical reactions, due in the most part to enzymes [11].

account the intellectual as well the social aspects – the zymase controversy 'can best be treated as a socio-cultural event' [91].

The activity in scientific research stimulated by the *new discipline of biochemistry* was significantly due to the guidance of a new paradigm [86, 92, 93]. The great advantage of enzymology was that it presented a new world of opportunities for biochemical work and for research which resulted in the central objective of biochemical enquiry. Biochemistry and physiological chemistry became more generally adopted, and biochemistry was established as a scientific discipline during the first decade of the twentieth century [11, 21]. It was mechanistic and reductionist in nature, relating biological events causally to chemical events on the basis of many findings on enzymatic reactions (cf. 94, p. 151). In defining the *scope of biochemistry*, its aim was stated to be the description, in the language of chemistry and physics, of physiological phenomena observed in living matter[16] [1].

By the late 1920s, it had become obvious that fermentation, both aerobic and anaerobic, and respiration were complex phenomena which although the details remained unclear, followed a common biochemical mechanism in their initial steps. Since then 'its *unity through life* became clear' [11, p. 23], [21]. The great number of investigations into *glycolysis* led to different hypotheses, both correct and erroneous, which were established and discussed over nearly 50 years thus creating a sequence of *epistemic things* that came to the fore during progress. Thus, Harden's 'coenzyme' was found to be a coenzyme system, a shuttle for redox equivalents and energy. Finally the elucidation of the complete sequence of reaction steps in glycolysis in the 1930s represented a triumph for biochemistry, showing that enzymatic reactions provided the energy for metabolic processes as well as mechanical (muscle) work, and that the metabolites contributed to the synthesis of the organic material of cells, namely carbohydrates, proteins, nucleic acids and so on [11].

The TCA (tricarboxylic acid) cycle formulated by Krebs was at the time essentially a theory, the chief aim of which was 'to describe, step by step, the chemical changes of the carbohydrate molecule leading to the formation of carbon dioxide and water' [11, p. 277–279]. It finally provided the synthons for the synthesis of biological material, and energy for biological processes. Surprisingly, nearly a century after Berzelius' speculation on enzymes synthesizing all natural compounds, this question had remained open. Thus, in his book Oppenheimer continued to speculate that there were phenomena which suggest that there are specific synthesizing enzymes [95, p. 33].

---

16) In more detail *'biochemistry* deals with the study of (1) the nature of the chemical constituents of living matter and of chemical substances produced by living things, (2) the functions and transformations of these chemical entities in biological systems, and (3) the chemical and energetic changes associated with these transformations in the course of activity of living matter'. This definition formed the basis for the successful establishment of enzymology, the identification of the intermediate steps of a microbiological phenomenon and alcoholic fermentation during the first decades of the 20th century. Physiological chemistry, established mainly in medical faculties in Germany, tended to solve biochemical problems *in vivo* by purely physiological methods.

The establishment of glycolysis and the citric acid cycle did not have an immediate impact on the development of biotechnological processes, apart from the research of Fernbach. However in the long term it provided the basis for optimization via advanced kinetics (e.g. citric acid and penicillin production, in this case with precursor strategies) and metabolic engineering strategies which had been undertaken since the 1990s to improve yields and make new (intermediate) products available.

The *nature of enzymes* and the *structure of proteins* required more than 40 years to be established (see also Chapter 9). From about 1900 E. Fischer had worked on a synthetic route for proteins with considerable success and compared the polypeptides he obtained composed of up to 18 amino acids with natural proteins. The products were similar to 'peptones' and proteins [96]. Nevertheless Fischer expressed a certain scepticism when estimates for natural proteins gave molecular weights of up to 15 000 Dalton [97, p. 8]. He was convinced that enzymes were proteins. However, the peptide theory was only a hypothesis advanced by E. Fischer and F. Hofmeister. Protoplasm, including proteins, was attributed with mystical and even magical properties and was widely thought to 'lose its virtue and disintegrate into mundane proteins when extracted'. In 1927 Willstätter was still denying that enzymes were proteins [28].

Establishing the fundamental protein structure thus was another *transition from mysterious substance* to the acceptance of the concept of a pure *chemical structure*, although of a complex three-dimensional order. It required many arguments to overcome the criticisms of the simple peptide bond linking amino acids and the macromolecular structure of which even E. Fischer was critical [97, p. 8]. The peptide bond as the structural element was not accepted as certain for decades. There were several competing theories which even differentiated 'living' and 'death' groups among proteins [28]. In their book, Oppenheimer and Kuhn [95] continued to maintain that the chemical nature of ferments (which was considered to be synonymous with enzymes) remained unclear, and that, almost certainly, these materials would not be members of any of the known chemical groups and would not be carbohydrates or proteins. A major problem with this assertion was that the results obtained by methods such as osmometry and ultracentrifugation indicated that the component molecules were of very high molecular weights. Staudinger had synthesized polymers and in 1926 advanced the theory of macromolecules which however, was not accepted and positively refuted in serious discussions by the established representatives of the scientific community as were the results indicating the macromolecular nature of proteins. It took about a decade for the theory of macromolecules to be accepted.

It was not until around 1940 that the structure of proteins was finally established and this was only after evidence had been provided by many different scientific disciplines including biochemistry and enzymology, physical chemistry, physics, biophysics which included the crystallization of enzymes, new synthetic methodology, proteolytic studies, the concept of the hydrogen bond introduced by Pauling, the ultracentrifuge method developed by Svedberg and various chromatographic methods [28]. Finally, X-ray analysis provided a method for analysing the crystal structure of complex proteins unequivocally (see Section 4.2.1 and Chapter 9). Thus, in contrast to preceding breakthrough events which were due to singular accomplishments of outstanding scientists, the remarkable feature of this innovative step was the *coordinated character of multidisciplinary contributions* which were necessary to elucidate the nature of protein structure.

Remarkably the different approaches are obvious from the historical examples discussed. E. Fischer's work essentially was *explorative* in nature. In contrast, that of Buchner and Pasteur was basically *theory guided* work. Buchner's work was based on an *experimentum crucis*. The work of E. Fischer, Buchner and Pasteur was essentially that of singular, outstanding scientists. In contrast, establishing the structure of proteins and developing penicillin production required the cooperative efforts of different scientific specialities and of different organizations, respectively, including academic, government and industrial laboratories.

A few *institutions* played key roles in research. In addition to the first 'Kaiser Wilhelm Institutes' (today Max Planck Institutes) which was established in 1911, a number of German schools housed in academic departments participated in analysing the pathway of fermentation and muscle glycolysis, that is the molecular aspects of metabolism. The expulsion of leading scientists by the Nazi regime terminated the pioneering role of German research groups. However, earlier established research institutions such as the Pasteur Institute in Paris, the Carlsberg Institute in Copenhagen and an institute in London that later became the Lister Institute continued to progress the research [11, p. 15, 16, 91].

New journals were established at that time: the *Zeitschrift für physiologische Chemie* (produced by Hoppe-Seiler in 1877 and devoted to the study of the 'chemische Lebensvorgänge', the chemical life processes), the *Journal of Biological Chemistry* and the *Biochemical Journal*, both set up in 1905, the *Biochemische Zeitschrift* established by Neuberg in 1906, the subtitle *Zeitschrift für die gesamte Biochemie* for the journal *Beiträge zur chemischen Physiolgie und Pathologie* established by Hofmeister in 1901, in addition to the abstract journal *Biochemisches Zentralblatt* created by Oppenheimer in 1903, and the first issue of the *Bulletin de la Société de Chimie Biologique* which was published in 1914. Furthermore, a vast biochemical literature became available which included student textbooks, monographs and encyclopaedias [19, 20, p. 187, 188]. Later, extended *series* on proteins and enzymes were published: *Annual Reviews in Biochemistry* (since 1932); *Ergebnisse der Enzymforschung* (since 1932); *Advances in Enzymology* (since 1941); *The Enzymes* (since 1950). The first *books* on ferments written in terms of enzymes included that by Effront (*Les Diastases*, 1898) and the most comprehensive volume by Oppenheimer and Kuhn [95]. The books by Nord and Weidenhagen [27] and Hoffmann-Ostenhof [98] illustrate the level of progress over the years from 1925 onwards.

4.4.2
Microbiology – Normal, Postparadigmatic Science

The research schools and books which referred to bacteriology or microbiology worked on pure empirical concepts. The work was not guided by the immediate use of biochemical principles as is obvious from the development of the processes for acetone, butanol and glycerol production, mostly proceeding in terms of *black box* or *trial and error* approaches. This was notably the case at the leading institutes in Copenhagen, Berlin and Vienna, the exception being the work of Fernbach at the Pasteur Institute in Paris (see below and Sections 4.2.2 and 4.3.3). The director of the Vienna institute had written a two-volume textbook, *Handbook of Technical Mycology* in 1896 (English translation 1998), that was followed by a five-volume second edition (1904–1914) [99]. A book on physiological chemistry (*Lehrbuch der Physiologischen Chemie*) by Hammerstein was published in 1926 [109], and a handbook entitled *Technische Mikrobiologie* was issued from 1897 to 1903, followed by a second edition comprising five volumes from 1904 onwards [100].

Thus, work on microbiology or bacteriology continued mainly following the routes established by Pasteur and expanding the field continuously in terms of *normal, postparadigmatic science* (see Section 3.4.3). Biology became a normal science and following the concept of the *unity of sciences* became integrated into the sciences of physics and chemistry nearly 100 years after the establishment of scientific chemistry. No forces or general principles other than those established in physics and chemistry were considered to be valid in biology. Microbiology was orientated around phenomenological topics, types and microscopic characterization of microorganisms, their fermentation products, optimal substrate transformation into products, fermentation conditions and so on. The general need for such work is obvious from a 'Conference on recent developments in the fermentation industries' (C. Chapman [101]) which dealt with 'the employment of micro-organisms in the service of the chemical industry'.[17] The 'important fact from a technical point of view is that through the agency of selected microorganisms, or more probably through the agency of the enzymes, which they elaborate, a yield which falls not very far short of that indicated by … [theoretical] equations can, under favourable conditions, be obtained'. Chapman highlighted 'the ease and completeness with which these complex transformations are effected at ordinary temperatures', the great progress in biochemistry, successful and profitable manufacturing, and international competition [101, 102].

Fernbach, an exceptional scientist who worked both on the biochemical elucidation of alcohol and other fermentations and on industrial applications at the Pasteur Institute, filed and obtained patents for the production of acetone and higher alcohols [44, 45]. Buswell [79] might also be cited as another exceptional scientist who formulated the mass balance equations for a range of anaerobic transformations of different classes of substances, including carbohydrates and fat which were of pure chemical nature, suggesting that he was thinking in terms of biochemical conversion.

---

17) Contributions to the conference directed attention to the inadequate provision made in the UK for systematic instruction in industrial microbiology. 'It will be necessary for this country to strain every nerve to meet foreign competition'. Chapman also discussed the desirability of 'systematic prose- cution of original research in connexion with any industry in which microorganisms or enzymes play an important part' F.L. Lloyd discussed the 'great difficulties in obtaining and retaining pure cultures for technical purposes', [101, 102].

4.4.3
Influences External to Science: War Requirements – the Case of Penicillin

The penicillin story reveals another *epistemic thing* – the significant and even unknown problems associated with the isolation, identification and testing of penicillin activity (Sections 4.2.3 and 4.3.4). It furthermore reveals a new research strategy – the coordination of many laboratories, state as well as university institutes, and industry laboratories which became necessary in view of limitations of the Oxford team. The dramatic success of microbial work in increasing the activity of penicillin in fermentation solution from the 3 units/ml obtained by the Oxford group, by several orders of magnitude to about 1500 units/ml at government and university laboratories illustrates how engineering work and technical development triumphed in the face of apparently insurmountable problems. The strategy of industrial development and manufacture had been applied earlier during World War I in Britain for the production of acetone and ethanol, but it now became a more sophisticated operation due to the number and diversity of laboratories and companies involved together with government coordination.

The development of penicillin production brought new problems with the efficient organization and coordination of the multidisciplinary teams and the multi-level-tasking that was required. Microbial, chemical and engineering work in addition to that of other disciplines, including chemical testing, had to be established at a new level of integration. Problems needed to be solved in parallel, not consecutively and within a very short period of time. This network of innovations also bred personal and national pride, resentments, jealousies, and stories which would themselves be an important component of the penicillin brand. 'The injection of funds, people, companies' and government interest meant a transformation in the ways of doing science' – at the *political level*. Severe problems due to the development of antibiotic-resistant microorganisms were recognized only later during the 1950s [56, p. 23–74, 116–139].

The motive of Florey and Chain to find a new drug for treatment of infections was *external to science*.[18] This was clear from their efforts to find partners both in government and industry. The nature of their work was *scientific* and was aimed at the identification of an *epistemic thing*, the drug that had escaped prior investigations. The work of the Peoria team and the cooperating institutions was clearly initiated externally, and was advanced using established microbial methods-at the highest level. (For general theories concerning technological development including government science and technology policies, and their success or failure, see [107, p. 266–290].

The external initiative did not lead to new basic discoveries, but it prompted the efficient development of new technologies and their management. A range of other

---

18) For a discussion and description of the aspects and problems of the motivations and initiatives internal or external to science see Van den Daele *et al.* [103]. Krohn and Van den Daele [104] distinguish three levels of analysis when external goals become internalized: the level of functional rationalities, that of programming decisions, and that of personal interaction. An example of external initiative in biotechnology such as the big government program (Studie Biotechnologie) initiated in Germany in 1974, has been analysed by Buchholz [105, 106] (see Section 5.2.2).

antibiotics and chemical derivatives, that is other biopharmaceuticals for example modified steroids, were discovered and produced prompting new activity in the pharmaceutical industries which had earlier relied mainly on chemical synthesis which had proved efficient in the manufacture of sulphonamides.

References

1 Fruton, J.S. and Simmonds, S. (1953) The scope and history of biochemistry, *General Biochemistry*, John Wiley & Sons, New York; (a) 1–14, (b) 17–19, (c) 199–205.

2 Fischer, E. (1891) Über die Konfiguration des Traubenzuckers und seiner Isomeren. II. *Berichte Deutschen chemischen Gesellschaft*, **24**, 2683.

3 Fischer, E. (1909) *Einfluß der Konfiguration auf die Wirkung der Enzyme. Untersuchungen über Kohlenhydrate und Fermente*, J. Springer, Berlin, p. 836–844.

4 Fischer, E. (1894) Einfluß der Konfiguration auf die Wirkung der Enzyme. I. *Berichte Deutschen chemischen Gesellschaft*, **27**, 2985.

5 Buchner, E. (1897) Alkoholische Gährung ohne Hefezellen. *Berichte der Deutschen Chemischen Gesellschaft*, **30**, 117–124.

6 Buchner, E. (1898) Ueber zellfreie Gährung. *Berichte der Deutschen Chemischen Gesellschaft*, **31**, 568–574.

7 Buchner, E. and Rapp, R. (1898) Alkoholische Gährung ohne Hefezellen. *Berichte der Deutschen Chemischen Gesellschaft*, **31**, 209–217.

8 Buchner, E. and Rapp, R. (1898) Alkoholische Gährung ohne Hefezellen (5.). *Berichte der Deutschen Chemischen Gesellschaft*, **31**, 1084–1090.

9 Buchner, E. and Rapp, R. (1898) Alkoholische Gährung ohne Hefezellen (6.). *Berichte der Deutschen Chemischen Gesellschaft*, **31**, 1090–1094.

10 Buchner, E. and Rapp, R. (1898) Alkoholische Gährung ohne Hefezellen (7.). *Berichte der Deutschen Chemischen Gesellschaft*, **31**, 1531–1533.

11 Florkin, M. (1975) The discovery of cell-free fermentation, in *A History of Biochemistry, Comprehensive Biochemistry*, vol. 31 (ed. M. Florkin), p. 23–37.

12 Buchner, E. and Meisenheimer, J. (1906) *Berichte der Deutschen Chemischen Gesellschaft*, **39**, 3201.

13 Buchner, E. and Meisenheimer, J. (1910) Chemical processes in alcoholic fermentation. *Chemical Abstracts*, 4, 2844; *Berichte der Deutschen Chemischen Gesellschaft*, **43**, 1773–1795.

14 Buchner, E., Meisenheimer, J., and Schade, H. (1907) Fermentation without enzymes. *Chemical Abstracts*, 1, 1041; *Berichte der Deutschen Chemischen Gesellschaft* **39**, 4217–4231.

15 Buchner, E. and Haehn, H. (1909) The action of the enzymes in the "press fluid" of yeast. *Chemical Abstracts*, 3, 2820.

16 Buchner, E. and Duchavek, F. (1909) Fractional precipitations from the press juice of yeast. *Chemical Abstracts*, 3, 1539.

17 Buchner, E. and Klatte, F. (1908) The co-enzyme of the yeast press-juice. *Chemical Abstracts*, 2, 2568.

18 Fernbach, A. (1910) Sur la dégradation biologique des hydrates de carbone. *Comptes rendus des séances de l'Académie des sciences (Paris)*, **151**, 1004–1006.

19 Florkin, M. (1972) A History of biochemistry, in *Comprehensive Biochemistry*, vol. 30 (eds M. Florkin and E.H. Stotz), Elsevier, p. 30–33, 1973.

20 Florkin, M. (1972) (a) The nature of alcoholic fermentation, the 'theory of the cells', and the concept of cells as units of metabolism. (b) The rise and fall of Liebig's metabolic theories. (c) The reaction against 'Analysm': Antichemicalists and physiological chemists of the 19th century. In: A history of biochemistry. Florkin, M, Comprehensive Biochemistry Vol. 30; (a) 129–144; (b) 145–162 (c) 173–189.

21 Kluyver, A.J. and Donker, H.J.L. (1926) Die Einheit in der Biochemie. *Chemie, Zelle, Gewebe*, **13**, 134–190.

22 Cori, C.F. and Cori, G.T. (1941) Carbohydrate metabolism. *Annual Review of Biochemistry*, **10**, 151–180.

23 Nord, F.F. (1940) Alkoholische Gärung, in *Handbuch der Enzymologie* (eds F.F. Nord and R. Weidenhagen), Akademische Verlagsgesellschaft, Leipzig, p. 968–1011.

24 Parnass, J.K. (1940) Glykogenolyse, in *Handbuch der Enzymologie* (eds F.F. Nord and R. Weidenhagen), Akademische Verlagsgesellschaft, Leipzig, p. 902–967.

25 Sumner, J.B. and Myrbäck, K. (1950) *The Enzymes*, vol. 1, Part 1, p. 1–27.

26 Sumner, J.B. and Somers, G.F. (1953) *Chemistry and Methods of Enzymes*, Academic Press, New York, p. XIII–XVI.

27 Nord, F.F. and Weidenhagen, R. (1940) *Handbuch der Enzymologie*, Akademische Verlagsgesellschaft, Leipzig.

28 Fruton, J.S. (1979) Early theories of protein structure, *The Origins of Modern Biochemistry* (eds P.R. Srinivasan, J.S. Fruton, and J.T. Edsall), New York Academy of Sciences, p. 1–18.

29 Waldschmidt-Leitz, E. (1932) Enzymes. *Annual Review of Biochemistry*, **1**, 69–88.

30 Hodgkin, D. (1979) Crystallographic measurements and the structure of protein molecules as they are, *The Origins of Modern Biochemistry* (eds P.R. Srinivasan, J.S. Fruton, and J.T. Edsall), New York Academy of Sciences, p. 121–145.

31 VanDemark, P.J. and Batzing, B.L. (1987) *The Microbes*, The Benjamin/Cummings Publishing Company, Menlo Park, California.

32 Fuchs, G. and Schlegel, H.-G. (2007) *Allgemeine Mikrobiologie*, Georg Thieme Verlag, Stuttgart.

33 Bud, R. (1993/1995) *Wie wir das Leben nutzbar machten*, Vieweg, Braunschweig/Wiesbaden, original version: 1993, *The Uses of Life, A History of Biotechnology*, Cambridge University Press (citations follow the 1995 edition).

34 Dellweg, H. (1974) Die Geschichte der Fermentation, *100 Jahre Institut für Gärungsgewerbe und Biotechnologie zu Berlin 1874–1974*, Institut für Gärungsgewerbe und Biotechnologie, Berlin, p. 17–41.

35 Frey, J.F. (1930) History and development of the modern yeast industry. *Industrial and Engineering Chemistry*, **22**, 1154–1162.

36 Vasic-Racki, D. (2000) History of industrial biotransformations – dreams and realities, *Industrial Biotransformations* (eds A. Liese, K. Seelbach, and C. Wandrey), Wiley-VCH, Weinheim, p. 3–29.

37 Buchholz, K. and Seibel, J. (2008) Industrial carbohydrate biotransformations. *Carbohydrate Research*, **343**, 1966–1979.

38 Tauber, H. (1949) *The Chemistry and Technology of Enzymes*, John Wiley & Sons, New York.

39 Fernbach, A. (1913) L'acidification des moûts par la levure au cours de la fermentation alcoolique. *Comptes rendus des séances de l'Académie des sciences (Paris)*, **156**, 77–79.

40 Fernbach, A. and Schoen, M. (1913) L'acide pyruvique, produit de la levure. *Comptes rendus des séances de l'Académie des sciences (Paris)*, **157**, 1478–1480.

41 Perkin, W.H. Jr. (1912) The production and polymerization of butadiene, isoprene, and their homologues. *J. Soc. Chem. Ind.*, **31**, 616–623.

42 Fernbach, A. (1919) Discussion, following the article by A. Gill. *J. Soc. Chem. Ind.*, **38**, 280T–281.

43 Speakman, H.B. (1919) The production of acetone and butyl alcohol by a bacteriological process. *J. Soc. Chem. Ind.*, **38**, 155T–161.

44 Fernbach, A. and Strange, E.H. (1911) Fermentation process producing amyl, butyl, or ethyl alcohol, butyric, propionic, or acetic acid, etc. Brit., 15,203 June 29.

45 Fernbach, A. and Strange, E.H. (1913) Fermentation process for producing acetone and higher alcohols from starch, sugars and other carbohydrates. US 1,044,368. Nov. 12 (CA 1913, Vol. 7, 206).

46 Gill, A. (1919) The acetone fermentation process and its technical application. *J. Soc. Chem. Ind.*, **38**, 278T–280T.

47 Nathan, F. (1919) The manufacture of acetone. *J. Soc. Chem. Ind.*, **38**, 271T–273T.

48 Connstein, W. and Lüdecke, K. (1919) Über Glyceringewinnung durch Gärung.

Berichte der Deutschen Chemischen
Gesellschaft, **52**, 1385–1391.

49 May, O.E. and Herrick, H.T. (1930) Some
minor industrial fermentations.
Industrial and Engineering Chemistry, **22**,
1172–1176.

50 Roehr, M. (1996) Citric acid,
Biotechnology, vol. 6 (ed. M. Roehr), VCH,
Weinheim, Germany, p. 307–345.

51 Roehr, M. (1998) A century of citric acid
fermentation and research. *Food
Technology and Biotechnology*, **36** (1),
163–171.

52 Roehr, M. (2001) personal
communication.

53 Schlegel, H.G. (1999) Geschichte der
Mikrobiologie, in *Acta Historica
Leopoldina*, Deutsch Akademie der
Naturforscher Leopoldina, Halle, Saale,
Germany, p. 173.

54 Clifton, C.E. (1966) Microbiology-past,
present, and the future. *Annual Review of
Microbiology*, **20**, 1–12.

55 Coghill, R.D. (1970) *The Development
of Penicillin Strains*, AIChE, p. 14–21.
(see [57])

56 Bud, R. (2007) *Penicillin – Triumph and
Tragedy*, Oxford University Press, Oxford.

57 AIChE (1970) *The History of Penicillin
Production, Chem. Eng. Progr. Symp. Ser.
No. 100*, vol. 66, American Institute of
Chemical Engineers, New York.

58 Pieroth, I. (1994) Penicillin – a survey
from discovery to industrial production,
in *50 Years of Penicillin Application* (eds H.
Kleinkauf and H. von Döhren),
Technische Universität Berlin/Public
Ltd., Prague.

59 Ohno, M., Otsuka, M., Yagisawa, M. *et al.*
(2002) Antibiotics, in *Ullmann's
Encyclopedia of Industrial Chemistry*,
Wiley-VCH, Weinheim, New York, www.
mrw.interscience.wiley.com/emrw/.

60 Demain, A.L. and Fang, A. (2000) The
natural functions of secondary
metabolites. *Advances in Biochemical
Engineering*, **69**, 1–39.

61 Greene, A.J. and Schmitz, A.J. Jr.
(1970) *Meeting the Objective*, AIChE,
p. 79–88 (see [57]).

62 Neidleman, S.L. (1991) Enzymes in the
food industry: a backward glance. *Food
Technology*, **45** (1), 88–91.

63 Uhlig, H. (1991) *Enzyme arbeiten für uns*,
Hanser Verlag, Munich.

64 Poulson, P.B. and Buchholz, K. (2003)
History of enzymology with emphasis on
food production, *Handbook of Food
Enzymology* (eds J.R. Whitaker, A.G.J.
Voragen, and D.W.S. Wong), M. Dekker,
New York, p. 11–20.

65 Trommsdorf, E. (1976) *Dr. Otto Röhm -
Chemiker und Unternehmer*, Econ,
Düsseldorf.

66 Weizmann, C. (1917) Fermentation
process for the production of Acetone and
butyl alcohol. E.P. 150,360. 25.1.17.

67 Weizmann and Hamlin (1920)
Fermentation process for the production
of Acetone and butyl alcohol. U.S. Patent
1,329,214. 27.1.20. Appl. 27.3.18.

68 Ullmann, F. (1953) *Enzyklopädie der
technischen Chemie*, vol. 4, Urban &
Schwarzenberg, München, Berlin,
p. 781–795.

69 Ullmann, F. (1928) *Enzyklopädie der
technischen Chemie*, vol. 2, Urban &
Schwarzenberg, Berlin, p. 709–715.

70 (1919) Review: The manufacture of
fermentation glycerin in Germany
during the war. *J. Soc. Chem. Ind.*, **38**,
287R.

71 Connstein, W. and Lüdecke, K. (1919)
Preparation of glycerol by fermentation.
J. Soc. Chem. Ind., **38**, 691A–692.

72 Northrop, J.H., Ashe, L.H., and Morgan,
R.R. (1919) Fermentation process for the
production of acetone and ethyl alcohol. *J.
Soc. Chem. Ind.*, **38**, 786A–787.

73 Gouthiere, H. (1910) The industrial
manufacture of lactic acid by
fermentation. *Chemical Abstracts*,
4, 1648.

74 Frey, J.F. (1930) Lactic acid. *Industrial and
Engineering Chemistry*, **22**, 1153–1154.

75 Ullmann, F. (1915) *Enzyklopädie der
technischen Chemie*, 1st edn, vol. 1,
Urban & Schwarzenberg, Berlin, p.
636–795.

76 Fulmer, E.I. (1930) The chemical
approach to problems of fermentation.
Industrial and Engineering Chemistry, **22**,
1148–1150.

77 Blank, F.C. (1930) Fermentations in the
food industries. *Industrial and
Engineering Chemistry*, **22**, 1166–1168.

78 Boruff, C.S. and van Lanen, J.M. (1947) The fermentation industry during world war II. *Industrial and Engineering Chemistry*, **39**, 934–937.

79 Buswell, A.M. (1930) Production of fuel gas by anaerobic fermentations. *Industrial and Engineering Chemistry*, **22**, 1168–1172.

80 Ullmann, F. (1928) *Enzyklopädie der technischen Chemie*, 2nd edn, vol. 1, Urban & Schwarzenberg, Berlin, p. 62–87.

81 Elder, A.L. (1970) *The Role of the Government in the Penicillin Program*, AIChE, p. 1–11 (see [57]).

82 Silcox, H.E. (1970) *The Importance of Innovation*, AIChE, p. 74–77 (see [57]).

83 Langlykke, A.F. (1970) *The Engineer and the Biologist*, AIChE, p. 89–97 (see [57]).

84 Perlman, D. (1970) *The Evolution of Penicillin Manufacturing Process*, AIChE, p. 24–30 (see [57]).

85 Demain, A.L. (1971) *Overproduction of Microbial Metabolites Due to Alteration of Regulation*, Adv. Biochem. Eng., 1, p. 129.

86 Buchholz, K. and Poulson, P.B. (2000) Introduction/overview of history of applied biocatalysis, *Applied Biocatalysis* (eds A.J.J. Straathof and P. Adlercreutz), Harwood Academic Publishers, Amsterdam, p. 1–15.

87 Bornscheuer, U. and Buchholz, K. (2005) Highlights in biocatalysis – historical landmarks and current trends. *Engineering in Life Sciences*, **5**, 309–323.

88 Rheinberger, H.-J. (2001) *Experimentalsysteme und epistemische Dinge*, Wallstein –Verlag, Göttingen, also: Rheinberger, H.-J. (1997) *Toward a History of Epistemic Things*, Stanford University Press, Stanford.

89 Buchner, E. and Klatte, F. (1909) The characteristics of expressed yeast juice and zymase formation in yeast. *Chemical Abstracts*, **3**, 187.

90 Maumené, E. (1897) Extrait des procès-verbaux des séances. *Bulletin de la Societe de Chimie Biologique*, **17**, 769–770.

91 Kohler, R. (1975) The History of Biochemistry – a Survey. *Journal of the History of Biology*, **8**, 275–318.

92 Barnett, J.A. and Lichtenthaler, F.W. (2001) A history of research on yeast: Emil Fischer, Eduard Buchner and their contemporaries, 1890–1900. *Yeast*, **18**, 363–388.

93 Demain, A.L. (1981) Industrielle Mikrobiologie. Spektrum der Wissenschaft, November, 1981, 21–26.

94 Gale, G. (1979) *Theory of Science*, Mc Graw-Hill, New York; (a) 143–168; (b) 169-L 277.

95 Oppenheimer, C. and Kuhn, R. (1927) *Lehrbuch der Enzyme, Einleitung*, Georg Thieme, Leipzig, p. 3–10.

96 Fischer, E. (1907) Proteine und Polypeptide; (a) (1907) *Z. Angew. Chem.*, **20**, 913; (b) in Fischer, 1923, Gesammelte Werke, Untersuchungen über Aminosäuren, Polypeptide und Proteine II., Springer, Berlin 43–52.

97 Fischer, E. (1907) Die Chemie der Proteine und ihre Beziehungen zur Biologie, (a) Sitzungsberichte d. Königl. Preuß.Akad. d. Wissenschaften zu Berlin, 1907, 35; (b) in Fischer, 1923, Gesammelte Werke, Untersuchungen über Aminosäuren, Polypeptide und Proteine II., Springer, Berlin 1–21.

98 Hoffmann-Ostenhof, O. (1954) *Enzymologie*, Springer, Wien.

99 Roehr, M. (2000) History of biotechnology in Austria. *Advances in Biochemical Engineering*, **69**, 125–149.

100 Metz, H. (1997) Nutzung der Biotechnologie in der chemischen Industrie, in *Technikgeschichte als Vorbild moderner Technik*, vol. 21, Schriftenreihe der Georg-Agrikola-Gesellschaft, Göttingen, p. 105–115.

101 Chapman, C. (1919) The employment of micro-organisms in the service of industrial chemistry. A plea for a national institute of industrial microbiology. *J. Soc. Chem. Ind.*, **38** (14), 282T–286.

102 (1919) Review: Conference on recent developments in the fermentation industries. *J. Soc. Chem. Ind.*, **38**, 261R.

103 Van den Daele, W., Krohn, W., and Weingart, P. (1979) Die politische Steuerung der wissenschaftlichen Entwicklung, *Geplante Forschung* (eds W. van den Daele, W. Krohn, and

P. Weingart), Suhrkamp Verlag, Frankfurt, p. 11–63.

104 Krohn, W. and Van den Daele, W. (1998) Science as an agent of change: finalization and experimental implementation, in *Symposium: 'Revisiting the theory of "Finalization in Science"'. Social Science Information*, **37** (1), SAGE Publications, London, p. 191–222.

105 Buchholz, K. (1979) Die gezielte Förderung und Entwicklung der Biotechnologie, in *Geplante Forschung*

(eds W. van den Daele, W. Krohn, and P. Weingart), Suhrkamp, Frankfurt, p. 64–116.

106 Buchholz, K. (2007) Science – or not? The status and dynamics of biotechnology. *Biotechnology Journal*, **2**, 1154–1168.

107 Weyer, J. (2008) *Techniksoziologie*, Juventa Verlag, Weinheim, Germany.

108 Berhauer, K. (1936) *Gärungschemisches Praktikum*, Springer, Berlin.

109 Hammersten, O. (1926) *Lehrbuch der Physiologischen Chemie*, Verlag von J. F. Bergmann, München.

5
Outlook, from 1950 Onwards: Biotechnology – Science or What?

5.1
Introduction

By the 1950s large scale production of for example, beer, cheese, citric acid, pharmaceuticals and other products of particularly high social and economic relevance, for example antibiotics had become well established. However, biotechnology did not exist as a scientific discipline and there were no books, journals, curricula or scientific conferences devoted to the subject. More importantly, there was no common language and understanding between scientists in biology and biochemistry or engineers involved in research and development. This chapter provides a short résumé of how Biotechnology (BT) developed over the period from the 1950s to the 1980s and also describes how politics came to recognize BT as a field of potential innovation and how it became integrated into science and technology policy. This led to significant funding programmes which widely promoted expansion of the field.

Growing recognition of economic relevance followed the success of penicillin and the manufacture of other antibiotics based on traditional BT, applied microbiology and the emergent discipline of biochemical engineering. An extended range of established products was being marketed, for example antibiotics, steroids, amino acids, organic acids, carbohydrates, vitamins and solvents. Work in the field of BT proceeded in various disciplines with a low level of coherence and little integration until the 1960s and 1970s. During that time BT became a political issue and benefited from major public funding and increasing economic impact. With the perspective and aim of further innovation, two important and influential studies were undertaken by scientific and government agencies in Germany in the early 1970s and in the USA in the early 1980s. The first dealt with traditional BT whereas the second focused on 'New BT' [1].

Major technical accomplishments and progress were achieved during the late 1970s and 1980s, notably due to genetic research and recombinant technologies (see Chapter 6). A milestone was the model of DNA as the molecular basis of heredity derived by Watson and Crick with the aid of data provided by Rosalind Franklin who worked in Maurice Wilkin's X-ray crystallography laboratory in 1953 [2]. Their theory

Concepts in Biotechnology: History, Science and Business. Klaus Buchholz and John Collins
Copyright © 2010 WILEY-VCH Verlag GmbH & Co. KGaA, Weinheim
ISBN: 978-3-527-31766-0

did not however, stimulate immediate technical innovations or put forward perspectives but led to a new understanding of genetic material. Thus, the 'DNA Revolution' as Hotchkiss termed it, progressed or penetrated slowly into technology, initially having little effect on traditional processes and products [3].

A significant change in BT only occurred during the 1980s and 1990s with common approaches being used in many different disciplines and with the merging of molecular biology and biochemical engineering. Industrial interest and the range of products expanded significantly and many new companies, mainly in the USA were founded. This new era of fundamental biology, genetics and molecular biology, as well as BT, is dealt with in Part II of this book.

5.2
Traditional Biotechnology and the Dechema Report

5.2.1
Traditional Products

As a result of the technology developed for penicillin fermentation and further progress in fermentation technology in general, a new class of additional high value products, the secondary metabolites, became available. These included additional antibiotics such as streptomycin and steroids obtained by biotransformation. Other major products of growing market relevance were amino acids, organic acids, carbohydrates and derivatives (hydrolysates, isomers), vitamins and solvents. Enzymes were also being produced for various novel applications. Economically the most important enzymes were hydrolases, amylases and amyloglucosidases for starch processing and proteases for detergent formulations. The new generation of biocatalysts, involving immobilization techniques developed in the academic field, led to a breakthrough in the processing of food and pharmaceutical compounds. Large scale processes were established using biocatalysts for penicillin hydrolysis (for the synthesis of semisynthetic β-lactam antibiotics) and glucose isomerization, market requirements being the driving force behind technical development. In addition, a range of further processes utilizing biocatalysts were introduced and these are dealt with in more detail in Section 15.3 [4, 5].

5.2.2
Political Initiatives: The Dechema Report

Starting in the 1970s and 1980s BT attracted the attention of government agencies in Germany, the UK, Japan, the USA and other countries as a field with innovative potential and capable of contributing to economic growth. A first enthusiastic report by the German chemical technology organization DECHEMA was produced in 1974 for the German Ministry for Education and Science (Bundesministerium für Bildung und Wissenschaft, BMBW). It was the first systematic approach for BT research funding, emphasizing classical BT and aiming to develop a unified research and development strategy. The German government wanted to accelerate BT research and development to identify and encourage innovations in industry [6–8, p. 183–212], [9]. The report mainly referred to Germany, but similar activities were observed in other European countries engaged in BT and even in the USA where approaches tended to change more rapidly.

Essential topics and aims of the *Dechema study* which reflect the main scientific and applied fields of BT (although not of genetics and molecular biology) at that time are summarized in Table 5.1: established fields were addressed in *applied microbiology*. Remarkably basic biological work was understood to comprise investigations following trial and error approaches such as random mutation aimed at increased productivity of microorganisms. In *biochemical engineering* emphasis was placed not only on the development of standard operations but also on the development of new engineering products, such as new types of fermenters. Most emphasis was placed on the *biological processes* that led to products of economic interest, whereby novel products were envisaged (Table 5.1). The importance of *education* was also emphasized and included detailed recommendations, for example for skills that biologists and engineers would require.

Table 5.1 Fields of biotechnology addressed by the Dechema study (1974).

Field of activity	Topic	Aim
Applied microbiology	Establishment of collections of microorganisms	Availability of strains for production
	Random mutation of microorganisms	Increased productivity of microorganisms
Biochemical engineering	Development of standard bioreactors, scale up	Knowledge about and optimization of mixing, aeration, mass transfer in large-scale reactors
	New types of fermenters	
	Unit operations for downstream processing	Simplified, efficient operation
Biological processes	High molecular weight products including biomass, enzymes and enzyme inhibitors	Products of economic interest
	Low molecular weight products including organic acids, vitamins and growth factors	
	Screening for new antibiotics, antifungal metabolites, antitumour metabolites	Search for new products
Microbial transformations and enzymatic reactions	Regio- and stereoselective modification of organic molecules, steroids and so on	Efficient production of fine chemicals and drugs
	Enzyme immobilization	Efficient enzyme processes

This study has been an intriguing example of interaction between politicians, industrialists and scientists and was termed a corporatist approach by Jasanoff [10]. It prompted major initiatives both in academic, institutional and industrial research and development in Germany. Subsequent studies in other European countries produced similar results. Thus studies on BT were initiated in the UK, Japan and France [8, 9].[1] To summarize, most topics were classical or conventional. Recombinant DNA methods were not mentioned since they were not available at the time of writing the study (1972–74) in contrast to the OTA study which was produced 10 years later and which had a fundamentally different focus (see below).

5.2.3
The Scientific Status of Biotechnology up to the 1970s

From the content of the study some general conclusions can be drawn with regard to the *scientific status* of BT and the character and type of research and development (R&D) work in the field of BT during this period. Biotechnological problems were complex, and remain so. They involve genetic and metabolic processes as well as regulation phenomena. They also include engineering aspects such as mass and heat transfer phenomena, mixing and scale up. The focus of interest was on the identification of new products and new or improved products, processes and services, such as antibiotics, citric acid, steroid drugs and environmental processes. Research problems in general were solved empirically by *trial and error methods* during the period from the 1950s to the 1970s and progressed on a *phenomenological* rather than the molecular level in microbiology. Black-box approaches were essentially used in applied microbiology as well as in biochemical engineering.

The *basic disciplines* involved in BT research and development work were microbiology, cell biology, biochemistry, and – to a limited extent – molecular biology and genetics in addition to chemical engineering. However, interdisciplinary communication amongst microbiologists, chemists, and engineers was found to be difficult due to the different working procedures, specialized scientific and technical languages and different approaches to this new field.[2] The proponents of one discipline, for example microbiology, distrusted those of the other disciplines such as biochemistry and chemical engineering. The traditional competition between microbiologists and chemists fuelled the problems of mutual trust and collaboration.

The major *industries* in Germany as well as in the rest of Europe, notably the chemical, pharmaceutical, and food industries, were conservative in their attitude

---

1) A second major programme paper, the Spinks Report, coordinated by the Royal Society (UK) was published in 1979 and followed a similar concept [8], as did a report for the European Commission entitled 'Biotechnology in Europe' [11].

2) The type of research or the phase of BT work can be described by phase one of the model advanced by Van den Daele *et al.* [12] which identified three phases: the explorative, the

paradigmatic, and the postparadigmatic research phases (see also Section 2.4.3). The type of research in BT taking place during the 1970s and 1980s corresponds to the explorative, preparadigmatic phase, applying trial and error methods, in contrast to theory-guided paradigmatic research which developed during the 1980s and 1990s. This was certainly the case for Germany, but seems to apply at the international level in general.

towards the new field of BT. With few exceptions they did not establish relevant research capacities with interdisciplinary teams. The management of large pharmaceutical companies, mostly chemists, and notably in Germany, was not convinced of the innovative potential of applied microbiology in manufacturing drugs, since the success stories of the early twentieth century, for example those of the Hoechst and Bayer companies were a result of chemical synthesis [7, 13, 14].

University education essentially remained bound to the traditional disciplines, faculties, and curricula. Exceptions to this were the specialized research units at a few UK and American universities that offered special courses.[3] University College London offered a curriculum granting a Master of Science in Biochemical Engineering in the 1960s, and another BT curriculum was established in the 1970s at the Technical University of Berlin [7, p. 69, 71].

The first BT journal of high reputation was established in 1958 by Elmer Gaden and was entitled the *Journal of Microbiological and Biochemical Engineering*. It later became *Biotechnology and Bioengineering* and is still a leading journal in the field. A few other journals were launched in the 1950s and 1960s.[4] As can be seen from Figure 5.1, few journals specializing in BT were established during this period as compared to the number that appeared later. Early *textbooks* on applied microbiology [15, 16] and biochemical engineering [17, 18] were published signifying increased interest in the field. The first encyclopaedia and a series on BT were produced by Rehm and Reed [19] and included the fundamentals of applied microbiology and biochemical engineering but with the focus on products and processes.

5.3
The Changing Focus in BT in the USA in the Early 1980s

(For a detailed discussion of the development of genetics and molecular biology see Part II).

5.3.1
The Role of Politics in the USA: The OTA Study, Promoting New BT

The 'DNA Revolution', as Hotchkiss termed it, progressed slowly with respect to innovation and technical application [3]. Nevertheless, major accomplishments

3) Thus courses were offered at the University of California, Berkeley from 1947, and later at Columbia and Pennsylvania. Graduate courses were offered at MIT by A.L. Demain and D.I.C. Wang from 1970. Such courses were also established in the UK, first in Manchester in 1958 and then in Birmingham, Imperial College, and University College London in the 1960s [7, p. 69, 71], [8, p. 104, 132, 202, 203]. In Birmingham a Bachelor of Science course was established in 1965.

Graduate courses were offered in 1963 in Tokyo by Aiba and Humphrey and at University College London [8, p. 104, 132, 202, 203].

4) For example *Applied Microbiology*, renamed *Environmental and Applied Microbiology*, *Applied Microbiology and Biotechnology*. The first biochemical engineering symposium took place in 1949 in the USA. The International Fermentation Symposium was first organized in the USA in 1964 [7, p. 69].

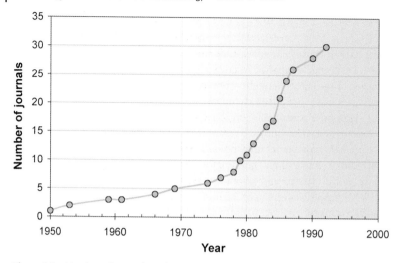

Figure 5.1 Number of journals on biotechnology produced over a period of 50 years.

and progress, notably due to genetic research and recombinant technologies became prominent. Berg and Cohen and Boyer introduced recombinant DNA technology in 1972 when they constructed the first recombinant virus and plasmid, respectively, which after their introduction into foreign host cells remained in a stable form [20, 21]. The field's progress is reflected by the rising number of journals on this theme that were established in the late 1970s and early 1980s (Figure 5.1).

In the USA, the perception of BT was rather different in the 1980s as compared to that in Germany and Europe in the 1970s. This can be differentiated with respect to academia, the industrial sector and at the political level (see Chapters 6 and 7) and is further evidenced by the emergence of new biotech companies during the 1970s and 1980s and by an OTA (Office of Technology Assessment, USA) study ([22]; all subsequent page numbers refer to this study). This difference in perception refers to the methods that arose from recombinant DNA methods, a milestone that allowed a novel definition of what could be envisaged as possible or practicable. In contrast to the reports mentioned previously, the emphasis in the OTA study was on genetic engineering and recombinant DNA (rDNA) technology resulting in commercial opportunities and support for the rapid commercial exploitation of scientific results and was closely associated with the business world: 'This report focuses on the industrial use of recombinant DNA (rDNA), cell fusion, and novel bioprocessing techniques'. 'In the past ten years, dramatic new developments in the ability to select and manipulate genetic material have sparked unprecedented interest in the industrial uses of living organisms' (p. 3).

The report summary gave more details on its focus: The industrial sectors addressed were pharmaceuticals, animal and agriculture, speciality chemicals and food additives, environmental areas, commodity chemicals and energy production. The emphasis was placed on the pharmaceutical sector which has been most active in commercializing BT. Thus the first BT products, such as rDNA-produced human insulin, interferon, and monoclonal antibody (mAb) diagnostic kits, were a direct result of the basic research that led to these new technologies (p. 5). Further products mentioned were antibiotics, vaccines, and hormone growth factors to be produced in the next five to 10 years (from 1984 onwards).

The highlights of the US competitive position were a well-developed life science base, the availability of financing for high-risk ventures, and an entrepreneurial spirit that had led the USA to the forefront of BT commercialization. The political framework made it possible for industrialists and scientists to rapidly capitalize on the results of basic research (p. 7). US companies which pursued BT applications included many of the established pharmaceutical companies and a large number of small new BT firms (NBFs) (p. 6). Based on this analysis, the report presented some strong general recommendations: US efforts to commercialize BT were subsequently the strongest in the world, derived from the unique complementarity and competition that existed between NBFs and established US companies in an open competitive market (p. 11).[5] The factors most important to the commercial development of BT were financing and tax incentives for firms, government funding of basic and applied research, venture capital and personnel availability and training (p. 12).

To summarize, the OTA report, in contrast to preceding studies, highlighted the impetus derived from genetic research which was transforming 'old BT' into 'new BT'. Its approach, with the emphasis on commercial issues, is basically different from that of others in that it focuses on new opportunities (see also [8, p. 183–212], [9], [10], [14]).

5.3.2
New Products and Companies

(A short overview has been given in this section but the topic is dealt with in detail in Chapters 9, 10 and 17).

The industrial breakthrough of recombinant technology came with drugs for pharmaceutical and medical use. The first recombinant industrial product was human insulin, developed by Genentech in cooperation with Ely Lilly in 1978 and approved by the US Food and Drug Administration in 1982. The insulin story is an example of great breakthrough but also of the delayed recognition of the potential of genetics and recombinant technologies by established pharmaceutical companies. In 1974, early in the history of BT, John Collins and later Herb Boyer in 1976 presented

---

5) To compare the US position in the BT field, OTA identified five foreign countries as the major potential competitors of the US in BT commercialization: Japan, Germany, the UK, Switzerland, and France [22, p. 9, 505]. Among the firms leading Japan's drive to commercialize BT were large established companies such as Takeda Pharmaceutical, Mitsubishi Chemical, Sumitomo Chemical, and Ajinomoto, leaning strongly towards BT in an effort to make it a key technology for the future. The Japanese government, which fell behind in starting to form a national support structure, had embarked on building a foundation for R&D (the total expenditure for rDNA research was $38.1 million in 1981) [22, p. 505].

their viewpoint on the impact of genetics on the synthesis of human insulin at Novo in Denmark which supplied 30% of the insulin world market. Their views were taken seriously by young scientists but initially not by management since they believed that there was no possible solution to the problems involved in the industrial development of recombinant bacteria. In contrast, Genentech in the US was convinced of the potential of insulin production by recombinant bacteria; Eli Lilly, which supplied 85% of the American insulin market, signed a research and development contract with Genentech and started to work on an empirical concept which went to scale-up in 1979, and after approval, to production in 1982/83. Novo 'didn't believe it could be done' and followed a competing strategy for synthesizing human insulin by enzymatic modification which led to production in 1982. However, in the 1980s the management of Novo changed its mind and initiated the development of a production process for insulin based on genetic engineering[6] (J. Collins, personal communication [26, 27]).

The breakthrough for human insulin was followed by a series of further recombinant products, mostly drugs, which in general could not be produced by other technical means, and some of which are of great medical interest as well as economic importance. Manufacture of insulin (produced by recombinant DNA) was followed by the production of human growth hormone in 1983, β-interferon and a hepatitis B vaccine in 1986, tissue plasminogen activator (tPA) in 1987, and erythropoietin in 1989 (product approval), all of major economic importance to the pharmaceutical industry. This was followed by the production of a further series of recombinant products, mainly drugs, which in most cases could not be produced by alternative procedures or from natural sources [28, 29]. In addition, from the mid-1970s onwards enzyme technology benefited significantly from recombinant methods. Overexpression in fast-growing host organisms with high protein productivity was competitive. Many enzymes which were not readily accessible and/or too expensive became available. The design of enzymes exhibiting new properties such as increased stability under extreme conditions or modified specificity increased the scope of synthesis, available substrates and products accessible by biocatalysis [5] (see Section 15.3).

The remarkable growth of Biotechnology in recent years is certainly based on the expanded and much greater synthetic potential of recombinant technologies compared to classical Biotechnology. Examples, in addition to those mentioned previously, are the production of mammalian proteins by microorganisms and/or cell culture which would not otherwise have been accessible, most notably monoclonal antibodies as potent drugs (see Sections 7.4, 9.7.2 and 10.8); the design of proteins, notably enzymes, and the design of new pathways to make products accessible using Biotechnology via easier, economic and/or ecological processes

---

6) Cloning of the first industrial enzymes, α-galactosidase and penicillin amidase, both in *E. coli*, was achieved by Boehringer Mannheim and Wagner, Mayer and Collins in 1979 [23]. The production of the first recombinant enzyme of large dimensions by Boehringer Mannheim followed in 1981 [24, 25]. In 1980 genetic engineering of amylase production was a test case at Novo for food enzymes, going to the approval process, and finally being marketed in 1984 [26, 27].

(metabolic engineering). This greatly expanded potential is due to the rapid progress in accumulating basic knowledge together with the development of new or improved methods and techniques which can be automated thereby enhancing their potential.

Large-scale investment by multinational companies, the founding of many small new companies, and state funded large-scale research characterized the early period of rapid growth from the mid 1970s to the 1990s. This was a type of *gold rush* to support the 'New Biotechnology' as recombinant technology was known in the USA. The transfer of science to the economic sphere resulted in the continued founding of new BT companies; in 2000 there were over 1200 in existence in the USA and some 1500 in Europe. A total of 77 recombinant biopharmaceutical products were available on the market in 2000 and that number had increased to 165 by 2006 [30, 31] (see Sections 9.7.2 and 17.4.2).

During this period the combination of basic scientific research and mechanization in the form of robotics led research to a novel form of *industrial quality*. Although discoveries such as restriction enzymes and the polymerase chain reaction (PCR) were scientific in nature, their practicality was nevertheless firmly rooted in their use in automated systems thus considerably increasing their potential and significantly accelerating research [1, 32]. In this context a number of consortia have formed worldwide since 1999 to pursue the goals of structural genomics.[7]

5.4
Conclusions

The integration of applied microbiology, biochemical engineering and molecular biology has led to the creation of biotechnology as a new scientific discipline in its own right with a common paradigm at the level of molecular research. Subdisciplines such as genomics, transcriptomics, proteomics, *metabolic flux analysis* with quantitative analysis of complex metabolic pathways and finally biochemical engineering and bioinformatics, have merged to create *biosystems engineering* [35–37]. A remarkable increase in the number of citations in *Biological Abstracts* between 1985 and 2001 is a testament to the increasing research activity and dynamics in these fields (Figure 5.2).

The basic discipline underlying new BT has been molecular genetics. This introduced a new paradigm based on knowledge of the molecular biology involved. Basic engineering concepts also became orientated towards molecular processes. Taken together this implies that theoretical concepts have guided progress through an understanding of molecular aspects. Biotechnology encompasses the study of highly complex genetic and structural properties in addition to reactions, biocatalysis and regulation phenomena. It further integrates biochemical engineering,

---

7) The Sanger Institute, Wellcome Trust, Cambridge, UK, founded in 1993, and Celera (USA) are outstanding examples of industrially organized research [33, 34].

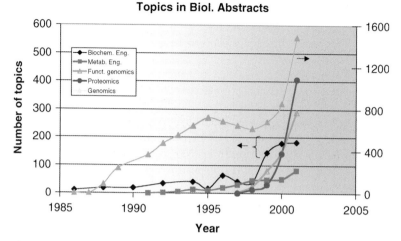

Figure 5.2 Numbers of research papers cited in *Biological Abstracts* on Biochemical Engineering, Metabolic Engineering, Functional Genomics, Proteomics (numbers of citations given on the left) and Genomics (right).

transport phenomena in reactors, their interaction with biological structures, and processes at the molecular and cellular level. It also aims at modelling and understanding these phenomena at the quantitative level [1] (see Sections 13.6 and 15.6).

References

1 Buchholz, K. (2007) Science – or not? The status and dynamics of biotechnology. *Biotechnology Journal*, **2**, 1154–1168.

2 Watson, J.D. and Crick, F.H.C. (1953) *The Structure of DNA*, in Cold Spring Harbor Symposia on Quantitative Biology, p. 123–131. Cold Spring Harbor, L.I., New York.

3 Hotchkiss, R.D. (1979) The identification of nucleic acids as genetic determinants, in *The Origins of Modern Biochemistry* (eds P.R. Srinivasan, J.S. Fruton, and J.T. Edsall), New York Academy of Sciences, New York, p. 321–342.

4 Buchholz, K. and Poulson, P.B. (2000) Introduction/overview of history of applied biocatalysis, in *Applied Biocatalysis* (eds A.J.J. Straathof and P. Adlercreutz), Harwood Academic Publishers, Amsterdam, p. 1–15.

5 Bornscheuer, U. and Buchholz, K. (2005) Highlights in biocatalysis – historical

landmarks and current trends. *Engineering in Life Science*, **5**, 309–323.

6 Dechema (1974) *Biotechnologie*, Dechema, Frankfurt, Germany.

7 Buchholz, K. (1979) Die gezielte Förderung und Entwicklung der Biotechnologie, in *Geplante Forschung* (eds W. van den Daele, W. Krohn, and P. Weingart), Suhrkamp Verlag, Frankfurt, p. 64–116.

8 Bud, R. (1993) *The Uses of Life, A History of Biotechnology*, Cambridge University Press, Cambridge;Bud, R. (1995) *Wie wir das Leben nutzbar machten*, Vieweg, Braunschweig/Wiesbaden.

9 Bud, R. (1994) In the engine of industry: Regulators of biotechnology, 19970–1986, in *Resistance to New Technology* (ed. M. Bauer), Cambridge University Press, Cambridge, p. 293–309.

10 Jasanoff, S. (1985) Technological innovation in a corporatist state: The

case of biotechnology in the Federal Republic of Germany. *Research Policy*, **14**, 23–38.

11 EFB (1983) *Biotechnology in Europe – A Community Strategy for European Biotechnology*, European Federation of Biotechnology, Dechema, Frankfurt.

12 Van den Daele, W., Krohn, W., and Weingart, P. (1979) Die politische Steuerung der wissenschaftlichen Entwicklung, in *Geplante Forschung* (eds W. van den Daele, W. Krohn, and P. Weingart), Suhrkamp Verlag, Frankfurt, p. 11–63.

13 Marschall, L. (1999) Industrielle Biotechnologie im 20. Jahrhundert. *Technikgeschichte*, **66** (4), 277–293.

14 Giesecke, S. (2000) The contrasting roles of government in the development of biotechnology industry in the US and Germany. *Research Policy*, **29**, 205–223.

15 Rehm, H.J. (1967) *Industrielle Mikrobiologie*, Springer, Berlin.

16 Pirt, J. (1975) *Principles of Microbe and Cell Cultivation*, Blackwell Scientific Publications, Oxford.

17 Aiba, S., Humphrey, A.E., and Millis, N.F. (1965) *Biochemical Engineering*, Academic Press, New York.

18 Bailey, J.E. and Ollis, D.F. (1977) *Biochemical Engineering Fundamentals*, 2nd, 1986, McGraw-Hill, New York.

19 Rehm, H.J. and Reed, G., *Biotechnology*, VCH, Weinheim, Germany, from 1981 on.

20 Cohen, S., Chang, A.C.Y., and Hsu, L. (1972) Nonchromosomal antibiotic resistance in bacteria: Genetic transformation of *E. coli* by R-factor DNA. *Proceedings of the National Academy of Sciences of the United States of America*, **69**, 2110–2114.

21 Cohen, S. and Boyer, H. (1979/1980) Process for producing biologically functional molecular chimeras. US Patent 4,237,224.

22 OTA (US Office of Technology Assessment) (1984) Impact of Applied Genetics, Washington.

23 Mayer, H., Collins, J., and Wagner, F. (1980) Enz. Eng. Conf (eds H.H. Weetall and G.E. Royer), Plenum Press, New York, 5, p. 61.

24 Mattes, R. and Beaucamp, K. (1981) Verfahren zur Herstellung eines Mikroorganismus, welcher α-Galactosidase, aber keine Invertase bildet, so erhaltener Mikroorganismus und seine Verwendung. Offenlegungsschrift (German patent application) DE 3122216; Umschau Wiss. Technik 1983, 83, 534.

25 Behrend, U. (2006) personal communication.

26 Poulsen, P. (2001) personal communication.

27 McKelvey, M. (1996) *Evolutionary Innovations*, vol. **128**, Oxford University Press, p. 129.

28 Demain, A.L. (2001) Genetics and microbiology of industrial microorganisms. Molecular genetics and industrial microbiology – 30 years of marriage. *Journal of Industrial Microbiology & Biotechnology*, **27**, 352–356.

29 Demain, A.L. (2003) Personal communications.

30 Ashton, G. (2001) *Nature Biotechnology*, **19**, 307.

31 Aggarwal, S. (2007) What's fueling the biotech engine? *Nature Biotechnology*, **25**, 1097–1104.

32 Buchholz, K. (2003) Historical highlights and current status of biotechnology–From mystery to the lock and key hypothesis to the design of recombinant enzymes. *Medicinal Chemistry Research*, **11**, 399–421.

33 Sanger (2002/2009) www.sanger.ac.uk/ Wellcome Trust Sanger Institute; www.wellcome.ac.uk/.

34 Venter, J.C. (2007) *A Life Decoded*, Penguin Books Ltd., London.

35 Sinskey, A.J. (1999) Foreword. *Metabolic Engineering*, **1** (3), iii–V.

36 Stephanopoulos, G. (1999) Metabolic fluxes and metabolic engineering. *Metabolic Engineering*, **1**, 1–11.

37 Reuss, M. (2001) Editorial. *Bioprocess and Biosystems Engineering*, **24**, 1.

Part Two
The New Paradigm Based on Molecular Biology and Genetics

Concepts in Biotechnology: History, Science and Business. Klaus Buchholz and John Collins
Copyright © 2010 WILEY-VCH Verlag GmbH & Co. KGaA, Weinheim
ISBN: 978-3-527-31766-0

6
Broadening of Biotechnology through Understanding Life, Genetics and Evolution

The ability to identify genes and gene products and to control the production of large quantities of pure products developed rapidly from 1973 onwards. This was enabled largely through the development of *gene cloning technology*. The accelerated analysis of gene structure and function which ensued revolutionized modern biotechnology to such an extent that today's biotechnology has become almost synonymous with the advancement of molecular biology and molecular medicine. Its progress has become largely dependent on rapid specific DNA amplification (PCR) and high-throughput DNA analysis procedures combined with large-scale electronic data analysis (bioinformatics).

One of the major areas of novel product development relates to monoclonal antibodies, in general now produced by gene technology-driven processes. The versatility and relative ease of developing recombinant antibodies has become a development area in its own right in which commercialization of the products plays a major role. The early Biotech start-ups of the 1990s pioneered by Cambridge Antibody Technology (CAT) in Cambridge, England which was founded by *Sir Gregory Winter*, have become well established with multi-billion dollar sales (see Section 10.8). The problems of antigenicity, half-life in serum and delivery have been and still are the major limitations to developing these products in common with all large protein molecules to be injected into the body. In spite of such problems the world market for recombinant DNA pharmaceuticals has grown to over $60 billion. This new Biotech industry grew in the wake of a small number of visionaries and pioneers who established the first molecular biology-based Biotech 'start-up' companies: *Cetus* (1971; *Ron E. Cape, Peter Farley* and Nobel laureate *Don A. Glaser*), *Genentech* (1976 *Herb Boyer and Bob Swanson*), *Biogen* (1978 *Wally Gilbert, Phil Sharp, Charles Weissmann, Bernard Mach, Walter Fiers, Heinz Schaller, Peter Hofschneider, Ken Murray and Brian Hartley*), *Amgen* (1980 *Ray Baddour*) and *Chiron* (1981, later bought by *Cetus; Bill Rutter*).

The fact that this new development is based on a deeper understanding of biology allowed an alteration in the general direction of development of products for medical application. Since understanding grew with respect to the cause of disease new emphasis was given to *preventive* rather than *corrective* medicine. Assessment of the genetic predisposition for disease which allows accurate and highly sensitive diagnosis of disease states is playing an ever increasing role in preventive medicine.

Concepts in Biotechnology: History, Science and Business. Klaus Buchholz and John Collins
Copyright © 2010 WILEY-VCH Verlag GmbH & Co. KGaA, Weinheim
ISBN: 978-3-527-31766-0

The type and potential malignancy of cancer cells or the presence of a virus can be detected and determined with high sensitivity. Recently it has become possible to detect environmental effects on gene regulation. These latter studies have given rise to the new scientific discipline of epigenetics. The predictive element of the analysis of predispositions is a new aspect which opens up novel possibilities for personal medicine.

In spite of the successes in most areas of recombinant DNA, the challenge of developing somatic gene therapy has not been met. Progress has been slow. Clinical trials have suffered major setbacks. Somatic gene therapy is a type of therapy in which genes are targeted to body cells. Although germ-line therapy is used in transgenic animals and plants, it is currently not allowed to be applied in humans since the long-term consequences of genetic alterations which can be passed on to future generations are uncertain. The main uncertainty is due to the inability to absolutely restrict the site of gene integration or exchange to the desired location.

A somewhat different outcome had been envisaged by *Paul Ehrlich* in his first ideas relating to drug discovery. He believed, on the basis of the ability to distinguish different microorganisms by their ability to bind different dyes, that empirical screening of chemicals specifically binding disease organisms might deliver a 'magic bullet', that is an antibiotic specific for the disease microbe.

In contrast, the more recent strategy is to try to understand the disease process and to target the development of new medicines directed to defined bottlenecks, or unique metabolic routes within a pathogen. One example is the targeting of peroxyredoxins in the tropical disease pathogens *Trypanosoma* and *Leishmania* spp., which lack the more efficient seleno-enzyme equivalents of most eukaryotes. The flavoenzyme trypanothione reductase is essential to enable the parasites to survive oxidative stress. This has been recognized as one approach which appears particularly promising in developing treatments for African sleeping sickness, *Chaga's disease* and *Leishmaniasis* [1].

'Gene mining' is a novel aspect of biotechnology arising from the enormous data pool provided by rapid inexpensive *whole genome sequencing* of microorganisms. It thus becomes relatively easy to compare the repertoire of protein-coding genes of parasites and their host. Currently thousands of entire genome sequences are available. Figure 6.1 shows data for the genomes sequenced by mid-2008. As discussed under the heading 'Future trends', DNA sequencing is expected to become 10 000-fold cheaper and faster between 2009 and 2013.

Recombinant DNA technology or rather molecular genetics has been so successful that for many it is almost synonymous with modern biotechnology. Although global players in the pharmaceutical industries were initially sceptical in accepting the potential of this new discipline, it is noteworthy that *Jurgen Drews*, then President of international R&D at Hoffmann-La Roche and later chairman of SAGB, the pan-European biotech industry lobby Organization, said in 1988 (personal communication to John Collins): *'In thirty years there will be no new pharmaceutical product coming onto the market which is either not produced by, or whose development did not involve recombinant DNA technology'.*

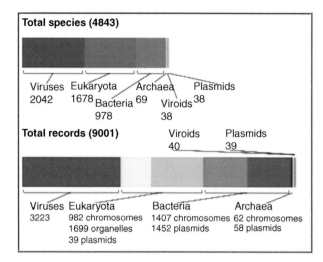

Figure 6.1 The internationally accessible DNA database already represented a wide range of species by 2008.

Recombinant DNA techniques have played a major role in the development of screening technologies, diagnostics and the identification of drug targets in disease as well as in the production of protein and peptide drugs. However, the development of somatic gene therapy has experienced a number of notable setbacks and is still in its infancy. Recombinant DNA technology has massively increased the potential of high through-put screening.

'High information screening' can inject new life into screening programmes for low molecular weight compounds and their semi-synthetic or synthetic chemical derivatives. *High information screening* gives better insight into the mode of action of a new drug and its potential side-effects and is expected to give a higher success rate for clinical phase I trials by greater selection during the preclinical phase. Recent breakthroughs in this area can be seen for example in the discovery of both *Epothilon*, a potential blockbuster as a replacement for the anti-cancer agent *Taxol* (*Paclitaxel*) and *Agyrin A*, a potential anti-cancer drug with a novel mechanism, originating from screening programmes of the secondary metabolites of the gliding bacteria (Genus *Myxobacteriaceae*).

References

1 Flohé, L. *et al.* (2003) *Biofactors*, **19**, 3–10.

7
The Beginnings of the New Biotechnology

7.1
Introduction

The following overview of the basic concepts and methods needed for the biotechnological revolution are presented in a largely historical perspective.

During the early twentieth century biological science matured to a more precise scientific discipline as compared to the previous more collative and descriptive activity of the landed gentry. The transition to explain causality had its origins with *Pasteur, Koch, Mendel* and others some 50 years earlier. Chronologically, explosive developments only occurred after the rediscovery of Mendel's laws of inheritance in 1905 and the movement towards the quantitative study of living organisms, their metabolism and inheritance and the development of a basic understanding of mutagenesis and variation. The new science was broadly based on the incorporation of chemistry, particularly biochemistry, and the migration of physicists and their mathematical tools to the study of biologically relevant topics (see also Section 4.2.1).

Pasteur's refutation of spontaneous generation, *Darwin's* concept of evolution and *Mendel's* laws of genetics allowed scientists to more readily accept the possibility that chemical substance(s) may be responsible for inherited characteristics, and furthermore that finding this in one type of organism might well be extended as a general rule to all organisms. The contentions of Pasteur and Darwin contained the germ of the idea that there was a common thread, later understood to be the *double helical thread* (= deoxyribonucleic acid, DNA) that had made it possible for all life forms to be related to one another as a result of common ancestry.

The initial characterization of phosphate-rich chemicals isolated from the nuclei of white blood cells (nuclein) was carried out by *Friedrich Miescher* (Basel; 1844–1895). The chemical composition was elucidated in 1919 as a result of the subsequent work by *Phoebus Levene* who identified the base, sugar and phosphate nucleotide unit which he correctly suggested, consisted of a base–sugar moiety, the nucleotide, linked together in short chains and separated by phosphates. *Albrecht Kossel* (Rostock, 1853–1927) continued to work on the structure of the purines and pyrimidines. The history of the development of X-ray crystallography for the analysis of biological molecules is treated separately later in this chapter with particular emphasis on

Concepts in Biotechnology: History, Science and Business. Klaus Buchholz and John Collins
Copyright © 2010 WILEY-VCH Verlag GmbH & Co. KGaA, Weinheim
ISBN: 978-3-527-31766-0

proteins. It should be noted here, that the study of nucleic acids using this technique had already commenced in the 1930s: in 1937 *William Astbury* (Stoke on Trent, UK, 1898–1961) produced the first X-ray diffraction patterns which showed that DNA had a regular structure. However 'seeing is believing' and 20 years later *Alex Rich* visualized the helical structure in high resolution using tunnel electron microscopy, apparently to the great relief of *Jim Watson* [1]. *Erwin Chargaff* established the parity rules for base composition: there was always the same amount of adenine (A) to thymine (T) and the same was also true for the ratio of cytosine (C) to guanine (G). The DNA composition of different organisms was variable but the ratios of (A + T)/ (C + G) was species specific. Chargaff reported this observation (Chargaff's parity rule) without any further suggestion that this in itself suggested a structural feature of the DNA molecule. Chargaff thought that DNA had a simple structure and did not believe that it could serve as a carrier for genetic information. Chargaff's parity rule served as the basis from which *Jim Watson* and *Francis Crick* modelled the DNA double helix in 1953. This structure suggested for DNA, a base-paired double helix, was also based on *Rosalind Franklin's* X-ray crystallography data for DNA and her estimate of the amount of water in the crystal. This latter information implied that the phosphates were located on the outside of the molecule as indeed does its property of being an acid.

The model had immediate obvious implications for how it could encode genetic information. The model showed a linear sequence of paired bases held together by weak hydrogen bonds between the anti-parallel strands. The weak bonds can be broken during replication with each strand acting as a template. The simple beauty of this hypothetical structure led to its acceptance long before more rigorous physical proof was available. It was assumed that when DNA was carefully isolated from a cell or a virus the molecules would be very long, since DNA solutions were found to be highly viscous. In summary DNA had all the characteristics required of a molecule which carries complex hereditary information from one generation to another in fairly stable form. It is capable of repair by semi-conservative replication and also able to mutate when an error occurs during replication. This latter characteristic is just as important as stability with respect to evolution. *Variability is the key parameter* which allows species radiation and speciation to occur and on which selection can act.

Early in the twentieth century, evidence for DNA being the genetic material was not absolutely conclusive. This changed due the following chain of discoveries:

- Microscopic methods for examining cells included basophilic aniline dyes which stained what we now call chromosomes. These investigations started in the mid-eighteenth century. In 1912 the human chromosome set was visualized ('karyotyping'). It was not until 1955 that *Joe Hin Tjio* and *Albert Levan* concluded that humans have 22 chromosome pairs and X and Y sex chromosomes and that karyotyping could be used as a reliable diagnostic tool. Purified chromosomes contained protein and DNA, although as mentioned before the best bet was that the protein was the genetic component. Giant (polytene) chromosomes from the salivary glands of flies could more easily be observed compared to normal chromosomes. 'Chromosome' puffs were thought to correlate with altered activity of genes. Loss of function was correlated with loss of polytene bands implying that chromosomes are the seat of the genetic material.

- Surprisingly aneuploidy (an alteration in the copy number of a particular chromosome) was not found to be associated with hereditary traits in humans until 1959 when the genetic basis of Down's syndrome was identified (*Jerome Lejeune*) even though this latter syndrome had already been described in 1866 by *John Langdon Down*. Note that alteration of a protein associated with a disease was first defined by Linus Pauling in 1949 [3]: "'Sickle cell anemia', a molecular disease", (see Section 11.2).

- In 1944 *Oswald Avery, Colin McLeod, and Maclyn McCarty* showed that the 'transforming principle' from virulent *Streptococcus pneumonia* which they obtained by carefully repeating an observation originally described by *Frederick Griffiths* in 1928, was DNA and not protein. Griffiths had shown that mice could be killed if they were infected with a mixture of the live non-virulent 'R' or rough strain, and a boiled extract of the virulent 'S' (smooth) strain. Mice were indifferent to either the boiled S-extract or the R-strain injected alone. Avery and his colleagues purified different types of molecules from the cells: protein, nucleic acid and lipids. *Alfred Hershey* and *Martha Chase* showed in 1952 that it was the DNA not the viral coat protein that entered the cell when *T2 bacteriophage* entered the *E. coli* cell.

- In the area of virology tobacco mosaic virus (TMV) was shown to consist of nucleic acid (actually RNA) and protein. It was purified separately and then reconstituted to give infectious virus (*H. Fraenkel-Conrat* and *R.C. Williams*, 1955 [2]). By labelling with radioisotopes it was shown that it was mostly the RNA that entered the cell. There was no immediate interest from those studying 'higher organisms' in much of this early work on plant viruses and bacteria and their viruses. However, it was through the 'phage group' with *Max Delbruck, Salvador Luria, Sydney Brenner* [4], *Francois Jacob* and *Jacques Monod* at its centre, that concrete genetic evidence emerged confirming that DNA was the genetic material [5].

Having thus arrived at the point where the biochemical components of living systems had been defined, we can now look at the developments in biology, genetics, biochemistry and chemistry which led to the explosion in recombinant DNA-driven biotechnology and medical research.

7.2
The Beginnings of Evolution Theory and Genetics

In describing the basis of speciation during a process in which the time axis is on a geological time scale of some 3.5 billion years, it was proposed that slight variation from one generation to the next would cause new species to arise, particularly under selection pressure to survive under competition. Higher fecundity of those individuals which were 'better' able to obtain nourishment and a mate(s) was taken into account as well as the ability to avoid predators and survive pathogenic infections. These ideas incorporated a vague concept that a mechanism existed to ensure fairly stable hereditary, that is that any particular species is preserved in a more or less recognizable form. This latter assumption was not entirely trivial at a time when for many spontaneous generation and transmutation of species was considered to be commonplace.

The evolutionary theories of *Charles Darwin* and *Alfred Russel Wallace* were first presented officially to the scientific community on 1 July 1858 by the Reader of the Linnaen Society [6], London, although the communication initially attracted little attention. The President of the Linnaen Society summarized the achievements of 1858 as not having brought anything worthy of particular mention in contrast to Daniel Dennet's later evaluation that 'evolution' is the single most important idea that mankind ever produced [7].

Both Darwin and Wallace had been impressed by *Thomas R. Malthus'* book *An Essay on the Principle of Population* (Thomas Robert Malthus, 1766–1834; Darwin read this book in 1838) which emphasized that competition for resources can always be considered to be the ultimate limitation to growth in the size of a population. Unregulated population growth culminates in widespread famine, disease (plague), and conflict (war) [8]. This provided the background scenario for constant selection in which only the fittest and most robust individuals will survive and produce off-spring.

It should be noted that the *survival of the fittest* concept in these first evolutionary theories is oversimplified compared to the way we understand it today: we now see that major roles are played by the accumulation of massive diversity in populations, accumulating variants which essentially confer equal fitness. Selection is very often exerted on communities of interacting interdependent species (commensals), for example in biofilms where interspecies cooperation or altruism can often be highly advantageous [9]. A radical suggestion has recently been made supported by mathematically modelling that selection may not be obligatory for speciation to occur [10]. In this scenario random drift provides a sufficient mechanism even in the absence of selection, particularly in small populations. This does not detract from the idea that selection is in fact constantly taking place and is a significant driving force in evolution.

Both Darwin and Wallace had initially been influenced by the earlier writings of *Jean Baptiste Lamarck* on evolution, but both eventually rejected Lamarck's ideas related to the acquisition and inheritance of acquired characteristics, for example strong muscularity developed as a result of an occupation as in the case of a blacksmith, being passed on to the blacksmith's son. Data has been accumulating which indicates that, to a very limited extent, there is the potential for an organism to 'learn' from its experiences in a particular environment. This information which adapts the response to these challenges can be passed on to the offspring for at least for a limited number of generations. Furthermore, it seems that mechanisms can be visualized for ways in which this information may be immortalized in the organism and passed on to future generations [11].

In Darwin's time there was no understanding of any laws of inheritance, only a feeling that there were gradual changes taking place in populations which over long periods of time would represent major changes in adaptation finally emerging in the form of a new species. It should be noted however, that we now not only understand that nucleic acid (DNA and RNA) is the heritable material and that the genes are the basic unit of inheritance but that additional mechanisms exist. Specific chemical modification of the DNA and the histone proteins which coat and package the DNA, are now recognized to be responsible for the long-term effects to environmental insults such as the effect of famine being passed on for several generations, for example in terms of small stature. Such an event which occurs independently of any sequence alteration (mutation) in the DNA is termed an *epigenetic effect*. The study of such effects is known as *epigenetics*. For instance, methylation of tumour-suppressor genes is now recognized as an effect of environmental insults such as cigarette smoking, which is likely to be a major factor in potentiating suscep-tibility to cancer.

The classic concept of evolution theory as a long fairly slow process of variation based on mutations accumulated in the population on which selection acts, is however still valid and considered to be the main force in action. The most important factors are the relative fecundity of the individual and the relative survival rate of its offspring. This is not a simple brutal selection principle as often portrayed in social Darwinism. Selective advantage may also arise from and in fact usually involves cooperation or altruism between species within commensal groups. The importance of this latter phenomenon was debated in letters between Charles Darwin and Alfred Russel Wallace and in Darwin's *Descent of Man* published in 1871, in which it was recognized as an important although unresolved topic at that time. The following publications can be regarded as the seminal works which expanded on these ideas and established them in current scientific thought: *Peter Alekseyovich Kropotkin*, 'Mutual aid: a factor in evolution', 1902; *J.B.S. Haldane*, 'Kin altruism', 1930; and *Robert Axelrod*, 'Evolution of cooperation', 1985.

7.2.1
Robust Cultures barely Addressed in Biotechnology

The reader may wonder why the authors have included a lengthy description of the concept and mechanisms of evolution, particularly placing such emphasis on the role of mutual cooperation or altruism. It should be pointed out that before Pasteur and back into antiquity, vinegar, wine, saki and sauerkraut production, the main repre-sentatives of biotechnological fermentations, in fact involved complex mixtures of micro-organisms in which the metabolic activities of one interacted with those of the others. These are intact robust communities that persist in open competition with other micro-organisms in the environment. This reflects the majority of real-life microbial communities in nature.

Pure mono-cultures, the maintenance of which requires expensive sterile fer-mentation conditions and media, are essentially a human artefact. The largest branch of biotechnology, namely the treatment of waste water also belongs to the category of open complex microbial fermentations. In spite of the complexity of the interplay between various species to deal with the degradation or conversion of essentially all conceivable organic substrates to CO_2, H_2, N_2 or CH_4 and water, the systems remain stable. Microenvironments are created when facultative aerobes remove all traces of oxygen to allow coexistence with obligate anaerobes. The broad parameters of mass action in some very large scale processes have in fact been analysed and used to create efficient units for waste treatment on a very large industrial scale (see Section 16.6).

It seems that the idea of controlling the metabolism of complex microbial communities for biotechnological processes has barely been addressed. An exception is a modern process developed in China for the production of vitamin C using two organisms, *Gluconobacter oxydans* and *Bacillus thuringensis* to convert sorbitol to sorbose and subsequently to 2-keto-L-gluconic acid with a yield of 85%, the latter being converted to ascorbic acid in two further chemical steps [12].

New methods of analysing the compositions of complex communities are discussed in other sections under the headings of *metagenomics* and *metabolomics*. The examples given are limited but show the potential of such approaches. Researchers are still a very long way from being able to manipulate the properties or composition of such complex systems. Natural diversity is discussed in this book as a source of genes for novel enzymes in 'gene mining' (see Section 7.6.9).

The problem of maintaining strain collections of such stable communities has also not yet been tackled. Taking a lesson from evolutionary principles perhaps it would be wise to invest some thought in how this potential could best be approached by developing multi-step conversions in a single fermentation rather than by using serial single-step conversions with individual strains or isolated enzymes or chemical reactions. This appears to be a novel area with a wealth of potential that might well profit best from the holistic ideas inherent in systems biology (see Section 13.6).

Gregor Mendel (1822–1884) analysed the segregation of parental 'elements' in peas, and derived hereditary laws based on segregation of such elements, later called genes, into the gametes (pollen and ovules). He proposed that the characteristics of the progeny were determined by redistribution/combination of these parental genes which came together in both parental and novel combinations in the progeny. This was the first time that a consistent model had been proposed and was supported by extensive evidence. He explained how the characteristics of the progeny could be different from those of their parents but determined by them in a *quantitative* fashion.

Mendel published his data and his interpretation in 1865 but it remained in obscurity for a further 35 years. During Darwin and Wallace's lifetimes, Mendel's work remained obscure and was only rediscovered and popularized in 1900 by *Hugo De Vries* and *Carl Correns*.

The inherited 'elements' were eventually given the name *genes* by *William Bateson* (1861–1926). In 1905 he proposed that the study of how an organism passes on its essential characteristics from one generation to the next should be named *genetics*. Slowly an understanding was developed for why progeny although closely resembling their parents, also differed. Evolution theory proposed that slowly over long periods of time selection for advantageous (= counter selection for disadvantageous) inherited characteristics was exercised at the level of these small variations.

It is the combination of these two insights, namely genetics and evolution through gradual change that can be designated as the event which really heralded in the modern age of biology and the new biotechnology. In particular, when Mendel's laws of inheritance were integrated with the chromosome theory of inheritance by *Thomas Hunt Morgan* in 1915, they became the core of classical genetics [13].

7.3
The Origin of Recombinant DNA Technology

The development of novel biotechnology processes and products was accelerated and broadened immeasurably by the advent of cloning, sequencing and identification of genes. The following text describes how gene cloning and DNA manipulation was developed by combining tools discovered in nature.

7.3.1
The Cradle of Gene Cloning: Bacterial and Bacteriophage Genetics

In 1964 at University College, London University, during the introductory genetics course which was part of the first Bachelor of Science course dedicated purely to Microbiology, the students (one of the authors included) took delight in hearing the following: 'There is an organism, which spends five times its normal life cycle in sexual conjugation'. The statement referred to the studies of *Bill Hayes* on the kinetics of gene transfer in crosses between two strains of bacteria. He showed that this process required a 'male' F (fertile) or Hfr (*High frequency of recombination*) strain and a recipient 'F-minus' strain. *Luigi Luca Cavalli-Sforza* had first described sexuality in bacteria in 1950 [14]. *Escherichia coli* which indeed in broth has a doubling time of some 20 minutes, was shown by Hayes to require during the act of conjugation about 100 minutes before the whole genome was transferred. Using interrupted mating *Francois Jacob* and *Elie Wollman* were able to map the order of the gene transfer. It seemed that the DNA transfer occurred from a fixed point in that there was a fixed time before a particular gene was transferred to the recipient cell. The results were consistent with a continuous closed genetic linkage map. It was concluded that there was a single circular chromosome. DNA could not be physically visualized at that time but the science of bacterial genetics was underway. Shortly thereafter, the circular nature of the bacterial chromosome was confirmed by *John Cairns* in 1963 using electron microscopy [15].

The 'PaJaMo' experiment [16] named after *Arthur Pardee, Francois Jacob* and *Jacques Monod* was carried out in 1959. They used these conjugation techniques to study how the genes encoding the lactose-metabolizing enzymes caused protein (enzyme) production within a minute of entering a cell. This led directly to the *very first models of gene regulation* and the concept of the operon, that is genes grouped together on the microbial genome which could be co-ordinately regulated in their expression. Although it transpired that these models were not so easily applicable to regulation in eukaryotes, it established the principle of how living organisms could be understood in terms of a complex interaction of self-regulating circuits. This recently led to the science of systems biology which attempts to develop a complete mathematical description of these interactions (see Section 13.6).

The initial description of specific gene regulation in response to external stimuli removed some of the enigma about how differentiated cells, which all carried the same genetic information, were able to produce different products. In a retrospective, *Pardee* and *Reddy* [17] make the following statement: 'In the early 1950s, biochemistry was mainly a study of structures, metabolism, and enzymology of small molecules. It was intellectually separate from bacteriology, genetics, and cell biology. Molecular biology did not exist; later it provided a unifying motif. Homeostasis was appreciated in physiology, but not in biochemistry'.

Monod also developed the concept of allosteric regulation by which molecular interactions, for example between proteins or between small molecules and proteins could change their properties. This change could be seen initially in the altered activity of enzymes. Later in 1966, *Walter Gilbert* and *Benno Müller-Hill* showed that allosteric characteristics were also responsible for altering the binding affinity of other molecules to DNA thus controlling gene expression [18]. *André Lwoff* [19] followed up on the initial work by *Eugène* and *Elisabeth Wollman* on viruses which used bacteria as hosts (bacteriophage). *Lwoff* shared the Nobel Prize with *Jacob* and *Monod* for their insights into molecular genetics. He developed an easily amenable system for studying virus production with the bacteriophage (phage) λ.

Max Delbruck was a prime mover in the phage field and his laboratory became an intellectual melting pot, initially in Pasadena and later in Cologne at the newly built Institute of Molecular Biology (IMB). *Benno Müller-Hill* joined Delbruck at IMB after leaving *Jim Watson's* laboratory in Cold Spring Harbor where he had just isolated the lambda repressor, the first protein characterized that was directly involved in gene regulation. Variant F-prime (F′) plasmids were also identified (*Adelberg* and *Wollman*) which after integrating into the chromosome via non-homologous recombination had formed a loop and detached from the bacterial chromosome to form a separate mini-chromosome carrying a piece of the host chromosomal DNA.

7.3.2
Antibiotic Resistance in Hospitals: Genetic Flux in Plasmids

By the mid-1970s in hospitals which used large quantities of antibiotics to treat both resident and ambulant patients it was recognized that among the *Enterobacteriaceae* family which includes the human gut flora *Escherichia* species, promiscuous host bacteria such as the *Erwinia* species, which are plant saprophytes and are often carried into the hospital environment on fruit and flowers, became resistant to five major classes of antibiotics. The *Erwinia* spp. was then able to become an opportunistic pathogen as the sole survivor of broad spectrum antibiotic treatment. The rapid spread of antibiotic resistance in hospitals was soon recognized to be associated with 'resistance' or R-factors. *Stanley Falkow, Stanley N. Cohen, Don R. Helinski, Richard P. Novick, Roy C. Clowes, Heinz Saedler, Peter Starlinger, Hans-Peter Hofschneider* and *Noemi Datta* were pioneers in recognizing the molecular basis of this health problem associated with mobile genetic elements: plasmids, IS elements and transposons [20]. These elements allowed micro-organisms to rapidly acquire useful genes and to adapt in a strongly selective environment. Promiscuous conjugation was able to take place among distantly related organisms, for instance between nearly all the members of the *Enterobacteriaceae* including *Escherichia* and *Salmonella* as well as plant saprophytes such as *Erwinia* species. It was discovered that this latter species sometimes acted as a reservoir for whole batteries of antibiotic resistance genes that had been accumulated often by transposition onto conjugative plasmids.

Antibiotic resistance is still a problem today. Currently there is an increase in particularly aggressive multiple drug resistant *Staphylococcus aureus* (MRSA; also methicillin-resistant) strains which alarmingly originate not only from hospital environments but also seem to have a selective advantage in the general population outside of health-care centres.

7.3.3
'Restriction and Modification': Finding the Enzymatic Tools for DNA Recombination

During studies on bacterial conjugation it was discovered that related strains differed in their ability to mate with one another. Type C might act as a recipient for genes from C- and B-type strains but conjugation with B-strains acting as the recipient were only successful with B-type donors. This correlated with the susceptibility of the strain to bacteriophage lambda which in turn depended on the strain in which the bacteriophage had been produced (*Salvador Luria*, 1952 [21, 22]; *Werner Arber* and *Daisy Dussoix*, 1962). This phenomenon was known as host-controlled 'restriction and modification'. It was discovered that the mechanism involved modification of the DNA itself, often a methylation of an adenine or cytosine nucleotide within a particular DNA sequence. In 1968 *Stewart Linn* and *Werner Arber* discovered a class of nucleases which were able to cleave DNA on both strands thus causing a double-strand break, however this was only possible when the nuclease bound to specific 'unmodified' regions [23].

These 'restriction' enzymes were found in pairs often in physical association with other proteins one of which was the modification enzyme (usually a DNA methylase) which actually catalysed the methylation of the sequences where the enzymes specifically bound.

The first enzymes (type I) cleaved randomly when activated, leading not only to degradation of the incoming DNA but also initiating the sacrificial death of the host by degradation of its own chromosome. Another type of enzyme (type II), actually *Hinc*II, was discovered in 1968 in the bacterium *Haemophilus influenzae* by *Hamilton O. Smith, K.W. Wilcox,* and *T.J. Kelley.* The Nobel Prize for Medicine was awarded to *Daniel Nathans, Werner Arber* and *Hamilton O. Smith* in 1978 for their discovery of restriction endonucleases.

5' NNNG|<u>AATT</u> CNNN 3'
3' NNNC TTAA|GNNN 5'

Figure 7.1 The letters represent the base sequence in a section of double-stranded DNA including the binding and cleavage site for the restriction endonuclease EcoRI. 'N' denotes that the base does not affect binding of the restriction enzyme; that is it can be any sequence. Note that 'A', adenine, pairs with 'T', thymine on the other DNA strand, and that 'G', guanine pairs with 'C', cytosine via hydrogen bonds. The enzyme cleaves one strand between the G and A (vertical lines) with inverse mirror symmetry on the other. Since the left and right side of the molecule are no longer covalently joined they will separate (melt) at ambient temperature. This creates a 5'-extension of four bases, AATT on both DNA strands. These can pair again at low temperature and are covalently sealed in the presence of DNA ligase. Restriction enzymes are now known which are able to cleave DNA at several hundred different sequences, some producing complementary ends as shown here, others flush ends and others cutting asymmetrically to generate unique protruding ends.

7.3.4
Gel Electrophoresis for DNA Analysis and Isolation of Specific Fragments

A classic paper by *K. Danna* and *D. Nathans* published in 1971 was the first to describe the use of gel electrophoresis for determining the size of DNA fragments produced by enzyme cleavage. They immediately realized that this could be used for analysing the deletion mutants of the DNA of the virus on which they were currently working and that physical maps could be constructed of the whole circular genome [24]. *Herb Boyer* isolated and characterized the *EcoRI* restriction enzyme which was later used in the first plasmid vector gene cloning experiment in bacteria, undertaken in collaboration with *Stanley N. Cohen* [25]. This latter experiment was the prototype for the gene cloning explosion which followed and was the harbinger of the new gene-orientated Biotech industry. Both EcoRI-DNA methylase (modification enzyme) and the EcoRI restriction enzyme (endonuclease) recognize the same DNA sequence (Figure 7.1). Such enzymes often have inverted mirror symmetry at the target site. A single strand cleavage occurs on opposite strands at the same sequence position, for example the EcoRI restriction enzyme binds the sequence shown which is then cleaved at the sites shown by the arrows.

Cleavage of a particular DNA (genome) with such enzymes yielded defined DNA fragments of particular sizes. Moreover, each fragment has a cohesive protruding end which allows it to bind in either orientation at low temperature to any other fragment generated by cleavage with this enzyme. Oligonucleotide synthesis *in vitro* had already been pioneered in the laboratories of *Gobind Khorana* and *Arthur Kornberg* in the mid 1960s using chemical synthesis or DNA polymerase respectively. In 1967 *Martin Gellert* described a form of enzymatic activity that was able to covalently join DNA (and RNA) strands on a double-stranded template; this enzyme was designated DNA ligase. The activity of this enzyme is also responsible for the covalent bonding of the annealed restriction fragments. Type II restriction enzymes usually work as single protein moieties. By 2008 over 3000 restriction enzymes had been studied in detail and more than 600 of these are available commercially and used routinely for DNA modification and manipulation in laboratories around the world. In early 1976 there were very few laboratories able to work with the technology and only some four restriction enzymes were routinely being used. They were usually made either by the laboratory which was using them or exchanged with other laboratories. A major effort was made by hundreds of laboratories worldwide to collate the development of these important tools and to identify, purify and characterize further enzymes. In the early days it was a very exciting time and cooperation, exchange of materials, strains and protocols was the normal procedure.

At many institutional centres individuals took the initiative to search for and characterize new restriction enzymes. They acted as an important source of enzymes and strains. One such individual was *Noemi Datta* who working at Mill Hill in London, UK, collected and characterized large numbers of micro-organisms many of which were clinical isolates. The largest repositories were the American Type Culture Collection (ATCC, Bethesda Maryland) and the German Collection of Micro-organisms which was originally located at the GBF (The German Biotechnology Research Centre, in Braunschweig) but then became a separate institute for patent purposes(= DSMZ (Deutsche Sammlung v. Mikroorganismen und Zellkultur).

Rich Roberts [24] at Cold Spring Harbor became a central figure in the collection and distribution of information and bacteria which could be used to make the restriction enzymes. This initiative eventually grew to provide the technology for the establishment of the New England Biolabs, New Haven, USA of which Roberts was a founder and which in 1975 started to produce restriction enzymes on a commercial basis. Today it provides a large selection of enzymes to the scientific community. At the time it was the first encounter that many biologists had had with industrial enterprises which 'collected' commercially valuable material from academic campuses. This rapidly changed as did the willingness of academics and industrial groups to exchange information. This hindered scientific openness but led to a reproducible industrial production standard.

During the 1960s it was thus discovered that many of the elements which allowed genes to spread rapidly through a population of bacteria by conjugation or even by viruses via the mechanism of transduction, involved the transfer of plasmid DNA. Plasmid DNA is present as double-stranded circular structures separate from the bulk of the cell's DNA, which although originally known as the 'chromosome' (by analogy to the eukaryotic cell chromosome which could be stained) is now referred to as the genome. It was not obligatory for the bacterial cell to carry these plasmids or viruses in addition to its core genome.

Genetic engineering developed as a result of the combination of the knowledge accumulated about both restriction enzymes and plasmid/virus biology and a third factor, namely the ability to introduce naked DNA back into the microbial cell. The restriction enzymes cleaved the DNA from a donor organism into manageable sizes for further analysis. The cleaved DNA is annealed (pairing the sticky ends created) and ligated (covalently sealed to) to plasmid or virus DNA which had previously been similarly cleaved (opened). The plasmid or virus which carried all the necessary DNA replication functions ensured that the hybrid molecule would be maintained once it was back in the cell. In cloning jargon the plasmid or virus used for this purpose is termed the vector. Antibiotic resistance genes described previously were used as selective markers to isolate bacterial colonies carrying the (recombinant) plasmids.

7.3.5
'Handling DNA'

Techniques for handling DNA were developed during this period. Hydroxyapatite columns were used in the early 1960s for separating double-stranded (which bound) from single-stranded DNA (which did not bind). However, it was difficult to obtain material of reproducible quality.

Acrylamide was initially introduced as a gel material for zone electrophoresis in 1959 by *Samuel Raymond* [26]. This was developed for protein analysis in various forms by *Hermann Stegemann* [27] and for the analysis of RNA by *Ulrich Loening* [28]. The technology thus became available and was applied by *Daniel Nathans* in 1971 for the high resolution separation of linear restriction fragments of the SV40 virus. Prior to this, centrifugation was the method of choice, initially employing sucrose or dextran gradients. Following the development of high speed ultracentrifugation (an exceptional engineering feat which involved developing rotors which would not explode under the huge *g*-forces generated) it was possible to separate DNA on the basis of its specific density in caesium salt isopycnic gradients. An experimental run using this technique would typically take 2 to 3 days and only produced separation in cases where the two DNAs to be separated had quite different AT:GC ratios or were glycosylated as in the case of DNA from a few viruses. This did, however, lead to the discovery that many genetic elements which were readily transferred from one cell to another were physical entities separate from the main genomic DNA. These were initially referred to as extra-chromosomal elements or episomes. Finally these 'independent' genetic elements were designated plasmids (a term proposed by *Joshua Lederberg* in 1952).

The closed circular structure of plasmids, viruses and mitochondrial DNA was confirmed by electron-microscopy (EM) of single molecules spread onto a water surface with cytochrome c and lifted onto a carbon film or mica surface which was then shadowed at a low angle with tungsten (*Alfred K. Kleinschmidt* [29], 1959). A reproducible method for preparing any circular plasmid DNA was developed by *Don Clewell* and *Don R. Helinski* (1969) based on the work of *Jerome Vinograd* who in 1966 had studied the differential affinity of constrained and unconstrained DNA (linear and open circular) for intercalating dyes such as ethidium bromide (Figure 7.2).

The method of *Don Clewell* and *Don Helinski* for separating the plasmid DNA thus worked fortuitously, since in opening the cells sufficient shear forces are generated to break nearly all of the very large chromosomal DNA but not the smaller tightly coiled plasmids. Separating chromosomal from plasmid DNA thus remains a technical

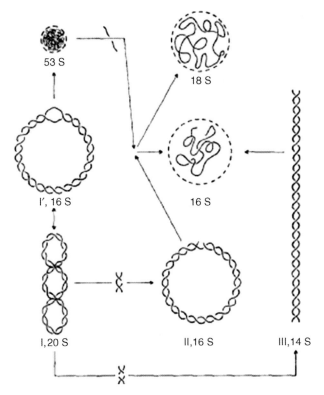

Figure 7.2 Diagrammatic representation of several forms of polyoma virus DNA. From Vinograd, J and Lebowitz, J. (1966) J. Gen. Physiol.49, 103–125.

problem in working with very large hybrid molecules such as BACS (see later) or YACS which are then separated by other techniques such as pulse-field electrophoresis.

7.3.6
New Tools to Study Single Molecules

Electron microscopy was one of the first methods to allow the visualization or *measurement of a single molecule*. Using electron microscopy, scanning electron microscopy or force field microscopy, the dimensions of single molecules with respect to their size and shape and even elastic properties when stretched can be determined. The patch clamp method of *Erwin Meyer* and *Bert Sakmann* is capable of measuring single ion channels. The development of highly fluorescent dyes as well as confocal laser microscopy enables minute volumes and single molecules to be visualized *in vivo* in living cells. Diffusion time and viscosity in the cytoplasm can also be estimated. The mode of action of proteins which regulate the central division of microbial cells can be elucidated (see for example work of *Petra Schwille* [30]).

Developments in the nanotechnology field and these aforementioned improvements of fluorescence-based photon-capture are being combined in a single molecule sequencing technique which is expected to revolutionize DNA sequencing. This is discussed in Section 13.5.

Up to this point tools had been developed for the physical characterization and manipulation of DNA molecules which involved the cleavage and rearrangement of DNA molecules. The next important step to consider was how to transfer intact genetic material into the living cell. It was known since the classic experiments of *Avery, McCleod* and *McCarty* in 1945 that pure DNA could be incorporated into bacteria thus transforming them to the virulent phenotype, hence the name transformation given to this process.

In 1970 *Morton Mandel* and *Akiko Higa* increased the natural efficiency of bacteria to incorporate DNA with techniques that involved washing the cells with calcium chloride. Various improvements followed.

Eukaryotic cells can often readily take up DNA when it is provided as a co-precipitate with calcium phosphate, a method developed by *F.L. Graham* and *A.J. van der Eb* [31]. This was a prerequisite for *Paul Berg's* later DNA cloning experiments with SV40 virus. The term 'transformation' has altered over time with respect to eukaryotic cells. Early on it referred to the event of immortalization which often follows the entry of an oncogenic (cancer-causing) virus into the cell.

Thus all the elements for the new recombinant DNA technology, at least for bacterial and animal cells were available:

- Methods to prepare DNA,
- which, following restriction cleavage (i.e. treatment with restriction endonucleases) could be covalently joined to a 'vector' with a DNA ligase
- A 'vector' (plasmid or virus) to ensure maintenance in the cell
- A method to prepare 'clean' vector DNA
- An efficient method to incorporate DNA into the cell
- Culture techniques to isolate single clones carrying a single recombinant hybrid molecule, including selective techniques to enrich for the cells 'transformed' with the vectors, for example selection for antibiotic-resistance genes.

With this technology the function of the genes isolated in the individual clones could be characterized. Part of such a strategy may involve transferring the DNA back into another host where the gene may be expressed and the resulting change in phenotype analysed. This strategy is referred to as 'reverse genetics' since it examines a gain of function rather than the approach used in classic genetics where the search was for loss of function after mutagenesis.

7.3.7
The First Recombinant DNA Experiments

Stanley N. Cohen (Stanford University) and *Herb Boyer* (UCSF), while at a meeting in Hawaii in 1972, conceived an experimental approach for gene cloning in bacteria which they published in 1973 [25]. They were the first to describe a simple protocol for cloning restriction fragments in bacteria, a method that was used primarily in the early years of gene technology by the majority of laboratories. This method was also patented by them, the applicant being the University of California, San Francisco. *Boyer* and *Bob Swanson* founded the first Biotech Company based on gene cloning in 1976, namely Genentech, with the immediate goal of producing pharmaceutical products. Genentech's first product manufactured using gene technology was human insulin which was approved for clinical use in 1982. For these latter activities which are considered to be the origin of the multibillion dollar/Euro 'New Biotech' industry, *Boyer* and *Cohen* shared the National Medal of Science in 1988, an honour bestowed on outstanding scientists and engineers by the President of the United States.

Paul Berg, Herbert Boyer, Stanley Cohen and *A. Dale Kaiser* also shared the Lasker Award in Medicine in 1980 for the development of gene cloning methodology. *Dale Kaiser* worked on the development of cohesive ends in DNA molecules. This work had already begun in 1957 using the bacteriophage lambda and continues to the present day with the enzymatic synthesis of DNA tails using terminal transferase, as employed in the first cloning of cDNA from mRNA. Cloning in animal cells employing viral vectors followed the development of gene cloning in bacteria using plasmid and viral vectors.

Berg and his student, *Stephen Goff*, initiated such experiments with viral vectors in animal cells in 1976, based on work by *Renate Dulbecco* who characterized the DNA of small animal viruses (polyoma) in the mid-1960s. In 1980 together with *Richard Mulligan, Berg* introduced a bacterial gene (XGPRT) into human cells which lacked this metabolic ability (HGPRT-minus) and demonstrated that the bacterial gene was able to 'repair' a metabolic defect. The selective medium (HAT-medium) was developed by *Waclaw Szybalski* in 1962 [32]. This provided a selective marker for vectors for the transformation of animal cells carrying a viral thymidine kinase gene into mutant cells lacking this enzyme.

Once the tools for gene cloning in the Gram-negative *Escherichia coli* had been established it became easy to develop gene cloning vectors which could be transferred to other species. This involved the identification of plasmids that replicated in other hosts and genes (promoters) that could be expressed and used for selection in the new host. A large part of this early work is described in detail in [33]. The initial work with yeast (*Kevin Struhl* in *Ron Davis'* lab, 1976, [34]) involved difficult protocols to make the yeast cell wall permeable to DNA using digestive enzymes extracted from snails and the 2μ (micron) DNA plasmid [35]. The initial development of systems for studying cloning and gene expression in insect cell lines involved the use of Baculovirus as vector. Since it was easier to clone and manipulate large DNA (see 7.3.8) in *E. coli*, vectors were often used that could replicate and be selected for in both *E. coli* and the second host organism, be it a bacterium, animal, human, plant, insect or yeast cell. These are referred to as 'shuttle vectors'.

Genetic engineering in plants began with the study of Crown Gall, a type of plant tumor. *Jeff Schell* and *Marc van Montagu* discovered that the Gall contained genes transferred into the plant cell from the saprophyte and symbiont soil bacteria *Agrobacterium tumefaciens* [36]. They adapted this mechanism to transfer other foreign and plant genes into plants. Although other mechanisms for transferring DNA into plants were later developed this system pioneered and facilitated genetic engineering in plants. Use of strong gene transcription promotors and viral vectors often facilitated good gene expression and gene propogation in the foreign host cells, e.g. the Cauliflower Mosaic Virus (CaMV) vector in plant cells [37].

7.3.8
DNA Sequence and Recognizing Genes and Gene Function

During this early period (1970s) methods were being developed for DNA sequencing, initially by partial chemical degradation (*Wally Gilbert*) or by base-specific strand termination (*Fred Sanger*). In both cases the products were analysed on long acrylamide gels. *Paul Berg, Walter Gilbert* and *Frederick Sanger* shared the Nobel Prize in Chemistry in 1980 'for fundamental studies of the biochemistry of nucleic acids, with particular regard to recombinant-DNA'. The complete sequencing of entire microbial genomes, even without the aid of cloning, has become a routine and affordable practice at every medical or life science research centre. With this surge in data it is evermore possible to recognize a gene function in a given DNA sequence by homology to known genes with known function. In the end, however, *functional tests are required* before the actual role of the gene can be determined with certainty. It may indeed have a very different role or even display several functions and interactions with diverse target molecules/partners (e.g. the proteins making up the lens of our eyes or those of the humming bird are essentially two different enzymes which have now clearly been co-opted to play a structural role; their roles are as completely different as those of metabolic enzymes in other differentiated tissues).

Various sophisticated techniques can be used to analyse the DNA- and histone-modification patterns at specific locations on the DNA which may be associated with malignant states. These latter modifications can often be accumulated slowly in response to environmental challenges such as bad diet or smoking. Gene expression patterns can be readily and rapidly detected in different tissues or in cancer cells and in response to pathogens. This enriches the possibilities available to practitioners in the human, animal and plant health-care fields, particularly with respect to the diagnosis of malignant states.

The isolation of known gene products which after functional testing leads to the recognition of novel functional gene products and hence to the production of novel (pharmaceutical) products through recombinant DNA technology is discussed in separate sections (see Section 7.5 and 7.6).

7.3.9
Cloning Larger DNA Fragments

In the early days of cloning (1973–1977) there was a strict limit to the size of DNA fragments which could be incorporated and propagated in a stable condition in the cell. Cloning fragments larger than 1 kb was extremely difficult. In 1977 *Noreen and Ken Murray*, and *Barbara* and *Thomas Hohn* worked on the development of *lambda bacteriophage* as a *vector* which led to an increase in the efficiency of cloning larger DNA molecules, although the size of the fragments was initially still relatively low (1–5 kb) and even later was still limited to less than 15 kb.

The efficient cloning of longer fragments, up to 40 kb, was first made possible by the development of the *cosmid cloning* system by *John Collins* and *Barbara Hohn* in 1978 [38]. They used small plasmids packaged *in vitro* into lambda bacteriophage coats. It was not immediately obvious how this could be achieved since efficient packaging can only be carried out on long concatameric substrates in which the *lambda cos*-site was regularly placed at 40–48-kb intervals in the same orientation. This difficulty was eventually overcome by using concatemers of restriction enzyme-cleaved vectors and large insert DNA (the inserts were partially cleaved and size fractionated) ligated at very high DNA concentration.

Further improvements to the cloning protocol were made in particular by *David Isch-Horowitz* and *John F. Burke* (1981) [39] to ensure that two fragments originating from different regions of the genome would not be present in the same clone. This was important for the use of cosmid 'banks' (DNA fragment libraries) to assist in the *mapping of DNA segments* in the early construction of 'contig' assemblies from overlapping clones. A further advance was the modification of cosmid vectors, for instance by substituting the high copy number plasmid replication origin(s) with the F-plasmid low number replication origin (fosmids). These were considered more stable for cloning eukaryotic DNA-containing repetitive sequences. *John Collins* [40] described mutations in the *E. coli* host's DNA recombination and repair system which contributed to at least partially stabilizing DNA repeats and palindromic DNA which are very unstable and had been considered to be unclonable in standard *Escherichia coli* hosts. The cosmid (and fosmid) cloning system was the most frequently used tool in genome cloning and mapping over some 12 years (1978–1990) and is still the tool of choice in metagenomic approaches to gene mining [41]. Cosmid cloning is a method based on functional expression and was first used to identify and isolate the penicillin acylase (=amidase) gene in order to produce acylase super-producing strains which were important in the production of semi-synthetic penicillins (*Hubert Mayer, John Collins*, and *Fritz Wagner* (1979); Patent Nr. DE2930794) [42]. This was one of the first examples of genetic engineering for which a patent was issued and which actually achieved commercial application. This paper also showed the general utility of cosmid cloning as a method for isolation of genes on the basis of *functional selection*. This served as a paradigm for gene mining using cosmids (and later fosmids and BAC gene libraries; see Section 7.6).

7.3.10
Further Tools for the Assembly of Physical Genome Maps

This step by step progressive accumulation, ordering and extension of clone collections was a widespread activity and is known as 'genome walking'. In this latter process DNA probes from the ends of one clone were used as hybridization probes for colony blots in order to identify cosmid clones carrying inserts which overlapped the original clone. Eukaryotic genomic fragments obtained by cosmid cloning were large enough to contain whole non-spliced genes along with most of the *cis*-acting elements required for their regulation. These could be transferred into and expressed in heterologous cell lines in which a human gene for instance could be shown to still be highly inducible (e.g. 10 000-fold for human interferon β gene in mouse fibroblast cells [43]).

7.3.11
Uniting Fragment, Sequence and Chromosome Maps

Much human genetic data was available in terms of mapping an inherited trait to a region on a particular chromosome. It was therefore of interest to map the position of particular DNA sequences and consequently genes on the chromosome map. Cosmid DNA was also found to be an ideal material for chromosome painting (Figures 7.3 and 7.4), a method used to enhance the information obtained from karyotyping including information on gene copy number and chromosome rearrangements, information which would otherwise be difficult to obtain. This latter approach has also been applied to total genome sequencing projects where it can be used to confirm the relative position of contig-sets (non-overlapping groups of overlapping clones) which have not yet been ordered or orientated on the genetic map. A further major advance was the combination of

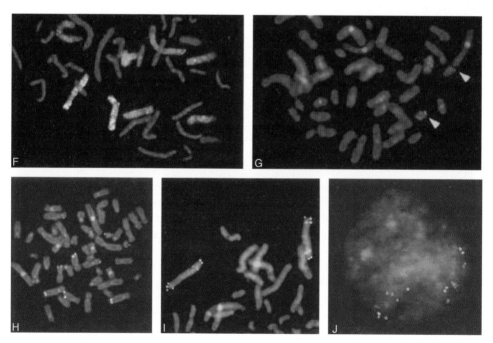

Figure 7.3 Chromosome Painting. From Ried, T. *et al.* (1992) 'Simultaneous visualization of seven different DNA probes by *in situ* hybridization using combinatorial fluorescence and digital imaging microscopy' Proc. Natl. Acad. Sci. USA, 89, 1388–1392. (David C. Ward's laboratory at Yale).

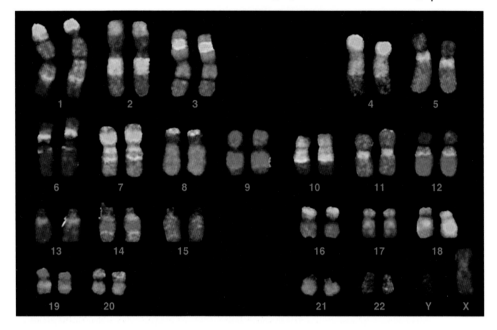

Figure 7.4 Fluorescent In-Situ Hybridization (FISH) identification of human chromosomes also known as 'Chromosome Painting': DNA probes specific to regions of particular chromosomes are attached to fluorescent markers and hybridized with a chromosome spread. The figure shows a computer-generated 'false colour' image in which small variations in fluorescence wavelength of the probes are enhanced as distinct primary colours. The combination of probes that hybridize to a particular chromosome produces a unique pattern for each chromosome. This makes it particularly easy to detect segmental deletions and translocations in the chromosomes [44].

previous methods and 'jumping libraries', now usually referred to as 'paired-end' libraries, which contain fragments from the ends of much larger size-fractionated material. This technique can deliver the unambiguous order of the unlinked contigs.

In the early 1990s two other powerful cloning methodologies were established, namely the Bacterial Artificial Chromosome (BAC) libraries developed by Hiroaki Shizuya [45] in 1992 and Yeast Artificial Chromosomes (YAC) initially developed by *Maynard Olson* in 1988 to 1991. The latter methodology was effectively used in 1993 in *Alan Coulson's* and *Sir John E. Sulston's* (Nobel Laureate 2002) laboratory in Cambridge in association with *R. Waterston* to map the genome of *Coenorhabditis elegans* (the round worm) (draft genome map presented at Cold Spring Harbor in 1989). In addition, the same methodology was also used in 1993 in *Jean Weissenbach's* laboratories in Paris (*Genethon* at Evry; in collaboration with the Centre de l'étude des polymorphisms humain (CEPH)) to map parts of the human genome. CEPH was established by *Jean Dausset* who received the Nobel Prize for his discovery of the histocompatibility phenomenon and HLA (human leukocyte antigens) now used for tissue-typing. He had established this institute with the intention of creating resources for human chromosome mapping in the form of human mouse chimaeric cell lines. After fusing mouse and human cell nuclei the hybrid starts to lose chromosomes, preferentially the human chromosomes. After months or years of cultivation, fairly stable cell lines usually carrying only one or a few human chromosomes can be obtained in reasonably large quantities. These were frozen and stored as reference material. Using non-repetitive DNA from human genomic cosmid, YAC or BAC clones to hybridize DNA from this reference material, the clones could be classified with respect to their chromosomal origin. The libraries generated by these methods in either *E. coli* (BAC) or yeast (YAC) hosts offer the possibility of cloning *larger* DNA inserts using BAC of up to a hundred to a few hundred kb and YAC of up to 1 Mb [24]. However, the advantage of the larger-size insert in YAC libraries may be outweighed by the lower frequency of chimeras and ease of clone handling observed with the BAC/PAC system [47]. As illustrated in the cited works, this further facilitated the creation of contigs and determination of their order on chromosomes (Figure 7.5). They were later also used as a source of DNA for megasequencing as it was known that essentially all the sequence data obtained from one set probably originated from a single contig which was the size of the BAC insert. The use of human cells as acceptors for the accessory chromosomes led to the development of human accessory chromosomes (HACs) as cloning vectors by *J.J. Harrington* in 1997 [46].

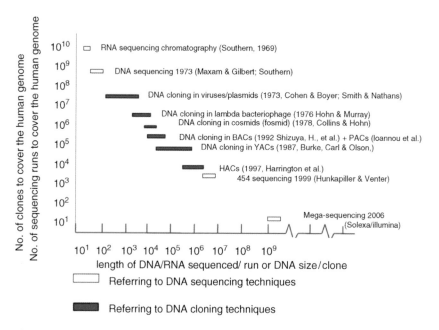

Figure 7.5 DNA sequencing, cloning and mapping technologies. With respect to the vertical axis, the filled bars represent the number of clones needed to cover the entire human genome and the open bars (sequencing methods) represent the number of runs required to give 99% coverage of the genome [47].

7.3.12
Understanding Gene Regulation in Eukaryotes: A Branch of Reverse Genetics

This in turn accelerated research into the mechanism of gene regulation in eukaryotes particularly with respect to regions adjacent to the genes which bound the transcription factors and enhancers. Many of these *cis*-acting sequences were located thousands of bases away from the site at which transcription was actually initiated. *Russel F. Doolittle* started cataloguing information on codon usage which enabled true gene translational coding regions to be recognized, and assisted in the identification of other often highly variable or degenerate sequence motifs of biological significance [48]. The first detailed and extensive collection of data relating to weighted sequence motifs correlated with the binding sites for proteins involved in regulation of gene transcription was established by *Edgar Wingender* (the TRANSFAC database). Combinations of such binding motifs near a gene could be predictive of how the gene may be controlled in a particular type of differentiated cell.

7.4
Oligonucleotide Synthesis Leads to Protein Engineering

During the initial work by *Marshal W. Nirenberg* and *Heinrich Mathei* and subsequent work carried out in association with *Gobind Khorana, Dieter Söll* and *Philip Leder*, the ability to synthesize DNA not only revealed the genetic code but also opened up new avenues in biotechnology. The synthesis of DNA probes made the rapid analysis of the genetic material possible. Genes could be mapped and sequenced. In the early days of genetic engineering synthetic DNA had already been used to build entire genes and their adjacent regulatory elements. Combined with gene cloning it then became possible to produce novel protein pharmaceuticals of high purity using reproducible production strains. By 1984 this revolution, initiated in 1974, was in full swing with several thousand laboratories involved worldwide.

In the short period between 1953 when the first structural model of DNA became available and about 1980 when proteins of a particular design were first produced using recombinant DNA technology, not only was a thorough understanding of the basis of heredity acquired but this information was utilized to create a novel discipline, namely *protein engineering*. This novel approach was not simply an extrapolation of the observation of the effect of *random* mutations and the phenomenon of *loss of function*. Using the new technology known as '*reverse genetics*', DNA could be reintroduced with or without *defined* modification into either the same species or across species barriers often resulting in *gain of function*.

One of the most dramatic findings of this period was the *universality of the genetic code*, that is, the rules used in the conversion of the linear sequence of bases in DNA into the linear sequence of amino acids in the protein product. The close relationship between all life forms is in line with their having been a common origin from which all life evolved. In the present context it was a serendipitous event for modern biotechnology that genes from one organism could be transferred and expressed in cells of another species. Being able to do this in a controlled fashion depended on the work carried out during the 1970s and the 1980s by hundreds of laboratories resulting in the accumulation of information relating to gene structure, gene regulation, operon structure and vector construction.

A detailed discussion of gene expression systems is not within the scope of this book. A good overview can be obtained by perusal of the catalogues of companies supplying laboratory materials for the life sciences.

However, one particularly noteworthy example is the widespread use of the tetracycline (Tet) promoter system for tightly controlled and highly inducible induction of gene expression using vectors which incorporate this promoter system. This system can be used in bacteria, yeast and animal cell culture, plants and whole animals using tetracycline derivatives such as doxycycline as inducers. It is the preferred system for controlling the production of factors required for the induction and maintenance of pluripotency in stem cells (iPS) or induction of even conditionally lethal functions in transgenic animals in a tissue-specific manner [49]. Hermann Bujard (Figure 7.6) initially developed this system at the University of Heidelberg and subsequently further improved it for use in a wide range of organisms.

Cloning of synthetic DNA (genes) led to the synthesis of the novel products in micro-organisms or in cell culture and later in transgenic animals. Computer models of protein structures were used to predict possible molecular interactions and potential alterations in the specificity of enzyme activity. The design of novel protein and gene structures with a predicted function would in principle then be possible over a period of years or months rather than millions of years of trial and error.

Figure 7.6 Hermann Bujard, the pioneer of inducible gene expression systems based on the tetracycline promoter.

The Doctrine of the Triad.
The Central Dogma: "Once information has got into a protein it
can't get out again". Information here means the sequence of
the amino acid residues, or other sequences related to it.
That is, we <u>may</u> be able to have

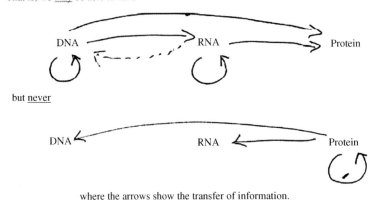

but <u>never</u>

where the arrows show <u>the transfer of information</u>.

Figure 7.7 The central dogma illustrated by Francis Crick in 1955 in this original draft.

This was the first time that Crick's *central dogma in molecular biology* (Figure 7.7)
had been challenged.

It should be noted that in conservative sections of the pharmaceutical industry these
novel tools were initially greeted with sober reservation and scepticism. Clear
advances, it was argued, had after all been made in massively increasing production
of antibiotics, for example by random mutagenesis of production strains and selection
for variants that produced higher yields. It still remained to be demonstrated that the
new technology was capable of producing a competitive product which would be
approved despite the nebulous but deep-seated fears related to the danger of
recombinant DNA 'manipulation' before it was finally released onto the market.

However, progress was relentless and unforgiving to the over-conservative. The
increased control and understanding of biological systems that was achieved using
reverse genetics and genetic manipulation initially in micro-organisms was decisive.
To give just a few examples illustrating the breadth of application: bacteria and yeast
were transformed to produce plasma proteins for medicine; transgenic animals were
constructed to provide improved models of human disease; selfish-gene-containing
transgenic insects were created to decimate populations of mosquitoes which are
vectors for malaria; and transgenic plants were developed which were better suited to
today's industrialized agricultural methods.

This was not just another example of human ingenuity but represented *a radical
advance in the mechanisms at work in evolution – a paradigm shift*. Man had implemented breeding programs over millennia but the outcome had relied on the genetic
variability already present in the breeding population. Combining protein analysis
which included the modelling of protein structures, the understanding of protein–
protein interaction and the mechanism of enzyme function with the ability to

synthesize the gene for a specifically designed protein, the central dogma on information flow in biology needed to be drastically modified (Figure 7.7). Information regarding the structure–function relationship of proteins could now be applied to the design of genes which would generate novel products with predicted novel function. Thus not only had the nature of human genetic material been elucidated but it was now also possible to alter it in a specific manner. The resulting euphoria amongst the visionaries led venture capitalists to invest in the 'New Biotech' industry with *Genentech* in California being the first to benefit in 1976.

The 'central dogma' in biology which was first formulated in 1955 by *Francis Crick*, states that genetic information flow is unidirectional in living organisms. It passes from one linear information-carrier to the next and originates in DNA from whence it is transcribed into RNA and finally translated into protein. This simple formulation of the dogma lasted only 15 years. In 1970 it was discovered that there were RNA viruses that encoded reverse transcriptase (*Howard Martin Temin* and *Renato Dulbecco* together with *David Baltimore* shared the 1975 Nobel Prize for Medicine 'for their discoveries concerning the interaction between tumour viruses and the genetic material of the cell'), an enzyme which produced DNA from RNA. This led to the first dramatic change in the acceptance of the central dogma as it demonstrated that in natural systems information was able to flow in the both directions between RNA and DNA (Dogma 2; in fact in an early draft the possibility of reverse transcription had been included as a possibility! Figure 7.10).

Around 1981 experiments with a radically new character emerged under the heading '*protein engineering*'. They involved gene synthesis for novel proteins which had been *designed* on the basis of models of structure and function. This may be derived, *a priori*, on completely theoretical grounds or by analysis and comparison of functional variants, for example homologues or paralogues from different species. The information flow diagram had come full circle for the first time in the history of life.

7.4.1
Chemical Synthesis of Macromolecules: Peptides and Oligonucleotides

The very beginnings of peptide chemistry in the early twentieth century, from *Emil Fischer* onward, are described in Chapter 4. Peptide and protein synthesis relevant to modern biotechnology, that is those compounds which have biological activity, began with the synthesis of an octapeptide, oxytocin, by *Vincent du Vigneaud* who published his findings in 1953. He was awarded the Nobel Prize for Chemistry in 1955 for this achievement. By 1958, the longest peptide which had been synthesized was a 23-residue fragment of adrenocorticotrophic hormone (ACTH). *Robert B. Merrifield* [50] synthesized a tetramer in 1958 using a technique known as solid phase peptide synthesis which he had pioneered. However, before this a Chinese group at the Shanghai Institute of Biochemistry was already working towards the total synthesis of a mammalian protein, namely bovine insulin. They had started this work shortly after the chemical structure of bovine insulin had been elucidated by *Fred Sanger* in 1955 (leading to his first Nobel Prize in 1958). Three years later in

1961 ribonuclease became the next protein to be sequenced. ACTH which contains 23 amino acids was the largest peptide to have been synthesized before 1963 and this was accomplished by the pharmaceutical division of CIBA-Geigy in Basel, Switzerland [51]. The project group led by Zou Chenglu published the synthesis of the B chain and semi-synthesis of insulin in 1964. The full synthesis was completed and the activity of the final product tested in mice in 1965 [52].

Although the total chemical synthesis of variant proteins was a theoretical possibility, it was entirely uneconomical and unfeasible for synthesis of a large number of different variants. As for most chemically synthesized products, each variant needed to be purified in small yield from a large background of contaminants with similar physical characteristics. Consequently, this could not be considered as a methodology for continuous production. This view later changed following the development of *ligation protein synthesis* by *Steve Kent* at the University of Chicago (see Figure 7.8) which led to the production of completely synthetic erythropoietin (EPO). The latter was brought to Phase I clinical trials by Gryphon Therapeutics, San Francisco [53]. Although the trials were terminated because of the immunological reactions to the polymers used as substitutes for glycosylation, it was shown that PEGylated derivatives were biologically active and not immunogenic. In principle this was a validation of the hypothesis that proteins for clinical use can be completely synthesized using fusion peptide chemistry and could be obtained both in high purity and with low immunogenicity at an apparently reasonable production price. Since

Figure 7.8 Stephen B.H. Kent who developed the process of peptide ligation which then facilitated the feasible synthesis of clean large proteins and the possibility of creating mirror image proteins.

this is a very interesting development it may be resurrected by other researchers at a later date, particularly in view of the improvements in reducing costs for chemical peptide synthesis which ensued during the synthesis of the AIDS drug Fuzeon (T20; see below) [54]; see Figure 7.9).

This development in methodology was like no other as it could also be used to synthesize 'mirror-image' tautomers which are entirely composed of D-amino acids instead of the normal L-amino acids which comprise 95% of normal proteins. Such proteins are resistant to proteases and are expected to be of low antigenicity since they cannot be easily processed by the proteasome for the presentation by the antigen-presenting cell. The presentation of such peptide fragments on MHC

Table 1	Peptide pharmaceuticals manufactured by chemical synthesis		
Peptide	**Length**	**Manufacturing method**
Adrenocorticotropic hormone (1–24)	24	C
Bivalirudin	20	C
Growth-hormone-releasing factor (1–29)	29	SP
Integrelin	7	C
Oxytocin	9	C
Atosiban (oxytocin antagonist)	9	C
Thymopentin (TP-5)	5	C
Thymosin α-1	28	SP
Thyrotropin-releasing hormone	3	C
Vasopressin analogues		
Desmopressin	9	C, SP
Felypressin	9	C
Glypressin	12	C
Lypressin	9	C
Pitressin	9	C
Corticotropin-releasing factors		
Human	41	SP
Ovine	41	SP
Angiotensin-converting enzyme inhibitors		
Enalapril, Lisinopril	2	C
Somatostatin and analogues		
Somatostatin and analogues	14	C, SP
Octreotide	8	C
Lanreotide	8	SP
Luteinizing-hormone-releasing hormone	10	C, SP
LHRH agonists and antagonists		
Leuprolide	9	C, SP
Goserelin	10	C
Triptorelin	10	C
Buserelin	9	C
Nafarelin	10	C
Cetrorelix	10	SP
Ganirelix	10	C
Calcitonins		
Human	32	C
Salmon	32	C, SP
Eel	32	C, SP
Dicarba-eel (elcatonin)	31	C, SP

C, classical solution-phase chemical synthesis; LHRH, luteinizing-hormone-releasing hormone; SP, linear solid-phase peptide synthesis.

From Bray, B.L.(2003) Large scale manufacture of peptide pharmaceuticals by chemical synthesis, Nature Rev. Drug Disc., 2, 587-593

Figure 7.9 Pharmaceutical peptides that can be successfully produced by chemical synthesis in direct competition with biotechnological production.

molecules is one of the steps involved in eliciting an immune response to the original antigen.

7.4.2
Combining Chemistry and Biology to Develop Novel Pharmaceuticals

A novel approach to developing D-peptide pharmaceuticals is to synthesize mirror image targets and then screen normal peptide libraries or phage-display libraries (of L-peptides) that bind to the all-D-protein target. Once a strongly binding variant has been found, an all-D version of the peptide may be synthesized and used as a novel protease-resistant pharmaceutical drug that will bind the natural all-L protein target.

The principle was worked through by *Stephen Kent* for the HIV protease, in creating the all-D-version of the enzyme, which was shown to be fully active on an all-D version of the normal all-L peptide substrate. Vice versa, the normal enzyme cleaved only the normal all-L substrate peptide [55]. *Peter Kim* used the same principle on gp41 to develop an all-D peptide inhibitor which had an activity in the 250 pM (picomolar) range to prevent entry of the HIV virus into the cell [56].

Smaller peptides for pharmaceutical use are now synthesized chemically on a large scale in competition with biotechnological processes. A dramatic turning point was the drive to produce *enfuvirtid* (also known as Fuzeon or T20) the 36 amino-acid peptide derived from HIV gp41 which inhibits entry of the HIV virus into the cell. Using classic solid-phase peptide synthesis (SPPS) three peptides containing between 9 and 16 amino acids were synthesized and then assembled in solution.

7.4.3
Biotechnology's Competitor: Cheaper Chemical Synthesis

This work commenced around 1997 under conditions that were initially very unfavourable for a successful commercial production procedure. Building blocks and resin ($15 000/kg) were very expensive and in very limited supply. Since the average therapeutic dose is about 80 g/patient/year, this creates a global demand for several tons per year. The final assembly step is carried out in batches of 300–500 kg of each of the three peptides. The purification steps are intensive, there being some 106 steps in the total synthesis. The chemistry is solid and robust with efficiencies of >99% per step for SPPS. Combined with the dramatic reduction in the cost of materials the process became a feasible proposition. The FDA approved the final product in 2003.

This is a success story where demand changed the limiting parameters (cost of raw materials). Although the material and target were found and characterized via genetic engineering it was a chemical synthesis procedure which finally dominated the production process [57].

Figure 7.9 shows a list of biologically active peptides where chemical synthesis now competes with biotechnological production. As outlined for EPO above, the size of peptides for which it will be economically feasible to develop a production process by total chemical synthesis, will probably continue to increase beyond those shown in the figure (maximum 41 amino acids).

7.4.4
'Muteins' and the Beginning of Protein Engineering

The term 'muteins' for an expressed protein altered by changing the coding sequence of its gene may have first been used in 1981 [58]. This represents a closure of the information transfer cycle previously discussed with respect to Francis Crick's central dogma. Knowing the protein structure enables the synthesis of the DNA (gene) (Figure 7.10). This can be seen as a new paradigm in science in which the design principle can be applied to making novel proteins.

The creation of muteins became a reality. Gene cloning provides the final link to close the circle. This aspect was certainly included in the work *John Collins'* group on the design of human PSTI mutein in 1984–85 which produced low picomolar affinity inhibitors of leukocyte elastase (patent application 1987). The meeting in June 1985 at the Royal Society in London caught the mood of the time. Some of the pioneers including *Max Perutz, Ian Fersht* and *Greg Winter* were present. In his introduction *D. M. Blow* summarized the advances which came together to generate the new discipline of *protein design*, by which point mutations could be introduced into cloned expressed genes and their products tested for the expected property changes. Cycles of prediction, mutation, mutein production and testing would yield more refined predictive models and a better understanding of protein structure and function [59].

The discussion highlighted the structural elucidation of protein structure which was limiting the process of *rational* protein engineering. Furthermore there was scepticism about the purported advantages of defined point mutation compared to gene-specific but nevertheless random mutagenesis. This brings to mind a statement made by Professor *E. von Wasielewski* then Head of the board of Directors of Hoechst AG, Frankfurt am Main who was of the old school which had for generations been successfully increasing yields of secondary metabolites by microbial fermentations.

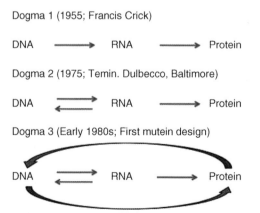

Dogma 1 (1955; Francis Crick)

DNA ⟶ RNA ⟶ Protein

Dogma 2 (1975; Temin. Dulbecco, Baltimore)

DNA ⇄ RNA ⟶ Protein

Dogma 3 (Early 1980s; First mutein design)

DNA ⇄ RNA ⟶ Protein

Figure 7.10 Refinement of the Central Dogma of Molecular Biology of 1955 up to about 1985.

In 1977 at the 'fifth Workshop Conference Hoechst: Pancreatic beta cell culture' [60] von Wasielewski was asked by Walter Gilbert: 'What would you pay for an *E. coli* clone producing human insulin, and what would you then do with it?'. von Wasielewski replied: 'I would pay $5 million and then irradiate it to produce variants that made more insulin!'.

It was *Michael (Mike) Smith* in 1979 who had been the first to show that site-specific mutation with synthetic oligonucleotides was feasible [61]. He was awarded the Nobel Prize for this in 1993. In the opening remarks of his lecture on that occasion in Stockholm, he pointed out that he had '... had the good fortune to arrive in the laboratory of Gobind Khorana, in September 1956, just one month after he had made the accidental discovery of the phosphodiester method for the chemical synthesis of oligodeoxyribonucleotides, a synthetic approach whose full exploitation led to elucidation of the genetic code and the first total synthesis of a gene' [62].

The work of *Jim Mark's* group at *Genentech*, South San Francisco, on the detailed structure–function relationship of a protein which entailed making a myriad of mutant Subtilisins [63] and analyzing their structures, was an early highlight in this area of fundamental research. It was unusual in that it was carried out in an industrial setting. The high quality of the research being carried out and the number of researchers entering the new biotech enterprises was characteristic of their innovative potential and perhaps a hallmark of this period.

Michael Smith quoted *Joshua Lederberg's* statement made in 1959: 'The *ignis fatuus* [will o' the wisp] of Genetics has been the specific mutagen, the reagent that would penetrate to a given gene, recognize it, and modify it in a specific way'. Neither could have imagined just how soon this dream would be realised. Subsequent to the discovery of 'silencing RNA' (siRNA) and micro RNA (miRNA) (see Section 13.5) synthetic DNA can be used to create silencing RNA directed to specific genes or to create genes for such RNAs which on specific induction can lead to breakdown of the specific mRNA of the target gene(s). This is already in use in transgenic animals and plants and has clear potential for clinical application.

The combination of chemical DNA synthesis with DNA cloning thus released enormous potential for the essentially unlimited specific alterations of the sequence and hence structure and function of proteins. This combined with the demonstration of the pragmatic success of Genentech in obtaining approval for recombinant human insulin in 1984, perhaps more than anything else generated the euphoria and the sense of a 'new age' dawning.

For the Biotech industry this euphoria waned late in the period 1990s to 2002, as investors lost confidence in the new biotechnology following a series of failures of products entering clinical trials, generating concerns about the relevance of targets and disease models and the immunogenicity of recombinant proteins.

There were also worries about gene patentability growing in the wake of vehement attacks on the NIH administration and on some of the pioneers of genome sequencing and annotation, such as *Craig Venter*. *Jim Watson*, the highly visible Nobel laureate, and for some time leader of the US genome programme was particularly damaging in this role as he was the most vitriolic critic.

2002 was described on the stock market as the 'nuclear winter of biotech'. The period 2002–2004 was critical for small Biotech companies. It saw many becoming insolvent or for those that were slightly luckier, being bought out at under-value prices by larger companies.

7.5
Synthetic DNA, Reverse Transcriptase: Isolating Genes

It was *P. T. Gilham* in the 1960s who developed the method of affinity chromatography using oligonucleotides covalently attached to cellulose. This led to the procedure for enriching specific eukaryote RNAs for example tRNAs, as a result of the work of Gilham and *Dieter Söll* (aiding the determination of the genetic code). More important in terms of isolating the genes was the use of immobilized random oligo-thymidylates to isolate poly-A-tailed messenger-RNA (mRNA). This latter type of RNA represents the majority of mRNA, the RNA which is translated into protein (*Haim Aviv* and *Philip Leder*, 1972 [64]).

The identification of mRNA as being the material which encodes the desired gene product was, however, extremely arduous and very indirect. One of the favourite methods used in the mid to late 1970s was the expression of the enriched mRNA on injecting it into the eggs (oocytes) of the three-toed South African toad, *Xenopus laevis*. The animal could simply be kept in a drawer in the laboratory and had few requirements other than an occasional piece of meat and change of water. The supernatant of the treated oocytes or even a lysate of the cells was examined after a day or so for the presence of the product which it was hoped had been encoded for in the fractionated mRNA. The investigation method for the final product was often a bioassay which usually took a further few days. This was true for example for the initial cloning of the highly active interferons and lymphokines, where insufficient protein was available for sequencing or inducing antibodies.

The following quote from Michael Smith in 1993 (Figure 7.11) puts the work of this time into perspective and enumerates the key developments which led to the simpler isolation of genes by hybridization of mRNA to synthetic oligonucleotides that were designed on the basis of short known protein sequences:

> '. . . the development of the full panoply of DNA cloning technologies in the early 1970s made it clear that the prime target for work with synthetic deoxyribo-oligonucleotides should be a method for *monitoring gene isolation*' [author's emphasis].

> Discussions with Dr Benjamin D. Hall made me aware of the important studies of F. Sherman on double-frameshift mutants of yeast cytochrome c. Knowledge of the amino-acid sequence of these mutants allowed the unambiguous prediction of the N-terminal coding sequence and, hence, provided a specific target for an oligonucleotide of defined sequence. [. . . . .]

> The required probe was synthesized using the enzymatic method (Gillam *et al.*, 1977) and was used to isolate the gene, albeit with some difficulty (Montgomery *et al.*, 1978), thus clearly validating the principle that a synthetic oligonucleotide could function as a probe for gene isolation. [. . . . .]

At the time when the iso-1-cytochrome c gene was isolated, the chemical method for DNA sequence determination (Alan M. Maxam and Walter Gilbert, 1977) and the chain terminating enzymatic method (Fred Sanger *et al.*, 1977) had newly been developed, the former being applicable to double-stranded DNA and the latter to single-stranded DNA. [. . . . .]

I decided to investigate the possibility of directly sequencing double-stranded DNA by the enzymatic method using synthetic oligonucleotide primers, based on my conviction that an oligonucleotide should hybridize at low temperature with a complementary DNA strand of a denatured DNA double helix in solution, even though the second DNA strand was present and capable of producing a thermodynamically much more stable structure. The critical question was whether the second DNA strand would displace a short oligonucleotide from a complex with the complementary DNA strand under the conditions of the Sanger enzymatic sequencing method (Sanger, 1981).

The first experiment was completely successful, an enormously exciting event. As a consequence, the sequence of the gene was completed by "walking" along the sequence using a series of short oligonucleotide primers 9 or 10 nucleotides in length (Smith *et al.*, 1979). I believe that the full potential of this approach to gene sequencing has yet to be realized'.

Figure 7.11   Michael Smith, 1993 Nobel Laureate and pioneer of site-specific mutagenesis using synthetic oligonucleotides.

However, the use of synthetic nucleotides as primers for DNA sequencing and introduction of point mutations is only one aspect. Once oligonucleotide synthesis became efficient and automated it was also possible to synthesize entire genes. These methods were developed by *Mike Smith* for the production of gene variants (Figure 7.11). Following the development of the technology for the polymerase-chain-reaction (PCR), see Section 7.7.2, it became possible for any researcher possessing the oligonucleotide primers specific for a particular gene, to amplify and isolate it directly from a minute amount of cell material, indeed even a single cell would suffice. With the knowledge of all the expressed gene sequences a novel type of diagnostic and research tool became available: a selection of gene-specific oligonucleotides that are immobilized on a chip and to which samples of other DNAs or RNAs can be hybridized. A quantitative evaluation of the bound material can be used in a number of different ways, for example for determining differential gene expression or analyzing variation in gene copy number. The data provided by this branch of biotechnology has enabled the identification of specific 'biomarkers' of particular disease states. Chemistry and engineering have delivered the DNA-CHIP-technology which is used to monitor these biomarkers (see Section 11.4.5).

The matrix-based technology requires the synthesis of large numbers of oligonucleotides and their ordered distribution on a two-dimensional surface. Part of the chain of developments included the development of methods and chemistries to facilitate simultaneous synthesis of large numbers of oligonucleotides. Amongst the pioneers were *Ronald Frank* and *Helmut Blöcker* who from 1982 to 1983 developed a synthesis method based on paper segmental supports (Figures 7.12 and 7.13) [65].

Figure 7.12 Helmut Blöcker.

Figure 7.13   Ronald Frank.

This was a general approach for the simultaneous chemical synthesis of large numbers of oligonucleotides on segmental solid supports which later found direct application in semi-automated and finally fully-automated oligonucleotide synthesis. This novel principle was also applied to peptide synthesis.

Alternatively beads were used to immobilize the growing chains of either oligonucleotides or peptides. In the synthesis of DNA specific nucleotides were added individually. Specific regions in the growing chains could be randomized by assorting a collection of beads into separate reactors, each specific for the addition of either adenine (A), cytosine (C), guanine (G) or thymine (T) nucleotides or all four.

The aforementioned method, the forerunner of combinatorial chemistry using solid-phase synthesis, became known as the 'tea-bag' technique, and was developed by *Richard A. Houghten* (Figure 7.14) in 1985 *for peptides* [66].

The use of gene technology thus allowed the isolation of genes in a form which allowed the controlled synthesis of the gene products. Gene technology was carried out initially in micro-organisms but was then applied in eukaryotic tissue culture (cell lines) and later even in whole animals and plants (see Chapter 12).

7.5.1
Reverse Genetics: Providing Cheap Renewable Resources for 'Germ-Free' Products

With respect to the production of products of biotechnological importance which could be used as pharmaceuticals it was important that this novel production route provided material which was free of possible contamination from pathogens. Contamination had always been a hazard associated with isolating pharmaceuticals

Figure 7.14 Richard Houghten the pioneer of peptide library synthesis.

from organs (e.g. growth hormone from the pituitary; insulin from the pancreas; many different hormones from the placenta) and blood (e.g. coagulation factors Factor VIII, Factor IV, Factor X, human serum albumin). For the 'second generation' products this was essentially the only method of producing pure products in the quantities required for therapy (interferons, erythropoietin, and lymphokines). On another level the long-term production of proteinaceous substances of controlled quality could be ensured, for example production of a particular monoclonal antibody previously isolated from an (unstable) hybridoma cell line.

Early applications in biotechnology focused on producing known products more cheaply and with a higher purity, for example insulin and human growth hormone, or penicillin acylase over-producing strains for the production of semi-synthetic penicillins and cephalosporins. In particular the large scale production of these pharmaceuticals, free of possible contaminating pathogenic material from human or animal sources (first generation products) was also a qualitative advance which can be seen as a major contribution to public health.

Apart from the economic advantages these developments ensured:

- Reproducible, storable production sources which could be readily tested for product quality. These are amenable to scale-up with known biotechnological engineering techniques and are not dependent on other market influences related to variables in agricultural production (drought, floods, foot and mouth disease, etc.).
- The products intended for medical or veterinary use could be produced from sources free of pathogenic viruses. Prior to 1978 tests for contamination of products with viruses such as the causative agents of hepatitis (HCV, HBV) or in

the case of acquired immune insufficiency disease the AIDS virus (HIV) and prions (e.g. the causative agent for variant Creutzfeld-Jakob disease; vCJD) was poor or non-existent.

7.5.2
Gene Cloning contributes to the Discovery of and Characterization of New Viral Pathogens: Risk Prevention for Biotechnological Products

There has always been the possibility of viral contamination during the isolation of pharmaceutical products. Just how hazardous this might be was often only estimated retrospectively after the discovery of widespread virus contamination in a product. The spread of Acquired human immunodeficiency syndrome (AIDS) caused by the HIV virus is a dramatic example of this, as hundreds of thousands of individuals were needlessly infected by blood-derived products prior to the use of gene-based technology to detect viral contamination.

A sensitive test for contaminants could not be developed until the disease agent was clearly identified. Identification of DNA (also reverse-transcribed copy DNA (cDNA) cloning where RNA viruses are involved) also contributed to the development of tests for such agents as contaminants in conventionally produced material which often originated from blood and was frequently produced in developing countries, for example erythropoietin, blood coagulation factors, serum albumin. This also applied to material extracted from other human tissue, for example growth hormone from the pineal body of the human brain; lymphokines and interferons from blood cells (e.g. buffy-coat fractions); placental hormones from human placenta in addition to material obtained from animal tissue, for example insulin from swine and sheep pancreas.

7.5.3
1979–1990: A Rapid Advance in Characterizing Viral Agents

The following list includes some of the major advances in the identification of viruses in biological material, food and in the body. These rapid advances were due to the development of gene cloning and DNA sequencing which revealed hitherto unnoticed contamination of the natural products used in medicine:

1979 Hepatitis B virus (HBV; *Galibert, F. et al.* [67] [*Pierre Tiollais'* laboratory, Pasteur Inst. Paris]). *Simon Wain-Hobson* was also intimately involved in both this discovery and in the cloning and sequencing of the HIV genome (see below).

1981–2 human papilloma virus (HPV; from warts and laryngeal papilloma; lead to vaccination and diagnosis of cervical cancer; *Harald zur Hausen*, DKFZ, Heidelberg; Nobel prize for medicine in 2008).

1981 Epstein-Barr virus (EBV; HHV-4, is the causative agent of Burkitt's lymphoma) (*Arrand, J.R. et al.* [68]).

1982 Cytomegalovirus (HCMV; this is the largest pathogenic virus genome and was the first genome to be mapped with cosmid cloning; 230 kb) (*Fleckenstein, B.,* *et al.* [69]). It often causes blindness in immune suppressed (e.g. HIV) patients in whom it is important to distinguish this virus from toxoplasma. It was often found associated with Kaposi Sarcoma which apart from being found in AIDS patients is an extremely rare cancer.

1984/5: Lymphadenopathy virus (LAV later known as HIV), *Simon Wain-Hobson* [70]. *Francoise Barré-Sinoussi*, and *Luc Montagnier* received the Nobel Prize for medicine for discovering HIV in 2008. However it should be noted that *Robert Gallo* who had spent two decades searching for a viral causative agent for human cancer shared both the *Albert Lasker Award for Clinical Medical Research* 1986 (the second time for Gallo!) with Montagnier, for this achievement, as well as sharing in the commercial development of an HIV-diagnostic test for human donor blood and blood products.

1986 Hepatitis Delta virus (HDV) when present with HBV causes 20% mortality. This was cloned and characterized in *Michael Houghton's* laboratory at *Chiron Corp.*, Emeryville, California [71], and

1989 Hepatitis C virus (HCV) which was also cloned in *Michael Houghton's* laboratory at *Chiron* Corp., Emeryville, California [72]

1990: Hepatitis E virus (*Reyes, G.R. et al.* [73]) was found to be the causative agent of what was previously known as non-A or epidemic hepatitis, it was given the latter name because of its rapid spread through faecal contamination. The low availability of antigen hampered the development of convenient serological assays until the viral genome was cloned and sequenced by Reyes's group.

Other viruses have been discovered in human tissue but their aetiology is at present unknown, for example GBV-C which was previously known as Hepatitis F virus (*Leary, T.P. et al.* [74]).

7.5.4
Sera and Antivirals: Biotechnological Products in their Own Right

The initial cloning and characterization of pathogenic viruses was a major breakthrough in its own right, leading to either recognition of the sole causative agents of some diseases, or at least to an understanding of their role in the pathology. This was the beginning of the detailed molecular study of virulence mechanisms and a route to both diagnostics and in some cases better vaccination against a particular disease. In a French article the discovery and cloning of Hepatitis C and E viruses was referred to as the 'miracle of molecular biology' which culminated in The Clinical Medical Award [75].

Understanding the genetic material and even characterizing all the components of a pathogenic virus or micro-organism is of considerable importance in identifying targets for drug intervention. This may be an antibiotic or a vaccine.

There is a long history of using a closely related cousin to a pathogenic organism as a vaccine to induce resistance to the disease. *Edward Jenner* initially demonstrated this in 1796 using cowpox as a smallpox vaccine. In other cases an attenuated (weakened) strain of the original pathogen was used as *Louis Pasteur* showed for Rabies virus. He weakened it by growing the virus for many generations in animal nerve tissue.

Effective vaccines have been developed for Hepatitis B virus (HBV) and for cervical cancer-inducing variants of Papilloma virus. For most viruses and micro-organisms this has proved difficult because variants appear rapidly and therefore change the nature of the antigen recognized by the antibodies induced by the vaccine of the day. A vaccine for HIV is still not on the horizon in spite of extensive research.

Influenza virus vaccines are produced on a yearly basis to combat the next wave of widespread infection (epidemic or pandemic (=world-wide epidemic)) by mainly targeting the neuraminidase (N) and haemaglutinin (H) functions. The latest pandemic, *swine 'flu* has the variant composition H1N1. Its origin seems complex as it incorporates regions from related viruses such as avian flu virus and other sources. Molecular genetics is key to providing a rapid response in terms of detection and containment as well as vaccine development in the case of out-breaks of a new virus. This was demonstrated for the SARS (Severe Acute Respiratory Syndrome) pandemic, where concerns were also raised about the rapid dispersal of pathogenic organisms in an era of rapid and extensive air travel. The rapid spread of the highly infectious Norovirus which causes debilitating diarrhoea but is of low mortality, has recently been recognized as a major hazard on ocean liners.

All the above examples illustrate the rapid spread of new viruses around the world which can create a pandemic within a few months. Recombinant DNA technology has facilitated the rapid development of efficient methods of molecular diagnosis leading to the identification and containment of infected individuals. The corre-sponding limitation in the rate of spread gives a period of grace of about 6 months which is the minimum time required to develop a new vaccine and to mount a vaccination programme.

As illustrated in these examples, and as we will see repeatedly throughout the key developments in molecular biology, major breakthroughs were often made in industrial, albeit often small biotech settings, particularly in the USA, where collaboration with university departments often played a key role. Key personalities from academia were either the founders or played major roles on the advisory boards of all the prominent biotech start-ups. See also Sections 10.4 and 10.5.

7.5.5
Gene Cloning and Production of Pure Material to Characterize Activity: 'second Generation Products'

The application of gene technology to products such as the interferons (second generation products) had a different dimension since it was only after gene cloning and the production of pure products that the actual properties of a protein could be distinguished from those of many similar highly active compounds, which had often been co-purified from complex mixtures present in bodily fluids or tissue-culture supernatants. Establishing the actual number of interferons and later other lymphokines such as the interleukins was *only possible* after the advent of this technology.

A further consequence of gene technology is the ability to produce for the first time sufficient quantities of such highly biologically active compounds. This was true of human growth hormone, for treating dwarfism, and erythropoietin (EPO) for stimulating red-blood cell synthesis after dialysis in kidney patients. Prior to the recombinant DNA era there had been an inadequate supply of these compounds. Many therapeutic compounds had long been isolated exclusively from human tissue or fluids: human growth hormone from the pineal body surgically removed from the brain of corpses; androgens from urine; and a number of hormones (progesterone; somatomammotropin (placental lactogen); oestrogen; relaxin, and chorionic gonadotrophin (β-hCG) from placenta. Organizing the collection, storage and preparation of such material as well as the aforementioned problem of contamination with pathogenic viruses, made the use of such human material very expensive and problematic, or as was seen in the case of contamination with the AIDS virus in the 1970s and 1980s, extremely hazardous.

7.5.6
Enzymes for Replacement Therapy

The same is true for the production of human enzymes: glucocerebrosidase (Cerezyme; Imiglucarase) used to treat Gaucher's disease; α-galactosidase A (Fabrazyme, Agalsidase-β) for treating Fabry disease; tissue plasminogen activator (Alteplase; + derivatives: Retavase, TNKase) to combat thrombosis, for example by dissolving blood clots which have formed in arteries of the heart; and DNAse I (Pulmozyme, Dornase-alpha) employed in the treatment of cystic fibrosis where it is used to reduce the high viscosity of lung mucous caused by the release of bacterial DNA during *Pneumococcus* infection (Table 7.1).

Some examples of human enzymes for which the genes have been cloned and the enzyme developed for either replacement therapy of lysosomal storage diseases (LSD), treating heart attack, or in relieving pulmonary distress in cystic fibrosis patients are shown in Table 7.1. Sales of Cerezyme and Fabrazyme alone accounted for some $ 1.7 billion in 2008. The variants of human tPA are examples of muteins where variants were made to improve certain therapeutic parameters.

The first successful application of enzyme replacement therapy for an inherited disease was the treatment of adenosine desaminase (ADA) deficiency using the modified enzyme. Enzon's (USA) first approved PEG product (ADAGEN®: (pegademase bovine) approved by the FDA in 1990) is used to treat patients afflicted with a type of severe combined immunodeficiency disease (SCID) who are unable to receive bone marrow transplants. SCID is caused by the chronic deficiency of ADA. Injections of unmodified ADA are not effective because of its short circulation life (less than 30 minutes) and the potential for immunogenic reactions to a bovine-sourced enzyme. The attachment of PEG to ADA allows ADA to achieve its full therapeutic effect by increasing its circulation life and masking the ADA to prevent immunogenic reactions [76].

In the case of the lysosomal storage diseases and ADA-deficiency these products represented the first possibilities for treatment of these rare 'orphan diseases' (see Section 14.1). Pulmozyme was the first new drug in 30 years to be approved for use in CF patients.

Table 7.1 Examples of human enzymes produced by gene cloning and used for human therapy.

Enzyme	Trade name	Application	Notes
α-Galactosidase	Fabrazyme; Afgalsidase	Fabry disease (1/40000); LSD=lysosomal storage disease)	Cloned, 1987; Orphan drug; FDA approved in 2003; sales $494 Mp.a. 2008
Glucocerebrosidase	Cerezyme; Imiglucarase	Gaucher's disease (1/12000; Aschkenazi jews 1/13 heterozygous); LSD	2 cloned 1988; (sales $1.2B p.a. 2008)
α-Glucosidase	Myozyme; rhGAA; alglucosidase α	Pompe disease; <1/40000; LSD	Orphan drug; FDA approved 2006; Limited application; high cost
DNAse I	Pulmozyme; Dornase alpha; rhDNAse I	Cystic fibrosis; to aid lung clearing	Cloned 1990, FDA approved 1997
Tissue plasminogen activator (tPA)	Altaplase; Activase	Thrombolytic	Cloned 1983; hu recombinant tPA;
Tissue plasminogen activator (tPA)	Reteplase; Retavase; Rapilysin	Thrombolytic e.g. heart attack; good clot penetration	Non-glycosylated hu tPA; FDA approved 1996
Tissue plasminogen activator (tPA)	Tenecteplase; TNKplase	Thrombolytic e.g. heart attack with reduced non-cerebral bleeding	Modified shortened hu tPA; FDA approved 2000
α-L-Iduronidase	Aldurazyme; laronidase	Mucopolysaccharidosis= MSP; LSD	Cloned 1990; FDA approved for MPS I, 2003

It should be noted that in spite of the costs of testing for contamination with human pathogens, *blood plasma is still a major source* of serum albumin, total immunoglobulin fractions and blood clotting factors. Production of these compounds from blood plasma is carried out in purification plants around the world, including India, Egypt, Iran and Australia. Pasteur-Mérieux, now Sanofi-Pasteur ceased producing placental hormones from collected placentas in 1990 because of the HIV (AIDS) risk. Foetal blood and placental/umbilical cord are now being used as alternative sources of stem cells for cell culture [77].

7.5.7
The Emergence of the 'New Biotech' Industry

The founding of *Genentech*, the first small recombinant DNA-based Biotech Company by *Robert Swanson* and *Herb Boyer* on April seventh 1976 marked the beginning of the new recombinant DNA (rDNA) era of Biotechnology. The demonstration that approval and worldwide marketing of insulin could be achieved in only 4 years after the insulin gene had been cloned (1978, insulin gene cloned; 1982, NIH approval and marketing), belied the considerable scepticism that prevailed within the pharmaceutical industry.

Genentech is also associated with another 'first': *Herb Boyer* who worked with *Arthur Riggs* and *Keiichi Itakura* from the Beckman Research Institute, were to be the first to successfully express a human gene (for somatostatin) in bacteria in 1977. Their achievement was not greeted with huge acclaim since the tiny peptide hormone somatotropin (somatostatin) could already be synthesized chemically. In contrast, the establishment of a bacterial source of human growth hormone, marketed by Genentech in 1985 was a different matter. In particular the demonstration that a small bioreactor of some 10 litres was sufficient to provide enough material for the entire world market was a dramatic step, and went a long way in stimulating an enthusiasm for investment in the new biotechnology.

7.5.8
'Third Generation' Products and the 'Post-Gene Cloning' Era

A 'third generation' of products can be distinguished which comprise novel variants (muteins) of known natural products that are characterized by novel (favourable) characteristics (e.g. storage half-life, pharmacokinetic properties, antigenicity).The modified products may be altered at the level of *the protein sequence or post-translational modification*. In the first case this involves gene mutation and modification in the second is usually achieved by using an alternative cell line for production. The first class of altered products were the *muteins*.

The extensive sequencing projects will be covered in more detail in later sections. Large-scale gene sequence comparisons between species known as *metagenomics* and *gene mining*, opened the way to new approaches for identifying novel enzymes and pharmaceutical targets and the isolation of 'lead compounds'. The development of such leads and their variants could in principle be carried out without gene cloning. It could be based purely on sequence data and synthetic chemistry. Gene cloning would however, still be required at the level of functional testing.

This could perhaps be referred to as the 'post gene cloning era of biotechnology' which commenced about 1998 with the widespread distribution of highly efficient sequencing automats. The emphasis was then transferred more towards bioinformatics approaches which have a greater reliance on the deeper understanding of many intracellular interactions and networks.

This type of holistic approach is encompassed by the new disciplines of *bioinformatics* and *systems biology* (see Section 13.6). This approach involves computer modelling of networks of interacting gene products, gene regulation- and metabolic-pathways and its aim is to finally predict how a cell or whole organism will react to a certain environmental challenge. In particular, understanding how changes in the disease state take place, for example on infection or during the development of a cancerous growth are challenges that are at the forefront of this branch of *molecular medicine*.

7.6
Biodiversity and Gene Mining

7.6.1
Recognizing Groups within Biological Diversity

It is important to be able to recognize particular types of animals and plants and to distinguish important characteristics. Are they edible or poisonous, or do they have some other interesting property which might be of use? Part of this awareness was documented by *Aristotle* (2300 BP) and in the *Book of Genesis* (>3000 BP?). The actual extent of biodiversity on the planet has been under scientific investigation since *Carl Linné* (*Carolus Linnaeus*; *Systema natura*, published 1735) started to catalogue, name and classify specific species. His use of two words in Latin to describe a species is still in use today and is referred to as binomial scientific nomenclature. He put species into groups, imagining that through his work man would be enlightened and would finally be able to see how these species fitted into a pattern of god-given and imperturbable order. The microbial world was not included in this systematic ordering until *E. Haeckel* included it in 1866 [78]. *Carl Woese* introduced a further modification in 1990 differentiating three main groups (Domains); *Eukarya, Bacteria* (= Kingdom of Eubacteria) and *Archaea* (= Kingdom of Archaebacteria) [79]. According to the 1977 nomenclature the Eukarya Domain comprises the Kingdoms known as *Protista* (single celled microbes containing a cell nucleus encased in a membrane and other organelles), *Fungi, Animalia* and *Plantae* (plants and multicellular algae).

7.6.2
Knowledge of Microbial Diversity was Limited by the Lack of Pure Cultures

It has recently been realized that probably more than 99% of microbial species on the planet are not cultivatable in pure culture, free of other life forms. The dependency of cultivation on the presence of (an)other organism(s) is exhibited to varying degrees. It may vary from commensalism, where organisms just share the same food source, through to mutualism where they may exchange metabolites to finally, either a completely interdependent symbiosis or a one-sided dependency as in a parasitic/pathogenic relationship. Thus one of the most exciting topics for many biotechnologists is the potential of untapped resources present in the as yet uncharacterized microbial world.

Microbes show extraordinary metabolic diversity which allows them to grow in extreme habitats under extremes of pH, concentrations of metal ions or salts, aerobically or anaerobically and at temperatures above 100 °C and below 0 °C. The fact that the enzymes they contain often show high stereo-selectivity but low substrate side-chain specificity makes them particularly attractive for use in the development of novel reaction pathways for novel or known reagents.

Viruses make up at least half the entities described as dependent on other organisms for their growth. Their existence is completely dependent on a host cell/organism. Their taxonomy now includes a classification system initiated by *David Baltimore* in 1971 which is based mainly on the nature of their nucleic acid content, RNA or DNA, single stranded or double stranded and the nature of the nucleic acid replication, which for retroviruses (RNA viruses) includes a DNA intermediate thus necessitating the involvement of the enzyme reverse transcriptase. The recent discovery of large viruses of amoeba implies that the taxonomic distinction between viruses and other organisms may in fact be blurred.

7.6.3
How Much Diversity is There?

So just how large is genetic (biological) diversity? An attempt to address this question follows and is solely concerned with the genomes of micro-organisms or virus species:

Estimating the global number of individuals in a number of classes

- 4–5×10^{30} soil microbes
- 3.6×10^{29} microbes in fresh or sea water
- 10^{30} (?) viruses in fresh and sea water ($>/=10^7/$ml)
- Plankton, nematodes and Rotifera in soil and other small life forms probably represent only a fraction of the diversity of the other classes.

These are huge numbers. Nevertheless genetic information is amazingly compact. In spite of the enormous diversity of life forms, a crude calculation estimates that the fraction of DNA contained in this biomass is only about 10^{-12}, that is one millionth-millionth of the Earth's estimated total mass of about 6×10^{27} g. Just how much real diversity does it represent? An insight has been obtained in the last 2 years as a result of large scale DNA sequencing projects.

7.6.4
Genetic Diversity can be studied by Sequencing Microbial, Organelle and Viral Genomes

DNA sequencing has transformed microbial taxonomy, our view of the detailed mechanisms of evolution and the possibilities for accessing novel biological diversity for biotechnological purposes. A brief history of the landmarks in the sequencing of total genomes follows:

1977 *Fred Sanger's* DNA sequencing method predominated in the early work on the sequencing of entire genomes of the bacteriophage, φX174 (**5368 bp**) [80] and in **1982** coliphage λ (**48 502 bp**) [81]. Animal viruses followed: **1986–1990**, for example the **229-kb** genome of cytomegalovirus (CMV) [82] and the **192-kb** genome of Vaccinia virus (**191.6 kb**) [83] (see also Section 7.5.3) as well as organelle DNA: **1986** chloroplast DNA (**121 024 bp**) [84] and in **1992** mitochondrial DNA of the liverwort *Marchantia polymorpha* (**184 kb**) [85].

In **1989**, under the leadership of *André Goffeau* a European consortium of 74 laboratories was set up to sequence the genome of the budding yeast *Saccharomyces* *cerevisiae* (**12.5 Mb**). This was completed in **1997** as a result of over 600 scientists in over 100 institutions working together. This was the largest collaboration that had ever taken place in molecular biology [86].

It was only in **1995** that the genome of a bacterium, *Haemophilus influenza*, [**1 830 137 bp**] [87] was sequenced in its entirety by Craig Venter's group. Many other complete genome sequences followed rapidly. High throughput DNA sequencing essentially started with the introduction of *Hunkapillar's* 454 sequencing machines in **1997**.

This accelerated the completion of the human genome sequence of 3 billion compiled bases by **2000**. This entailed collecting at least 15 times as much raw sequence data (i.e. more than 45 billion).

The exponential accumulation of sequences available in the public domain through GenBank is shown in Figure 7.15. The rate of accumulation of DNA sequence data doubled every 1.5 years from 1982 onwards. This represents a 10 000-fold increase over a period of 20 years or a 1000-fold increase in 15 years. This is faster than the rate of increase in computer speed as represented by 'Moore's Law' which increases twofold every 2 years. It is more than likely that this will continue, supported by the development of ever faster mass sequencing technologies (see Section 13.5). This can be extrapolated to predict a total of some 10^{15} base pairs by 2029. This might not include the sequence of the entire genome of every animal, plant, microbial and viral *species* on Earth, but it would be getting close. It would not cover, by far, the entire diversity of variants present in the *entire population* of each species (globally perhaps slightly less than 10^{31} individual genomes; see above). It is, however, not unlikely that the entire diversity of genetic variation in the human species will have been analysed by then.

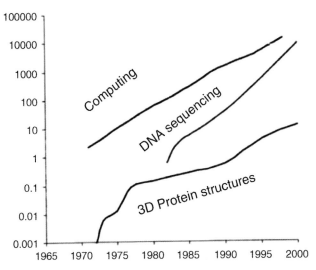

Figure 7.15 Relative rates of growth of technology from 1965 to 2000.

7.6.5
Microbial Genome Analysis and Bioinformatics to Identify Druggable Targets

The identification of novel 'druggable' targets for the development of antibiotics via a combination of genome analysis, network analysis and computational chemistry is another exciting possibility, which in the light of at least one recent publication has real potential [88].

The strategy involved:

- Metabolic bottlenecks are identified that involve non-redundant enzymes which when blocked will block biomass increase and/or synthesis of virulence factors.
- A large number of low molecular weight chemical compounds are examined in *in silico* docking analyses with respect to acting as inhibitors of the bottlekneck enzymes of known structure (in fact involved in bacterial-specific fatty acid synthesis). In the case of enzyme redundancy, compounds are selected which inhibit both enzymes.
- Candidates are tested in clinically relevant tests on *Escherichia coli* and *Staphylococcus aureus* strains.

From approximately a million chemical entites 41 chemicals were finally tested. A hit rate of some 40% for high inhibitory activities is significantly higher than had previously been achieved from virtual screening attempts and implies that the field is reaching a certain maturity. This is an attractive approach particularly for organisms which are otherwise refractory to laboratory experimentation such as *Plasmodium falciparum* and *Mycobacterium tuberculosis*.

7.6.6
Recognizing and Utilizing Useful Functional Diversity

So what are the best strategies for utilizing this information in the novel cottage industry of *gene mining*? How can *useful functional* diversity be recognized and harnessed for human purposes?

There are two main approaches as seen in studies to develop novel enzymes:

- Annotate entire genome or metagenomic sequences and compare them with gene sequences of known function, making inferences as to similar function
- Functional screening of libraries of random/'shot-gun' clones using vectors which allow the cloning of large intact DNA fragments, for example in cosmid, fosmid or BAC (PAC) cloning vectors.

Following either of the former routes the families of interesting related genes will be expressed and fragments assorted into new combinations. This provides a *focused diversity* which, in combination with functional selection may well lead to optimization of certain properties of the enzyme, for example heat stability or altered substrate specificity. This strategy has been demonstrated in principle and in practice, for example by the employment of 'sexual' PCR, also referred to as 'DNA shuffling' or even 'molecular breeding' [89]. The method was developed in 1994 by *Willem 'Pim'*

Stemmer who was then working at Affymax; it formed the basis for companies specializing in mining the benefits of gene diversity, for example *Maxygen*, set up to optimize proteins for therapeutical purposes; the latter spun-off *Codexis* for the production of industrial enzymes and *Vernia* who specialized in agricultural applications (see also *Diversa*, etc.). This led to the discovery and creation of novel combinations of mutations in the gene required for the product in question and has been extended to the optimization of gene expression and to whole organism optimization. Iterative cycles of selection, (high-throughput) screening, amplification and characterization using clone sequencing have been employed to identify improved strains.

These are empirical approaches that have been successful at least in selecting for altered properties of enzymes.

The laboratory of *Manfred T. Reetz* at the Max-Planck-Institute for Coal Research in Mülheim has been particularly innovative in this respect. They have developed the following strategy:

- Identify several individual amino acids in the protein on the basis of modelling structural features, which are likely to affect the properties that are to be modified.
- Introduce mutations at these codons with PCR primers containing codon mixtures thus essentially introducing saturation mutagenesis at the desired site.
- Take the best mutants from this first selection which have the desired 'improvements', and repeat the saturation mutagenesis in a second, defined region and repeat selection. The iterative process is referred to as Iterative Saturation Mutagenesis (ISM).

This has been shown to be successful for instance in developing enzymes with altered selectivity for one or other of the enantiomers (mirror image structures) in reactions that yield chiral products [90] and changing the thermostability of enzymes [91]. These examples contain both the strategies of random mutagenesis, with *a posteriori* rationalization of structural alterations relating to function, and *a priori* modelling guiding the definition of regions to be mutated both involving lipase from *Pseudomonas aeruginosa*.

There are a number of approaches which take advantage of natural diversity using gene mining with or without randomized or site-directed mutagenesis and with or without the prediction of structure-function or gene shuffling. The examples that originated in *Reetz's* laboratory demonstrate that in experiments where a novel change in the chirality of an enzyme is required, rapid advances can be made in a few screening steps using only a minimal number of mutants. Using recombination of the selected mutations, final products containing three or four mutations which exhibit alterations in chirality preference of four orders of magnitude can be obtained. A different impression is obtained from examining another case [92] where a considerable change in activity was achieved by converting ethyl(S)-4-chloro-3-hydroxybutyrate to ethyl(R)-4-cyano-3-hydroxybutyrate using a bacterial halohydrin dehalogenase. The final enzyme had accumulated 35 of some 47 detected 'positive' mutations to yield a 4000-fold improvement in enzyme activity. This cyanation process is involved in the production of atovastatin

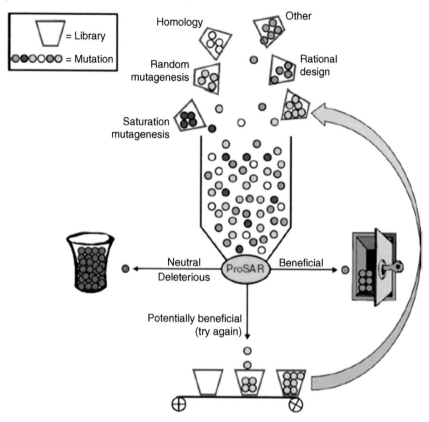

Figure 7.16 Use of multiple sources of diversity in combinatorial biology strategies. From Fox *et al.* [92].

(Lipitor), a drug designed to lower levels of blood cholesterol. This involved the use of an activity screen (ProSAR in Figure 7.16) on more than half a million strains. It is particular noteworthy that at 85% of the altered sites mutations involved only conservative or moderately conservative amino-acid exchanges and only 2% radical exchanges. Furthermore, *no single mutation was seen to cause more than a threefold effect.* Eighteen cycles (!) of enrichment were necessary at an increment rate of 1.5-fold per round. The latter investigation availed itself of the whole range of diversity generation and selection (Figure 7.16) in which homologous enzyme sequences from metagenomic studies, random mutagenesis, local saturation mutagenesis and rational design elements were incorporated into random mixing experiments. The enrichment of this 'melting pot' by addition of potentially beneficial and beneficial mutations enriched from a previous round represents an iterative evolutive principle. This seems to be a central and expected feature common to successful projects aimed at designing and improving enzyme specificities and activity. The latter project is a good example of the large-scale industrial application of screening methods to correlate structure and function. It links empirical combinatorial and selection methods. It is to be hoped that more channelled strategies can be discerned in the near future that will bring this type of project into the realms of academic budgets.

The work of *Dan S. Tawfik* and co-workers seems to be particularly important in understanding what we can expect from natural diversity and how to use it meaningfully [93]. Their work shows that natural variants of genes are quite likely to be excellent starting materials for laboratory evolution experiments. This can be interpreted as showing that present day genes are the product of long periods of high mutagenesis and neutral drift which have *maintained essential features in a robust form* while *permitting variability in non-essential regions*, thus providing a reservoir of useful diversity. This is a *sine qua non* for the gene mining approach to be productive when combined with combinatorial approaches for the study of laboratory evolution as described above.

The recent huge influx of novel DNA sequence data, particularly from metagenomics studies provides opportunities for identifying an enzyme that can recognize (altered?) substrates for a ligation reaction in order to create novel reagents. This has been demonstrated well with the isolation of genes involved in the synthesis of polyketides [94], Figure 7.17, which are produced as secondary metabolites by bacteria, fungi, plants and animals and which have widespread application, for example in the form of antibiotics, cytostatics, animal growth promoters and insecticides [95].

7.6.7
Functional Screening of 'Shotgun' and Metagenomic Libraries

The feasibility of functional screening of cloned genes began with efficient high-representation cosmid cloning of the entire genome (so-called 'shotgun' library) of a rare *E.coli* strain and the selection of the penicillin acylase (amidase) gene in a laboratory strain in 1979 [96]. This gene was subsequently used in the commercial preparation of semisynthetic penicillins and cephalosporins. This was shortly followed by other applications in which even entire functional operons were isolated and characterized, for example the entire nitrogen fixation operon (nif) from *Klebsiella* spp. Vectors such as cosmids (also low-copy number cosmids known as fosmids) and BACs (or PACs), which can accommodate large genomic fragments are still used preferentially for the creation and screening of metagenomic libraries. Since the efficiency of cloning is limited and the number of clones thus limited, it is often unclear whether the metagenomic library produced reflects the complexity of the microbial population sampled. Populations of microbes from extreme environments may be a particularly interesting source of enzymes which possess the following properties:

- Increased thermostability
- Increased activity in organic solvents
- Altered substrate specificity
- Altered enantio-selectivity

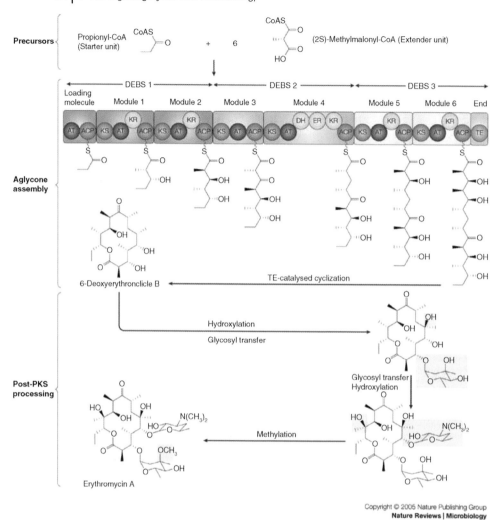

Figure 7.17 An example of polyketide synthesis using specialized enzyme complexes. Using genetic engineering to transplant enzymes from such pathways into related species may lead to the synthesis of polyketide variants.

- Increased gene expression
- Increased enzyme activity

These properties would address some of the unmet needs of large scale application in an industrial bioreactor.

Some authors prefer to consider the large fragment libraries more as initial archives of the metagenomic genome population sample, carrying out screening with smaller fragments where transcription is powered by read-through from promoters located on the vector (e.g. lambda or standard plasmid vectors).

7.6.8
Entire Genome Sequence of Microbes of Special Interest

The sequencing of an entire microbial genome sometimes turns up surprising information. The gliding bacteria, *Myxobacteria*, are a group of bacteria which produce particularly complex secondary metabolites, e.g. *Sorangium cellulosum* produces the new anticancer drug Epothilon (see Section 17.2.3).

The graph (Figure 7.18 [97]) shows that the newly sequenced genome of *Sorangium cellulosum* has the largest microbial genome so far sequenced with over 13 million base pairs (larger than fission yeast and normal yeast). This and another previously sequenced Myxobacterium, *Myxococcus xanthus*, also contains a high number and density of genes coding for protein kinases of the type normally associated with eukaryotic cells. The number and density of genes is even greater than that found in simple eukaryotes. The ORFs for the genes involved in secondary metabolite synthesis are clustered and orientated in the same direction. A take-home lesson is that we must not assume that all bacteria are less complex than all eukaryotes (Figure 7.19).

Myxobacteria exhibit particularly complex social interaction between cells: aggregation, filament building and spore body formation. Complex intercellular communication/behaviour is associated with an increase in genetic complexity. In the gliding bacteria this is clearly related to the large variety of protein kinases, enzymes which are involved in signal transduction from cell surface receptors, often massively and rapidly amplifying signals in response to activators and inhibitors.

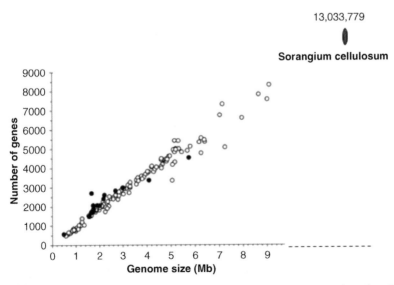

Figure 7.18 The complexity of bacteria and Archaea is shown by the estimated number of genes and the size of the genome. A gene has an average size of some 1.3 kilobase. Modified from Gregory and DeSalle (2005).

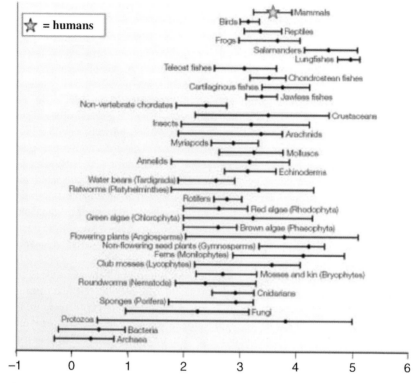

Figure 7.19 Diversity of the complex genomes of Archaea, bacteria and eukaryotes (e.g. total DNA size/2 for diploid organisms) as currently established from the analysis of the genomes of more than 10 000 species. The horizontal axis shows the $\log_{10}$ megabase pair scale (i.e. 0 = 1 million bases; 3 = 1000 million).

7.6.9
'Gene Mining' Commercialization and Terminology

With reference to this technology, a number of new terminologies are often used synonymously and as such confused with one another. Alternatively, one or other of the following terms are sometimes used to denote different activities: *gene mining*, *bioprospecting* (the term *biopiracy* has been coined by critics who see this as robbing ethnic cultures of their ancient medicines; see Chapter 14) and *biopanning*. *Gene mining* is sometimes used simply to mean the annotation of genomes in which the likely locations of genes are listed. In another section on personal medicine, gene mining has been used to refer to the search for quantitative trait linkages (QTL; see the HapMap project). However a slightly different aspect of gene mining is at the root of a new branch of biotechnology in the sense used by Ferrer *et al.* [98]. A better term would perhaps be *metagenomic prospecting*.

All these endeavours have the common goal of establishing novel and useful findings among the huge diversity of genetic information being made available by the

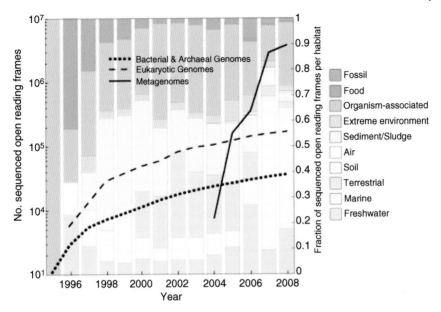

Figure 7.20 Trends in the increase of genomic data and represented habitats. The number of sequenced ORFs continues to increase exponentially, accompanied by an increase in the number and complexity of represented habitats. Raw habitat and sequence data were collected from the Genomes Online database, and habitats were classified into the categories mentioned above using the Habitat-Lite terms of Environment Ontology. Note that (i) dates reported for each sequence are publication dates, even if the genome was released to the public earlier in database form, and (ii) the numbers for 2008 represent the available data until June 2008 [99].

large scale sequencing of not only individual genomes, but also by the *simultaneous sequencing of the DNA of entire populations of microorganisms*. This latter approach is usually focussed on a particular biological niche such as the digestive tract of a cow or a soil or water sample (Figure 7.20).

Diversa is a company dedicated to this type of research and development. They paint the following picture of their aims: '... one of an increasing number of Companies hoping to make blockbuster biotech products from the estimated 99% of the Earth's microbial biodiversity that cannot be cultured in the laboratory and that has thereby eluded previous scientific study'.

Although this latter statement may sound astonishing to any biologist who was trained in the 1960s or the 1970s and has subsequently specialized in other fields, it is now accepted as a fact that the biological material available for detailed investigation from pure cultures of organisms and isolated as single colonies, represents a minute fraction of the actual number of life forms. The diversity of uncultured microbes was approached in the late 1990s using techniques that employed PCR amplified detection of specific ribosomal-RNA (rRNA) segments which were potentially diagnostic for the presence of a particular bacterial species (e.g. see [100]), thus by 2002 a range of specific rRNA signatures had already been catalogued for some 23 000 bacterial species using this methodology.

Until recently ignorance of the diversity and characteristics of microbial flora surprisingly also included a lack of knowledge relating to the major constituents of the biomass of the planet. An organism first isolated as SAR11 from the Sargasso Sea in 2004, was found to be *the most abundant organism* on the planet. It is now known as *Pelagibacter ubique.*

A new challenge thus arises which was previously unsatisfactorily addressed, namely, how can type culture collections of members of interdependent species (from commensals to true symbionts) from specialized niches be established and maintained.

Such sudden revelations of our ignorance motivate caution in our abilities to predict where science is going. At least they may make us wonder how the entire scientific community could have been so preoccupied with other details that these extraordinary gaps in our knowledge remained for so long. The positive side of the story is that this is now a burgeoning area for expansion and a potential cornucopia for the biotech industry.

7.6.10
Dealing with the Data Flux

In the public domain *Alfred Pühler,* (Figure 7.21) of Bielefeld, Germany was one of the first to specifically address the sequencing and annotation of bacterial genomes from

Figure 7.21   Alfred Pühler, a pioneer in the analysis of the genomes of industrial microorganisms.

organisms of biotechnological interest. He was recently awarded the Bundesverdienstkreuz in recognition of this pioneering work.

This work led directly to a number of open access tools for genome annotation now centred at CeBiTec.

This work highlights the complexity of dealing with many different types of information. Self explanatory user interfaces and compatibility between formats are prime concerns (Figure 7.22). Dealing with the more than exponential growth in the volume of data is an additional concern that is being addressed by massive parallel computing strategies including GPUs and high speed graphic processors.

The scientist has many requirements which need to be addressed: gene annotation and designation of gene products into functional categories; evolutionary comparisons; whole genome comparisons, comparison of metabolic pathways; and even constructing metagenomic analyses. The tools should deal with data mining, data integration and should bridge the gap between the fields of genomics, transcriptomics, proteomics and metabolomics (see also Section 15.6). Compatible data warehouses must be established.

Perhaps one of the most critical decisions relates to the best manner in which expert knowledge can be integrated into the entire system. This was certainly a critical

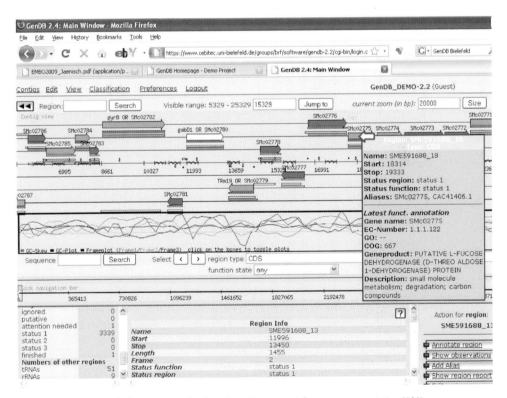

Figure 7.22 GenDB which is an example of an Open Source tool for genome annotation [101].

consideration in teaching chess-playing programmes and in the long run it may be the most important aspect for understanding the genomes.

The use of comparative metabolomics was nicely illustrated by the annotation of the *Corynebacterium kroppenstedtii* genome. This species is unusual for corynebacteria in that it lacks mycolic acid. Novel user-friendly tools rapidly identified the complete loss of a number of genes for enzymes of a particular section of a relevant metabolic pathway [102].

A recent interesting activity of this same group has been the analysis of the DNA sequences present in a biogas ferment, which although containing three dominant organisms showed the presence of over a thousand different species.

With the advent of the novel superfast and inexpensive single molecule DNA sequencing methods outlined in Section 13.5 and known as flash sequencing, it can be envisaged that such projects will deliver a complete picture of the microorganisms present in such complex communities.

Once again we emphasize that with present algorithms and computing hardware there would be a huge bottleneck at the level of data analysis.

7.7
Creating New Diversity by Design or Empirically

7.7.1
Limitations in Predicting Protein Structure and Function

A major long-term goal is to define the basic structure–function rules for inferring which alterations are most likely to result in a given alteration of (say) enzyme specificity. In the absence of such clear rules empirical selection methods are often favoured. Examples of different approaches are given below. Purists from the protein modelling camp deride such an approach as 'irrational', compared to an approach in which a purely theoretical analysis of the essentially chemical problem should in theory, deliver an optimized answer to any problem in enzymology.

The details of protein structure and function were elucidated during the latter part of the twentieth century. For some well understood classes of enzymatic reactions progress was made for *de novo* design of enzymes for simple hydrolytic reactions which would work with novel substrates. This usually makes use of a model for the structure of the activated substrate–enzyme intermediate, that is the transition-state intermediate. One problem is that although geometry is relatively easy to calculate, the position of water molecules is not always apparent from X-ray crystallography. The flexibility of the molecules cannot always be inferred from NMR studies although flexible and rigid regions can usually be differentiated. One of the successes relates to the development of 'ligand field theory' by *Leslie Orgel* as it relates to transition complexes with metals. This facilitated the prediction of catalytic centres and the creation of some simple *de novo* designed hydrolytic enzymes for novel substances which are of potential biotechnological relevance. Seen quantitatively, there are two arguments against such an approach and in favour of a pragmatic empirical search for natural variants and combinatorial studies of their recombinants.

Can we evaluate which approach actually delivers the best results? Catalytic activities of *a priori* designed enzymes are in general very low as are the turnover numbers. The vast majority of new enzymes have been isolated from natural sources. Recombination between related variants delivers new active enzymes, sometimes with altered properties. There seems to be a type of inherent '*hybrid vigour*', such as that already recognized as a general phenomenon by *Gregor Mendel* and again supported by the finding of *Danny Tawfik* as discussed above in relation to gene mining.

In this context it is of interest to note the Salomonic observation of *Kira Weissman* [103] 'As our understanding of structure–function relationships improve, it should be possible to launch a more focused attack on particular regions of the protein landscape, and so optimize evolution experiments. In the end, it may be a "*rationally random*" approach that finally enables engineers to create novel enzymes that are able to perform any reaction we desire'.

7.7.2
Genetic Recombination to Create New Assortments of Gene Fragments: PCR

Combinatorial biology, gene fragment shuffling or 'sexual PCR', developed by *Pim Stemmer*, the founder of *Maxygen* for creating novel variants of related genes, can generate seemingly endless variety (Figure 7.23). One of the early developments at Maxygen was to develop a heat stable para-nitrobenzyl esterase. This was obtained fairly readily and involved 13 amino-acid substitutions. Although the selection system functioned well it was not a simple task to explain *a posteriori* how each of these alterations had contributed to the stability of the final enzyme. Properties selected for in these combinatorial experiments involved the use of techniques discussed in 7.6.6.

7.8
'Genetic Fingerprinting'

We recognize individuals by their appearance more or less in the order of the following categories: sex, age, skin colour and other features that we may associate with their ethnic origin.

In addition it has been noticed that the whorls on our finger tips are highly variable and can also be used to identify individuals. As such fingerprinting has become a standard tool in forensic procedures. It is also true that every individual has a unique constellation of genes. One of the major surprises brought to the fore after the 'completion' of the human genome sequencing project in 2007, was the discovery of the large difference in copy number of genes between individuals [104] and the finding that this seems to account for nearly 20% of the differences in gene expression between individuals. That the most variable regions are relatively short

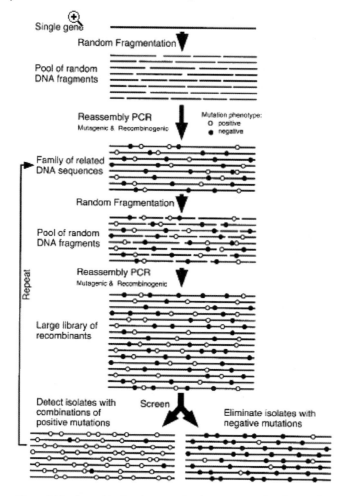

Figure 7.23 Recombination between closely related gene variants or mutants using PCR-mediated 'DNA shuffling' also referred to as 'sexual PCR'. From Zhang, J.H. *et al.* (1997)

'Directed evolution of a fucosidase from a galactosidase by DNA shuffling and screening', *Proc. Natl. Acad. Sci. USA*, **94**, 4504–4509.

and occur hundreds of thousands of times in the genome often in blocks of repeats, had been discovered earlier. During the formation of sperm and egg the chromosome pairs line up with each other and undergo recombination, often through a crossover in just such variable regions. This changes the lengths of the repeat blocks at that exchange point.

Alec Jeffreys, now Sir Alec (Figure 7.24), of Leicester University, UK, applied the PCR DNA amplification method to the analysis of regions containing these 'minisatellite' highly variable repeated DNA regions which he had serendipitously discovered within the intron of the human myoglobin gene while comparing it with

Figure 7.24   Sir Alec Jeffreys who developed PCR-based DNA finger-printing.

seal myoglobin. The variation between individuals was so high that blood from one individual could clearly be distinguished from that of any other individual. On the other hand the satellite-DNA bands originating in the father and mother will most often both still be seen in the genetic make up of the child. Therefore this method is used to determine paternity [105].

The method became internationally recognized after Jeffreys applied this technology to help in the conviction of a rapist and murderer in Leicester, England, also leading to the exoneration of the prime suspect who had been wrongly accused. The method was also used to confirm the identity of the corpse of the Nazi War Criminal Josef Mengele who had died in 1979. Sir Alec has received notable distinctions for this development other than his knighthood: the Albert Lasker Award for Clinical Medical Research (2005); the Royal Medal awarded by the Royal Society (2004) and the Great Briton Award, Greatest Briton of 2006.

The method delivers a type of 'genetic fingerprint' essentially unique to a particular individual. The pattern thus obtained is a mixture of some of the patterns present in each parent, as well as a few new ones caused by the recombination as discussed previously.

Applied in legal cases concerning immigration it has been used to speed up the correct processing of claims to enable families to stay together, thus avoiding both the problem of separating parents and children in the absence of documentation and the resulting emotional distress that this process had previously entailed.

A clear result may even be obtained from a single cell. Although bloodhounds may have an excellent ability to trace persons on the basis of the scent of the clothing, this test is many million-fold more sensitive than any existing test with regard to whether or not a person was present at a certain location, for example by analysis of a hair or a flake of skin found at the site of a crime. The method has become routine as a forensic tool. It can also be extended to detect other specific genetic traits which are more likely to occur in particular ethnic groups, as well as health-related factors such as the probability of obesity, medical problems such as diabetes or to give an indication of eye and skin colour.

The 'DNA Fingerprint Act' was passed by the USA Congress in 2005, thereby allowing federal authorities to collect DNA samples from anyone who is arrested or detained. More than a million samples per year are added to the US national DNA database NDIS [106].

The use of this technique has given rise to some debate regarding its social and ethical impact. Often screens are made of whole groups in order to find a rapist, for example. Since the data can give insight into the personal details of innocent citizens, there has been considerable concern about such fingerprint data being kept indefinitely and the potential of its misuse if privacy is not ensured. According to Jeffreys, at the time of writing, data from over a million people in Great Britain are being kept in the databases.

The European Court of Human Rights ruled in 2008 that it is a violation of human rights to keep the DNA of innocent individuals in a national DNA database based on the European Human Rights Convention (Chapters 8 and 14).

The new DNA-based tests are powerful tools and the way in which they are applied and the way in which the data are made available or restricted, are important decisions which require good counsel and careful handling. Other countries see such databases as an important tool in combating terrorism and conclude that it is the state's obligation to protect their citizens from violence, this having priority over considerations of privacy.

Analysis of DNA samples can provide insight into the genetic background of an individual and therefore information relevant to their predisposition or succeptibility to disease.

7.9
Inherited Predisposition to Disease

As outlined previously an individual can obtain an analysis of his 'personal genome'. At the moment, for an affordable price several thousand regions can be analysed for *haplotypes* that will show certain factors which may already be known to the individual such as intolerance to large amounts of coffee, or traits that may be of more relevance to *pharmacogenetics*, that is the ability to rapidly break down certain drugs. This is information that may be useful to a doctor in determining the appropriate dosage of a new medication for a particular patient.

As a result of their family history, some couples are concerned about the possibility of passing on adverse genes to their offspring. In the past, in the absence of an alternative (i.e. lack of a genetic diagnosis), they would often just have avoided having children altogether. Others consulted a genetic councillor.

The combination of advances in fertility medicine, for example using *in vitro* fertilization and pre-implantation DNA analysis on a single cell from either 8- or 16-cell embryos, allows selection of embryos for implantation which do not carry the disease allele. This *pre-implantation diagnosis* has for example been applied in cases of devastating and untreatable syndromes such as Huntington's chorea or the Lesch-Nyhan syndrome.

7.9.1
Heritable Factors Affecting Probabilities of Developing Disease: Predisposition

Just before Christmas 2008 a couple in England were reported to be the first to have used the pre-implantation diagnosis for counter selection of a 'potential susceptibility' gene which was found to be present in the family of one of the parents. The gene variant increases the likelihood many-fold for the carrier to develop breast cancer. This is called a *genetic* or *inherited predisposition* for developing cancer. It is a statement of likelihood that is dependent on many other factors. Therfore this type of reproductive decision has a quality that is different from that of previous practice, in that the child-to-be who carries the gene variant which predisposes cancer, would not *necessarily* develop the cancer. The child would however, be at a very elevated risk of doing so.

Such procedures are still illegal in other parts of Europe such as Germany. Concerns that led to such restrictions are based on fears that widespread application would cause a trend for 'designer babies', a type of resurrected *Eugenics movement*.

The benefits of personal predisposition trait testing as is now offered by a number of companies such as *Navigenics*, Redwood, California, *23-and-Me*, Mountain View, California, or *DeCODEme*, Reykjavik, Iceland, is however, highly questionable. Psychosocial scientists, who have been studying how individuals react to risk assessment, came to the conclusion that the reaction of individuals to such data is likely to be neither well informed nor sufficient motivation for a change in lifestyle [107].

Most predispositions, will involve many different genes and specific alleles at each genetic locus. Each contributes a very small risk factor. In a recent summary of the impact of genomic analysis on a large number of individuals in a search for genes more frequently associated with particular diseases, *Walter Bodmer* and C. *Bonilla* [108] concluded that very large numbers will need to be analysed before any useful conclusions can be drawn. They argue that the most useful conclusions will not be, as originally intended, the identification of the most frequent variants of a particular gene which have an effect on the overall likelihood of disease occurrence but it would instead be more advantageous to identify those much rarer variants that have a *high penetrance*. 'High penetrance' means that the effect of the gene is largely independent of the environment and independent of the presence or absence of genetic variation in other genes. As such high penetrance genes would have a relatively high chance of contributinging to a health problem. Such information would therefore have significant value in preventive medicine.

7.9.2
Services for Genome Analysis as Part of Personal Medicine

Companies offering such services have responded to some extent to the criticism as exemplified by *Knome Inc.*, Cambridge, Massachussets, who at the time of writing offered complete genome sequencing with analysis of predisposition traits for $99 500 (about € 70 000) and include a round table discussion with expert geneticists, clinicians and bioinformatics experts at the Harvard Club in Boston. A complete genome sequnce without analysis of the genetic information is being offered in July 2009 by *Illumina Inc.*, San Diego, California, for €34 000. Future trends and further expectations for continued dramatic price reductions in addition to increasing sequencing speed are discussed in Section 13.5.

In terms of genes affecting longevity, very little correlation was found between the results of a number of studies on populations around the world. However, one constant association was found for the *Sir1* gene which effects the sequestering of DNA in chromatin and the refreshing of the chromosome ends via telomerase. This responds to *epigenetic methylation* signals in particular. It also represents a possible link to the effects of diet in that there is a metabolic link mediated by NADP interaction. It has been acknowledged for some time that high sugar intake and modern lifestyles result in high blood sugar levels and negatively affect longevity. Very low food intake and a high level of telomerase in the cells (comparing different species and individuals) are correlated with longevity. This is the first clear genetic trait found that might be directly involved with this phenomenon following on a number of studies comparing genetic haplotypes of octogenarians and older individuals with the general population.

References

1 Crick, F. (2004) *Of Molecules and Men*, Prometheus Books, New York.

2 Fraenkel-Conrat, H. and Williams, R.C. (1955) *Proceedings of the National Academy of Sciences of the United States of America*, **41**, 690–698.

3 Pauling, L. *et al.* (1949) Science, **110**, 543–548.

4 Brenner, S. (2001) *My Life in Science*, Biomed Central Ltd., London, ISBN: 10: 0954027809; 13: 978-0954027803.

5 Cairns J., Stent G.S., and Watson J.D. (eds) *DNA and the Origins of Molecular Biology*, Cold Spring Harbor Laboratory Press, New York.

6 The Linnaen Society, founded in 1788, is the oldest biological society which is still active.

7 Dennett, D.C. (1995) *Darwin's Dangerous Idea: Evolution and the Meanings of Life*, Penguin Books, London.

8 The 'Malthusian catastrophe'.

9 Axelrod, R. (1992) *The Evolution of Cooperation*, Penguin Books, London.

10 de Aguiar, M.A.M. *et al.* (2009) *Nature*, **460**, 384–387.

11 See: Mattick, J.S. (2009) *EMBO Reports*, vol. 10, p. 665 (Opinion) Nature (publ.).

12 BASF AG, Takeda Chemical Industries Ltd. (2001); Rueckel, M. (2000) EPO

0972843. Hoffmann La Roche, Switzerland.

13 Stubbe, H.(revised 2nd edition 1965; English translation 1972) *History of Genetics*, From prehistoric times to the rediscovery of Mendel's laws. MIT Press, Boston.

14 Cavalli-Sforza, L.L. (1950) La sessualità nei bacteri. *Bollettino Dell Istituto Sieroterapico Milanese*, **29**, 281–289.

15 Cairns, J. (1963) *Journal of Molecular Biology*, **6**, 208–213.

16 Pardee, A.B., Jacob, F., and Monod, J. (1959) *Journal of Molecular Biology*, **1**, 165–178.

17 Pardee, A.B. and Reddy, G.A. (2003) *Gene*, **321**, 17–23.

18 Gilbert, W. and Müller-Hill, B. (1966) *Proceedings of the National Academy of Sciences of the United States of America*, **56**, 1891–1898.

19 Morange, M. (2005) *Journal of Biosciences*, **30**, 591–594.

20 This is just one example of the way in which homologies between plasmids were analysed via melting DNA of both plasmids and letting them anneal together. Examination of spreads showed up regions of homology as well as IS elements and transposons which had inverted symmetry regions: Gorai, A.P., Heffron, F., Falkow, S., Hedges, R.W., and Datta, N. (1979) Electron microscope heteroduplex studies of sequence relationships among plasmids of the W incompatibility group. *Plasmid*, **2** (3), 485–492.

21 Luria, S.E. and Human, M.L. (1952) *Journal of Bacteriology*, **64**, 557–569.

22 Luria, S.E. (1973) *Life: The Unfinished Experiment*, Charles Scribner Son's, New York.

23 Arber, W. and Linn, S. (1969) *Annual Review of Biochemistry*, **38**, 467–500.

24 Roberts, R.J. (2005) *Proceedings of the National Academy of Sciences of the United States of America*, **102**, 5905–5908 (Review); Danna, K. and Nathans, D. (1971) *Proceedings of the National Academy of Sciences of the United States of America*, **68**, 2913–2917; Smith, H.O. and Wilcox, K.W. (1970) *Journal of Molecular Biology*, **51**, 379–391.

25 Cohen, S.N., Chang, A.C.Y., Boyer, H.W., and Helling, R.B. (1973) *Proceedings of the National Academy of Sciences of the United States of America*, **70**, 3240–3244.

26 Raymond, S. and Weintraub, L. (1959) *Science*, **130**, 711; Hermann Stegemann was a pioneer (Braunschweig & Göttingen) who organized an international Protein Analytical Symposium with practicals to optimize and promulgate the methods for rapid protein and tissue comparisons starting in 1960. His laboratory was also instrumental in developing 2-D gel electrophoresis of proteins, particularly those of the potato (Mitteilung der Max-Planck-Gesellschaft 1–2 (1963) 72–76; see also Raymond, S. and Aurell, B. (1962) 2-D gel electrophoresis. *Science*, **138**, 152–153.

27 Siepmann, R. and Stegemann, H. (1967) Enzyme electrophoresis in embedding polymers of acrylamide. A: Amylases, phosphorylases (German publication). *Zeitschrift für Naturforschung. Teil B: Chemie, Biochemie, Biophysik, Biologie*, **22**, 949–955 (in German). Stegemann's group published an atlas of potato 2-D protein gels of potato crosses from around the world three decades before the initiation of the human 'proteome' project.

28 Loening, U.E. (1967) *The Biochemical Journal*, **102**, 251–257.

29 Kleinschmidt, A.K. *et al.* (1962) *Biochimica et Biophysica Acta*, **61**, 857–864.

30 Petrásek, Z., Hoege, C., Mashaghi, A., Ohrt, T., Hyman, A.A., and Schwille, P. (2008) *Biophysical Journal*, **95**, 5476–5486.

31 Graham, F.L. and van der Eb, A.J. (1973) *Virology*, **52**, 456–467.

32 Szybalski, W., Szybalska, E.H., and Ragni, G. (1962) Genetic studies with human cell lines. In: Analytic cell culture. *Journal of the National Cancer Institute. Monographs*, **7**, 75–87.

33 Collins J. (1977) Gene cloning with small plasmids. *Current Topics in Microbiology and Immunology*, **78**, 122–170.

34 Struhl, K., *et al.* (1976) Functional genetic expression of eukaryotic DNA in

Escherichia coli. *Proc. Natl. Acad. Sci. U.S.A.*, **73**, 1471–1475; Struhl, K., *et al.* (1979). High frequency transformation of yeast: Autonomous replication of hybrid DNA molecules. *Proc. Natl. Acad. Sci. U.S.A.*, **76**, 1035–1039.

35 Berg, D. E. and Howe, M. M. (eds.) (1989) 'Mobile DNA', American Society of Microbiology, ASM Press (publ.), Herndon, USA, ISBN-10: 1555810055.

36 Schell, J. and van Montagu, M. (1977) The Ti-plasmid of Agrobacterium tumefaciens, a natural vector for the introduction of nif genes in plants? *Basic Life Sci.*, **9**, 159–79.

37 Lebeurier G., *et al.* (1980) Gene, **12**, 139–46.

38 Collins, J. and Hohn, B. (1978) *Proceedings of the National Academy of Sciences of the United States of America*, **75**, 4242–4246.

39 Isch-Horowitz, D., John, F., and Burke, J.F. (1981) *Nucleic Acids Research*, **9**, 26.

40 Collins, J. (1981) *Cold Spring Harbor Symposia on Quantitative Biology*, **45**, 409–416.

41 Luen-Luen Li *et al.* (2009) Biotechnology for biofuels, **2**, 10–17.

42 Mayer, H., Collins, J., and Wagner, F. (1979) Patent Nr. DE2930794.

43 Hauser, H. *et al.* (1982) *Nature*, **297**, 650–654.

44 http://www.mun.ca/biology/scarr/FISH_chromosome_painting.html.

45 Shizuya, H. *et al.* (1992) *Proceedings of the National Academy of Sciences*, **89**, 8794–8797.

46 Harrington, J.J. *et al.* (1997) *Nature Genetics*, **15**, 345–355.

47 BAC reference: Shizuya, H. *et al.* (Mel Simons' laboratory) (1992) *Proceedings of the National Academy of Sciences of the United States of America*, **89**, 8794–8797; PAC reference: (P1 vector): Ioannou, P.A. *et al.* (1994) *Nature Genetics*, **6**, 84–89; YAC reference; HAC reference: Harrington, J.J. *et al.* (1997) *Nature Genetics*, **15**, 345–355;Smith, L.M. *et al.* (1986) *Nature*, **321**, 674–679; Lewis, E.K. *et al.* (2005) *Proceedings of the National Academy of Sciences*, **102**, 5346–5351; Margulies, M. *et al.* (2005) *Nature*. doi: 10.1038/nature03959; Shendure, J. *et al.*

(2005) *Scienc Express*. doi: 10.1126/science.1117389.

48 Doolittle, R.F. (1986) *Of URFs and ORFs: a Primer on How to Analyze Derived Amino Acid Sequences*, University Science Books, Mill Valley California.

49 Blau, H.M. and Rossi, F.M. (1999) *Proceedings of the National Academy of Sciences of the United States of America*, **96**, 797–799; Markoulaki, S. *et al.* (2009) *Nature Biotechnology*, **27**, 169–171.

50 Merrifield, R.B. (1963) *Journal of the American Chemical Society*, **85**, 2149.

51 Schwyzer, R. and Sieber, P. (1963) *Nature*, **199**, 172–174.

52 http://www.hlhl.org.cn/english/shownews.asp?newsid=379, for reminiscences from this period in China, by Zou Chenglu himself.

53 Kochendoerfer, G.G. *et al.* (2003) *Science*, **299**, 884–887.

54 Kent, S.B. (2009) Total chemical synthesis of proteins. *Chemical Society Reviews*, **38**, 338–351.

55 Milton, R.C., Milton, S.C., and Kent, S.B. (1992) *Science*, **256**, 1445–1448.

56 Welch, B.D. *et al.* (2007) *Proceedings of the National Academy of Sciences of the United States of America*, **104**, 16828–16833.

57 Bray, B.L. (2003) *Nature Reviews. Drug Discovery*, **2**, 587–593.

58 Shepard, H.M. *et al.* (1981) *Nature*, **294**, 563–565.

59 Entire meeting's report (1986) *Philosophical Transactions of the Royal Society of London. Series A: Mathematical and Physical Sciences*, **317**, 293–453.

60 von Wasielewski E. and Chick W.L. (eds) Pancreatic beta-cell culture: Workshop Conference Hoechst, Vol. 5 (1977) Excerpta Medica, Amsterdam.

61 Gillam, S. and Smith, M. (1979) *Gene*, **8**, 81–97; 99–106.

62 Khorana, G. (1979) *Science*, **203**, 614–625.

63 Carter, P. *et al.* (1989) *Proteins*, **6**, 240–248.

64 Aviv, H. and Leder, P. (1972) Proceedings of the National Academy of Science of the United States of America, **69**, 1408–1412.

65 Frank, R. *et al.* (1983) *Nucleic Acids Research*, **11**, 4365–4377; Blöcker, H. and Frank, R. (1984) *Bio/Technology*, **2**, 694–695.

66 Houghten, R.A. (1985) *Proceedings of the National Academy of Sciences of the United States of America*, **82**, 5131–5135.

67 Galibert, F. *et al.* (1979) *Nature*, **281**, 646–650.

68 Arrand, J.R. *et al.* (1981) *Nucleic Acids Research*, **9**, 2999–3014.

69 Fleckenstein, B., Müller, I., and Collins, J. (1982) *Gene*, **18**, 39–46. CMV is the largest pathogenic virus genome and was the first genome to be mapped using cosmid cloning; 230 kb.

70 Wain-Hobson, S. *et al.* (1985) Nucleotide sequence of the AIDS virus, LAV. *Cell*, **40**, 9–17; Alizon, M. *et al.* (1984) *Nature*, **312**, 757–760.

71 Wang, K.-S. and Choo, Q.-L. *et al.* (1986) *Nature*, **323**, 508–514.

72 Choo, Q.-L. *et al.* (1989) *Science*, **244**, 359–362 and 362–364.

73 Reyes, G.R. *et al.* (1990) *Science*, **247**, 1335–1339.

74 Leary, T.P. *et al.* (1996) *Journal of Molecular Virology*, **48**, 60–67.

75 Alter, H.J. and Houghton, M. (2000) *Nature Medicine*, **6**, 1082–1086.

76 Behrendt (2009) personal communication; see: http://www.enzon.com/index.php?id=36.

77 Plasma Products Therapeutics Association: http://www.pptaglobal.org Re: therapeutics: http://www.pptaglobal.org/plasma/therapies.aspx as well as personal communication with Karl Simpson who personally inspected one such purification plant in Iran.

78 Haeckel, E. (1866) *Generelle Morphologie der Organismen*, Reimer, Berlin.

79 Woese, C. *et al.* (1990) *Proceedings of the National Academy of Sciences of the United States of America*, **87** (12), 4576–4579.

80 Sanger, F. *et al.* (1977) *Nature*, **265**, 687–695.

81 Sanger, F. *et al.* (1982) *Journal of Molecular Biology*, **162**, 729–773.

82 Chee, M.S. *et al.* (1990) *Current Topics in Microbiology and Immunology*, **154**, 125–169.

83 Goebel, S.J. (1990) *Virology*, **179**, 247–266 and 517–63.

84 Ohyama, K. *et al.* (1986) *Nature*, **322**, 572–574.

85 Oda, K. *et al.* (1992) *Journal of Molecular Biology*, **223**, 1–7.

86 Mewes, H.W. *et al.* (1997) *Nature*, **387**, 7–65.

87 Fleischmann, R.D. *et al.* (1995) *Science*, **269**, 496–512.

88 Shen, Y. *et al.* Proceedings of the National Academy of Sciences of the United States of America, 0909181107 (on-line early edition 2010).

89 Minshull, J. and Stemmer, W.P.C. (1999) *Current Opinion in Chemical Biology*, **3**, 284–290.

90 Reetz, M.T. *et al.* (1997) *Angewandte Chemie*, **109**, 2961–2963.

91 Reetz, M.T. *et al.* (2008) *Biotechnology and Bioengineering*, **102**, 1712–1717.

92 Fox, R.J. *et al.* (2007) *Nature Biotechnology*, **25**, 338–344.

93 Bershtein, S. and Tawfik, D.S. (2008) *Current Opinion in Chemical Biology*, **12**, 151–158; Bershtein, S., Goldin, K., and Tawfik, D.S. (2008) *Journal of Molecular Biology*, **379**, 1029–1044.

94 Weissman, K.J. and Leadlay, P.F. (2005) *Nature Reviews Microbiology*, **3**, 925–936; See also: Sherman, D.H. (2005) *Nature Biotechnology*, **23**, 1083–1084; and Schirmer, A. *et al.* (2005) *Applied and Environmental Microbiology*, **71**, 4840–4849.

95 Review: Weissman, K.J. and Leadlay, P.F. (2005) *Nature Reviews Microbiology*, **3**, 925–936.

96 Mayer, H., Collins, J., and Wagner, F. (1979) *Plasmids of Medical, Environmental and Commercial Importance* (eds K.N. Timmis and A. Pühler), Elsevier/North Holland Medical Press, Rotterdam, pp. 456–459.

97 Gregory, T.R. *et al.* (2006) *Nucleic Acids Research*, **35**, Database issue D332-338: in 2006 the Animal Genome Database contained 5677 records from 601 sources, covering 2953 species of vertebrates and 1323 species of invertebrates.

98 Ferrer, M., Beloqui, A., Timmis, K.N., and Golyshin, P.N. (2009) *Journal of*

Molecular Microbiology and Biotechnology,
16, 109–123; Lorenz, P. and Eck, J. (2005)
Recommended further reading. *Nature
Reviews Microbiology,* **3**, 510–516.

99 Bork, P. (2009) *J. Bacteriol.,* **191** (1),
32–41.

100 Wilson, K.H. *et al.* (2002) *Applied and
Environmental Microbiology,* **68**,
2535–2541.

101 Meyer, F. *et al.* (2003) *Nucleic Acids
Research,* **31**, 2187–2195.

102 Tauch, A. *et al.* (2008) *Journal of
Biotechnology,* **136**, 22–30.

103 Weissman, K.J. (2004) *Chemistry World,* **1**,
28–33. online://www.rsc.org/
chemistryworld/Issues/2004/July/
national.asp

104 Strange, B.E. *et al.* (2007) *Science,* **315**,
848–853.

105 Jeffreys, A.J. *et al.* (1985) *Nature,* **314**,
67–73.

106 *The Economist,* August 1–7, 2009 pp. 37.

107 see *Nature* http://www.nature.com/
news/2008/080528/full/453570a.html.

108 Bodmer, W. and Bonilla, C. (2008) *Nature
Genetics,* **40** (6), 695–701.

8
Ethical Aspects Related to Genome Research, and Reproductive Medicine

8.1
Negative Public Reaction to Gene Technology

A prominent Director of a large German pharmaceutical Company gave his opinion during a Board meeting of the German National Centre for Biotechnology Research (GBF – Gesellschaft für Biotechnologische Forschung; privileged communication JC) in Braunschweig, that 'there will never be a recombinant DNA product approved for human use'. This statement was made only a few weeks before the approval of human insulin in 1982. We, the authors, consider this as representative of the hesitance apparently combined with ignorance that prevailed in a large part of European industry, to actively incorporate modern molecular genetics into product development programmes. In the early days announcements of successful product development based on the new technology were often greeted with scepticism, particularly by the conservative European pharmaceutical industry. This continued for some time even after the first rDNA pharmaceutical product had successfully passed through clinical testing and had been approved for clinical use (recombinant human insulin, initially produced as two separate (A and B) chains and approved for human use in 1982) [1]. Irving S. Johnson describes the strong public reaction which followed the scientists' own concern about certain unpredictable dangers which might arise from genetic engineering. This was taken up by professional public interest groups, particularly the group headed by Jeremy Rifkin in the United States and by the 'Green' political movement in Europe. Arguments brought against the use of rDNA were technical and ethical, and often ideological, in nature.[1]

1) One example of the latter was the fact that in 1984 in Berlin the Green party declared that Genetic Engineering was a misogynous (frauenfeindliche) technology, in that it would broaden the areas for which prenatal examination would be applicable. At the International Women's Conference in 1986 and 1991 in Brazil it was clearly established that the actions against Gene Technology had mostly been of a campaign nature with the intention of stagnating development in this area by 'throwing sand in the works' rather than by working towards altering legislation in a democratic fashion (see e.g. http://www.kuzeb.ch/karnikl/04-15.htm).

Concepts in Biotechnology: History, Science and Business. Klaus Buchholz and John Collins
Copyright © 2010 WILEY-VCH Verlag GmbH & Co. KGaA, Weinheim
ISBN: 978-3-527-31766-0

Similar experiences, especially in Europe, are not uncommon, for example the insulin production plant at Hoechst, near Frankfurt: although planning had started in the 1980s the pilot-plant was not operative until 1993 and the production plant not until 1995. The product came first onto the German market in 1999.

Apparently in addition to public resistance, there were problems of management perception and strategy. *Günther Eisbrenner* who was responsible for the development of the process, points out that this represents *a delay of over 15 years* in a highly competitive market (the worldwide market was worth about $3.6 billion in 2006).

The last 30 years of the twentieth century saw the rapid continuing decline of the German pharmaceutical industry which had previously dominated the market for decades. This involved tens to hundreds of billions of dollars in lost sales, inevitably resulting in tens if not hundreds of thousands of job losses when this decline is extrapolated to the whole range of products affected by the new technology over this time period.

The situation is reflected perhaps even more dramatically in the area of recombinant (green) plants and microorganisms for release into the environment, where Europe essentially no longer played a role in development. It has been inferred that the latter also caused the loss of thousands of jobs per year, for instance by *Michael Kock* from BASF.

The number of field trials for genetically modified (GM) crops in Europe illustrates this disparity between European and American research and development. Starting in 1991, the number of field trials in Europe peaked at some 250 per annum decreasing to some 35 in 2001. In 2001 there were some 1500 trials being carried out p.a. in the USA alone.

Novel biotechnological developments which go beyond classic plant breeding include the areas of *nutriceuticals* which involves the idea of increasing the concentrations of important beneficial nutrients, and *pharming*, applied to both plants and animals that act as genetically modified hosts for the production of pharmaceutical products. Such radical concepts were greeted with scepticism.

The concept of large-scale production in the order of tons of highly purified product and low cost has not been achieved so far, but is still a goal. The production of human serum albumin from cow's milk can be taken as a fairly recent example. It was pioneered by Pharming Inc., together with Genzyme Transgenics in 1997 and more recently with Fresenius AG (Bad Homburg) [2]. A further example of ultra low-cost products is the production of human insulin from recombinant Safflower seeds [3].

(*Foot note 1 continued*)

A personal encounter between one of the authors with the German militant arm of this movement the 'Rote Zora', involved their placement of 9 kg of plastic explosive at the entrance to the Biocenter at the University of Braunschweig in 1986. Although the detonator fired, the bomb failed to cause any damage. This was only one of several bomb attacks organized in the 1980s directed against gene technology institutes.

8.1.1
Initial Lack of Acceptance for Gene Technology in Europe

In the case of gene technology in Europe the cost/benefit of the drawn-out discussion and stringent regulation of recombinant DNA research, appeared to be very negative and a technology gap developed within a very short time. The loss of these crucial years in the development of rDNA technology in Europe, particularly in Germany, can be seen to be due largely to regulatory wrangling which included having to defend civil court cases brought by the Green faction whose intention was to ban recombinant DNA technology altogether. This blockade could only be enforced with the support of a popular majority that was negatively disposed to novel technology, and as we will also see accompanied by a militant fringe activity which was partially tolerant.

In Germany anything to do with human genetics was regarded with particular suspicion. There are reasons for this which stem from the misuse of genetics during the Third Reich as is well documented by *Benno Müller-Hill* (1985) in 'Tödliche Wissenschaft/Deadly Science'. In addition, mistrust in many novel applications of new technology was reignited by the *Chernobyl disaster* and the '*Schweizerhof' accident* at Sandoz in Basel, both in 1986. The latter involved the release of some 500 tons of chemicals which were washed into the Rhein by water used by the fire-fighters. These two incidents were considered to be the largest man-made environmental disasters that ever occurred before the 'Deepwater Horizon' BP Oil Spill in 2010. They shook public confidence in the safety standards of the pharmaceutical (chemical) industry and in applied science in general. An example of the strength of public feeling following these incidents can be judged from the result of a referendum amongst the population of Basel,[2] Switzerland which rejected the building of a production plant for rDNA products in Basel. The plant was later built a few kilometres away on the other side of the French-Swiss border but only after the resignation of *Kaspar von Meyenburg* who had initiated the project, and who now continues his career in a venerated biotechnological tradition, namely viniculture, on the hillside above the Lake of Zurich.

The self-imposed rigour for setting high standards in safety which ensued immediately after the public debate also caused an inhibition of research and development within the pharmaceutical industry around the world. This is an event that has rarely been commented on.

George B. Rathman presents a case in point through his own experience as health safety officer in 1979 at *Abbot* where research and development were in his opinion essentially crippled by this effect.

The positive result of this was that Rathman, later revered as 'the Golden throat' by the pharmaceutical industry, left Abbot to head the greatest success story of the new Biotech industry. He was the first CEO of *Amgen* (founded in 1980) and within a few years under his influence it became the largest biotech company in the world.

At about this time the Mayor of Cambridge (Boston), *Alfred Vellucci* in his now infamous statement threatened to 'turn *Harvard* into a parking lot' if it continued to

---

2) Home of the pharmaceutical giants, Hoffmann La Roche and Novartis which is itself a fusion of Sandoz and Ciba-Geigy that took place in 1996.

carry out hazardous (recombinant DNA) work. *Rathman* believed that this repressive attitude was a crucial element in enabling *Genentech* to win the race to develop the first recombinant DNA products at a time when *Wally Gilbert's* laboratory and others in the Boston area had been considered as the favourites for this distinction.[3]

8.1.2
International Scientists' Concerns Led to Guidelines for Gene Cloning

Paul Berg was the first to call for a public discussion on the possible hazards of recombinant DNA research. He addressed an open letter to the public following the Gordon conference in 1973. *Berg* himself deferred an experiment to introduce bacterial sugar-metabolizing genes into animal cells using his SV40 virus vector, until a further evaluation of the risks inherent in recombinant DNA experiments had been made.

He informed the President of the United States of his concerns and wrote a letter in 1974, which was published in *Nature, Science* and the *Proceedings of the National Academy of Sciences (USA)*, on the possible risks involved in recombinant DNA experiments. He proposed that although most experiments would not represent any risk some should perhaps rather be avoided in the absence of further risk assessment. He proposed a *moratorium* on the cloning of antibiotic resistance genes, toxin genes and cancer genes.[4]

The entire scientific community both academic and in industry, in fact observed this moratorium. Berg, along with *David Baltimore, Maxine Singer, Richard Roblin and Sydney Brenner* [4] organized the *Asilomar* conference in Pacific Grove, California, in February 1975 to evaluate the conditions under which the moratorium could be lifted [5].

The meeting was attended by 140 scientists, lawyers, journalists and government officials. This led to the idea for defining the risk level of individual experiments and designing physical containment conditions suitable for each risk category. To limit the risk of spread outside of the laboratory special weakened bacterial host strains were proposed for biological containment. These recommendations were refined in countless meetings, exchange of drafts and discussions.

This eventually formed the basis of the *guidelines for recombinant DNA work*. This was rapidly implemented and was already in force in the USA by July 1976. Following a number of revisions it is still functioning in country-specific form internationally. It contains rules for documenting and categorizing experiments for examination and approval by local safety officers. Higher risk categories usually require approval from external regulatory authorities. This unique self-imposed cautiousness was also intended to have a reassuring effect on the public.

---

3) Researched in databases available on-line from the University of California Berkeley; Regional Oral History Office University of California; The Bancroft Library Berkeley, California; Program in the History of the Biological Sciences and Biotechnology.

4) www.dnai.org/text/81_the_moratorium_letter_regarding_risky_experiments_paul_berg.html.

It is true that a necessary discussion had been initiated leading to meaningful changes in laboratory procedure. In the opinion of the authors Berg's contention that this calmed public outcry may be applicable to the USA but in contrast, it certainly did not have the same effect on the public in Europe, in fact it had exactly the opposite effect.

Are there lessons to be learnt for the future? Can scientists and government adequately prepare the public for the impact of new technologies? In Paul Berg's recent review of the effects of the Asilomar conference, 33 years later he is pessimistic about this: '. . . so many issues in science and technology today are beset by economic self-interest and, increasingly, by nearly irreconcilable ethical and religious conflicts, as well as by challenges to deeply-held social values. A conference that sets out to find a consensus among such contentious views would, I believe, be doomed to acrimony and policy stagnation. That said there is a lesson in Asilomar for all of science: the best way to respond to concerns created by emerging knowledge or early-stage technologies is for scientists from publicly funded institutions to find common cause with the wider public about the best way to regulate – as early as possible. Once scientists from corporations begin to dominate the research enterprise, it will simply be too late'.

The impact on European recombinant plant biotechnology was especially dramatic, even though the technology had originated there, pioneered by *Jozef (Jeff) Schell* (1935–2003) and *Marc van Montagu* (Figure 8.1).

The technical issues involved in risk assessment were inherently complex and involved probability estimates involving complicated systems. The public fear of a new disease agent is overwhelming set against a lack of understanding of the myriad steps required to turn a harmless bacterium into a pathogen.

The following questions are an example of the appraisal according to an average microbiologist:

Figure 8.1   Marc van Montagu and Jeff Schell (archiv 1986) who were the pioneers of genetic engineering in plants.

- How can a non-pathogenic recombinant organism survive in the environment?
- How can it overcome entry barriers into the body and then avoid or inactivate the immune system?
- How can it scavenge sufficient iron while in the bodily fluids?
- How can a wide range of antibiotic resistance be acquired?

These are just a few examples of a multitude of factors, for each of which there would be a vanishingly low probability of occurrence.

In spite of this quite simple argument for the lack of imminent danger of this type of threat, scientists are trained never to say that there is NO danger. This is immediately misunderstood as a weakness since the general public lacks an understanding of chance and probabilities. It is sometimes hair-raising for a scientist to hear that his recounting '*a theory*' is placed in the general consciousness as being as good as a neighbour's dreamed-up scenario rather than what it really is, namely a rigorously tested hypothesis which is as close to the truth as we may come in the foreseeable future.

So what can be done? Unilateral national initiatives have been highly repetitive and thus wasteful. They require ongoing refinement to remove incongruous international differences.

The best scientists will want to work where their research possibilities are unfettered. The 'brain drain' therefore starts almost immediately in a direction away from the restricted areas. The long term costs may be considerable. They may develop over longer periods to complete loss of dominance in whole sectors of global markets.

8.1.3
Towards a Solution: Increase Scientists' Involvement in Technology Assessment

We see international effort towards establishing a *modus vivendi* (*operating rules*) which is aimed at reinforcing the status of the scientist in the public eye. Is it conceivable that the public can recognize the scientists' endeavours to be not only to develop novel ideas and techniques which may lead to valuable products, but also to be clearly involved in predicting the consequences of these innovations? This could be supported by having highly competent groups of scientists advising government along the lines of earlier Offices of Technology Assessment.

The formation in 2008 of the *European Research Council* (ERC) involving the 27 countries of the EU as well as seven other associated states including Switzerland, Israel and Turkey, can be seen as one important step in this direction, with a precedence function. The 22 members of the *Scientific Council of the ERC* are elected on the advice of members of a representative group of scientific societies, including *FEBS, EMBO, ESF* and *EURY* (see Appendix on European funding organizations) which in turn represent the vast majority of scientists in Europe. The ERC could thus be seen as *the ideal body*, beyond reproach, *to deal with technology assessment advice for Europe*. It was created as a research funding organization with a budget of $1 billion annually and will have an enormous impact on shaping the direction of European research.

The establishment of the ERC is an attempt to rectify the gap in funding which has opened between Europe (1.8%; % GDP as of 2005), the USA (2.6%) and Japan (3.6%). After the 2001 European Council Summit the following aim of improving research funding, evaluation and technology transfer was to make Europe the 'most dynamic and competitive knowledge-based economy in the World by 2010'. After the Summit meeting in Barcelona in 2003 it was declared that 3% of Europe's GNP will be spent on science research by 2010. In view of the current economic crisis this may be difficult to achieve, however the establishment of the ERC was a significant step in the right direction.

It should be noted that the optimism about gene technology in the 1990s was not an improvement on that in the 1980s. In 1991 polls showed that 47% of the population in highly developed countries were optimistic and only 17% pessimistic about the impact of biotechnology. This dropped to an all-time low in about 1999 following the discussions about the possible misuse of the human genome sequencing project. The situation again improved, back to the 1991 level, as many positive results became apparent: better characterization of infectious disease agents, cancer diagnostics and understanding of basic biological phenomena.

8.2
Ethical Aspects: Animal Cloning and Fertility Research

A prize-winning animal of whatever species, be it a race horse, racing camel, a disease-resistant bull, a fast growing pig or a hardy sheep, has always been worth a small fortune for the short period during which it could be used as a stud in breeding programmes. In natural breeding programmes the genetic traits which were seen in the original animal are usually diluted out in the first generation offspring. In contrast, a *cloned animal*, one in which the new born animal foetus is generated from a body cell (e.g. skin of the 'single (parent) adult'), has the identical genetic make-up.

In 1973 *Ian Wilmut's* team successfully resuscitated a frozen embryo and implanted it into the womb of a surrogate mother animal, leading to the birth of the first animal from a frozen embryo (see Chapter 12). Such a manipulation can be seen as an extension of *reproductive* or *fertility medicine* which is becoming more efficient year on year. At the time of writing, some 100 000 humans are on the Earth as a result of *in vitro* fertilization: fusion of the sperm and egg outside of the human body.

A single cell from the 8-cell stage in the development of the embryo can be analysed to determine part of the genetic composition of the future individual without harm to the embryo. This type of manipulation, which is carried out before the embryo has been implanted into the (surrogate) mother is referred to as *pre-implantation diagnostics* (see Section 7.8). It is usually carried out to establish the presence of a (dominant) or highly 'penetrating' gene for a severely debilitating and usually untreatable disease such as Lesch-Nyhan or Huntington's chorea. This type of test is usually undertaken with the prior understanding that in the case of the adverse gene being present, the embryo will be discarded.

The ethical discussion centres on human rights. At the European Human Rights Convention in 2001, chapter 11 of these rights was extended explicitly for the human embryo. Many would argue that if the embryo is not yet implanted it has not yet achieved this status, and as is analogous to the case of a woman having a contraceptive spiral in her womb so that the incoming embryos will not implant and will be lost. Empiricists will argue that the principle is to prevent suffering of the future individual and its family and to give the potential parents the privilege of a positive reproductive choice. Previously a couple had been faced with a narrower choice. In the absence of such diagnostics but with the knowledge of there being close family relatives who had been affected by a devastating hereditary (familial) disease, they would usually reject the idea of procreation completely.

The most recent example of a critical pre-implantation diagnostic again resurrects the ghost of the Eugenics movement and critical slogans such as '*designer babies*'. The analysis referred in this case to the *potential to develop* a type of breast cancer with a certain statistical probability (see Section 7.9 'Inherited predisposition to disease'). Below we discuss the development and impact of the analysis of genes which carry such predispositions, designated through *haplotype* mapping. This is becoming increasingly more accurate in combination with information relating to short sequences of the genomic regions or genes which may be involved.

If the current trend for reduction of the cost for DNA sequencing continues one can expect that within a few years one can have a complete personal genome sequence which will most likely be affordable for less than the cost of a television screen. (See Section 13.5 Flash sequencing DNA: a human genome sequence in minutes?)

8.2.1
The Origin of Cloning Animals

The idea of cloning an animal was suggested as a real possibility in 1938 by *Hans Spemann* (Nobel Prize in Medicine, 1935). It was attempted unsuccessfully in 1951 by *Robert W. Briggs* and *Thomas J. King* at the Cancer Research Institute in Philadelphia. *John B. Gurdon* achieved the first success in 1963 at Oxford University with the three clawed South African Toad, *Xenopus laevis* which produces an unusually large and robust egg cell. The first successful cloning of a mammal, the Dorset lamb 'Dolly', using a nucleus from a normal body cell inserted into a nucleus-free egg cell was carried out by a team of scientists led by *Ian Wilmut* in 1996 which employed a novel starvation protocol for the egg cells before the nuclear transfer [6]. Ian Wilmut received the Nobel Prize for Medicine in 2005 (see Section 12.2 for further developments).

In the meantime Wilmut has left the field of applied animal cloning and is highly critical of the limitations of the technique, particularly with regard to the malformations in the embryo that he observed in his own experiments. This may be dependent on the type of cell used as the nuclear donor. Ian Wilmut now dedicates his research abilities to studying how and why these abnormalities arise. As long as there is an uncertain outcome to this technology it would be irresponsible to attempt to apply cloning to humans.

Figure 8.2 Clone of the original *Brucella-, Salmonella-* and *Mycoplasma*-resistant bull produced in 2001 at Texas A&M from frozen fibroblast culture cells. The animal retained all the desired properties of disease resistance. Westhusin *et al.* [7].

8.2.2
An Example of Cloning in Animal Husbandry

In 2001 *Mark Westhusin* and his team at the Texas A&M University reported the successful cloning of a rare bull which had been found to be naturally resistant to a number of bacterial pathogenic bacteria, including *Brucella abortus, Mycobacterium bovis* and *Salmonella typhimurium* (Figure 8.2). These organisms are common contaminants of meat and can be transferred to humans, particularly farm hands [7]. This was pioneering work in its field not only because it led to the further study of a disease resistance that was poorly understood but also because it demonstrated that an animal could be cloned from tissue that had been frozen for some 15 years. The tissue had been taken from the original bull after its death from natural causes in 1996 and since no semen was available for artificial insemination, this valuable genome would have been lost had fibroblast cell cultures not been established in 1985, cryopreserved, and stored in liquid nitrogen for future genetic analysis.

This work has also inspired others to freeze and store animal (or plant) tissue of endangered species or rare breeds with the possibility of perhaps later saving a species on the brink of extinction.

References

1 Johnson, I.S. (2003) *Nature Reviews Drug Discovery*, **2**, 747–751.

2 Eichner, W. and Sommermeyer, K. (1999) Animal cell technology: challenges for the 21st century. Proceedings of the Joint International meeting of the Japanese Association of Animal Cell Technology (JAACT) + the European Society of Animal Cell Technology (ESACT) 1998, Kyoto, Japan, pp. 199–202.

3 Owen, M.R.L. and Pen, J. (eds) (2006) *Sembiosys Genetics, Calgary Canada, Announced Viable Production Cultivars*, John

Wiley & Sons, Weinheim, ISBN: 978-0-471-96443-8.

4 Brenner, S. (2001) *My Life in Science*, Biomed Central Ltd., London, ISBN: 10: 0954027809; 13: 978-0954027803.

5 Berg, P. (2008) *Nature*, **445**, 290–291.

6 Wilmut, I., Cambell, K. and Tudge, C. (2000) *The Second Creation: The Age of Dolly and the Age of Biological Control*, Harvard University Press.

7 Westhusin, M.E. *et al.* (2007) *Journal of Animal Science*, **85**, 138–142.

9
Elucidating Protein Structure: The Beginnings of Rational Protein Design

Studies in biochemistry were underway in the 1940s and 1950s that showed how organic molecules interacted with each other, culminating in a picture of enzymes as flexible protein structures which interacted by binding their substrates and stabilizing transition state intermediates. This early work was reviewed by *Crosland* [1] and the principles and modern understanding of enzyme action elegantly explained by *Fersht* [2].

9.1
Cambridge England, the Cradle of Structural Analysis of Macromolecules

The powerful tool of X-ray diffraction had been developed by the father–son team of *Sir William Henry Bragg* (1862–1942) and *Sir William Lawrence Bragg* (1890–1971) and applied to elucidate the molecular structure of ordered crystals (minerals) for which they had jointly received the Nobel Prize in 1915. *John Desmond Bernal* and *Dorothy (née Crowfoot) Hodgkin* were the first to investigate X-ray diffraction of protein crystals in 1937. *Bernal* was mentor to *Dorothy Hodgkin*, *Rosalind Franklin* (who was the first to show the alpha-helical structure of DNA), *Aaron Klug* (Nobel Prize 1982 for pioneering crystallographic electron microscopy; compiling images from different angles to create 3 D images) and *Max Perutz*, who in turn was mentor to *Jim Watson* and *Francis Crick* (Nobel Prize for modelling a DNA structure in 1962). The discovery of mRNA and the triplet-based genetic code were largely attributed to Crick's associate, *Sydney Brenner* [3].

This 'scientific lineage' described above is an interesting example of a centre of excellence attracting talent. It maintained a leading position with respect to the development of methods and as a source of cutting-edge instrumentation for the investigation of molecular structure which was unsurpassed for decades [4].

Dorothy Hodgkin (Figure 9.1) was the first to use X-ray crystallography to analyse and determine the structure of organic molecules (cholesterol, 1937; penicillin, 1945

Concepts in Biotechnology: History, Science and Business. Klaus Buchholz and John Collins
Copyright © 2010 WILEY-VCH Verlag GmbH & Co. KGaA, Weinheim
ISBN: 978-3-527-31766-0

Figure 9.1 Dorothy Crowfoot Hodgkin, pioneer of X-ray crystallography of organic molecules, died 1994 and was posthumously recognized on a British postage stamp issued in 1996.

and vitamin B12, 1959) for which she received the Nobel Prize in 1965, she also elucidated the structure of insulin in 1969.

It was, however, the work of *Max Perutz* (Figure 9.2) and *John Kendrew* (Nobel Prize in Chemistry in 1962) which perhaps first caught the imagination of the scientific community and animated the field for modern molecular biologists as a result of their elucidation of the structure of the first complex protein, namely myoglobin.

The pivotal writings of *Erwin Schrödinger* ('What is life?', 1944) and *Jacques Monod* ('Chance and necessity', 1972) [5] had great influence in attracting scientists from other disciplines to find intellectual fulfilment of the highest order in biology. Perutz' and Kendrew's early work on haemoglobin and myoglobin, elegantly explained the molecular pathway for uptake and transport of oxygen in the blood (bound to haemoglobin) and its delivery to the muscle (with stronger binding to myoglobin). In 1961 *Jean-Pierre Changeux* had shown how reversible changes in the structure of an enzyme, so-called *allosteric* effects, caused by the interaction with an 'activator' or an 'inhibitor' binding at a remote site, could alter enzyme activity [6].

Such insights into the fine tuning of molecular interactions in biological systems produced great hopes for a new paradigm of 'rational drug design' (see *D.E. Szymkowski*, 2005) [7].

9.1.1
Using Knowledge of Structure for Protein Design

Thus we discern two approaches: one depends on the prediction of the physical structure of the target molecules involved in an interaction of interest for pharma-

Figure 9.2 Max Perutz pioneer of protein structure analysis.

ceutical intervention; the other approach is more pragmatic, screening for compounds which will fulfil the functional criteria. Limitations to the first approach basically derive from the uncertainties in predicting exactly how the protein will now fold and how a particular change will affect activity.

As we will see later, it is in fact quite productive to screen randomly created libraries for some specific functions. This is due in part to the fact that the number of residues actually involved in a particular interaction is limited, often less than a dozen atoms in each molecule. In addition, it had already been noticed during the early determination of crystal structures that many proteins shared similar folds, that is, spatial geometry, despite a high degree of sequence divergence. It seems that in evolution only a limited set of possible protein folds has exhibited sufficient robustness (plasticity?) to be of use in evolving functional proteins: perhaps less than 2000 of some 4000 basic geometries (R.A. Goldstein, 2008) [8]. The presence of such a natural selection among folding patterns is proposed as a fundamental explanation of why metagenomic gene mining should be more fruitful in delivering useful design elements than say designing a functional protein from basic principles.

The following paraphrases Szymkowski's ([7] ibid.) analysis, with the addition of more recent examples. He distinguishes alterations in protein structure which involve varying the primary amino-acid sequence with the addition of post-translational and post-production modifications.

Those recombinant DNA products that were produced as *first generation* substitutes for known highly active compounds derived from animal tissues such as human insulin, growth hormone (huGH) and human G-CSF, have also been most frequently investigated with respect to variant derivatives. This has resulted in there being a number of approved products in which primary sequence alterations have yielded products with altered, desired characteristics.

The design principles for variant proteins are discussed in the following sections.

9.2
Redesigning the Protein Core

Design via protein modelling has been extensively applied to human growth hormone and G-CSF, using in particular novel so-called 'dead-end elimination algorithms' (Looger and Hellinga, 2001) [9] which although they do not facilitate calculations of the actual folding consequences of each alteration (the backbone is maintained invariant throughout), do eliminate changes which would cause geometric clashes. Protein modelling filters out a few suggestions which might lead to more compact core structures from very large numbers of possible variants (e.g. in excess of 10^{21} combinations considering 10–14 positions within the 25–35 amino-acid core of G-CSF) based on energy minimization parameters. Active and more thermostable variants of huGH were obtained which contained six to ten alterations.

9.3
Redesigning the Protein by Altering Primary Sequence

A number of properties are subject to 'improvement' particularly in relation to proteins for pharmaceutical use. Shelf-life, solubility, serum half-life, uptake into the body and antigenicity are important factors. These have been addressed in a multitude of ways.

9.3.1
Multimer States

In the case of *insulin*, the design of muteins has focused on altering association of the molecules in multimeric form causing changes in solubility of the final product, producing either a rapidly soluble quick acting form (*Humalog*® or *Novolog*®) or a less soluble, slow release form (*Lantus*®; *Insulin glargine*). Both *Eli Lilly* (*Humalog*) and *Novo Nordisk* (*NovoLog*) have produced variants of *human insulin* which remain preferentially in the monomeric or dimeric form rather than the normal hexamer, which is only active once dissociated. These novel forms, in which a single amino acid charge has been altered, represent insulin which acts very rapidly after injection (peaking 45–120 min after injection) after subcutaneous injection. These products have been clinically approved.

9.3.2
Solubility and Stability

A Group at Sejong University analysed the structure of the *serpin protease inhibitor family* and found that most of the members had suboptimal folding characteristics, caused largely by their very strained structure and buried side chains. More than 30% of all substitutions improved the thermostability of these inhibitors.

9.3.3
Selectivity and Potency

It has been shown with several examples, of which a selection is given here, that biological activity of potent biological mediators can be modified to produce for example, a higher activity in a given assay. This does not categorically mean that these products may be more effective as pharmaceuticals. Their altered amino-acid sequence has the potential to be immunogenic. Immunogenicity has proved to be one of the major stumbling blocks in development of protein-based pharmaceuticals originating from the rDNA revolution. One basic problem is that even normal human proteins such as insulin, interferons or interleukins may become immunogenic after long periods of intravenous use.

9.3.4
Consensus or Combinatorial Products of Natural Gene Products

Fourteen human α-*interferon* genes have been discovered. A consensus sequence gene was constructed which produced an interferon with relatively high activity (*Infergen*®; *Amgen/InterMune/Yamanouchi*).

Similarly, work carried out at Maxygen on the recombination of sequences from seven mammalian *interleukin-12* molecules yielded an 'evolved' IL-12 that was more potent than the original human product (*EvlL12*). Some interleukins are particularly prone to endocytic degradation or recycling due to protein instability under acidic conditions after binding to the cellular receptor and being taken up into the cell. Variants of both *IL2* and *G-CSF* which are more robust and are thus recyclable and therefore of a higher apparent efficacy have also been produced [10].

9.3.5
Protein Fusions

Various IgG fusions to a number of proteins have altered the pharmacokinetic properties of the original products, altering the half-life in serum or altering tissue penetration:

> **Enbrel**® (soluble TNFR2 fused to IgG1) increased half-life.
> **Epo-Fc** (erythropoietin conjugate to IgG1-Fc) increased lung permeability.
> **Abatacept** (CTLA4-IgG4-Fc fusion) increased half-life; used in the treatment of rheumatoid arthritis as it acts as an antagonist to co-activators of TNF action.

9.3.6
Protein Truncations

> **Retavase**® (tPA, increased half-life of non-glycosylated material; solubilization of blood clots which lead to heart attacks);

Refacto® a form of the anti-haemophilic Factor VIII was more easily manufactured in the truncated form.

Chimaeric and humanized antibodies are discussed separately in Chapter 10 which includes a table of clinical antibodies (Table 10.2).

9.4
Post Translational Modifications

Many proteins found in nature contain post-translation modifications, of which more than 80 types are known. Some of these affect specific functions of the protein while others produce an effect on the general physical properties of the protein, particularly its stability in the natural environment.

9.4.1
Glycosylation

Glycoproteins are characterized by long flexible chains of sugar subunits which are attached after translation. The lengths of the sugar chains, the type of links between the sugars and the terminal modifications vary from species to species and from tissue to tissue and are also subject to micro-heterogeneity. The latter is often of unknown biological significance and difficult to analyse and control in bulk production processes.

In the early 1980s glycosylation was considered to be little more than a way of making more soluble protein products. Interferon-β is, for example, barely soluble in the absence of glycosylation. For a large number of products, particularly antibodies, glycosylation has been shown to reduce immunogenicity as an alternative to or in addition to using completely human antibodies as well as affecting complement-activation. Glycosylation also increases the molecular weight of small proteins and peptides considerably increasing their retention time in serum. Combined with a contribution to the resistance to degradation by proteases, this leads to an increase in the half-life of the glycoprotein in the serum compared to the naked polypeptide (see also Table 10.2).

The glycosylation machinery which is essential for glycoprotein production, is absent in *E. coli*. However, it is present in *Saccharomyces cerevisiae* and is of a simpler structure and in general the sugar polymer is of greater length than that in mammalian cells. Glycosylation is highly conserved and complex amongst mammalian cells and exhibits natural micro-heterogeneity and in some cases tissue-specific modification (Figure 9.3). Insect cells again differ in this respect.

The mammalian cells which are most frequently used for recombinant protein production include baby hamster kidney (BHK) and Chinese hamster ovary (CHO) cells. Although more technically complex and expensive than microbial cell production, they are amenable to the post-translational modifications that may be necessary for biological activity; examples include EPO, many interferons, blood factor VIII and others [12].

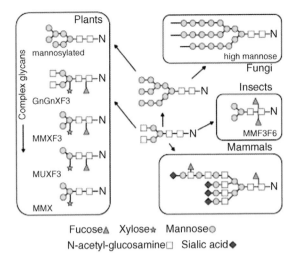

Figure 9.3 Simplified overview of the basic modified sugar composition of glycan side chains added to glycoproteins in different species as a result of post-translational modification [11].

Mammalian glycosyltransferases have been successfully cloned and expressed in other cell lines where they allow the production of glycoproteins with correspondingly more complex glycosylation side-chains. Even in insect cells which can scavenge sialic acids from glycoproteins added to the medium, such modifications have been successfully carried out. It would appear that in the near future there should be no limitations to finding appropriate host cells for the production of glycoproteins with any desired modification (Figure 9.4).

9.4.2
PEGylation

The addition of a hydrophilic polymer (polyethyleneglycol; PEG) which to some extent imitates glycosylation, has been used effectively to increase the half-life of therapeutic proteins in serum, particularly that of antibodies. In the case of *Somavert*® (*pegmisovant*), a commercial therapeutic growth-hormone antagonist, pegylation of human growth hormone caused changes in functional specificity, yielding at once both an antagonist to the growth hormone receptor as well as eliminating unwanted side-effects related to binding to the prolactin receptor.

9.4.3
HESylation: An Alternative to PEGylation

Although a number of PEGylated pharmaceutical products (e.g. PEG-interferon-α; Neulasta, PEG-G-CSF) are already on the market, it is noteworthy that a compound with similar properties to polethylene glycol is being evaluated as an alternative for use in post-translational modification. Hydroxyethyl-starch is approved for adults

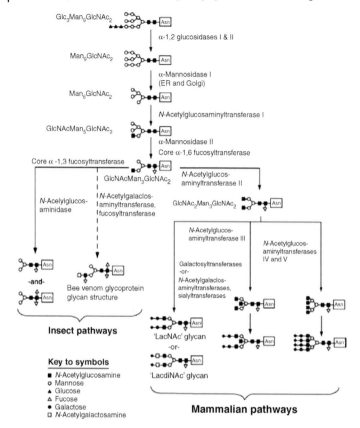

Figure 9.4 Sialyltransferases have been introduced into insect cells to allow mammalian cell-type glycosylation of glycoproteins. The common pathways and differences of insect and mammalian cells are shown here [13].

and infants as a 'blood volume replacement' and for decades has been used in very large doses for injections over long periods of time. Allergic reactions occur in <0.035% of patients, making hydroxyethyl-starch one of the least allergenic substance known. Apparently initial trials are underway and are looking sufficiently promising for Fresenius-Kabi to radically expand its capacity to produce such modified proteins.

9.4.4
An Alternative to Post-Translational Modification?

An interesting alternative to glycosylation and/or PEGylation is the 'XTEN' platform technology used by *Amunix*, a company founded by *Pim Stemmer* in 2005. The proprietory technology involves expressing an unstructured glycine-rich hydrophobic 330–200 amino acid sequence in the same translational reading-frame as the therapeutic peptide or protein. This massively increases the apparent molecular weight/size of the molecule and demonstrably increases serum half-life with apparently little evidence of any increase in immunogenicity. Concatamers in which such rPEG sequences alternate with copies of the therapeutic peptide sequence can in principle potentiate the activity of the compound to clustered target receptors and further increase the serum half-life.

9.5
Total Chemical Synthesis

In general it had not been considered a viable commercial proposition to produce compounds for pharmaceutical use by complete chemical synthesis if the chain length was more than some 30–40 amino-acids. Following the development of the peptide ligation method by *Steve Kent* [14], it became possible to obtain highly purified synthetic protein products by fusing smaller highly purified shorter peptides.

Since the mid-1990s it had been the aim of *Gryphon Therapeutics* of San Francisco to completely synthesize erythropoietin which consists of 166 amino acids, with synthetic non-peptide side chains. The biological activity of EPO normally depends on the surface charge. Using defined polymers the optimal charge could be studied by changing the charge on the added synthetic polymers. It was found that a slight negative charge gave the optimal activity; these molecules were tested in a phase I trial in the clinic and compared with the PEGylated material. The synthetic EPO with the 'optimized' charge was more active in trials compared to PEGylated material [15]. Unfortunately the trials were terminated, since a few patients produced low levels of IgM in response to the controlled size (controlled charge) polymers.

It should be noted that where the product is synthesized completely by chemical means we are really departing from the realms of biotechnology since no living cells or their components are being used. However, Peter G. Schultz's lab has developed a microbial system in which unnaturally modified amino-acids can be introduced at specific gene regions containing amber-stop codons, which can then be used as intermediates for the subsequent site-specific addition of a post-translational modification [16]. Random PEGylation for instance, often alters biological properties, for example human Growth Hormone (hGH) is converted into an agonist by random PEGylation.

Other examples of such alterations include terminal glycosylation of TNF after the replacement of five lysine positions using genetic means, to prevent random glycosylation at these sites (as well as changing the total surface charge). The terminally glycosylated molecule demonstrated a higher activity compared to the non-glycosylated or randomly mono-glycosylated material. In one final example reductive alkylation (via a Schiff base) was used to attach a single PEG molecule to the amino terminus of human G-CSF (*Figrastim*) to yield the product *Neulasta*. Neulasta has been approved for inclusion in products for clinical use, specifically to treat neutropenia.

9.6

Validation of Drug Design Based on the known Structure of the Target

In 2003 *Brian J. Druker* received the Braunschweig Prize for his involvement in the development of the drug *Glivec* used in the treatment of Chronic Myeloid Leukaemia (*Gleevec* (US spelling) approved in the USA in May 2001). The drug was also approved in 2008 as the only post-surgical treatment indicated to delay the return of a highly aggressive form of gastrointestinal stromal tumours (GIST), thus satisfying a major requirement for such patients.

The development of Glivec really began with the 1960 discovery of a specific chromosome aberration known as the Philadelphia chromosome (Figure 9.5) [17]. This aberration is found in 90% of CML leukaemia patients. The gene fusion that occurs, results in the creation of a novel tyrosine kinase which alters signal transduction in the leukaemia cells resulting in their cancerous growth. *Glivec* was developed to specifically inhibit this tyrosine kinase and is thus a strong and specific inhibitor of this type of leukaemia in which the enzyme in the leukaemic cells has been changed as a result of a recombination within the *Abl* gene. The development of this drug is the first validation of this direct approach based on the discovery of a specific target gene product. This process of gene fusion and the protein were first identified in the late 1980s. However, it took over 15 years before the drug was available in the clinic. This included the time needed for the development of an effective drug and the period of clinical testing.

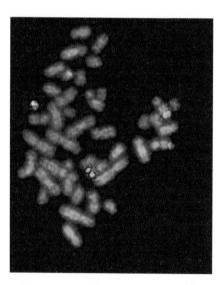

Figure 9.5 Philadephia chromosome karyotype: FISH coloration of the BCR and ABL1 gene segments shows up the fusion that has occurred during chromosome rearrangement which leads to the production of the micro 'Philadelphia' chromosome. This is found in some 90% of chronic myelogenous leukemias.

The development of Glivec is recognized as the first validation of a novel concept in drug development based on information about novel targets discovered by gene and gene expression analysis. The development of Glivec took 29 years to complete. In principle this was the same concept proposed a century earlier by *Paul Ehrlich* for the development of 'magic bullets' against specific pathogens. Antibiotics had been identified empirically as they were screened for functionality without knowledge of the target. Once a lead substance had been discovered its mode of action and target could also be established. In a number of cases this led to the development of families of drugs whose specific functionality was often screened using the purified target: β-blockers, β-lactamase inhibitors, β-amylase inhibitors, lymphokine agonists, neuraminidase inhibitors, reverse transcriptase inhibitors and proteasome inhibitors amongst many others.

Many different types of leukaemia and other tumours are characterized by chromosome rearrangements. Over 45 000 such rearrangements have been documented [18]. Many projects have been initiated in order to find specific tumour targets of similar nature and to develop specific protein kinase inhibitors for medical application.

The approval in 1998 of Herceptin, an antibody against the Her2 antigen, can be considered to be the first target-based recombinant DNA therapeutic to be approved, although it is not absolutely specific for the tumour cell. This has led to further target-driven anti-tumour drug development involving genomic analysis strategies which have recently centred on various receptor protein tyrosine kinases other than that discussed in the development of Gleevec, for example HER3, FGFR4, Afl/Ufo and Flk/VEGFR2.

9.7
General Considerations in Drug Development

9.7.1
rDNA Technology: Yields an Altered Drug Development Paradigm

Jörgen Drews had already predicted in 1989 that '...in 30 years there will be no pharmaceutical drug which is not either a recombinant DNA product or, has used rDNA at some point in its development'. It seems that even by 2009 we were approaching the fulfilment of this prophecy. Most of the potential targets for developing new drugs have been identified by gene cloning or sequencing. Using the recombinant DNA technology discussed in Section 7.5 a new cornucopia of biological diversity has become available to assist in the discovery of novel products or to guide the semi-synthesis of novel derivatives and secondary metabolites.

Recombinant monoclonal antibodies provide incisive insight into the quantification of *post-translational modification* of proteins which guides the manner in which eukaryotic cells respond to the environment. The latter is a qualitative advance in understanding how the cell functions. It allows a distinction and understanding of the disease- and healthy states of individuals which was previously unachievable.

9.7.2
Recombinant DNA has Generated Novel Biotech Products over the Last Decade

In 2007 the market for rDNA-derived pharmaceuticals exceeded $40 billion. This included recombinant erythropoietin (EPO), human growth hormone (hGH), insulins, interferons, and monoclonal antibodies. Perhaps the most successful development of novel products has been in the recombinant antibody field (see Section 10.8 'A survey of therapeutic antibodies'). According to *Wetzel* [19] there were, at the beginning of 2000, some 284 medically-related biotechnology-derived candidates *in some stage of development*:

- 28 gene therapy applications
- 62 vaccines for infectious diseases and cancer
- 14 cell-based therapies
- 60 monoclonal antibodies

According to the *Pharmaceutical Research and Manufacturers of America* (PRMA) in 2000 there were over 369 biotech medicines being utilized in the clinic and on the market for the treatment of a range of over 200 diseases, for example *protein or peptide rDNA products* in clinical development:

- 98 vaccines
- 12 interferons for 25 different indications
- 59 monoclonal antibodies
- some 52 other recombinant proteins
- some 75 biotech products had been approved for *diagnostics*

Walsh compiled data in 2006 on the extraordinarily high number of biopharmaceuticals in the pipeline [20]: over 600 for cancer, over 300 for AIDS, near to 300 for other infectious diseases, followed by more than 10 other classes of biopharmaceuticals and a range of 100 other projects. Development trends favour four classes of biopharmaceuticals, notably complex ones (number of industrial projects given) [21]: monoclonal antibodies, 160; vaccines, 60; gene therapeutics, 42; recombinant proteins (other), 41.

In April 2008 PRMA reported that in the area of clinical drugs alone 750 new medicines (over 300 specific to rare diseases; more than 275 medicines for heart disease and stroke; 109 being for HIV/AIDS) were in clinical trial. Nearly half were targeted at the most devastating forms of cancer in terms of mortality. Of the products in development 113 were for the treatment of lung cancer; 90 for breast cancer; 65 for colorectal cancer and 88 for prostate cancer.

This shows a steady trend over the last decade for novel products in the pharmaceutical pipeline driven by the genome revolution.

It was also reported that spending for research and development had grown in the US Biopharmaceutical Companies from 2006 to 2007 by over $3 billion, to a sum of $58.8 billion. The conclusion is that this branch of drug development is one of the most research investment-intensive industries, spending five times as much of their income on research and development as the average American Company.

9.7.3
The Complex Factors (ADMET) Important for Pharmaceutical Drug Development

In the development of a pharmaceutical drug the way in which the compound behaves in the human body can be a decisive factor determining its effectiveness. This starts with the kinetics of uptake, the distribution to the tissues and finally the overall half-life of the substance determined by breakdown and active or passive secretion of the compound. This is just as important to the overall effect as the affinity and specificity of the compound for the therapeutic target. These are called the 'ADMET factors';

- *Adsorption = uptake,*
  - *Resistance to breakdown in the stomach/intestine;*
- *Distribution in the body,*
  - *Tissue entry,*
  - *Crossing the blood–brain barrier;*
- *Metabolism in the body (new field of study known as Pharmacogenetics);*
- *Excretion/secretion;*
- *Toxicity.*

Engineering proteins to meet these requirements by changing their stability, tertiary structure, molecular size and in particular the effect of post-translational modification, where the latter is dependent on the cell-line used for production, are discussed in Sections 9.3, 10.6, 10.7 and 10.8.

9.7.4
The Problem of Antigenicity

Antigenicity is of major concern as it can occur with any recombinant protein. The potential for protein denaturation during storage and shear during administration, for example by intravenous injection, can generate immunogenicity in completely 'natural' proteins from the same host. This is perhaps due to the higher local concentration of partially denatured material at unusual sites or the formation of 'unnatural' complexes with other serum proteins.

9.7.5
Improving Drug Screening

Very early tests for toxicity particularly carcinogenicity of chemicals were developed by examining their mutagenic effects in bacteria. By first adding the compound to animal tissue extracts such as dog liver microsomal fractions, it was also possible to check for conversion of the original compound to other (more?) toxic compounds (*Bruce Ames*), imitating what might happen in the body.

The tragedy of *Contergan* (thalidomide; Contergan = trade mark in Germany) which caused retardation of arm formation in the embryo during clinical use between 1960 and 1961 in Germany, highlighted the need for extensive testing and extreme caution with respect to induced deformations in the growing embryo (teratogenic effects). This also showed the limits of animal models since this very specific effect of thalidomide is limited to man and a few primates. In the absence of a valid teratogenic test many drugs will be contraindicated for use during pregnancy simply due to the *potential* risk. With time, sufficient data will accumulate to determine risk levels, contingent on routine use in adults.

9.7.6
High Information Preclinical Screening

A new speciality branch of drug development is the provision of a screening service for as many variables as possible at a reasonable price to identify any problems before expensive and critical clinical trials are initiated for a particular compound. *High information screening* of large numbers of new compounds is the aim.

9.7.7
Animal Models

In animal trials, novel transgenic animals have been produced which can be used as models for particular diseases.

One interesting example is the generation of a transgenic mouse strain which carries the human internalin gene in which the human internalin acts as a receptor for the bacterium allowing it to cross the intestinal barrier, thus facilitating studies of Listeriosis [22]. Unfortunately there is no such mouse model for the Hepatitis C virus at present.

Perhaps the most famous example was the early 'oncomouse' developed for studying cancer treatment. Models exist for autoimmune disease, osteoporosis, obesity and many other diseases. There is no satisfactory animal model for septic and toxic shock which has had dramatic consequences for the clinical trials of anti-CD28 antibodies which are intended to treat these problems (see Section 10.7 'Two severe setbacks during clinical testing'). There is a fundamental problem with such animal models since they are only a rough approximation of a clinical situation; there may for instance be many different causes or genetic predispositions for a single clinical picture, for example only 10% of patients suffering from motor neurone disease (MND/ALS) lack a particular super oxide dismutase and there are several thousand distinct types of neoplastic diseases (cancers).

9.7.8
Animal Models to Identify Targets for Human Hereditary Disease

Genetically well characterized systems have recently been screened after intense (saturation) mutagenesis for phenotypes that might be of use in the identification of targets of human hereditary syndromes.

In *Caenorhabditis elegans* (*C. elegans*, the round worm) it is a simple matter to create knock down mutants of every gene in the genome which can be screened on a large scale for developmental and behavioural alterations.

The Zebrafish, *Danio rerio* [23], has been developed as an animal in which all early embryonic development stages are easily observable. Sophisticated mutagenesis and screening methods have been implemented to generate hundreds of mutants that may be suitable models of human disease. They can be used for high-throughput screening of potential drugs on a scale not feasible without great expense in other vertebrate systems.

Various transgenic animal models are commercially available or are used in the development of pharmaceuticals, for example at *Artemis Pharmaceuticals GmbH*, Heidelberg, a company co-founded by the Nobel laureates *Christine Nüsslein-Volhard* and *Klaus Rajewsky* along with ex-Bayer manager *Peter Stadler* in 1998 (see also Chapter 12).

9.7.9
Biomarkers: The New Buzz Word in Clinical Screening

The complimentary clinical approach therefore, is to use biomarkers and genetic analyses to differentiate patients into groups which may have to be treated differently. Although these individuals exhibit similar symptoms there may be a multitude of causalities. The development of biomarkers is a growing cottage industry (see Section 11.4 'The personalized genome and personal medicine') whereby both the diagnosis and the drug response can be monitored at the level of the specific gene response by assaying mRNA levels.

References

1 Crosland, D.E. (1959) *Journal of Cellular and Comparative Physiology*, **54**, 245–258.

2 Fersht, A. (September 15 1998) *Structure and Mechanism in Protein Science*, 1st edn, W.H. Freeman, ISBN: 10: 0716732688; 13: 978-0716732686.

3 Brenner, S. (2001) *My Life in Science*, Biomed Central Ltd., London, ISBN: 10: 0954027809; 13: 978-0954027803.

4 Perutz, M. (2003) *I Wish I'd Made You Angry Earlier* (Autobiography), Cold Spring Harbor Laboratory Press, New York.

5 Monod, J.(Reprint 1974; original in French, 1970) *Chance and Necessity*, Fontana/Collins, W. & Sons, Glasgow.

6 Changeux, J.-P. (1961) *Cold Spring Harbor Symposia on Quantitative Biology*, **26**, 313–318; Monod, J., Wyman, J., and Changeux, J.P. (1965) *Journal of Molecular Biology*, **12**, 88–118.

7 Szymkowski, D.E. (2005) *Current Opinion in Drug Discovery & Development*, **8**, 590–600.

8 Goldstein, R.A. (2008) *Current Opinion in Structural Biology*, **18**, 170–177.

9 Looger, L.L. and Hellinga, H.W. (2001) *Journal of Molecular Biology*, **307**, 429–445.

10 Ricci, M.S. *et al.* (2003) *Protein Science*, **12**, 1030–1038.

11 Malandain, H. (2005) *European Annals of Allergy and Clinical Immunology*, **37**, 122–128, 247-L 56.

12 Gellissen, G. (ed.) (2005) *Production of Recombinant Proteins*, Wiley-VCH, Weinheim, Germany; Knaeblein, J. (ed.)

(2005) *Modern Biopharmaceuticals*, vol. 1, Wiley-VCH, Weinheim, Germany.

13 Harrison, R.L. and Jarvis, D.L. (2006) *Advances in Virus Research*, **68**, 159–191.

14 Dawson, P.E. and Kent, S.B.H. (2000) *Annual Review of Biochemistry*, **69**, 923–962.

15 Kochendoerfer, G.G. *et al.* (2003) *Science*, 299, 884–887; Kochendoerfer, G.G. (2005) *Current Opinion in Chemical Biology*, **9**, 255–260.

16 Xie, J. and Schultz, P.G. (2005) *Current Opinion in Chemical Biology*, **9**, 548–554.

17 Nowell, P.C. and Hungerford, D.A. (1960) *Science*, **132**, 1497.

18 See: Hahn, Y. et al. (2004) *Proceedings of the National Academy of Sciences of the United States of America*, **101**, 13257–13261.

19 Wetzel, G.D. (2001) Medical applications of recombinant proteins in humans, *Novel Therapeutic Proteins: Selected Case Studies* (eds K. Dembowsky and P. Stadler), Wiley-VCH, Weinheim.

20 Walsh, G. (2006) Biopharmaceutical benchmarks. *Nature Biotechnology*, **24**, 769–776.

21 Schulte, 2007, Merck AG, Darmstadt, (Germany), Lecture, SFB-Colloquium, Braunschweig, 29.1.07.

22 Lecuit, M. *et al.* (2001) *Science*, **292**, 1722–1725.

23 Dooley, K. and Zon, L.I. (2000) *Current Opinion in Genetics & Development*, **10**, 252–256; Lieschke, G.J. and Currie, P.D. (2007) *Nature Reviews Genetics*, **8**, 353–367.

10

The Development of Antibodies as Pharmaceutical Products

10.1
An Introduction to the Immune System

Antibodies are produced in the blood by specialized B-cells. During the formation of an active antibody-producing gene, cassettes are formed by fusion of diverse gene segments. This results in the production of a unique antibody structure in each cell. The gene region that encodes the six peptide fingers which form the binding pocket for a potential antigen, represent the most variable regions in the final product (the V regions of the antibody's light and heavy chain; see Figure 10.1). The level of sequence variability allows an individual who produces about a million different antibodies, to deal with the challenges of recognizing *all* intrusive microbes. This is called the class II-mediated response

The system can respond quickly to the detection of novel invaders by extending the repertoire. There is an ongoing generation of new pre-B-cells that produce novel antibodies some of which can be further refined by selection from a random pool of mutant variants. This ongoing monitoring and *de novo* response depends on the continuous generation of B-cells carrying novel antigens and the selection and amplification of those that best fit their purpose. This involves a positive feedback loop created by an increased number of signals for B-cell clonal expansion. It should also be noted that additional alterations are introduced into the antibody genes during selection and amplification of the functional B-cells. The end result of this process, known as somatic maturation, is an increase in the affinity of the final antibody for the specific antigen.

The T-cells determine the type of immune response that ensues. This response can be either a clonal expansion of B-cells and an increase in antibody production over several days or months, or an inflammatory reaction involving production of interleukins such as TNFα or interferon γ and expansion of T-cells, often leading to the killing or apoptotic (programmed cell death) death of the antigen presenting cell (APC). This is the class I-mediated response.

An intrinsic or innate response also allows the T-cells to react directly and very rapidly to the presence of foreign molecules. The presence of invasive microbes is detected via a set of toll-like receptors (TLRs) which bind substances which normally

Concepts in Biotechnology: History, Science and Business. Klaus Buchholz and John Collins
Copyright © 2010 WILEY-VCH Verlag GmbH & Co. KGaA, Weinheim
ISBN: 978-3-527-31766-0

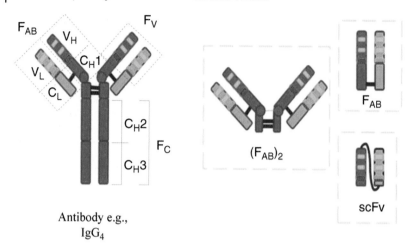

Antibody e.g.,
IgG$_4$

Figure 10.1 Antibody (immunoglobulin = Ig) structure. A typical antibody is shown on the left. Fragments generated by proteolytic cleavage (F$_{AB}$) or genetic engineering (scF$_V$) are shown on the right. The protein-folding domains derived from either the antibody light chain (subscript L) or heavy chain (subscript H), are shown in the fine-dashed boxes. Thick blue lines = S-S bridges which stabilize the dimers and link the light (grey green) and heavy chains (olive-green). scFv fragments are produced using rDNA methodology, with V$_H$ and V$_L$ fragments in the same ORF.

only occur during infection. This latter response also augments any ongoing MHC (Major Histocompatibility Complex) class I- or class II-mediated response.

10.2
The Beginnings of Applied Immunology

10.2.1
The Origin of Vaccination

In 1796 *Edward Jenner* bravely attempted to induce resistance to smallpox by injecting small amounts of material found in pustules on cows suffering from a superficially similar disease, now known as cowpox. His reasoning was simple. The girls who milked cows suffering from this disease were somehow immune to infection with smallpox from this source.

He was fortunately correct and the treatment, *vaccination* derived from the Latin *vacca* meaning cow, was highly successful, so much so that it resulted in the first complete eradication of a human disease, the declaration that it had been removed from the world's population was made in 1980. It had been a major scourge of humanity and was a key factor in eradicating entire ethnic groups including most of the indigenous population of South America.

10.2.2
Pure Culture, the Beginnings of Microbiological Science Leading to Production of Vaccines

The establishment of sterile technique and the production of pure cultures of bacteria by *Louis Pasteur* served to confirm his '*germ theory of disease*'. *Robert Koch* developed and implemented precise postulates (1891; Nobel Prize 1905) which had to be fulfilled to provide exact scientific proof that a particular bacterium was the causative agent of a particular disease. Pasteur fully appreciated the work of Jenner and the basis for vaccine production was established using pure cultures which had been killed or attenuated before use. Just how pure these vaccines were was not always certain, especially when tissue culture was used. The highly successful and widespread Salk Polio vaccine was later found to have been contaminated with the simian virus SV40, which luckily is not harmful to man. When the first transfer of genes into animal cells was being carried out by *Paul Berg* and colleagues, this was relatively well accepted as safe methodology due to the nature of the vector used, which was in fact SV40.

Products inducing immunity are known as *vaccines*. Their active constituents are the *antigens* against which the antibodies (immunoglobulin glycoproteins) are produced and to which they then bind. To enhance the immune response a cocktail of other components can be added to the vaccine and are administered together with the antigen. These substances are called *adjuvants* and act non-specifically.

Milestones in the history of immune biology can be illustrated by the Nobel laureates in this field. However, *Paul Ehrlich* is of particular importance as it was he who first proposed the existence of antibodies which bound to antigens ('side chain antibody interaction') and who developed the first toxin-neutralizing antisera (against Diphtheria toxin and snake venom). *Elie Metchnikoff* first distinguished between a humoral and a cellular immune response. This latter work essentially marked the beginnings of cellular immunology. Ehrlich and Metchnikoff jointly received the Nobel Prize in medicine in 1908.

10.3
Monoclonal Antibodies

10.3.1
A Critical Mass of Intellect and Technology Forms the Cradle for Monoclonal Antibodies

It was not mere chance that these developments had their origin in Cambridge, UK, where recombinant antibody production also began later with the work of *Sir Gregory P. Winter*. It had its roots in the fact that *Fred Sanger*, subsequent to being awarded his first Nobel Prize in Chemistry in 1958 for elucidating the structure of insulin, attracted *Cèsar Milstein* to join an already impressive group of scientists working on the most fundamental questions in molecular biology. This group included the best brains and technically gifted scientists in structural, particularly protein structural chemistry and biochemistry. Competence in the area of protein structure was paralleled only by *Linus Pauling* in the USA. As *Klaus Rajewsky* writes in his obituary to Cèsar Milstein: '(*Cèsar Milstein*) in 1963, took up Sanger's invitation to join the Medical Research Council's Laboratory of Molecular Biology (LMB), which had already attracted other top scientists from abroad such as *Max Perutz, Aaron Klug* and *Sydney Brenner*. It was on *Sanger's* advice that *Milstein* then began to study antibody structure and diversity: this was to be his life's work with LMB (Laboratory of Molecular Biology) in Cambridge, his scientific home'. The key to success appears to have depended on establishing a critical mass of bright people brought together at the *right* time and place in an incomparable academic atmosphere. They were able to address the most basic questions free of any ulterior constraints. 'Right' is defined here contextually with reference to the unique availability of investigative methods that could be used to answer these basic questions. This also includes the tradition of respecting academic freedom by supporting it in a tangible manner.

It had been realized in the 1970s that in a particular type of cancer, such as non-Hodgkin Lymphoma which involves the uncontrolled growth of differentiated (mature) B-cells, each individual's lymphoma cells produced a specific antibody. This is in fact monoclonal antibody production, but since it could not be guided to a particular specificity, was thus not of any immediate potential use. *Georges Köhler, Cèsar Milstein* and *Niels Kaj Jerne* worked together to establish a novel hybridoma technology for the production of *monoclonal* antibodies in 1975 (Nobel Prize in 1986). A key factor was the use of a technique to identify single cells that were producing a particular antibody. Köhler brought the method to Milstein's laboratory in Cambridge. The specific antigenic response was induced by injecting antigen into the mouse. Repeating doses for several weeks boosted the number of antibody-producing cells in the animal. These were then fused to immortal myeloma cells which were maintained in tissue culture [1]. After prolonged culture a relatively stable cell line was obtained that produced a single, or monoclonal, antibody.

Since specific antibodies could be induced in mice and production maintained almost indefinitely in the hybridoma cultures, this was a major advance over the previous use of polyclonal sera from (say) cattle or sheep where the exact specificity and cross-reactivity to other proteins varied from animal to animal thus making the sera unreliable.

10.3.2
High Specificity and Sensitivity of Antigen Detection

Using the hybridoma technique, monoclonal antibodies (mAbs) became a major tool for testing for the presence of a specific antigen often with an exquisite specificity that can differentiate a single amino-acid change, or distinguish between different positions and types of post-translation modification. Nucleic acid-based analytical techniques are not applicable for this purpose.

The method was made more sensitive, for example by using immuno-gold staining, the position of a single antigen molecule could be seen in a tissue section with either a light or electron microscope. Other highly sensitive methods involved using radio-actively labelled antibodies in radio-immuno assays, or fluorescent dye labelling via fluorescence techniques. Signal amplification was made possible by coupling an enzyme to the antibody (the *enzyme-linked immunosorbent assay* (ELISA)). This methodology led to the development of test kits for quantification or detection of the location of an antigen in a cell section or in a gel-blot. In combination with either two-photon methods or infra-red-labelled mAbs, antigens could be visualized in living organisms, for example just under the surface of the skin or in the body of a mouse.

10.3.3
The Antibody Molecule

The structure of a typical antibody (e.g. IgG) is shown in Figure 10.1, along with the structure of derivatives, which are screened or produced by recombinant DNA techniques. Antibody genes are put together during a somatic rearrangement of DNA cassettes which are present in the DNA of every cell, a process during which the DNA is further mutated (somatic maturation). At each step the cells that produce variant antibodies with the highest affinities are recognized and clonally amplified. The result is a specific response to foreign antigens within days. The antigen-binding site on the surface of the antibody comprises the regions of both the light- and heavy-chain variable domains which have the greatest variability (Figure 10.2).

10.3.4
rDNA Monoclonals: Reproducible Tools for the Life Sciences

The use of 'monoclonals' heralded in an era in which antibodies of exquisite and defined specificity could, for the first time, be produced indefinitely with the same quality as the original antibody batch. Their use as tools in molecular biology and as diagnostics in medicine had a qualitative effect enabling precise investigations of causality in biological systems and accelerating all areas of research in the life sciences.

10.3.5
Selection of Antibodies from Combinatorial 'Display' Libraries

Antibody expression in bacteria is not trivial. For instance it is difficult to correctly dose 1 : 1 expression of light and heavy chains and to have them correctly secreted into the periplasmic space and finally have them assemble into the light–heavy chain heterodimer. Phagemid libraries, therefore, often use a format in which the light and heavy chains are coupled in one fusion protein (single chain Fv (scFv)). If the linker between the V_L and V_H regions is long, a small variant can be obtained. Alternative structures are also possible in which for example a double specificity (two headed) molecule is created (Figure 10.6). Many antibody display libraries use synthetic DNA or natural segments, particularly from the regions that have the most dominant

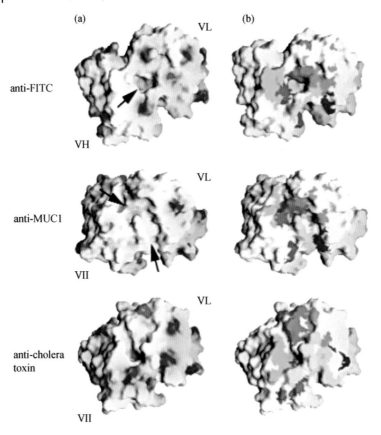

Figure 10.2 The variable structures of the surface of scFv selected from phage display libraries. The arrows indicate a pocket in the anti-FITC antibody and a groove in the anti-MUC1 antibody. On the left the red areas indicate positively charged and blue areas negatively charged surfaces. To the right the coloured regions indicate portions of the surface originating from different CDR regions [2].

effects on the interaction with antigens, namely the CDR-1, -2, and -3 regions, to give additional structural variation to the antibodies in the absence of natural *somatic maturation*. The constant regions of the light chain come in two classes, namely the *lambda* and *kappa* chains.

10.3.6
In Vitro Selection for Novel Monoclonal Antibodies

Alternative methods for producing commercial therapeutic and diagnostic antibodies, which make up a significant share of the 'new biotech' market, are discussed below. Recombinant DNA techniques with combinatorial antibody gene libraries replaced hybridomas as a lead discovery during the mid-1990s. The selection of antibodies of desired specificities were obtained using affinity enrichment techniques (see Section 10.4) on *in vitro* immobilized targets, displaying the variant proteins, on either a phage or phagemid vector, as described below. Other combinatorial methods such as yeast cell surface display or alternative *in vitro* techniques such as ribosome-display or RNA–protein-display methods have become more popular. This latter method formed the technology platform for the founding of the Company *Phylos*. In this method puromycin is coupled to the mRNA. The puromycin causes termination of translation near the end of the gene while being incorporated into the nascent peptide. In this way the peptide is covalently attached to the mRNA (or as a reverse transcript to the gene). Complexes of this sort can be enriched in cycles of affinity enrichment (on the target) and amplification (via transcription and PCR) in a process carried out entirely *in vitro*.

10.3.7
Transgenic Mice with a Human Immunoglobulin System

Transgenic mice have been created in which the entire human antibody generation machinery has replaced the equivalent mouse genes. Injecting an antigen causes the induction and selection of high affinity essentially human monoclonal antibodies in such mice. Although this does not provide the freedom to challenge specific targets *in vitro* where for instance selection can be guided to epitope-specific antibodies, it does have the advantage of being subject to *somatic maturation* steps that occur *in vivo*. This often results in higher affinity antibodies. Since the antibodies are very similar to those that would be produced in humans, this should mean that there is a reduced risk of generating an immunogenic or even an allergenic reaction to the novel antibody. This may increase the likelihood of avoiding antigenic or even allergenic products and thus of successfully completing clinical trials (see Section 10.6 for more details).

10.3.8
Antibody Classes and Antibody Functionality

During maturation of the antibody in the B-cell, the antibody gene not only undergoes recombination with in the variable F_V coding region of the antibody but also exchanges cassettes for the heavy chain constant region in a process known as 'class switching'. Table 10.1 illustrates the class types and the influence this has on the functionality of the final antibody. This relates not only to which other cells it may interact with but also to where it will end up in the body, for example it may cross the blood–brain barrier, it may terminate its journey in the skin or intestine, it may be capable of interacting with the complement cascade which will culminate in it penetrating a cell membrane and killing a cell or it may opsonize a target bacteria for ingestion by a phagocytic cell. All of these factors, particularly the half-life of the protein in the blood, will have a critical effect on the effectiveness of the antibody as a therapeutic.

10.3.9
A Summary of the Cell Biology in the Generation and 'Evolution' of a Natural Antibody

In the natural process of generating an antibody molecule, the repertoire of novel antibodies which could potentially be generated is astronomical. These processes

Table 10.1 Classes of antibodies: the carboxy terminal region of the heavy chain determines the functionality of an antibody.

Class/Subclass		IgG1	IgG2	IgG3	IgG4	IgM	IgA1	IgA2	sIgA	IgD	IgE
H chain		$\gamma1$	$\gamma2$	$\gamma3$	$\gamma4$	μ	$\alpha1$	$\alpha2$	$\alpha1/\alpha2$	δ	ε
Molecular weight (kDa)		146	146	170	146	970	160	160	385	184	188
Serum concentration (mg/ml)		9	3	1	0.5	1.5	3	0.5	0.05	0.03	0.00005
Half-life (day)		21	20	7	21	10	6	6	7	3	2
Complement binding		++	+	+++	–	+++	–	–	–	–	–
Fc receptor binding	Fc γ R	I, II, III	(IIa)	I, II, III	I	+++	–	–	–	–	–
	Fc α R	–	–	–	–	–	+++	+++	+++	–	–
	Fc ε R	–	–	–	–	–	–	–	–	–	I, II
Placenta permeability		+	+	+	+	–	–	–	–	–	–

IgE is involved in eliciting allergic responses and IgA is located in the skin and mucosal layers. IgM and IgD classes are produced during the early stages of development of the antibody-producing B-cells before extensive somatic maturation of the antibody variable regions has taken place. Class switching can occur by exchange of the gene segments encoding the constant region of the antibody heavy chain. The light chain constant regions derive from two families of genes, designated kappa ($\varkappa$) and lambda (λ).

include combining variable cassettes, error-prone cassette fusion and the somatic maturation process. The antibodies thus produced are considered to be sufficiently wide ranging to generate an antibody which can bind (or 'recognize') essentially any foreign substance. During the 'evolution' of a novel antibody an initially small pool of weakly binding candidates is subjected to mutation and clonal amplification. This is a critical step in the successful functioning of the system. Any cell generating an antibody of improved affinity will be subject to strong positive selection. This is ensured in an animal by the interplay of T-helper and B-cells via a positive feedback loop which depends on the efficient binding and uptake into the cell of the antigens by the antibodies. Peptides are generated from these antigens for presentation to the T-cells by binding the MHC (major histocompatibility) complex. The T-B-cell interaction in turn causes lymphokines to be produced which stimulate B-cell growth and differentiation.

10.3.10
Combinatorial Display Mimics Natural Antibody Generation and Maturation

Combinatorial display methods attempt to mimic this process, for example by using bacterial expression and phage-display in entirely *in vitro* compartmentalization strategies or in yeast cell surface display expression; iterative rounds of *affinity-based enrichment procedures* followed by *amplification* are always involved. Some systems also include interspersed *specific mutation* and *cassette recombination*. These are in sum analogous to the natural selection and maturation procedure and are discussed in more detail in the following section.

10.4
Producing Antibodies via rDNA and Combinatorial Biology

The rearrangement of gene cassettes as occurs in the mammalian immune system, serves as 'the model' for all man-made strategies for utilizing combinatorial biology techniques in biotechnology. The empirical approach via combinatorial display methods may be an anathema to purists who might prefer structure-prediction and protein design but it is, however, highly successful.

A number of biotech companies incorporated or were founded on this technology by pioneers in the field: in the case of antibodies, the pioneers and associated companies were *Sir Gregory P. Winter* (CAT, Cambridge Antibody Technology, Cambridge, UK; and *Domantis*, Cambridge, UK), *Hennie R. Hoogenboom* (*Dyax* and *Ablynx*, Ghent, Belgium), and *Simon Moroney* and *Andreas Plückthun* (*Morphosys*, Munich); and in the case of other smaller ligands (later antibodies also), *Willem 'Pim' Stemmer* (*Maxygen*, Redwood, California; based on DNA shuffling; later subsidiaries being *Codexis* and *Verdia*; also co-founded *Amunix, Evolva* and *Avidia*), *Bob Ladner* (Dyax, Boston, based on phage display), *John Collins* (*Cosmix*, Braunschweig, Germany; based on cosmix-plexing), *Peter Wagner* (*Phylos*, Frankfurt (later Boston), based on combinatorial RNA–peptide fusion *in vitro* technology), *Sven Klussmann*

(*Noxxon*, Berlin, based on SELEX selection on enantiomer targets), *Arne Skerra* (*Pieris Proteolab AG*, Freising-Weihenstephan, Germany; based on Anticalin® display) and *Andreas Plückthun* (*Molecular Partners AG*, Munich, Germany; based on DARPin® technology).

Combinatorial biology methods can be seen as having their origin in the development of a system in which the genetic material (DNA) is physically coupled to the gene product. This was first demonstrated by *G.P. Smith* in the USA and in parallel by *Valery H. Petrenkov* in Russia, with the filamentous bacteriophage (phage) M13. A cassette for a foreign protein was inserted into the gene region encoding the amino-terminal region of the virus' minor coat protein gpIII (gp3; Figure 10.3). It was shown that the hybrid protein was presented on the surface of the phage. Extending this finding to inserting variable oligonucleotide cassettes created collections of clones (the assemblage of clones being referred to as a 'library' of variants) where each phage particle carried one of the many variant proteins. Since each phage clone originates from the infection and replication of a single cell (compartmentalization), phage particles are produced which carry the gene for the specific variant which is displayed on the surface of the cell. This method is therefore known as '*phage-display*'.

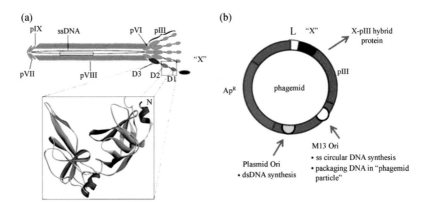

Figure 10.3 Phagemid display vector: (a) The M13 bacteriophage requires the minor pIII protein to infect its host, *Escherichia coli*. (b) A phagemid is shown schematically on the right (circular double stranded DNA); foreign DNA ('X') has been inserted between the leader sequence and D1 domain of the pIII protein, a hybrid X-pIII protein is thus produced [6]. The functional pIII-D3 domain allows the assembly of these hybrids into the bacteriophage particle so that the foreign protein or peptide is 'displayed' in low copy number on the surface of the particle. When the cell is super-infected with an M13 helper phage, all the normal bacteriophage products required for assembly of a virus-like particle are produced. The phagemid DNA is also replicated and packaged as if it were a viral DNA. Abbreviations: Ap^R, antibiotic resistance; L, leader peptide which transports the N-terminus of the X-pIII hybrid protein through the cell membrane [7]; pIII, pVI-IX, M13 coat proteins 3 and 6 through 9; 'X', foreign DNA; D1 to D3, the three domains of the pIII protein. The insert shows a ribbon diagram of the pIII-D1 and -D2 domain structure (from Holliger [8]).

10.5

Affinity Enrichment on Surfaces of Immobilized Target Molecules

Applying affinity selection where a specific target is immobilized on a bead or surface enables phage particles which display variant peptides with affinity for the target to be enriched. These can be grown up in large quantities in a new *E. coli* culture. Cycles of enrichment and amplification are iterated and after three or four rounds, peptides or proteins with affinities in the medium to low nanomolar range can be obtained (Figure 10.4). Using the monovalent phagemid-display system and introducing

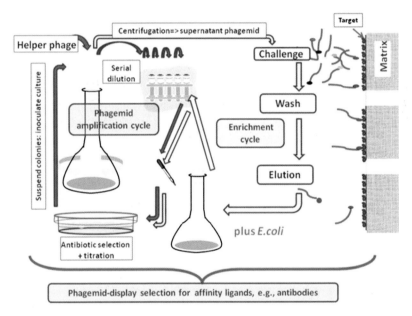

Figure 10.4 Phagemid display is achieved by iterative cycles of affinity selection on a target protein immobilized for example on Sephadex beads. Bound phagemid particles are released on addition of a competitor or by denaturing the target. The eluted particles can transfer (transduce/inject) their single stranded DNA into *Escherichia coli* (F-plus) where the phagemid DNA then continues to replicate as double stranded DNA and can be stably inherited as a plasmid. If the culture is infected with a 'helper bacteriophage' (e.g. M13 K07) the phagemid DNA is again packaged into phage coats and 'sweated' out of the cell into the medium. Phagemid particles are contained within the culture medium. Phagemid-containing colonies are selected on antibiotic-containing medium. After some four to six rounds of selection clones will usually be found which have a specific affinity for the target molecule in the mid-nanomolar range. Clones are sequenced and further characterized, for instance by plasmon resonance, to measure the actual affinity of the ligand. The success of the method depends on the initial population of phagemid being highly variable for a peptide or protein 'displayed' as an extension of one of the minor proteins of the bacteriophage coat. Such a collection of phagemids, known as a library, is constructed by recombinant DNA techniques to display either peptides or larger proteins such as antibodies. This methodological approach is also known as 'panning'.

further variation through site-specific mutation enables the selection of improved affinities. The development of strong protease inhibitor variants of various specificities by alteration of the human pancreatic secretory trypsin inhibitor using firstly protein design and later combinatorial phagemid-display libraries (PSTI; Collins' group, Germany [4]) is one early example of the use of this strategy. A similar project was carried out with bovine secretory trypsin inhibitor using combinatorial phage display (BSTI; Ladner's group USA [5]). Both projects yielded low picomolar inhibitors of human leukocyte elastase.

Fusion of a plasmid to a small viral fragment which carries the viral replication start (origin) produces plasmids which can be packaged into a viral coat and 'sweated' into the culture medium after the necessary viral functions have been provided by a helper-phage. This is used to infect the cell containing the plasmid. Such a plasmid, packagable in a bacteriophage coat is termed a phagemid. In a manner similar to that described for a phage, a pIII gene with variable cassettes can be inserted to produce a *phagemid-display* vector. Phagemid DNA replication is more robust and deletions are less frequently observed than with phage vectors.

The cell acts as a single compartment in which DNA of only one clonal type is produced. This yields particles in which the gene for a particular variant is physically linked to its variant gene product. Affinity selection of particles which can bind a particular target, selects both displayed product and the gene encoding it. This can be easily sequenced and the product readily defined. Protein modelling can be carried out and further variants designed.

It should be noted that the display requires secretion of the pIII–protein hybrid through the cell membrane and as such is dependent on a leader-peptide. The variant portion of the hybrid peptide is normally presented at the N-terminus of the hybrid.

10.5.1
Other 'Display' Scaffolds; Peptides to Knottins

Peptide display was the technological platform that led to the founding of three companies: *Dyax* in Boston by *Robert C. Ladner* in 1989 with the additional introduction of novel combinatorial DNA methodology, *Cosmix* in Braunschweig founded by *John Collins* and *Maxygen*, San Diego, founded by *Willem 'Pim' Stemmer* in 1997. The technology was also later implemented in larger pharmaceutical companies, notably by *Sachev Sidhu* at *Genentech* and *Peter S. Kim* at *Merck*. Ladner's controversial patents were defended aggressively by *Dyax*, Cambridge MA, contributing to some of the turmoil and uncertainty surrounding the new biotechnology in the late nineties (see more on intellectual property in a separate section).

The use of various peptide or protein scaffolds in combinatorial librairies has recently been reviewed (Figure 10.5) [9]. It can be seen that these represent a range of molecular structures mostly with a fairly rigid core or backbone and variable loops. The anticalin β-barrel structure is particularly suitable for developing high affinity ligands for small organic molecules. The only other molecules

Affibody   **ImmE7**   **Cytochrome b$_{562}$**   **Ankyrin repeat**

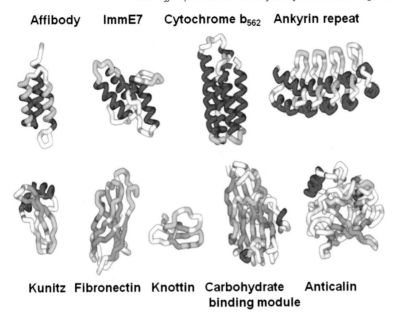

Kunitz **Fibronectin** **Knottin** **Carbohydrate**   **Anticalin**
binding module

Figure 10.5 Alternative combinatorial 'display'-library scaffolds.

capable of forming pockets for smaller molecules are the larger structures such as antibodies or antibody fragments containing 'six-fingered' loops.

10.5.2
Fusion to other Minor Coat Proteins

Other phage-display variants produced using an engineered viral gpIX protein can display peptides or proteins fused to its carboxy terminus. This is of course of interest not only for the study of those proteins that require a free carboxy terminus for function but also for more efficient screening to identify the functional binding of proteins encoded in cDNAs after random cloning of sheared fragments. The reason for this is that the cDNA insert is only required to be in the same (correct) open reading frame where it is fused to the gpIX protein (one third of the cases) as opposed to insertions (e.g. into gpIII) where the open reading frame must be the same at both fusion points (1/9 of cases) [10].

10.5.3
Combinatorial Library Affinity Selection: Alternatives to Phage Display

In vitro expression and selection systems have largely replaced phage-display technology in commercial settings despite the fact that they are more expensive, technically demanding and require a dedicated task-force to maintain operations; it is for these reasons that these systems are not generally used in academic laboratories. All the *in vitro* systems have the advantage of being able to produce a larger repertoire of different molecules compared to the phage-display system, the maximum library size of which is some 10^{10} variants.

One method, '*ribosome display*', relies on maintaining an intact complex of ribosome and mRNA with the nascent peptide still physically attached.

A second *in vitro* system developed at *Phylos* which was originally located in Germany, utilizes the covalent linkage of the nascent peptide to mRNA which is converted to an RNA–DNA hybrid or dsDNA molecule complex with the peptide attached.

Finally, *Andrew D. Griffiths* and *Dan S. Tawfik* developed an *in vitro* compartmentalization system (aqueous-oil emulsion) [11] in which during *in vitro* protein synthesis of a DNA-methylase peptide hybrid gene template, a covalent attachment occurs between the 5-flouro-cytosine-modified template DNA and the restriction methylase hybrid. The physical linkage of the genetic material to the protein product is analogous to phage-display. Peptides of specific affinities for defined targets can be obtained from cycles of affinity enrichment on immobilized targets followed by PCR amplification of the templates and repeated *in vitro* transcription/translation cycles.

Anaptys Biosciences Inc. (San Diego) uses a novel combination of rDNA technology combined with the natural process of site-specific recombination within the variable regions of the antibody variable gene segment, namely *somatic cell hypermutation*.

10.5.4
Pros and Cons of Microbial versus Animal Cell Systems for Selection and Production

Why use microbial systems rather than animals for the selection of the optimal binding partner for a target molecule? First, the requirements of efficient secretion through the cell membrane, correct folding and stability of the backbone scaffold need to be addressed. Once this has been achieved it is technically easier to produce libraries of larger numbers of variants in bacterial and bacteriophage (phage-display) systems than in yeast or in animal cell lines.

In vitro ribosome-display technology has become the dominant methodology partly because of the very large number of molecules that can be added to the library.

Transgenic mice in which the antibody gene repertoire has replaced the murine gene equivalents (see Section 10.6) is potentially an even stronger competitor since it can be used to rapidly produce a *highly effective* (high affinity) antibody of essentially human type. Before production the B-cell clones should either be identified and/or both the heavy and light chains cloned for optimized antibody production. The main variable in the choice of cell line for use in production is the degree and type of post-translational modification required, in particular glycosylation. This will affect the antigenicity, solubility and half-life of the antibody in the medium.

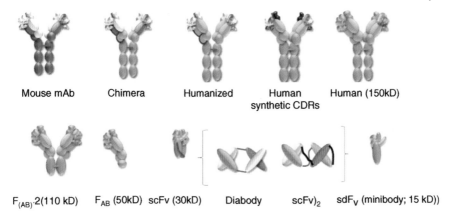

| Mouse mAb | Chimera | Humanized | Human synthetic CDRs | Human (150kD) |

F$_{(AB)}$·2(110 kD)   F$_{AB}$ (50kD)   scFv (30kD)   Diabody   scFv)$_2$ sdF$_V$ (minibody; 15 kD))

Figure 10.6 Forms of antibodies and antibody fragments developed for clinical use many of which are referred to in the compilation of antibodies (Table 10.2) which were still in clinical trials up until early 2009. Some are of murine origin, some of human origin and some hybrid or chimaeric. The theory is that the more 'human' the antibody is, the less likely it is to be immunogenic. It should however be noted that each antibody is a unique substance and that there is vast divergence in the human population which makes it difficult to predict how such a protein will be recognized by the immune system.

A number of alternative formats are used for antibody products, particularly for those which have been developed by selection in microbial systems. A selection of the main variants is shown in Figure 10.6.

10.6
Mice with Human Antibody Gene Repertoires

A project was commenced in the early 1990s at *Medarex* to replace the mouse antibody gene repertoire with human homologues. This was carried out in a series of steps including a close partnership with the pharmaceutical division of the *Kirin Breweries*, Japan. The Medarex 'HuMAb transgenic mouse' has a full repertoire of light and heavy chain variable regions with a limited range of non-variable human regions [12]. Kirin Breweries had produced a 'Kirin C mouse' in which the human variable cassettes replaced the mouse antibody genes and which was able to engage in the full range of class-switching with all human heavy chain isotypes.

The two companies collaborated in the crossing of these two transgenic mice to produce the '*KM mouse*' which essentially produces a humoral immune response which is very close to that which would be achieved in a human being. Since the sequences are derived from human DNA cassettes, the products should be less likely to provoke an antigenic or, worse, an allergenic response.

It seems to the authors of this book that this is a rather simplistic viewpoint. However elegant the technology may be, an antibody produced by one individual always has the potential to be antigenic when injected into the blood of another individual who may have a different MHC (HLA) repertoire.

The advantages of this technology are that the whole repertoire of somatic hypermutation during B-cell maturation and clonal expansion can be made available, thus facilitating the production of antibodies with very high affinities for their antigens (10^{-12} M has been cited for some products).

In addition, once the mouse strains have been well established, they can be maintained and used at low cost for the initial production and selection of essentially human antibodies of any desired specificity. As with the other technologies the final antibody production will still be carried out in either recombinant microorganisms or preferentially in immortalized human cell culture.

At the time of writing, one of the latest products of this technology, Golimumab (see Table 10.2) has been approved for the treatment of inflammatory diseases such as psoriasis, rheumatic arthritis and ankylating spondylitis. It joins four other recombinant antibodies which are directed against TNF-α. In 2008, the market value of the existing antibodies Infliximab, Etanercept, Adalimimab and Certolizumab pegol was a substantial $US 13.5 billion [13].

10.7
Two Severe Setbacks during Clinical Testing

One of the most dramatic developments in the monoclonal antibody field was the *withdrawal* of *Tysabri* in 2005 after it had been approved in 2003 for the treatment of multiple sclerosis. It is a humanized ($IgG_{4\varkappa}$) monoclonal antibody derived from murine myeloma cells that is directed against integrins containing $V\alpha4$ chains, which are involved in either transport of leukocytes across the blood–brain barrier or guiding the lymphocytes to sites of inflammation.

The withdrawal followed the occurrence of three cases of PML (progressive multifocal leukoencephalopathy) which resulted in two deaths. PML is a rare, but serious viral infection of the brain. The FDA put the clinical trials on hold in February 2005, but then allowed limited continuation under highly controlled conditions with intense monitoring for potential signs of PML. In January 2008 the FDA gave restricted approval for the use of *Tysabri* in the treatment of the inflammatory disease Morbus Crohn, whereby distribution was only allowed through TOUCH programme centres in order to identify risk factors which might be used for the early detection of patients at high risk of developing PML.

A further event which caused grave concern and public outcry over the lack of diligence in designing experimental evaluation before entering clinical trials occurred on 13 March 2006 in London. *Parexel*, a contract clinical testing company was carrying out the first routine Phase I 'toxicity' tests on low doses of the new humanized monoclonal anti-CD28 drug, TGN1412 which was the primary product manufactured by a small 12-man gene biotech company in Würzburg, Germany.

Table 10.2 Clinical applications of recombinant human or humanized/chimaeric antibodies.

Name(s)	Autoimmunity/inflammatory disease			
	Type	Company	Target/mode of action	Application	Clinical development

Name(s)	Type	Company	Target/mode of action	Application	Clinical development
Inflammatory disease					
ReoPro; Abciximab	F$_{AB}$	Centocor/Lilly	Gp IIb/IIIa, integrin	Complications in coronary angioplasty	FDA approval December 1994 *First Fab approved*
Remicade; Infliximab; anti-TNF-α-cA2	Chimaeric mu/hu	Centocor/J&J	TNF-α	Morbus Crohn	FDA approval 29 August 1998
				MTX resistant rheumatoid arthritis (+ MTX)	FDA approval 10 November 1999
Enbrel; Etanercept	TNF-receptor–IgG1 hybrid	Amgen (via Amgen (Immunex) acquisition)	Blocks TNF action	Autoimmune disease, arthritis; psoriasis etc.	FDA approved May 2006 (Later mandatory warning about latent tuberculosis)
Humira; Adalimumab; D2E7	Human IgG1	CAT (with Abbot)	Binds and neutralizes TNFα	Rheumatoid arthritis, reducing inflammation in autoimmune disease	First approved in December 2002 >51 countries (2005)
				Crohn's disease	Phase II, 2004
				Multiple sclerosis	Phase II, 2004
Tysabri; Natalizumab; Antegran	Humanized IgG4x from murine myeloma cells	Biogen Idec/Elan (PDL)	Vα4 integrin. Prevents leukocytes crossing the blood–brain barrier. α4β1 and α4β7 integrins guide lymphocytes to inflammation sites	Multiple sclerosis. (solid tumors??), inflammation	Approved 23 November 2004 in the US. *First mAb for Multiple sclerosis* ** (MS) ** Withdrawn following three cases of PML (two fatal); other trials terminated

(*Continued*)

Table 10.2 (*Continued*)

		Autoimmunity/inflammatory disease			
Name(s)	Type	Company	Target/mode of action	Application	Clinical development
Tysabri				Morbus Crohn	Failed first Crohn disease prevention/phase III for maintaining remission $+/-$ interferon β. Finally FDA approved January 2008 for Crohn's disease: restricted distribution through the TOUCH program.
Golimumab; CNTO 148	Human (from immune-humanized mice); IgG$_1$-κ	Centocor/Medarex	Anti-TNF-β	Inflammatory diseases	FDA approved April 2009 for treatment of rheumatoid arthritis, psoriasis, ankylosing spondylitis
Ustekinumab; CNTO 1275	Bivalent	Centocor/Medarex	Anti-IL12/IL23	Inflammatory diseases	FDA approved for treatment of psoriasis June 2008. EC Approved 1. May 2009
Certolizumab pegol; Cimzia	Humanized F$_{AB}'$ fragment double pegylated	USB Inc.	Anti-tumour necrosis factor alpha (TNF-α) (inhibitor)	Morbus Crohn. Inflammatory bowel disease	Approved April 2008 for Crohn's disease. *First pegylated antibody fragment approved*
Tocilizumab; Actemra; Roactemra	Humanized	Chugai/Roche	Anti-IL-6-receptor	Ulcerative colitis. Rheumatoid arthritis; Systemic lupus erythromatosis. Castleman syndrome	BLA approved in 2008 in the US; Approved 2008 in Japan. Approved in Japan in 2005

Cancer/Neoplasms

Name	Type	Company	Target	Indication	Notes
Rituxan; Rituximab	Chimaeric	Biogen Idec	B-cell surface CD20	Non-Hodgkin lymphoma (NHL) originally after failed chemotherapy now as monotherapy	FDA approval November 1997. Approved in over 70 countries; $1.5 billion in sales in 2003
Herceptin; Trastuzumab	Human rDNA	Genentech (PDL)	HER2-overexpressing cancer cells	Metastatic breast cancer combined with chemotherapy	FDA approval September 1998. *First target-based agent for breast cancer*
bH1-44	Variant of Trastuzumab	In development at Genentech	Double target affinity! Binds ERBB2 and VEGFA	On cultured tumour cells has similar effect to either Herceptin or bevacizumab (Avatin)[§§]	First engineered antibody showing double specificity, largely separated in the light and heavy Ig chains respectively
Mylotarg; Gemtuzumab ozogamicin	Humanized- linked to calicheamicin	Wyeth/American Home (PDL)	CD33 (on myeloid but not stem cells)	Acute myeloid leukaemia (patients over 60 years after failed chemotherapy	FDA approval May 2000. *First conjugated mAb approved*
Campath;Alemtuzumab; Campath-1H; LDP-03	Humanized IgG1κ	Millenium/Ilex/Berlex	CD52 on B- and T-lymphocytes (21 to 29 kDa glycoprotein)	Chronic B-cell leukaemia after failed chemotherapy	FDA approved July 2001
Zevalin; Ibritumomab tiuxetan	Mouse; radiolabelled Yt-90 or In-111	Biogen Idec/Schering AG	CD20 on B cells; tiuxetan is a linker chelator for attachment of the radionuclide	Non-Hodgkin lymphoma (NHL)	Approved for B-cell lymphomas 19 February 2002 (*First radio-conjugated-targeted reagent approved for cancer!*)

(Continued)

Table 10.2 (*Continued*)

Name(s)	Type	Company	Target/mode of action	Application	Clinical development
Autoimmunity/inflammatory disease					
Bexxar; Tositumomab	Mouse Ab labelled with I-131	Corixa	Anti-CD20 on B cells	Non-Hodgkin lymphoma (NHL)	Approved June 2003 for NHL refractory to rituximab
Erbitux; Cetuximab; αEGFmAb; IMC-C225	Chimaeric	Weizmann Institute, Israel ⇒ ImClone and BMS	EGFR-1 (flt-1); binds extracellular domain	Receptor present on colorectal, breast and other cancers	FDA approval for treatment of metastatic colorectal cancer: 12 February 2004
	IMC-11F8 derived from mouse mAb 225				Patent on antibodies to EGFR + chemotherapy
Avastin; Bevacizumab	Humanized mAb	Genentech (PDL)	VEGF-α binding blocks activity	Prolongs life about 5 months in metastatic colorectal cancer	FDA approved January 2005 for treatment of colon cancer. (*First antiangiogenic agent to show prolonged survival in cancer*).
1D09C3	Human	GPC Biotech/Morphosys	B-cell surface MHC-classII (HuCal library isolate)	Synergy with Rituxan in an animal NHL model	Phase I, started February 2005
				CLL	*First antibody from the HuCal library (Morphosys)* Orphan drug status
Vectibix; ABX-EGF; panitumumab	Human (ex-XenoMouse) IgG2κ	Abgenix + Amgen (Immunex)(Amgen)	Anti-EGF receptor	Anti-angiogensis; anti-solid tumour (prostate, colorectal; renal etc.	FDA approval September 2006
Infectious disease					
Synagis; Palivicumab	Humanized	Medimmune(PDL)	Respiratory syncytial virus	Treating RSV infections	FDA approved June 1998

TGN1412	Humanized	TeGenero. Boehringer Ingelheim (manufacturer)	AntiCD-28	Septic shock treatment	Phase I: six healthy probands became very ill. *Radical re-evaluation of preclinical testing*
Graft rejection					
Orthoclone OKT3; Muromonab-CD3	Mouse	Johnson & Johnson	CD3 on T 'killer' cells	Allograft rejection	Approved June 1986 First monoclonal approved
Zenapax; Daclizumab	Humanized IgG1	Protein Design Labs-> Hoffman La Roche	CD25(p55α, or Tac) subunit of the IL-2-receptor present on activated lymphocytes	Preventing acute organ rejection in kidney transplants	Approved December 1997 *First humanized mAb approved*
Other indications					
Lucentis; Ranibizumab (injection)	Humanized mouse mAb IgG1fragment for intraocular injection	XOMA/Genentech	Anti-VEGF-α	Wet age-related macular degeneration.	Approved July 2007
Bapineuzumab; AAB-001; 3D-6	Humanized mAb	ELAN, Wyeth	β-Amyloid	Degenerative brain disease (Alzheimer's) in non-ApoE4 patients	Trial stopped 2008, new phase III for treatment of Alzheimer's disease restarted June 2009

Tragically all six healthy subjects who had been paid to act as volunteers suffered disastrous illnesses within hours of administration of the antibody. It is unclear how many of them will suffer long-term effects. Pain and swelling occurred almost immediately in the first subject, but regardless of this reaction all six men were injected within a short period of time. This procedure cannot be considered to be in any way rational for the initial testing of a drug in humans and is totally reprehensible, and for this reason the company must be reprimanded.

In view of the analysis below, it can be said in retrospect that the evolutionary relatedness of man to the animals used in preclinical trials must be taken into account at the molecular level. Many human, dog, rat, horse, cattle, rabbit, monkey and mouse genomes have been completely sequenced and although mouse and rat CD28 (a receptor on the surface of T-helper cells that aid T-B cell interaction) share 93% sequence identity in the extracellular moiety of the molecule, human CD28 is only 65% identical to the mouse sequence. Tests on *human T-cells* do not seem to have been carried out in advance.

Other studies had already established that the stimulation of CD28 contributes to the lethality of super-antigen toxins which occurs via the deregulation of the T-helper-1 cells, leading to an overproduction of lymphokines and ultimately to organ failure. It was also already known that such a 'cytokine storm' could be induced in a rat model by certain antibodies that were agonistic to CD28 activation.

The CTLA4-ligand-IgG conjugate *Orencia* (manufactured by *Abatacept* and approved in 2005) had been validated and approved as an anti-inflammatory agent for autoimmune diseases such as rheumatoid arthritis and Morbus Crohn. It functioned by competing with CD28 for interaction with CD80 or CD86 on various T-cells. This latter reaction represents the second stimulus during the normal interaction between a T-cell and an antigen-presenting cell. There should, therefore, have been ample warning that there was a potential danger in developing an antibody to CD28 and particularly in the situation where sepsis was already involved.

Nevertheless, this example emphasizes the need for care in planning preclinical testing and the necessity of choosing animal models that have been adequately validated.

10.8
A Survey of Therapeutic Antibodies

Table 10.2 gives an overview of some of the antibodies which have been successfully developed for clinical applications. This gives an impression of the range of targets which have been addressed and documents the chronological development of milestones in this area. A first appraisal indicates that there was considerable competition for a relatively small number of targets, considering the wealth of information pertaining to potential targets that had been obtained during the post-human-genome-sequence era of molecular medicine. The large number of antibodies which have failed to be accepted for clinical use due to their lack of therapeutic efficacy or adverse reactions detected during clinical trials have not been listed in

Table 10.2, although the failure rate is perhaps no worse than the average for therapeutic agents in general.

The antibody therapeutics in Table 10.2 are presented with respect to three main aspects: (i) the type of antibody developed, which is limited or determined by the technology used; (ii) the application area and, if known, the specific target and (iii) the time-line for development during clinical trials.

At the time of writing there were 23 therapeutic monoclonal antibodies which had achieved FDA approval. In spite of some dramatic setbacks as described above, the development of monoclonal antibodies can still be regarded as one of the most successful and established applications of genetic engineering to the production of biotechnological products. This can be illustrated by the following figures: the global sales were estimated by one biotech advisor, Krishan Maggon, to be $33 billion (milliard) in 2008 as compared to $27 billion in 2007. *Remicade* was the market leader with sales of $6.5 billion in 2008, followed by *Rituxan, Herceptin, Avastin* and *Humira*. There were eight monoclonal antibodies with sales of $1 billion or more in 2008. That was a very conservative year for the approval of new antibodies: the FDA approved only *Cimzia* manufactured by *UCB* and *Tysabri* produced by *Biogen Idec* for an additional indication for Crohn's disease. *Genentech/Roche* was the market leader producing four blockbuster antibody drugs, followed by *J&J* and *Abbott*.

As shown in Figure 10.6 monoclonal antibodies can be classified according to their origin and the subsequent recombinant DNA alterations that may have been made, particularly with respect to the degree of homology that they have with human germ-line sequences:

- Mouse monoclonal antibodies originally produced in mice and isolated via hybridoma technology.
- Humanized monoclonal mouse antibodies, in which the original cloned mouse monoclonal antibody gene has had most sections encoding the non-variable portions of the immunoglobulin replaced by human genomic sequences.
- Fully human monoclonal antibodies, usually originally selected in transgenic mice in which the complete set of mouse immunoglobulin gene cassettes has been replaced with their human equivalents.
- Fragments of antibodies; scFv, F_{ab}, diabodies, single domain antibodies (dAbs™) and scFv dimers are examples of this group.

In addition hybrid molecules (chimaera) have been generated in which a non-immunoglobulin region determines the binding specificity and the cell killing action is mediated by the immunoglobulin-constant region.

A short summary of the clinical uses of recombinant antibodies:

- In 1986 the very first monoclonal antibody was approved for clinical use: the *OKT3* mouse monoclonal which recognizes CD3 on the surface of T-killer cells. The reduction of killer cell activity reduces the body's potential for allograft rejection. A total of four antibodies have been approved for this type of application, including the first humanized mAb, *Zanapax* which was approved in 1997.

- The first F_{AB}-fragment to receive approval for clinical use was *ReoPro* in 1994, for reducing the risk of heart attack during heart operations in atherosclerosis patients. This application is associated with the largest disease group for which monoclonal antibodies have been approved, namely that of autoimmunity and inflammatory disease.
- Another main area of application is the treatment of neoplasms (cancers, leukaemia) as exemplified by *Herceptin*, the first target-based antibody for cancer which was approved in 1998 for the treatment of breast cancer.
- In the area of infectious disease only one mAb has been approved. This is *Synagis*, approved in 1998 for the treatment of Respiratory Syncytial Virus (RSV).

References

1 Schwaber, J. and Cohen, E.P. (1973) *Nature*, **244**, 444–447.

2 Söderling, E. *et al.* (2000) *Nature Biotechnology*, **18**, 852–856.

3 Röttgen, P. and Collins, J. (1995) *Gene*, **164**, 243–250.

4 Collins, J., *et al.* (07.01.1987) Patent Nr. EGB8700204; Applicant, GBF, Braunschweig; Bayer AG, Wuppertal, Germany

5 Roberts, B.L., et al. (1992), Gene, 121, 9–15

6 Smith, G.P. and Petrenko, V.A. (1997) *Chemical Reviews*, **97** (2), 391–410.

7 Modified after Glökler, J. (Ph. D. thesis, 2000, TU Braunschweig).

8 Holliger, P., Riechmann, L., and Williams, R.L. (1999) *Journal of Molecular Biology*, **288**, 649–657.

9 Hosse, R.J., Rothe, A., and Power, B.E., (2006) A new generation of protein display scaffolds for molecular recognition. *Protein Science*, **15**, 14–27.

10 Weiss, G.A. and Sidhu, S.S., (2000) *Journal of Molecular Biology*, **300**, 213; Collins, J. (1997) Phage display. *Journal of Biotechnology: Reviews in Molecular Biotechnology*, **1**, 210–262 (Elsevier); Kehoe, J.W. and Kay, B.K. (2005) *Chemical Reviews*, **105** (11), 4056–4072. doi: 10.1021/cr000261r

11 Tawfik, D.S. and Griffiths, A.D. (1998) *Nature Biotechnology*, **16**, 652–656.

12 Lonberg, N. *et al.* (1994) *Nature*, **368**, 856–859.

13 Pappas, D.A. and Bachon, J.M. (2009) *Nature Reviews Drug Discovery*, **8**, 695–696.

11
Hereditary Disease and Human Genome Analysis

11.1
Introduction

Genes passed from parents to their children can influence their likelihood of developing or being resistant to disease. Morbus Crohn, Huntington's disease and various 'bleeding' syndromes in which there is a deficiency in blood coagulation factors are well known examples of a genetic burden passed down through the generations. Medical diagnosis including analysis of the genetic contribution which is provided by genome analysis is a rapidly growing branch of biotechnology. In addition it contributes to the discovery of novel targets and the development of pharmaceuticals for both intervention and prophylaxis.

11.2
Heredity Studies and Family Counselling

Genetic counselling services were made available to those who had a family history of some relatively rare disease suspected of being hereditary in origin. The results often determine reproductive decisions particularly in the case of debilitating disease. More recently prophylaxis has become an important factor where a propensity (predisposition) for a disease is involved. Prior to the gene sequencing (PCR) era, diagnosis could sometimes be carried out by blood protein analysis of family members, or particularly relevant to newborn or even prenatal examinations, by examining the chromosomes in the white blood cells (karyotyping) for large scale abnormalities, such as having three copies of chromosome 21 which is the main cause of Down's syndrome. *Linus Pauling's* discovery of the protein alteration in sickle cell anaemia in 1943 (Section 7.1) led to the first test for a "molecular disease" [1].

Concepts in Biotechnology: History, Science and Business. Klaus Buchholz and John Collins
Copyright © 2010 WILEY-VCH Verlag GmbH & Co. KGaA, Weinheim
ISBN: 978-3-527-31766-0

11.2.1
The Beginnings of the Study of Hereditary Factors: Altered Proteins

Some hereditary traits have been recognized for centuries, for instance the taste of honey in the urine of individuals suffering from *diabetes mellitus* and maple syrup disease, black urine in those suffering from phenylketonuria and the salty sweat produced from the forehead in cystic fibrosis patients. The first molecular studies examined alterations at the protein level. *Oliver Smithies* (see Figure 12.2) introduced zone electrophoresis using starch gels as an improvement over the previous paper chromatographic separations for protein analysis. In 1955 this was applied to the study of blood protein disease.

The association between protein variants and genetic disease was discovered in the following chain of events: *Linus Pauling* showed that sickle cell anaemia globin had an altered electric charge (pI) [1]. *Max Perutz* had shown that deoxy-haemoglobin in sickle-cell anaemia patients formed crystals in affected red blood cells [2] (this causes some of the symptoms of the disease). He showed this by using polarisation micros copy and comparing the data with his data on pure deoxyhaemoglobin crystals which he had used in his X-ray crystallographic studies. *Vernon Ingram* who worked in Perutz's laboratory in 1957, showed that the composition of a single peptide fragment which was generated on protease digestion of the globin and could be separated by 2-D paper chromatography (a method that *Fred Sanger* had used to analyse and sequence insulin a few years previously), was altered in samples of sickle cell globin originally brought to the laboratory by *Tony Allison*. Allison had shown that carriers of the sickle cell trait were resistant to malaria. These developments can be seen as the first steps in analysing the molecular mechanisms involved in heritable disease. *Smithies'* technology could be simply applied to genetic studies of protein alterations in thallasaemias and other inherited traits [3].

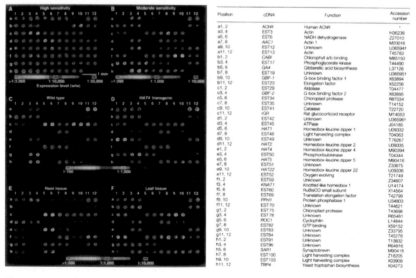

Figure 11.1 Early in 1995, gene specific cDNA-spot microarray was hybridized with fluorescein-labelled cDNA from a named source [18].

The method was improved by the development of polyacrylamide as a gel substance for protein electrophoresis (*S. Raymond* and *L. Weintraub*, 1959 [4]). In polyacrylamide gel electrophoresis separation of the molecules is based on size filtration in one dimension. Combining this method with a subsequent separation based on total protein charge (pI) led to the development of a powerful technique called 2-D electrophoresis which enables the separation of many thousands of proteins in a relatively short time. This is used for comparison of proteins from two individuals or for comparing protein composition in normal and diseased tissue. 2-D electrophoresis has been largely replaced by transcript analysis using DNA chips for this purpose (Figure 11.1); this methodology falls under the general heading of 'proteome and transcriptome research' (see Section 11.4).

11.2.2
Predisposition for Disease: Quantitative Traits

Genetic analysis was not initially considered to be necessary for most common diseases. *R. A. Fisher* pioneered the analysis of 'quantitative' inheritance based on Mendelian gene segregation and in 1918 had already introduced the term *'variance'* in its modern form. This is the concept that a gene may be found in a number of variant forms in a population, where each variant may have quantitatively differing effects on expression of a characteristic. In an extreme form this effect may be termed a familial disease or syndrome. Fischer's successor, *John Thoday* tackled the mathematics for mapping the position of genes starting with family data relating to a particular characteristic (phenotype), the final effect of which, for example appearance of each individual, was influenced by more than one gene (or more correctly by a pair of identical genes in a diploid organism). He coined the term Quantitative Trait Loci (or *QTLs*) in 1961 for genes mapped using his analytical method.

It is only recently that sufficient data has become available due to cheaper analytical methods. Large-scale international efforts to compile a genome-wide analysis of QTL locations has led to the identification of QTLs on the genome which may increase or decrease the probability of an individual suffering a common disease. This includes the realistic concept that most forms of illness are not dominated by the presence or absence of a single type of gene variant but are influenced by the interplay of many different genes, each contributing in a small way to the final effect. This also takes into account that the environment (behaviour, diet, smoking, exercise, neurotic or peaceful environment, etc.) may have a decisive role in the prevention or development of a disease state.

11.2.3
The Concept of a DNA-based Physical Map of the Entire Human Genome, about 1980

David Botstein and colleagues [5] demonstrated how DNA sequence alterations which resulted in the loss or gain of restriction sites (restriction fragments length polymorphisms, RFLPs), could be used as physical genetic markers to create a physical map of the human genome. Although this had been achieved for small viral genomes, the principle was in fact quite revolutionary for a genome some four to five orders of magnitude larger. Using Botstein's approach one could rapidly produce thousands of markers around the chromosome within a very short time. If used for analysis, on say the DNA from a collection of sperm, it could be used to give fine resolution to a physical map. Since individuals in the population differ in these RFLPs, genes involved in inherited disease could be seen to be either coupled or not to these physical loci. Rapid developments in PCR technology and the identification of single base pair mismatches within otherwise unique sequences known as 'single nucleotide polymorphisms = SNPs', pronounced 'snips', enables the presence or absence of such a specific sequence to be 'seen' using CHIP technology (Figure 11.3), thus increasing the number of recognized polymorphism loci from thousands (RFLPs) to millions.

11.2.4
Prenatal Diagnosis

In the mid-1980s the search for genes with clinical relevance was at its peak, particularly for the genes which were of relevance to debilitating disease and therefore of interest in prenatal diagnosis.

Prenatal diagnosis involves the analysis of:

- amniotic fluid extracted from the amnion with a syringe (invasive)
- foetal cells extracted from the chorionic villus (invasive)
- the limited number of foetal cells circulating in the mother's blood after enrichment with immobilized antibodies to the foetal cells (non-invasive)

The invasive methods were associated with a low although significant (a few percent) risk of spontaneous abortion. DNA from these cells could be amplified using PCR and then analysed for relevant DNA sequences that were expected to be associated with the probability of inheriting the trait.

11.2.5
Pre-implantation Diagnosis

As infertility research progressed relatively efficient methods of *in vitro* fertilization became routine for couples with unfulfilled desires for having their own children. In this case eggs obtained from the mother's ovaries are fertilized with sperm and the early development followed under the microscope. After a few hours the cells have undergone three doublings and consist of a small ball of eight cells known as the blastocyst. One of these cells is then removed for DNA analysis, a process dependent on DNA amplification via PCR. The remaining seven-celled blastocyst which will have been frozen for the duration of the DNA analysis can then be thawed out and implanted into the uterus where it will continue to grow and develop into a normal embryo, so that about 9 months later a normal child will be born.

This type of examination, called pre-implantation diagnosis, would normally entail refraining from implanting any embryos which carry a dominant deleterious gene or a non-complemented copy or (two) copies of an affected recessive gene allele. In the 1990s laws were passed in many countries which regulated the use of human eggs and embryos. As with any laws related to rapidly developing technologies these are now largely out of date and in some cases pre-implantation diagnosis is prohibited. In retrospect, some of the legislation lacked certain important provisos such as the fact that although the origin of the eggs was carefully documented, the origin of the donor sperm was not, making it impossible for a child to later trace their genetic father. Further developments in stem cell research are discussed in Chapter 12 and the expectations for the immediate future are reviewed in Section 13.4.

11.3
Early Attempts to Analyse the Human Genome

Mapping of hereditary disease genes on human chromosomes would have looked very different had it not been for the dedication and enormous productivity of *Victor McKusick* (1921–2008: recipient of many prizes including the Lasker award, the US National Medal of Science and the Japan Prize [6]).

His first book published in 1957 already addressed the genetics of connective tissue disorders. He and colleagues showed linkage of colour vision to the gene G6PD for glucose-6-phosphate dehydrogenase, the most prevalent gene variation in man.

Other enzyme changes were discovered to cause hereditary disorders. Depressed enzyme levels cause favism, a condition in which the red blood cells are sensitive to broad beans and which is prevalent in areas where malaria is widespread. From this grew further encyclopaedic volumes culminating in *Mendelian Inheritance in Man*, the most authoritative work on human genetics.

McKusick was innovative in implementing computerized databases in *1964*, making this book the first medical text to be published *after* being computer generated. This became the internet version OMIM, one of the very first medical uses of the worldwide web (www). He was the first to make the clarion call for an all-out attack on the human genome at the 1969 international meeting on birth defects held at The Hague; 'What we should know in full detail are the structure and geography of man: the full nucleotide sequence of all genes determining the amino-acid sequence of proteins and the location of each on the chromosomes of man'. Considering that four years later in 1973, *Gilbert* and *Maxam* reported the sequence of only 24 base pairs using a chromatography method known as 'wandering-spot' analysis, McKusick's vision is quite remarkable.

It took another 20 years before adequate methods had been developed which would eventually make the sequencing of the whole genome feasible (see below). It was, therefore, most appropriate that when the decision had been taken to start such a project Victor McKusick was named as the founding President of the international Human Genome Organization (HUGO; 'Victor's' HUGO, with alliterative association to another very prolific writer).

Perhaps one of McKusick's most significant activities was training generations of medical geneticists, a human resource of immeasurable value without which the human genome sequence would essentially be meaningless with respect to inheritance. This was a weakness in Germany where the number of human geneticists was depleted and the discipline had been in disrepute for decades following the atrocities carried out in the name of science, particularly human genetics, under the Third Reich (see Benno Müller-Hill's book *Murderous Science*) [7].

11.3.1
Human Geneticists give Meaning to the Physical Map

Just how important human geneticists were can be seen from the successful studies carried out, for example by *Leena Peltonen-Palotie* (1952–2010) who examined inbred populations in small villages in Finland and *Nancy Wexler* who studied large families in South America in which several generations had inherited Huntington's chorea. Genetic analysis of inherited disease, combined with DNA studies led to the rapid early mapping of the disease genes which was out of all proportion to the amount of sequence information available. Actually most of the genetic traits for heritable diseases were known and had been mapped before the genome had been entirely sequenced.

11.3.2
Organizing (and Funding) the Genome Effort

The creation of a concrete plan of action seems to have its origins in serious discussions that took place in the early summer of 1985, at a workshop organized by *Robert Sinsheimer* at the University of California Santa Cruz. A little later *Renato Dulbecco* (Nobel laureate) suggested that sequencing the human genome should be part of the USA's declared 'war on cancer' [8] and *Sydney Brenner* [9] at the Medical Research Council Centre (MRC) in Cambridge, England, was trying to animate the European effort.

However, with the technology then available the goal looked unachievable and many were sceptical and unclear as to how best to proceed. The random approach covering the whole sequence, most of which was considered to be repetitive 'junk', seemed like a waste of scarce research funding. At this time, in spite of the War on Cancer, there was no real tradition of large-scale funding for projects in biology; the scale of funding for biological and medical basic research was trivial compared to the funding for even a single experiment in space research, physics or military projects.

The American Department of Energy (DOE) became involved as a prime supporter of the genome project at the instigation of *Charles DeLisi*. Decades before, the DOE had become involved in studying the mutagenic effects of radiation due to *Thomas Hunt Morgan*, and was still responsible for understanding the effects of the atomic bomb on the survivors of Nagasaki and Hiroshima.

11.4
The Personalized Genome and Personal Medicine

11.4.1
DNA CHIP Technology and Haplotype Mapping

It is hoped that CHIP technology together with the data from the Human Genome Programme will enable scientists to analyse accurately the genetic basis for susceptibility to widespread health problems, in cases where many genetic and environmental factors appear to play a role and no single genetic variant is found to be highly dominant. This goal has been publicly addressed by the creation of the *HapMap* project, whose name is derived from the abbreviation for haplotype (the actual collection of variant alleles (detected for example as SNPs) present within or 'tagging' a gene region) mapping.

Once the sequence of the genome was known it became possible to rapidly diagnose variant genes associated with predisposition to a disease. This has also led to the development of the science of *pharmacogenetics* which relates the sensitivity to drugs or the ability to metabolize drugs to the personal genetic make-up of the individual, for example individuals carrying P450 liver enzyme variants have varying abilities to break down pharmaceutical compounds. Rapid diagnosis can be carried out by hybridizing labelled DNA from the patient to a DNA chip.

This does not mean that in all cases the SNP in question is the cause of the disease, but it may be found more frequently linked to a gene variant (mutation) which is causative for that condition (disease or predisposition). It should be emphasized that this analysis yields information which must be taken as a probability, that is, the statistical likelihood of it indicating the presence of a particular condition.

11.4.2
High-throughput Screening

In the last forty years an important change in laboratory methodology has taken place as a result of the development and introduction of sample preparations that can be addressed spatially. Combined with robotics, or later with optical surface analysis, and electronic data processing this led to high-throughput screening (HTS) for novel substances with defined properties [10].

An early and extraordinary innovation in *high-throughput screening* (HTS) followed immediately in the footsteps of droplet-deflection-based technology coupled to the measurement of particular size parameters using laser scanning of the droplets. This innovation was developed at the University of California Science Laboratory at Los Alamos in 1965 [11]. At a later stage, fluorescent-labelled antibodies which bind specific cell surface molecules were used to enable (analysis and collection) large numbers of different cell types to be sorted simultaneously (= fluorescent activated cell sorting = FACS). The droplet deflection device was improved to create a large machine which could be used to plate massive arrays of single cells for high-throughput isolation of rare temperature-sensitive mutations from bacterial cultures, or to study the effects of drugs on prokaryotic or eukaryotic cells. The inventor of this apparatus was *Don Glaser* [12] of the University of California in Berkeley who was already a Nobel Laureate (for the bubble chamber used to detect subatomic particles) [13]. Although the machine was used for a small number of NIH and US military financed projects in the early 1980s, it became obsolete. However, the possibilities it opened attracted investors, leading to the first modern biotechnology company, *Cetus* (founded in Berkeley, California in 1972) which was later incorporated into the *Chiron Corporation* (Berkeley, 1991).

Berkley became the cradle for the development of PCR for which *Karin Mullis* obtained the Nobel Prize in Chemistry in 1993. This in turn led directly to the large scale development of the biochip industry with *Affymetrix* being formed as a spin-off from *Affymax* which in turn was a spin-off from the *Chiron Corporation* [14].

These innovations ushered in the era of combinatorial chemistry in the late 1980s, or at least accelerated its widespread application[1] (see also Section 7.3).

During the late 1970s and early 1980s a large number of routine investigations employed ever more standardized modules and increasingly smaller sample volumes, beginning with small centrifuge tubes ('epis', 24-tube racks and centrifuge rotors) and extending to microtitre plates in 24-, 96- and 384-well standard robotic formats. These are still being used in the pharmaceutical industry for studying the effects of large (some 3 million compounds) collections of candidate small-molecular-weight drugs on tissue cultures in the search for anti-cancer agents, for example at *Bayer AG*, Leverkusen in 1996.

11.4.3
Ordered Micro-arrays and DNA-CHIPs

DNA probe hybridization to cancer cell-derived mRNA spotted onto nitrocellulose can be seen as a transition step from Northern blot analysis (which identified specific RNAs) to mass screening on micro-arrays. In 1987 for example, this methodology appeared in simple form for the analysis of the quantitative level of expression of a small number of genes supposed to be potentially involved in oncogenesis.

This led to the surprising finding that endothelial carcinomas which over-express the 'oncogene' *ras*, were in fact less malignant than others with a lower level of expression [15]. The diagnostic potential for this type of method became immediately clear. The concept that miniaturization would permit the simultaneous analysis of an ever-increasing number of samples with an increasing number of probes and with reduced material costs, drove further developments towards CHIP and other microarray technologies.

The following developments were an extension of the techniques that had already been established for the synthesis of oligonucleotides and peptides on solid supports (see Section 7.5). Some important transition stages and early milestones are listed below:

- Photolithographic techniques for making microarrays with immobilized specific synthetic oligonucleotides were established in 1991 [16]
- By 1993 densities of 10^5 spots/cm^2 were obtained, this was also the case for immobilized peptides [17]
- The birth of the first real quantitative microarray of cDNAs on glass slides with non-radioactive probes. This arose from a collaboration between the groups led by *Ron W. Davis* and *Patrick O. Brown* (Figure 11.2) at Stanford University in 1995 [18]
- Optical sensor arrays (1993–1998) [19]

---

1) For citations relating to the history of combinatorial chemistry, immobilized compounds on arrays, beads, and so on see URL: www.5z.com/divinfo/classics.

Figure 11.2   Patrick O' Brown, 'father of microarrays'.

Large numbers of samples can be evaluated in ordered micro-chambers (e.g. Evotech, which was founded on the basis of technology developed in *Manfred Eigen's* group at the MPI (Max-Planck-Institute in Göttingen)) to study molecular interactions. This requires channelling the liquids to be tested into micro-chambers where they are mixed with a second component. Free molecules and interacting molecules have different rotational rates and mobility. Essentially single interactions can be analysed in a tiny volume at the focal point of a concentrated laser beam.

The use of microgrids is also an essential feature of some of the third generation DNA sequencing machines (e.g. the *454 Life Sciences* (Roche Group)), which as discussed in Section 13.5, has resulted in a massive reduction in the price of sequencing a complete individual human genome (in 2010 the cost is apparently under $100 000, in December 2007 it was estimated at $1 million dollars and in 2000 *Craig Venter* estimated the cost of *de novo* complete sequencing of his genome which was essentially completed in that year, to be some $74 million).

11.4.4
Scannable Random Arrays: Basis for a New Generation of DNA Sequencing

Recent more powerful methodologies have employed the unordered immobilization of DNA samples which are either locally amplified or treated as single molecules. These processes use optical monitoring of each 'nano-spot', one of millions examined under a microscope during the sequencing reactions. The machines produced by the *Illumina* Company utilizing the *Solexa* DNA sequencing technology generate some 1.5 Gigabase in a few days and require only a single technician to operate the equipment. This enables 'light sequencing' of the human genome to be completed within a few weeks.

The process by which the genome of an individual, be it a bacterium or a person, is analysed using 'light sequencing' for the identification of the total distribution of SNPs and other DNA variations, is referred to as 'resequencing'. Combining this large array scanning method with the *single-molecule sequencing* technique of *Helicos*, allows a single technician to accumulate nearly 150 G bp of raw data on a single machine within a month. This was sufficient to allow a 28-fold depth resequencing of a single individual [20], revealing some 3 million single nucleotide polymorphisms (SNPs). The accuracy was sufficient to recognize regions of *copy number variance* (CNV) which was not the case in previously sequenced individuals such as J. Craig Venter and Jim Watson.

Such analysis may be carried out on a genome wide scale or be targeted to particular subsets of sequences, for example by amplifying with PCR primers or precipitation of subsets of DNA from chromatin fragments interacting with particular regulation factors (e.g. chromatin immunoprecipitation-mediated ChIP_CHIP sequencing)[2]. This is proposed to be the method *in ascendi* for the future of personalized medicine, where there is a shift from corrective intervention, once a disease state has established itself, towards preventive measures involving diagnosis of predispositions.

The reduced cost of this analysis allows the study of microorganisms following the systems biological approach. The altered phenotype of a number of variants can be examined at the level of the entire genome of each variant. In this way the causative genotype (mutation or mutations) should be unambiguously correlated with the resulting alteration in the observed characteristic (the phenotype). Earlier genetic backcrosses had been necessary in order to isolate single mutations or to create specific combinations.

11.4.5
DNA CHIPS for Pharmacogenetics and Cancer Diagnostics

One product that has been approved for use in the diagnosis of drug susceptibility is *AmpliChip CYP450*® (approved 04/2005) which functions by identifying the presence of the cytochrome P450 gene alleles 2D6 and 2C19 which play a key role in the metabolism of several toxins, nutrients and medications. The 15 000 oligonucleotides immobilized on the microarray chip can detect many variants as well as copy number changes for the genes CYP4502D6 (19 variants) and CYP4502C19 (two common variants (Figure 11.3)). These variants are of considerable clinical relevance particularly with regard to the rate of conversion of codeine

---

2) Gene expression involves binding of proteins to the chromatin which directly affect initiation of transcription. After cross-linking proteins with formaldehyde and shearing, complexes are precipitated by further cross-linking with antibodies specific for one of the regulatory proteins of interest (immunoprecipitation; IP). The DNA incorporated into these complexes can be analysed by hybridization to a DNA array especially designed for this purpose, either to give a broad overview of what is happening at the genome scale or to analyse the fine detail of what is occurring in a particular region (e.g. tumour suppressor gene region)

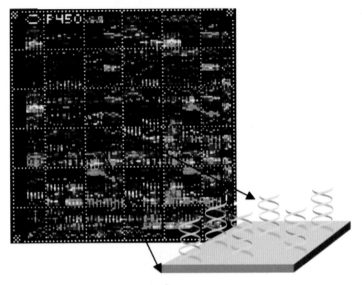

Each 20 µm² cell on the array can contain
10⁷ DNA fragments, or probes

Figure 11.3 Microarrays (CHIPs) with spots of specific DNA probes immobilized on the surface. After hybridization with labelled DNA samples and PCR amplification, samples of patient blood are hybridized to the CHIP. The presence of specific gene variants, in this case for AmpliChip CYP450 cytochrome P450, can be visualized by fluorescence microscopy. http://www.roche.com/med_backgr-ampli.htm.

to morphine, and the degradation of the anti-depressant Prozac, which if not metabolized accumulates in the blood.

A second CHIP product *MammaPrint*® which was approved in February 2007 is used for diagnosing the risk of metastasis in breast tumour tissue. This product can identify the expression of the metastasis-relevant RNA in cells obtained from a tumour biopsy. The steps involved are as follows:

- Isolation of RNA
- DNAse treatment of isolated RNA
- Reverse transcription into cDNA and then into cRNA
- Fluorescent-labelling of tumour and reference cRNA
- cRNA purification; hybridization of the cRNAs of the tumour and reference sample to the chip®
- Scanning of the microarray and data acquisition
- Calculation and determination of the risk of recurrence in breast cancer patients

Another chip, the *AmplichipLeukemia* which was developed earlier by Roche[3] is used to analyse the level of gene expression in blood or lymph nodes of leukaemia patients. Preparation and analysis of the samples require some 2 days. The

3) http://www.roche.com/pages/downloads/media/video/flash/roche_amplichip.swf.

differential gene expression profiles obtained allow 20 different classes of leukaemia to be distinguished. This should in principle, lead to a more rational decision-making process, with respect to (say) the usefulness of a chemotherapy regime.

The acquisition of instrumentation capable of reading the output of a DNA hybridization CHIP is still a considerable investment. Manufacturers are therefore offering a wide range of applications which can be carried out with the same basic format and indeed with the same microarray. One of the latest products aimed at personal medicine contains over a million markers incorporating the latest data on copy number variation and SNPs. This data was obtained from the '1000 genome project', an international cooperation aimed at providing more information about the genome and population-wide occurrence of particular haplotypes or SNPs and their possible association with predispositions for frequent disease syndromes.

Total sequencing and resequencing projects using CHIPs are realistic for microbial genomes, but are limited to targeted resequencing of small regions of the human genome. Figure 11.4 summarizes the multiple uses of DNA CHIPs for biological projects and medical applications. The upper part of the figure shows a diagram of the ordered samples on the microarray and the upper left area illustrates how a heterozygous (two variants) SNP variant shows up when bound to a fluorescently-labelled sample.

DNA sample (Genetic counselling)		mRNA cDNA (diagnosis)	DNA samples Comparative analysis

• Total sequencing
 • Resequencing
 • Targeted resequencing
 • Indels
 • SNPs
   – disease/cancer assoc^d
   – splice sites
	– AIMS
	 – Blood type
	 • miRNA binding sites
	 • introns
	 • Promoters
	 • ChIP
		• Coding SNPS (cSNPs)
		 • Expressed SNPs (eSNPs)
		 • Prostate androgen response (PAR)
		 ADME genes
			• DNA methylation
			 • Copy number variation (CNV)

Figure 11.4 A multipurpose DNA-CHIP can be used to analyse a wide variety of medical or research relevant topics, delivering information of use for genetic counselling and personal medicine as well as for research purposes. Samples may have to be prepared in different ways for the different applications, for instance for accurate read-out on copy number variance or DNA methylation patterns.

ADME (adsorption, dissemination, metabolism, excretion in relation to pharmaceutical drugs) genes shown in Figure 11.4 encode gene products relevant to pharmacogenetics. These includes the relevant flavin-containing mono-oxygenases (FMO), cytochrome P450 (CYP) and ATP-binding-cassette (ABC)-transporter superfamily variants. AIMS (ancestry informative marker SNP) is an abbreviation referring to the fact that these markers have a skewed distribution in certain ethnic groups and can therefore be used to give statistically relevant information about ancestral origin.

In summary it can be seen that sequencing technology and in particular the power of the fluorescent detection CHIP readers, has allowed companies initially involved in either sequencing or SNP analysis to broaden their methodologies to cover an extensive range of possible research and development activities based on nucleic acid sequences.

11.4.6
Analysis of Complex Systems in Personal Medicine

The complex assembly of information pertinent to the development of personal medicine is growing rapidly and requires an integrated or 'holistic' approach.

Systems biology in which the dynamics of interacting networks of proteins, RNA, metabolites, cells and organs are considered together are brought into play to understand health and the perturbations which we recognize as disease. A recent review clearly addresses the problem and why it is necessary to portray and analyse the interaction networks. *Erwin Eric Schadt* of *Pacific Biosciences*, a company involved in developing super high speed DNA sequencing, proposes that it is through this approach that we will arrive at an understanding of the complex association between the many gene variants and propensity for a particular disease [21].

11.5
Analysing the Effect of the Environment on the Human Genome: Epigenetics

An extension of personal medicine is discussed here under a separate heading since it involves analysis of changes in the genome which occur in each individual in response to his or her lifestyle.

Changes that occur at what is called the epigenetic level, often involve alterations in the methylation of DNA and/or methylation, acetylation and other modifications of the *histone proteins* which are directly associated with the DNA (the chromatin). These latter modifications change during an individual's lifetime and affect the level of local gene expression. However, these changes in response to the environment do not alter the gene sequence.

Since DNA-methylation patterns should all be reset during production of the gametes (eggs and sperms) and again specifically modified during early embryogenesis, they should not be inherited. In fact this is not entirely true. Some effects, for instance following periods of famine, seem to persist for one or even two generations. This is known as an epigenetic effect.

It has been found that these alterations may well be the main cause of differences between identical twins in response to different lifestyles. In particular, with reference to human health care, there are major differences in DNA methylation of tumour suppressor genes in tumour cells, these being associated with a higher cancer risk, for example in tobacco smokers or -chewers.

This indicates the potential for novel strategies for cancer prophylaxis based on inhibiting specific chromatin (histone) de-acetylation enzymes (HDAC inhibitors) and/or specifically inhibiting or activating DNA methylation: the company *Methyl-Gene*[4] is developing HDAC inhibitors incorporating demethylation agents to further improve cancer therapy. Drugs targeting DNA methylation in chromatin were first developed to treat a group of blood disorders called myelodysplastic syndromes. *Pharmion* developed *Vidaza*, the first demethylating agent approved by the FDA in 2004.[5]

References

1 Pauling, L. *et al.* (1949) *Science*, **110**, 543–548; Perutz, M. and Mitchison, I.M. (1950) *Nature*, **166**, 677–9.

2 Perutz, M. and Mitchinson, I.M. (1950) State of haemoglobin in sickle cell anemia. *Nature*, **167**, 1053–4.

3 Smithies, O. and Walker, N.F. (1955) *Nature*, **176**, 1265–1266; Smithies, O. (1955) *The Biochemical Journal*, **61**, 629–641; Smithies, O. (1955) *Nature*, **175**, 307–308.

4 Raymond, S. and Weintraub, L. (1959) *Science*, **130**, 711.

5 Botstein, D. *et al.* (1980) *American Journal of Human Genetics*, **32**, 314–331.

6 Rimoin, D.L. (2008) *Nature Genetics*, **40**, 1037.

7 Müller-Hill, B. (1981) Tödliche Wissenschaft, Rohwolt, Hamburg; Engish version (1989) "Murderous science: elimination by scientific selection of Jews, Gypsies and other in Germany 1933–1945". Cold Spring Harbour Laboratory Press. ISBN 0-87969-531-5.

8 Dulbecco, R. (1986) A turning point in cancer research: sequencing the human genome. *Science*, **231**, 1055–1056.

9 Brenner, S. (2001) My life in Science. Publ: Biomed Central Ltd., London; ISBN-10 0954027809.

10 Service, R.F. (1997) *Science*, **277**, 474–475.

11 Fulwyler, M.J. (1965) *Science*, **152**, 910–911.

12 Sevastopoulos, C.G., Wehr, C.T., and Glaser, D.A. (1977) *Proceedings of the National Academy of Sciences of the United States of America*, **74**, 3485–3489.

13 Aidells, B.D., Konrad, M.W., and Glaser, D.A. (1979) *Proceedings of the National Academy of Sciences of the United States of America*, **76**, 1863–1867.

14 Rabinow, P. (1996) *Making PCR: A Story of Biotechnology*, The University of Chicago Press.

15 Augenlicht, L.H. *et al.* (1987) *Cancer Research*, **47**, 3763–3765.

16 Fodor, S.P.A. *et al.* (1991) *Nature*, **251**, 767–773.

17 Fodor, S.P.A. *et al.* (1993) *Nature*, **364**, 555–556.

18 Schena, M. *et al.* (1995) *Science*, **270**, 467–470.

19 Walt, D.R. *et al.* (1993) *Applied Biochemistry and Biotechnology*, **41**, 129; (1995) *Science*, **269**, 1079; (1998) *Analytical Chemistry*, **70**, 1242.

20 Pushkarev, D. *et al.* (2009) *Nature Biotechnology*, **27**, 847–850.

21 Schadt, E.E. (2009) *Nature*, **461**, 218–223.

4) www.methylgene.com.

5) *Genetic Engineering News*, Editorial by Carola Potera, 1 June 2008, 28, (11); www.pharmion.com, now owned by Celgene.

12
Transgenic Animals and Plants

12.1
Introduction

It is possible to go beyond previously described advances in reproductive or fertility medicine which include *in vitro* fertilization (IVF), cryopreservation of embryos and reproductive acceleration by cloning from somatic tissue (as opposed to germ line) cell nuclei. It is possible to introduce foreign DNA into an animal or plant cell nucleus and using the tools available to *reproductive medicine* create *transgenic animal* (*or plant*) *clones*. This opens up the possibility of creating animals with novel characteristics which may be of obvious benefit for the animal, for example increased resistance to certain pathogenic viruses or bacteria. Equally well, an animal may become the source of a product which can be used in medicine, for example a cow that produces human serum albumin (HSA; required in tons per year) in its blood rather than its normal bovine serum albumin (BSA) or human alpha-1-antitrypsin inhibitor in its milk that can be used during heart surgery.

The key novel developments originating from recombinant DNA and particularly involving transgenic animals and plants, are summarized below:

- The ability to produce essentially any component of a living cell in large quantities and the analysis of the causes of disease at a fundamental level led to the targeted development of novel medicines.
- Transgenic plants facilitate acceleration of breeding programmes.
- The use of transgenic animals and plants enables the development of novel varieties beyond the previous limitations of natural crosses.
- Cloning allows the propagation of premium animals without dilution of the desired characteristics.

Plants that have been engineered to contain useful drugs can be categorized under the heading *farmaceuticals*. The particular advantage of products which require mammalian post-translational modifications to improve their solubility or activity for example, is the low cost of rearing animals or plants compared for instance, to that of maintaining a sterile fermentation plant for tissue culture. Highly active

Concepts in Biotechnology: History, Science and Business. Klaus Buchholz and John Collins
Copyright © 2010 WILEY-VCH Verlag GmbH & Co. KGaA, Weinheim
ISBN: 978-3-527-31766-0

compounds such as interleukin 2 and interferon-α which are not required in huge quantities and do not require glycosylation, can be more efficiently produced by recombinant microbial fermentation processes.

The danger inherent in most transgenic methods which involve altering the germ line, is the untoward effects of possibly disrupting essential genes during the integration of the novel DNA.

12.2
Stem Cells and Gene Targeting

12.2.1
Selective Gene Markers for Animal Cells

In 1977 Michael H. Wigler and Richard Axel (Figure 12.1) showed that a thymidine kinase gene could be used to compensate the loss of the host cell mutant in mouse cell fibroblasts [1]. They used this gene as a selective marker for vectors in eukaryotic cells. This is analogous to the way that metabolic complementation was used in the development of yeast vectors. The immediate advances were dramatic. Researchers were able to use this technique in animal cell cloning experiments to select for and isolate the first vertebrate genes and the first human oncogenes.

Oliver Smithies, Mario R. Capecchi and *Sir Martin J. Evans* jointly received the Nobel Prize in 2007 for 'for their discoveries of principles for introducing specific gene modifications in mice by the use of embryonic stem cells'.

Figure 12.1   Richard Axel.

Figure 12.2   Oliver Smithies.

12.2.2
Site-Specific DNA Mutation

It was the work of *Oliver Smithies* (Figure 12.2) which paved the way for gene cloning in animals via *predetermined alteration* of specific DNA regions. It should be noted that even before he started work on gene insertion into animal cells via homologous recombination, Smithies had already created by January 1954, a method for analysing proteins by gel-electrophoresis which initially used starch gels.

This led him to the discovery that certain proteins differed from one individual to another. Variants which could be separated on the basis of size and shape during their migration through the gel matrix under the influence of an electric field were found to be present in a population. The analysis of genetically-determined haplotypes of the alpha- and beta-globin chains had been achieved long before the development of DNA analysis (see Section 11.2) of globin. It is perhaps a consequence of this that Oliver Smithies subsequently became involved in development of methodologies which are used to study the functionality of gene alterations, transferring them by homologous recombination into animals.

Some 30 years after the initial work on protein variants, *Oliver Smithies* in 1985 was the first to show that 'gene targeting', that is exchanging a gene in the chromosome with a variant gene present on DNA introduced into mammalian cells in tissue culture, was indeed possible. This functioned in spite of the enormous complexity of the host cell genome [2]. Although the frequencies were very low, the results of an experiment recombining altered DNA into the globin gene via homologous recombination were unequivocal and a new age of genetic engineering in animal (in this case the human animal) cells began.

Figure 12.3 Mario R. Capecchi.

12.2.3
Microinjection into the Cell Nucleus

In 1980 *Mario R. Capecchi* (Figure 12.3) developed the methodology for microin-jection of DNA and other molecules, directly into the cell nucleus. This considerably increased the efficiency of producing recombinant cell lines compared to just allowing uptake of DNA into the cells or injecting it into the cytoplasm. About 30% of the cells, injected at a rate of nearly a 1000 per hour, incorporated foreign DNA. Capecchi relates in his Nobel acceptance speech how these genetically altered cells were implanted into the uterus of foster mothers thus generating transgenic mice, he also remarked that 'The generation of transgenic mice, in which chosen exogenous pieces of DNA have been randomly inserted within the mouse genome, has become a cottage industry'. Other improvements followed. Electroporation was used as a simpler method to transfer DNA into a very large number of cells simultaneously. Capecchi went on to develop methods for selecting knock-out cells where the HGPRT gene mutants could be selected for survival in the presence of 6-thioguanidine. The presence of a neomycin resistance gene allowed selection of this gene after the transfer of DNA cloned from the transgenic mouse back into bacteria.

12.2.4
Altered Embryonic Stem Cells

Martin Evans discovered that embryonic stem cells (ES) could be isolated from the inner trophoblast of early embryos (blastula). These could be maintained for long periods in culture without differentiating and then added back to a new blastula where they would contribute part of the tissue of the resulting chimaeric animal.

Capecchi and Evans demonstrated that by injecting DNA into ES cells it was possible to create knock-out ES cell lines, containing the 'genetically engineered' region.

12.2.5
Transgenic Animals from Engineered ES Cells Injected Into Early Embryos (Blastula)

Smithies injected DNA into the nucleus of such ES cells which could develop in the growing embryo as part of a tissue mosaic which made up the final animal. Sometimes the eggs or sperm would contain the altered gene and a recombinant 'transgenic animal' could be bred out. In the meantime stem cells have also been discovered in other tissues such as the intestine, where they can be induced to totipotency by inducing the LRR-receptor.

The scientific community was hardly prepared for these discoveries in the early 1980s. Grant proposals on this subject had been turned down as unrealistic and over-ambitious, but luckily these three researchers had the persistence to continue their work on low budgets and thus prevailed.

12.2.6
Specific Gene Alterations in Animals

The combined discoveries of *Smithies* and *Capecchi* together with *Evans*, led to an acceleration and qualitative alteration to the study of the functionality of genes, namely that the genes could now be studied in their actual normal environment in the animal. Smithies *et al.* [2] demonstrated that both random integration, and more importantly, homologous recombination which enabled site-specific exchanges at specific gene locations, were possible.

Initially many of these alterations were achieved by the destruction of gene function and were referred to as 'knock-out' mutants, usually in mice. Later, alteration of function or specific induction or repression of function could be used to study functionality in more detail. Animals carrying such alterations are referred to as 'knock-in' variants.

Some transgenic animals carry mutations which make them suitable models for human (inherited) disease, for example certain mutants of *Drosophila melanogaster* are considered to be appropriate for the study of Parkinson's disease [3]. More recently a method of cloning differentiated tissue from patients suffering from a particular inherited disease has been developed, thus providing more relevant material for studying the activity of potential therapeutics that are under development (see Section 12.1).

12.2.7
Study of Gene Functionality in Transgenic Animals with Conditionally Induced Genes

Tissue-specific promoters for fluorescent proteins (e.g. GFP = green fluorescent protein) have been attached to genes which are then introduced into 'knock in' transgenic animals. This technique can be used to determine at what stage the genes that are normally regulated by these promoters are turned on and off during embryogenesis.

12.2.8
Multicoloured Fluorescent 'Tags' to Analyse Complex Tissue Development

Using this methodology it became possible for the first time to study the development of individual brain cells in detail. In 2008, *Osamu Shimomura, Martin Chalfie* and *Roger Tsien* (Figure 12.5) were awarded the Nobel Prize in recognition of their contribution to the development of multicoloured labelling using fluorescent GFP-derivatives. These Harvard researchers had tagged brain cells in mice with some 90 colours in the 'spectacular experiment' for which they received the Nobel Prize. The technique is known as 'Brainbow' (see Figures 12.4 and 12.6). Shimomura first isolated and characterized the GFP protein from jelly-fish in 1961. Some 30 years later (1991), Chalfie showed that the GFP gene could make individual nerve cells in

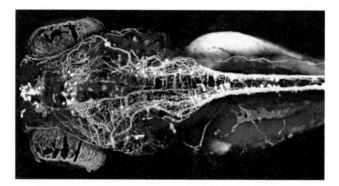

Figure 12.4   Fluorescence image of the brain of a 5-day-old zebra fish larva (Albert Pan, Harvard University) which won fourth place in the 2008 Olympus BioScapes Digital Imaging competition.

Figure 12.5   Left to right, Martin Chalfie, Roger Tsien and Osamu Shimomura.

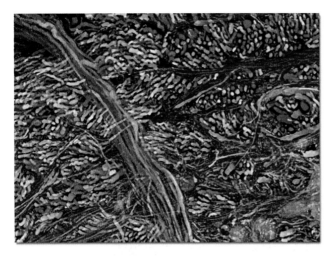

Figure 12.6 Image of the brainstem of a 'Brainbow' mouse by Jean Livet of Harvard University (Copyright Olympus).

the worm *C. elegans* glow bright green. Tsien's work provided GFP-like proteins that extended the scientific palette to a variety of colours.

12.2.9
Transgenic Model Organisms: A Tool for Biotechnological Development

Other organisms traditionally used as models in genetic studies may also be produced in transgenic form. In the fruitfly, *Drosophila,* the method of achieving this is much simpler, in that DNA can be injected into the polar region of the fly egg cell. This is also the case for Zebra fish (*Brachydanio*) which for example was used as a basic tool in transgenic studies undertaken by *Christiane Nüsslein-Volhardt* (Figure 12.7) who also received a Nobel Prize for her work in 1995 together with

Figure 12.7 Christiane Nüsslein-Volhardt.

her associate *Eric F. Wieschaus* who established the understanding on how gene function determined form during embryogenesis.

Most of their initial work involved the analysis of mRNA production and the interaction of the gradients of products of pairs of genes identified by classic genetics, which influence the formation of specific organs during embryogenesis. Later, however, transgenics played a central role in these analyses (see Nüsslein-Volhardt's book for a crash course in embryology; although without any original references [4]).

12.2.10
A New Type of Biotechnology Company Based on Transgenics

This work, apart from enriching scientific endeavours, led to the creation of a new type of Biotech Company, *Artemis Pharmaceuticals, GmbH*, founded by Nüsslein-Volhardt together with the former Bayer Manager, *Peter Stadler* and *Klaus Rajewsky* (also a Nobel Laureate; Figure 12.8) in 1998. This company specialized in defining novel targets for pharmaceuticals by examining the effects of gene alterations in transgenic Zebra fish. The long-term aim was to utilize this information to develop novel recombinant DNA products for use as pharmaceuticals. This was extended to the generation of transgenic mice, which in 2005 led to a strategic alliance with *Taconic Farms, Inc*, Germantown, USA.

12.2.11
Tackling a New Problem Where Cancer Potential is Sequestered as Stem Cells

Following an agreement with the Spanish company *CIBASA*, Artemis acquired a new technology, Oncostem® which addresses the recently discovered problem in cancer of some forms surviving treatment as stem cells. Stem cells which do not divide rapidly

Figure 12.8 Klaus Rajewsky.

in the absence of differentiation signals are not attacked by classic cancer treatments and therefore remain as sources for the formation of future metastases.

12.2.12
Vertebrate Clones from Somatic Cells: The Origin in Transplantation of the Cell Nucleus

In 1962 *Sir John Bertrand Gurdon* reported that he had taken a nucleus from the gut cell of a frog (actually from *Xenopus laevis*, a South African toad) and injected it into an enucleated mature egg (oocyte). The fact that this oocyte which contained the nucleus of a differentiated body (somatic) cell, had undergone apparently normal embryogenesis to produce a viable tadpole, caused a sensation. This contradicted the then prevailing dogma that once cells had differentiated, they were committed to form a particular type of tissue and this process could not be reversed. There were many specialized genes in the cell that were turned on and others that were 'permanently' turned off.

Gurdon had dramatically demonstrated that, in this case at least, the differentiated state of the body cell had been reprogrammed so that it could behave as a normal fertilized egg cell nucleus. What is not always emphasized is that the differentiation process was not a complete demonstration of cloning from somatic cells, since the tadpoles died before developing into adult toads! The reason for this is still not completely understood.

However, this key experiment kindled the imagination of scientists and challenged them to identify the conditions under which a body cell or its nucleus when injected into an oocyte, would undergo the entire, carefully orchestrated process of cell division and differential development (differentiation) into specialized cells (e.g. fibroblast, myoblast, chondrocytes, macrophage, erythroblasts, eggs or sperm). In addition the whole process required that the differential migration of cells within the growing embryo together with programmed cell death, should lead to organ formation and an entire animal clone which would be *an exact copy of the original animal*.

A cell with the potential to form an entire animal is termed '*totipotent*'. This is in contrast to a cell which may be maintained in culture and following induction by the addition of combinations of growth factors, is able to develop into one or other of the many different specialized cell types. This latter type of cell is referred to as '*pluripotent*'. Pluripotent cells can be used in the techniques described in the previous section to create cloned animals and also transgenic animals with the help of a blastocyst that is already in the process of embryogenesis. Even tumour cells can be 'tamed' by injecting them into blastocysts, where they then develop and become part of a somatic mosaic in the resulting animal.

12.2.13
Advances in Fertility Research

It had been shown that single cells could be taken from the ball of developing cells constituting the embryo at the eight-cell stage (morula) without inhibiting further development. The techniques described in the following sections can be categorized under the heading of 'infertility therapy' or 'enhanced reproductive medicine'.

12.2.14
In Vitro Fertilization and the 'Test Tube Baby'

For both animals and humans, *in vitro fertilization* (IVF) and implantation of the embryo into the womb of either the genetic or a surrogate mother has been a standard alternative in cases of infertility since 1978. On 25 July 1978, Louise Joy Brown, became the first 'test tube baby' to be born. On 21 December 2006, she gave birth to her own child conceived naturally. IVF was developed by *Robert Evans* and *Patrick Steptoe* in the UK. By Louise's 21st birthday in 1999, more than 300 000 babies had been born throughout the world using IVF. The methodology has been further enhanced by the development of cryo-embryology and cryopreservation of sperm and egg cells.

12.2.15
Prenatal and Pre-implantation Diagnosis

Prenatal diagnosis can be carried out by analysis of foetal cells isolated from the amniotic fluid (amniocentesis), the chorionic villus (at the mouth of the womb; CVS) or by the safer non-invasive method of isolating some of the few foetal cells (after enrichment with antibodies against the foetal cells) that have entered the mother's blood via the placenta. This analysis can be made highly sensitive even for point mutations by using PCR and DNA probes and extends beyond the analysis of chromosome abnormalities (karyotyping: aneuploidy; e.g. trisomy 21) or sex determination of the unborn for which international standards had previously been established in 1971. It should be noted that these analyses may be used in addition to biochemical tests for hormones and α-feto-protein.

In combination with *in vitro fertilization*, a diagnosis can be carried out on a single cell taken from the eight-cell morula *before it is implanted into the womb*. This is known as *pre-implantation diagnosis*.

12.2.16
Animal Clones and Accelerated Breeding

In animal breeding it may be desirable to produce a large number of particular clones as quickly as possible. This can be done by dividing the morula into individual cells and allowing each of the eight or 16 cells to develop as identical clones after implantation in the womb. This procedure is a form of '*clonal expansion*' *prior to implantation* and is used extensively in accelerated breeding programmes for farm animals.

Figure 12.9 Ian WIlmut, co-'engineer' of the first cloned mammal.

12.2.17
Creating Totipotent Cells with Somatic Cell Nuclei: The Breakthrough for Transgenic Animal Cloning

Most experimenters had assumed that a fast growing cell would be the ideal material for generating a 'totipotent' nucleus for implantation into an enucleated egg cell. The breakthrough came when *Keith Campbell* had the novel idea that in contrast to a fast-growing cell, perhaps a quiescent or resting cell may be in better condition for generating totipotent cells. Cells can be made quiescent by starving them in culture media. Together with *Ian Wilmut* (Figure 12.9) they carried out the definitive experiment and it worked. Their first such clone was the now infamous sheep 'Dolly' (*5 July 1996 → †14 February 2003) [5].

Recent advances in regenerative medicine resulting from the rapid progress in (protein) induced pluripotent/totipotent stem cell (iPS and piPS) research is discussed in Chapter 13.

References

1 Wigler, M.H. *et al.* (1977) *Cell*, **11**, 223–232.
2 Smithies, O. *et al.* (1985) *Nature*, **317**, 230–234.
3 Feany, M.B. and Bender, W.C. (2000) *Nature*, **404**, 394–398.
4 Nüsslein-Volhardt, C. (2006) "Coming To Life: How Genes Drive Development", Kales Press, Carlsbad, USA, ISBN 978-0-9798456-0-4 [Translation from German: originally Published 2004].
5 Wilmut, I., Campbell, K. and Tudge, C. (2000) "The second creation: Dolly and the age of biology control". Farrar, Straus and Giroux (Publ.).

13
Extrapolating to the Future

13.1
Summary of the Status Quo

The initial technical impact of the New Biotechnology was to allow more efficient production of certain known products, with dramatic qualitative improvements such as freedom from possible pathogenic contaminants which had been a potential hazard for products derived from animal tissue or human blood. The second step was to discover new products and targets for intervention particularly in medicine. In a third phase, a huge diversity of novel protein (and perhaps oligonucleotide) variants were developed which had a myriad of applications in medicine, diagnostics and classic biotechnology. The relevance of high-throughput screening tests then became the bottleneck, shifting the emphasis towards research, knowledge acquisition and knowledge management, high-content screening and translational science centres, the latter conceived to optimize the interaction between research and unmet medical needs. In animal husbandry and in agricultural the development of transgenic animals and plants allowed rapid development relevant to the economic demands of these industries.

13.1.1
Where are the Novel Tools and Interdisciplinary Studies Taking Us?

Some refer to the present period as the 'omics-era' since there has been a continuing specialization or drawing of boundaries around ever smaller areas of expertise, leading to terms such as *proteome* and *proteomics* (study of the proteome, the entirety of proteins which can be produced in a cell or organism), the *kinome* (knowledge centred on the complete collection of protein kinases), the *transcriptome* (information on all the RNAs produced), the *spliceome* (knowledge of all the mRNAs, miRNAs, and siRNAs produced from the transcriptome), the *methlyome* (knowledge of the distribution of methylation sites on the DNA in relation to gene expression/transcription levels in different cells and under different conditions) and the *interactome* in which

Concepts in Biotechnology: History, Science and Business. Klaus Buchholz and John Collins
Copyright © 2010 WILEY-VCH Verlag GmbH & Co. KGaA, Weinheim
ISBN: 978-3-527-31766-0

the molecular interactions between pairs and groups of proteins (dependent on their post-translational modifications) are defined.

13.1.2
Understanding the Whole

As a reaction to the generation of ever finer levels of 'omics', a new generalist approach has emerged, perhaps in analogy to the emergence of holistic medicine ('*Ganzheitsmedizin*' in Germany), known as *systems biology*. This has the ambitious aim of providing mathematically-based descriptors for all the variables involved in establishing a stable metabolism in a living system. A system component may comprise, in ascending order of complexity: a gene regulation switch; a signal transduction pathway; the whole cell; cell–cell interaction; organ formation; the whole organism; commensalism and symbiosis of different species; even extrapolation to an entire ecosphere or finally encompassing the balance between life forms and the environment of the entire planet, perhaps eventually even insights into limiting factors for evolutionary potential.

This is an approach that *Leonardo da Vinci* would have approved of. He considered that a discipline could only qualify as a science if its theories could be expressed or captured in mathematical terms. A development in this direction would then be considered as the final maturation of the life sciences.

13.2
Insect Control Through 'Sterile' Males (SIT)

In the past the screw-worm fly *Cochliomyia hominivorax* from North and Central America, the Mediterranean fruitfly (Medfly) *Ceratitis capitata* from California, and the tsetse fly* (*Glossina* spp.) have been effectively controlled by the dissemination of irradiated sterile males whose offspring were largely non-viable. Irradiation, however, reduced the competitiveness of males compared to untreated flies.

Based on the pioneering work of *Luke Alphey* (Figure 13.1) at Oxford University, Oxitec, a company located in Oxford, UK, has produced during the last five or six years, a number of transgenic insects for use as vectors in pest- and disease control. In particular a line of *Aedes aegypti* mosquito in which the males transmit a dominant lethal mutation to the offspring, making them non-viable [1].

A. aegypti is the main vector that transmits Dengue fever to humans. Some 100 million people worldwide are infected and the problem is becoming more severe. The transgenic mosquito requires tetracycline for the development of the larvae and thus can only be cultivated in controlled conditions and cannot survive in the wild. The males, which transmit this dominant lethal gene to their offspring, *effectively compete* with their wild-type brethren. At the time of writing it is expected this RIDL (releasing insects with dominant lethal) programme will be approved and implemented shortly to reduce the population of *Aedes aegypti* in Malaysia in an effort to reduce the spread of Dengue fever.

Figure 13.1   Luke Alphey.

A two-year RIDL open field test to control the cotton pest *pink bollworm* has been completed in the USA.

13.3
The Future of Gene Therapy

13.3.1
Slow Progress of Somatic Cell Gene Therapy

Although recombinant DNA has brought dramatic acceleration and a more rational approach to the development of novel medicines, *somatic gene therapy* trials, initiated in 1990, have been largely unsuccessful and have suffered setbacks due to deaths that occurred during the clinical trials and which appear to have been caused by the inherent methodology of the protocols used.

There are a number of factors frequently associated with this type of therapy that have limited its development:

- Somatic cells introduced into the patient as a replacement therapy have a *limited half-life* (as opposed to the self-replacing character of stem cells; see below).
- Immune rejection: *tissue incompatibility* may require accompanying immune-suppression treatment or may lead to immune rejection or inflammation.

The use of viral vectors has been shown in at least one dramatic incident in 2003, to have caused (perhaps by immune reaction to the adenovirus used) the death of the patient. There is ongoing work to develop episome or 'extra chromosomal'-type vectors which will be stable and work without integration into the genome and/or the production of foreign antigens.

The number of diseases and patients that can be treated using this methodology seems to be limited to those diseases caused by the functional loss of single genes. The majority of chronic disease is, however, recognized to result from the effects of a wide variety of genes.

A recent clinical study of gene vectors based on ancient transposons already present in inactive form in the human genome has now been approved. The transposons have been genetically altered by *Zoltan Ivics'* group at the Max-Delbrück Centre in Berlin, so that they can efficiently and specifically integrate by homologous recombination at a defined site [2]. It is intended to use cells taken from the patients (*ex-vivo*). The cells are transformed and reintroduced into the body after amplification in tissue culture. This method should enforce a specific immune reaction directed to the cancer cells.

13.3.2
Germ-Line Therapy

Although *germ-line therapy* is used in transgenic animals and plants, it is currently not allowed to be used in humans. This is because the long-term consequences are uncertain in view of the inability to absolutely restrict the site of gene integration or exchange to the desired location, which may in turn lead to the inactivation of normal genes. In the long term however, these concerns may be overcome by the same type of experiments, in which genes are guided to integrate into the genome via directed homologous recombination, or maintained without integration as auxiliary chromosomes [3].

13.4
Stem Cell Therapy

In January 2009 the FDA gave approval for a clinical trial using human stem cells. Approval was given to the *Geron Corporation* to test their product 'GRNOPC1, a human embryonic stem cell (hESC)-based oligodendroglial progenitor therapeutic' for its efficacy on patients with spinal cord damage. It had previously been shown that stem cells can be used to re-myelinate nerves and stimulate nerve 'sprouting' in damaged spinal cord in a rat model. At present there is no effective method of regenerating nerves and stimulating production of multiple nerve growth factors, that is, the proteins that promote the survival and regeneration of neurons damaged during spinal cord injury, as would be necessary for the treatment of spinal injuries (paraplegia) [4].

13.4.1
An Alternative to Human Embryos as a Source of Pluripotent Stem Cells

The introduction of stem cell therapy for the treatment of human disease had been beset with difficulties. Ethical considerations had overshadowed the use of human

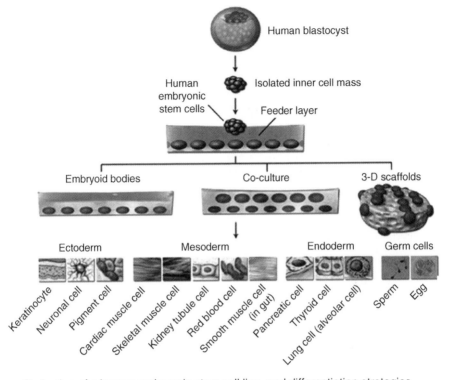

Derivation of a human embryonic stem cell line, and differentiation strategies

Figure 13.2 Until recently pluripotent cells could only be generated from embryonic stem cells thus involving the ethically questionable use of embryonic tissue.

embryos for research and as a source of the cells for therapy (Figure 13.2); in addition severe international restrictions had been placed on the use of human-embryo cell lines that had already been established. In America there have been indications that under the Obama administration such restrictions will be relaxed to a certain extent. As discussed in more detail above, alternative sources of stem cells have recently been established and are derived by using the correct cocktail of growth factors to reverse the differentiation of somatic cells (non-germ-line) thus leading to the generation of novel pluripotent or even totipotent stem cells. If the Geron product is successful, this may open the way for the development and approval of many other novel stem cell therapies – a new branch of biotechnology.

At the EMBO meeting in Amsterdam in August 2009, *Shinya Yamanaka*, (Figure 13.3) along with other pioneers in the field, such as *Rudi Jaenisch* (Figure 13.4), *Hans Clevers* (Figure 13.5) and *Austin Smith* discussed the development of the latest techniques of induced pluripotent cells (iPS).

In summary, major advances are expected in:

- *Production of specific cell lines as models for disease.* Particularly for neurodegenerative diseases such as Long QT syndrome, Parkinson's, ALS (= MND, motor

Figure 13.3 Shinya Yamanaka, pioneer of induced pluripotent stem cells.

neuron disease), spinal muscular atrophy, Alzheimer's and multiple sclerosis (MS), it is hoped to have an *in vitro* system derived from tissue from affected individuals which show a disease-specific characteristic. This could be used for screening drugs in completely human tissue that might reverse the effect. Initial screening in patients with associated risk factors would be avoided. This concept appears to have been validated recently in the case of *familial dysautonomia* (FD). It was shown that the FD patient-derived iPS cells showed phenotypic differences in the development of neuronal crest cells. These differences could be used as a basis for screening various substances. *Kinetin* did in fact have an effect by ameliorating the splicing and developmental defects, thus presenting a promising start for the screening of a broader range of candidate drugs [5].

Figure 13.4 Rudi Jaenisch.

Figure 13.5 Hans Clevers.

- *A source of tissue for regenerative medicine*, including the aforementioned diseases. In addition it has potential for the treatment of chronic diseases of an ageing population such as diabetes, cardiac damage and loss of bone mineralization (osteoporosis). Tissue regeneration has long been established and is based on older technology, for example that for skin grafts in the 1970s.
- This former category can be seen as a potential *form of personal medicine* if the cells used to establish the *iPS* cells are derived from the patient him-/herself. Yamanaka sees, however, a realistic alternative in the creation of iPS cell banks: he concludes that creating 50 cell lines representing different combinations of a set of homozygotic HLA A, B and DR haplotypes, would cover the needs of regenerative medicine for 90% of the Japanese population and could be completed by 2015. Using pre-amplified iPS cells from an iPS Cell Bank would cut down treatment time by some 2 months compared to having to establish novel iPS cells from individuals, as envisaged for creating a personal medicine regenerative treatment programme.

13.4.2
Vectors for Re-programming Somatic Cells to Become Stem Cells [6]

The following methods can be used to re-program normal differentiated cells to form pluripotent stem cells. These methods can be used to introduce the genes for the 'Yamanaka factors[1] into the cell where their expression is induced (e.g. by the addition of doxycyclin if the genes are under the control of the tetracycline promoter):

---

1) Proteins from the klf-4, c-myc, sox-2 and oct-4 genes.

- Lentivirus vector: modified self-inactivating HIV-1 [7]
- Lentivirus vector: site-specific integrase-deficient lentiviral vector (IDLV) [8]
- Baculovirus: modified insect Baculovirus with large cloning capacity [9]
- Electroporation with linear or circular DNA: homologous recombination of DNA transferred during electroporation [10]
- Integrase: pseudo-attP site-specific phiC31 integrase-mediated targeting ensures stable introduction of the introduced genes [11]

13.4.3
A Revolution in Regenerative Medicine

As shown in Figure 13.6, once the iPS stem cells have been established they can be used for regenerative medicine, that is, for tissue replacement, in drug screens or for research applications.

Where are the limiting factors to further development? What can we expect in the short term?

- Using retroviral vectors as part of a personalized medicine treatment would avoid immune reactions to the medication, and *ensure a later down-regulation of the oncogene c-myc.* This would reduce the risk of the iPS generating cancerous tissue.
- It is expected that the efficiency of inducing iPS stem cells will increase rapidly in the short term. Dramatic announcements have recently been made with respect to the accelerating effect of p53 and p21 which have already significantly reduced the time needed to generate a particular mass/number of iPS cells.
- Methods have been developed for inducing iPS cells from somatic progenitor cells without the use of retroviral vectors which employ for instance, plasmid vectors,

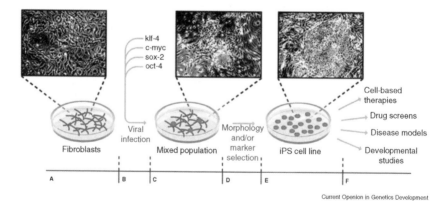

Current Openion in Genetics Development

Figure 13.6 Recently it has become possible to use somatic cells to generate pluripotent stem cells thus avoiding the use of embryonic tissue. Effective methods have been developed for the production of tissue-specific stem cells for a number of purposes including regenerative medicine. This involves either adding the Yamanaka factors to normal somatic cells or adding the genes for controlled production of these factors in the form of DNA or virus (see *Vectors* below).

mini-chromosomes, or just the addition of protein components (piPS). At the moment these are not particularly efficient but are undergoing improvement.

The fact that *iPS technology removes the need for embryonic material* (the ES cells) as a starting point is perhaps one of the most dramatic changes to be made in the generation of materials for this type of regenerative medicine.

Many questions regarding the best ways to use the cells in medical applications still remain unanswered:

- At what differentiation state should the cells be applied, for example as pluripotent stem cells or introduced locally as partially differentiated cells or even tissues?
- How should the cells be dispersed or introduced into the body?
- What are the best monitoring methods?

At the time of writing, exciting advances in this area are being made at a very rapid pace.

13.5
Flash Sequencing DNA: A Human Genome Sequence in Minutes?

Single molecule DNA sequencing of an entire human genome was shown to be a viable option with the recent publication of Pushkarev *et al.* [12], in which they demonstrate the accumulation of some billion bases of raw sequence data per day. Molecules of some 100–200 bases in length were end tagged with short DNA sequences to bind a universal primer. Molecules were randomly immoblized on a surface at some 100 million per cm^2 and several rounds of incorporation of fluorescently-labelled bases using DNA polymerase were scanned. At the time of writing this represented the fastest resequencing of a human genome, which was essentially complete within about 1 week.

However, recently *two* novel DNA sequencing methodologies were reported, both of which appear to be superior to those that have previously been developed. One at least appears to have reached a level of validation which suggests that it will be operational within the next 2 years. This represents a significant step in further accelerating DNA sequencing beyond the already amazing capabilities that have been developed in the last 15 or so years. This new technology is a combination of the advances in nanotechnology and sensitive single-molecule fluorescence detection [13].

Nanotechnology is used to create an observation system, a nanophotonic structure, known as the zero-mode waveguide (ZMW), (Figure 13.7) which can reduce the volume of observation by more than three orders of magnitude relative to confocal fluorescence microscopy, that is down to the zeptoliter (10^{-21} litre) range. This enables single-fluorophore detection. A single phi80-DNA polymerase moelcule is immobilized in each tiny well and acts as the site for the synthesis of a complementary strand for a single DNA molecule. Progress can be followed by changes in fluorescence as nucleotide dNTPs are cleaved and incorporated.

Since this is a single DNA molecule it is a stochastic process and the accuracy of the read remains the same throughout the period in which the polymerase is active which

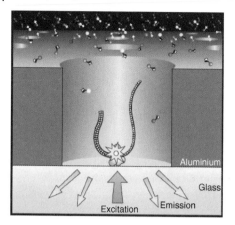

Figure 13.7 Nanotechnology, in this case the Zero-mode wave-guide (ZMW) technique makes fluorescent sequencing of single molecules a reality.

can account for more than a thousand bases per molecule. It requires no cyclic changes of materials. Once one DNA molecule is complete synthesis can then begin on a fresh DNA template.

Parallel sequencing using 14 000 wells could deliver sufficient, high accuracy, data to produce a single human genome sequence per day. The level of accuracy already achieved is sufficient for resequencing applications. This is equivalent to what can be achieved by the best current sequencing technology commercially available (say Illumina or SOLID), in which the high throughput relies on high parallel sequencing with some million samples being sequenced simultaneously. The speed of sequencing a single 'sample' with the ZMW method is over a 1000-fold faster than the current method, is far less expensive and because of the high accuracy resulting from sequencing longer sequences, requires less raw data to achieve the same aim. This method is being developed commercially by *Pacific Biosciences* in Menlo Park, California. It is estimated that in the mid-term raw data acquisition could be made some 10 000-fold faster.

A second methodology [14] uses a protein nanopore with a covalently attached adaptor molecule which can continuously identify incorporation of unlabelled nucleoside 5′-monophosphate molecules with accuracies averaging 99.8%. Methylated cytosine can also be distinguished from the four standard DNA bases and this method does not require preparation of the molecules.

When automats based on either or both of these techniques are finally developed and become routine methodology, the limiting time factor in genome research would have shifted from data collection to data handling.

13.5.1
Another Aspect of Personal Medicine becomes a Reality

This paradigm shift is already taking place. When the research is complete, personal medicine will become a reality for everyone, probably within the next

3 years. In principle nothing would stand in the way of cheap personal genome analysis. The genome analysis technology portrayed in the science fiction book/film *GATTACA* by Andrew Niccol, (film, 1997) would become reality, with the exception that the wait for an individual's complete genetic constitution and 'fitness' as mate, as portrayed in the film, would not be so long. A simple DNA-based 'almost instant' personal identity check will certainly become a possibility. We will be confronted with decisions as to how we wish to apply such powerful technologies sooner than had been expected.

13.5.2
Terrasequencing: Stimulating Development of Biotechnology and Molecular Medicine

It can be assumed that this further acceleration in man's analytical abilities with respect to his and other genomes will continue to stimulate biotechnology particularly as a result of the better understanding of molecular medicine in addition to accessing the cornucopia of natural gene variance that exists in the planet's living resources. This will also bring a better understanding of the interrelatedness and interdependency of all life on Earth. Hopefully we will use this new power wisely so as to maintain a stable long-term and harmonious development during the application of these new tools.

13.6
Systems Biology and Looking for 'Druggable' Targets

13.6.1
Understanding Mechanisms of Action and Finding 'Druggable' Targets

Part of the success in the development of monoclonal antibodies and other pharmaceuticals for autoimmune and inflammatory diseases and in the treatment of neoplasms, has been the definition of a range of new targets for medical intervention. To define the best strategy to continue this endeavour is not trivial.

This may be the area which can profit most from an input from *systems biology.*

A detailed analysis of protein kinases and the discovery of their role in disease led to the development of specific drugs as discussed previously in relation to the development of Glivec, the first example of target-based drug development.

In an attempt to more accurately predict suitable targets for drug development, so-called '*druggable targets*', companies such as Merrimack Pharmaceuticals, Boston, USA, place integrated systems biology at the fore of their product development strategy.

They ask:

- Which is the best target?
- Is there built-in redundancy that would affect the activity of the drug?
- Is there signal amplification or attenuation in the system that may affect the potency of the drug?

- How might heterogeneity, such as differences between cells in a tumour or between different patients, affect the activity of the drug?

In the case of their first drug candidate to enter Phase I trials, Merrimack have relied strongly on what is currently considered arguably, to be the most elegant example to date. The mathematical analyses of and insights into the signal transduction pathways downstream of the EGF receptor were established in 2002 and were largely derived from the work of *Birgit Schoeberl* who at that time was working in *Ernst Dieter Gilles'* group at the Max-Planck-Institute in Magdeburg, Germany, and that of *Doug Lauffenburger's* group, at MIT, Boston. The fact that predictions from mathematical modelling were indeed borne out by experimentation was particularly impressive [15, 16]. The role of different variants of the receptor which influenced the rate of uptake of the receptor into the cell could be correlated with the effectiveness of the EGF-receptor-binding drug, Iressa, which had entered clinical trials but had been producing very variable results (see Section 13.6.2 below: 'Industrial acceptance of systems biology').

Pioneering work has also been carried out by *Alex Levchenko's* group on the complex network interaction involving nuclear factor-κB (NF-κB) isoforms which influence gene regulation. The cell uses this system to either rapidly down-regulate an acute response after short exposure to a growth factor or to maintain a long-term steady level of response in the presence of longer stimulation by varying levels of growth factor [17]. The models show features of dampened oscillatory feedback loops, the mathematics for which will be familiar to those who have worked on controlling parameters in fermenters, where such oscillatory behaviour was first described in yeast in 1973 [18].

13.6.2
Industrial Acceptance of Systems Biology

Although by mid-2007 the pharmaceutical industry was estimated to be spending some $55 billion on research, systems biology was being treated with extreme scepticism by established industry and venture capitalists alike [19]. Dynamic biological systems seemed too complex to be subjected to mathematical analysis. In the course of evolution there has been strong selection for feedback control in living cells that enables metabolic and energy flow to be channelled through the organism. The system must also be robust to mutational change but must allow either rapid adaptation in a changing environment or the attainment of Nash-Pareto optimality in a more stable environment. Mathematical models to deal with such concepts originated in the 1940s with the game theory of *John Nash* and *John von Neumann*. Today it is applied to the analysis of the behaviour of economies and populations. It seems that only very few network topologies are highly successful in this respect [20]. Systems biology approaches will most likely considerably reduce the actual complexity involved in analysing living systems.

AstraZeneca had put *Iressa* into clinical trials in 2003 as an anticancer drug which antagonized the EGF binding to the ErbB1 receptor. There were problems of lack of response in some patients, but they were not able to pin down the common denominator. Following developments in the systems biological approach as mentioned previously, they decided to establish their own systems biology group in house to model the networking and kinetics of the system downstream of the receptor. It became clear that the problem was that Iressa was only able to function on those ErbB1 variants that were slow to internalize into the cell, that is, available to the extracellular drug molecule [21]. AstraZeneca could then establish new screening diagnostics targeted at a selected patient class where a higher degree of effectivity can be expected. Further drug development will address other intracellular downstream targets in the rest of the patient population.

This is one of the first examples of a classic pharmaceutical company openly embracing the concept of systems biology. A project involving the development of an antibody to target the extracellular target was then terminated saving the company an estimated $20 million.

13.6.3
Estimated Number of Potential 'Druggable' Targets

Various statements have been made over the last 20 years about the number of proteins which can realistically be targeted by pharmaceuticals. *Jurgen Drews*, then head of global research at Hoffmann-LaRoche, started the ball rolling by examining the actual targets of all known pharmaceutical drugs. He concluded that only 482 were actually targeted. Most drugs were small molecular weight compounds. A recent statement by bioinformatics personnel at CurAgen Corp. who had recalculated this figure, concluded that in fact there are only 272 'druggable genes' that had actually been targeted and that by 2002 this figure had risen to only 273. The question is then: Just how much more potential is there? Estimates as high as 5–8000 seem unreasonable even assuming that the current estimate of the total number of genes in the human genome increases from 20 000 to perhaps double that number by including micro- and silencing RNA genes [22]. It is unclear whether drug accessibility, functional redundancy and the side effects of involvement of a target in multiple functions, have been adequately considered.

13.6.4
Target Modulation rather than 'On/Off' Agonist Antagonist Effectors

Optimism with respect to a large number of novel targets seems to stem from systems biology as outlined above, combined with the concept that *modulating the function* of a target may be more effective in many cases than direct antagonistic or agonistic effects.

An example of an agonist might be a drug competing with a natural ligand for binding to a receptor. An example of an antagonist would be an inhibitor blocking an enzyme active site. The alternative is to take a leaf from nature's own lexicon where the function of many proteins can be *quantitatively* influenced by the binding of molecules at sites remote from a defined active site. This is termed an *allosteric* effect.

In contrast many drugs which act directly as agonists or antagonists are described as acting in an *orthosteric* mode.

It is known that many ion-channels and indeed a further whole class of trans-membrane proteins, the G-protein-coupled receptors (GPCRs) are susceptible to allosteric modulation. However, it has been difficult to work with such targets, and this can still be considered to be a growing field of research with considerable promise. The first allosteric-action drug to enter the market was Amgen's *Sensipar* (cinacalcet) in 2004, which is a modulator of the calcium-sensing receptor for use in hyperparathyroidism. At the time of writing 10 candidate drugs of this class are in an early stage of clinical development.

Where enzymes occur in *similar isoforms* the active sites are often very similar, as is the case with the superfamily of phosphodiesterases (PDEs). Viagra is a well-known inhibitor of PDE5. PDE4 and an isoform PDE4D play a role in respiratory diseases and inflammation and cognition respectively. It was thought that success in specifically targeting PDE4D activity could be achieved by finding an *allosteric inhibitor* rather than an enzyme inhibitor to block the active site. Modelling differential potential binding areas on these targets helped *deCODE* to develop the drug DG071, which is now in early clinical testing for the treatment of Alzheimer's disease.

A further level of complexity is imposed by the fact that many potential targets are involved in various stable complexes, heterodimers or heteromultimers, where any interaction confers a very different function. The novel structure at the interface of the heteromer provides a new function-specific target for drug development which has so far been little exploited, although for GPCRs in particular, this presents a wealth of opportunity.

13.7
Synthetic Biology

Chemists and biologists are now collaborating in multidisciplinary projects which have rightly or wrongly received the novel designation of *synthetic biology*. They use a combination of tricks accumulated from metabolic engineering and genetic engineering in which enzymes or pieces of whole synthetic pathways are put together in the same organism to carry out a multistep synthetic reaction (see also Section 15.6).

In 2005 the European Commission provided the following definition: 'Synthetic biology is the engineering of biology: the synthesis of complex, biologically based (or inspired) systems, which display functions that do not exist in nature. This engineering perspective may be applied at all levels of the hierarchy of biological structures – from individual molecules to whole cells, tissues and organisms. In essence, synthetic biology will enable the design of "biological systems" in a rational and systematic way'.

Another recent definition [23] emphasizes 'the employment of engineering principles to reprogram living systems' as the distinguishing feature of this burgeoning discipline involving elements of both biology and engineering.

13.7.1
Engineered and Synthetic Organisms

A recent publication discerns two general approaches used in synthetic biology [24]:

- Complex metabolic engineering but with increased input of mathematical modelling and knowledge of 'genetic circuitry'
- Organisms with genomes constructed chemically
    - –and as an extension of the last category, minimal cells in which 'superfluous' genes for growth have been removed
    - –An ultimate goal would be to create a life form completely from synthetic building blocks. Although it is not expected that any supernatural barriers will be encountered, the complexity of the undertaking is still daunting.

13.7.2
Synthetic Genomes and Minimal Organisms

The idea of creating 'novel life forms' has been with us since perhaps the first transformation experiments with purified DNA showed that a virulence gene could be transferred into a bacterial strain (*Avery, McLeod* and *McCarty*, 1945). Certainly with the demonstration by *H. Fraenkel-Conrat and R.C. Williams* in 1955 [25] that it was possible to reconstitute the tobacco mosaic virus (TMV) from its purified constituents (protein and RNA), the requirement for any mystic component of life forms was banished and the roots of a synthetic biology, based on chemical components, which could, at least theoretically, be synthesized in the laboratory, were established.

Knowledge of the DNA or RNA sequence of viral genomes and later entire bacterial genomes went hand-in-hand with the ability to synthesize and knit together ever longer DNA fragments, although, initially a time-consuming process: in 1977 the entire sequence of φX174 virus consisting of 5386 bases was reported [26]. Twenty-six years later in 2003 the virus was fully synthesized and transformed into bacteria where it functioned normally [27].

In 2005 an 'ancient DNA', the influenza virus responsible for millions of deaths during the 1918 Spanish 'flu pandemic was sequenced and the sequence published despite the fear of 'possible misuse by terrorists'. An entire synthesis of the viable virus was completed in 2007 [28]. In 2005 the full sequence of the smallpox viral genome (185 kb) was published. In 2006 a journalist from the *Guardian* newspaper had parts of the smallpox viral genome sent to his private adress simply by ordering synthetic oligonucleotides from a mail-order DNA synthesis company.[2] This occurrence both reignited a debate about the possibility of resurrecting a pathogen to which the human population no longer had any resistance and raised the question of how lax the safety control of recombinant DNA experiments had become.

---

2) *The Guardian*, Wednesday 14 June 2006.

Hamilton O. Smith and *Craig Venter* have undertaken the total synthesis of bacterial genomes with the intention of creating efficient organisms for the conversion of CO_2 to methane; an attempt to create novel fuel sources and at the same time to combat the problems of CO_2 as a causative agent in global warming. The idea was to reduce the genome by eliminating unnecessary genes in an organism that was growing in an energy efficient manner in a defined environment.

The model organism chosen was *Mycoplasma genitalium*. A patent has been taken out on the fully synthetic genome of this ideal organism, designated *Mycoplasma laboratorium* (US Patent) which in one formulation would contain a total of 381 genes, 101 genes fewer than the wild-type organism. This has attracted investment in projects at the *J. Craig Venter Institute* (Rockville, Maryland and San Diego, California) which grew out of the initial microbial sequencing projects at The Institute for Genome Research (*TIGR*; founded 1992 in Rockville Maryland) with the participation of the US Department of Energy (DOE).

Expectations were focused firstly on the remarkable finding that a complete genome replacement can take place in transforming *Mycoplasma capricolum* with the purified DNA from *Mycoplasma mycoides*. The mechanism of replacement involved is not well understood nor has it been repeated with other combinations of host and donor genomes. Oligonucleotide fragments comprising the complete genome for an 'ideal' non-virulent variant of *M. genitalium,* designated strain JCVI-1, were synthesized in the laboratory and knitted together into a single molecule. This was accomplished in a number of steps in *Escherichia coli* BAC clones, the final assembly being carried out with four or more fragments assembling themselves by homologous recombination into a *Saccharomyces cerevisiae* YAC/BAC vector [29].

13.7.3
Generation of a Bacterium with a Completely Synthetic Genome

At the end of 2009 the successful transformation of a restriction endonuclease mutant of *M. capricolum* with the entire synthetic genome of *Mycoplasma mycoides* JCV1-syn1.0 was reported [30]. Some surviving supporters of the notion that a *'mystic life force'* is required for life may feel that their goal posts have been moved. Others see this as a useful extension to the tools for genetic modification of an organism of which there had previously been a restricted supply.

Although many, including *George Church* at Harvard, do not see the necessity of using an entirely synthetic genome for the construction of a minimal genome, such an approach would allow the construction of safety strains, for instance with altered genetic codes (it should be noted that *Mycoplasma* does not use UGA as a stop codon), in perhaps a more controlled fashion than with any other approach. Whatever the criticisms may be with respect to 'hype', these projects illustrate the rapid growth of new capabilities in the field of synthetic biology.

13.7.4
Organisms with Alternative Genetic Codes

One aspect of synthetic biology is to consider developing microorganisms which use a *different genetic code*, not just variable with respect to incorporating one of the canonical 'natural' 20 amino acids (see variation discussed above with respect to encoding tryptophan in *Mycoplasma spp.* and other bacteria). The idea here would be to reserve a particular codon for encoding a particular novel amino acid not normally incorporated into protein during normal translation. Nature has found this necessary at least twice: in the case of *selenocysteine* replacing serine in some enzymes needed for oxidation at very high redox potentials, and in the example of *pyrolysine* which is incorporated into specific enzymes in the methane-producing metabolic pathway in methanobacteria belonging to the *Archaea*. These incorporations use an extended structure in the mRNA to allow recognition of the position in the protein (usually the terminal amino acid) where such an exceptional incorporation should occur. Although the *de novo* engineering of a microorganism variant able incorporate a novel amino acid at a desired position in a protein or proteins is not trivial, several groups have taken on the challenge [31]. One aim would be the production of novel protein variants containing the non-natural amino acids directly from a bacterial fermentation with or without further chemical modifications.

13.7.5
Understanding other Structural Complexities in the Cell

Petra Schwille (Dresden) and others are using novel nanotechnologies to understand the mechanisms of cell and membrane formation and division. In spite of the apparent *irreducible complexity* involved, it does not seem to be impossible to extrapolate to a near future where all the components of a living system may be synthesized from basic chemical building blocks and reassembled to form a living cell. Having once achieved this goal, it would seem highly likely, that having at last fully appreciated the complexity and the elegance with which living cells simply accomplish this self perpetuating feat, we would resort to classical biotechnology to meet daily demands with empirical simplicity.

13.7.6
Gene Mining Meets Synthetic Biology

The main expectations for improvements based on gene mining and synthetic biology, was recently reviewed by Ferrer *et al.* (2009), with an emphasis on biotechnology [32]. New catalysts and cell systems should be identified or developed to address the following (our summary):

- replace chemical reactions
- reduce energy consumption compared to present biological processes

- increase renewable resources, as opposed to generating environmental pollutants, by using processes which produce only biodegradable waste
- allow production of novel pharmaceuticals (e.g. novel polyketides; next generation myxobacterial secondary metabolites?) some of which will address unmet needs.

13.7.7
Synthetic Biology: The New Metabolic Engineering

An exemplary synthetic biology project is the synthesis of *artemesinin*, an effective anti-malarial drug, for which high yields of a precursor, artemisinic acid, are produced by fermentation in *E. coli* and finally converted in two further chemical reactions to yield the desired compound. This was accomplished by *Jay Kiesling's* group at the University of California Berkeley [33]. Artemesinin is naturally produced in the 'weed' Sweet Wormwood. Over a period of 4 years the yield of artemisinic acid obtained by fermentation increased by a factor of a 100-fold per year, finally to a million-fold increase over the initial levels.

The project involved obtaining expression of the plant genes and active enzymes in *E. coli* by reworking the codon usage and additionally imposing combinatorial selection to optimize expression. Initially the plant genes were not expressed at all. Some metabolic pathways were introduced from the yeast *S. cerevisiae* to deal with the complex toxicity problems that arose from the novel anabolic capability contributed by the foreign genes. This event was analysed and solutions proposed as to how to remove the small molecular weight intermediates that had been identified as the culprits. Detailed knowledge of promoter activity and control was employed to obtain a proper metabolic balance of the many interacting components. This work is considered to be one of the most extensive implementations of the principle of metabolic engineering to date and illustrated that an adequate understanding of the variables affecting gene expression and the modelling of interacting metabolic and anabolic pathways, had attained such a level of sophistication that even novel complex anabolic pathways could be introduced and optimized in organisms which were previously naïve to such synthetic capabilities. This was also made possible by the work of microbial geneticists over the last 30 years who have honed the tools for genetic engineering and controlling gene regulation and expression in *Escherichia coli*.

13.7.8
RNA Machines: Real Genetic Engineering

Perhaps more than any other recent work *Christine Smolke* (Figure 13.8) and *Maung Nyan Win's* creation of 'plug and play' *RNA machines* can be cited as an example of the direction in which synthetic biology is progressing.

They imagined creating an RNA device which is capable of detecting and responding to a unique biomarker in (say) a diseased liver cell where it would 'then give the go ahead to do its stuff'. These developments are based on the observation which had already been made in the early 1970s, that many bacterial mRNAs

Figure 13.8 Christine Smolke.

contained a secondary fold which could take on different structural forms during the interaction with other components in the cytoplasm. The first such structure to be recognized originated from the short translational open reading frame in the tryptophan operon of *Escherichia coli*. When tryptophan becomes limiting an RNA loop opens up allowing the ribosomes to bind downstream. The result is that, as soon as tryptophan becomes limiting, those enzymes required for tryptophan synthesis are immediately synthesized without initiating new transcription. Such RNA structures are now known as *riboswitches*. Many naturally occurring examples have been discovered which bind small molecular weight molecules. This binding induces structural changes in the RNA folding pattern which can be envisaged as a type of chemical sensor, for example variants have been found which can bind: adenosylcobalamin; thiamine pyrophosphate (TPP), flavin mononucleotide (FMN), S-adenosylmethionine (SAM), lysine, guanine and adenine.

More recently it has been shown that with the SELEX affinity selection technology small RNA or DNA molecules (*aptamers*) can be selected *in vitro* which will have affinity for almost any small molecule, thus extending the technology in an extremely flexible manner. The natural structures alter their conformation upon binding the small molecule, usually in the 5′-UTRs (upstream untranslated region of the mRNA) leading to an immediate alteration of gene expression. Those aptamers acting as riboswitches in metabolite-regulated genes are found to be highly conserved sequences which can essentially be considered as independently-acting *modules*. [34].

These modules can be treated as *sensors* that bind both the small molecular weight ligand, a *mediator* which significantly alters the secondary structure in the riboswitch, and an *actuator* which alters the sensor's interaction with a target sequence, for example by forming or disrupting a hammerhead nuclease RNA cleavage or forming or disrupting a translational initiation (access to a ribosome binding site, RBS) or termination (formation of a hairpin, blocking translation). Recent publications have shown how fine-tuning the ligand concentration can be achieved as well as establishing higher order co-operation of binding to the small ligand. Various architectures of module arrays produce regulatory elements in which all types of Boolean switch related to the binding of the two different types of molecule can be generated (e.g. AND, NOT, NAND, NOR; Figure 13.9; M. N. Win and C. D. Smolke) [35].

13.7.9
Summing up Synthetic Biology

In 1974, *Waclaw Szybalski* introduced the term '*synthetic biology*' as follows: 'Up to now we are working on the descriptive phase of molecular biology. . .. But the real challenge will start when we enter the *synthetic biology* phase of research in our field. We will then devise new control elements and add these new modules to the existing genomes or build up wholly new genomes'. In 1978 as Editor-in-Chief for *Gene* he wrote: 'The work on restriction nucleases not only permits us easily to

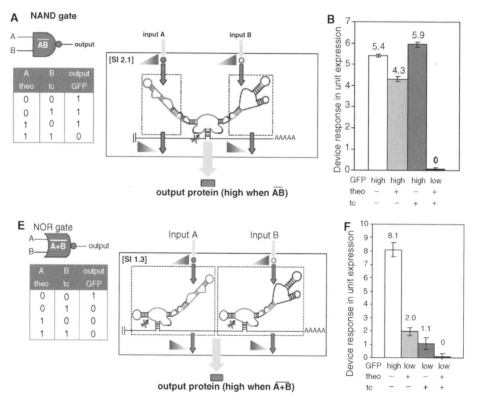

Figure 13.9 Various RNA-machine architectures for different sensor, mediator and actuator modules are shown. These can respond in various ways corresponding to the Boolean operators AND, OR or NOR with respect to the altered response to small molecules in the cell (e.g. in this case theophylline (theo), a caffeine-related compound and tetracycline (tc)).

construct recombinant DNA molecules and to analyse individual genes, but also has led us into the new era of *synthetic biology* where not only existing genes are described and analysed but also new gene arrangements can be constructed and evaluated'. Emil Fischer, who saw the limitations of the application of chemical synthesis to biological problems in his own time, was still optimistic about future developments. In 1917 he stated: 'I strongly believe that, through the products of organic synthesis, it will be possible to gain influence over the development of organisms and to produce changes that surpass all that can be achieved by conventional breeding'. Both visionaries have been proved correct, probably long before they would have predicted.[3]

Let us close by concurring with them, in that we see enormous possibilities in the application of these new developments, but take heed of *J. Robert Oppenheimer* to qualify our hubris[4]: 'Although we are sure not to know everything, and rather likely not to know very much, we can know anything that is known to man, and may, with luck and sweat, even find out some things that have not before been known to man'. Although this may sound like a truism for any scientific enterprise, it is particularly true for the rapidly growing science of biotechnology as an active branch of human enquiry. We see it delivering both the right questions, with respect to the careful development of our industrialization to meet as yet unfulfilled needs, and the tools to create the right solutions. We hope that this will occur mainly in a spirit of international cooperation to further improve our capabilities in the face of the unpredictability of nature and with due consideration for the planet's resources, atmosphere and biosphere.

References

1 Thomas, D.D. *et al.* (2000) *Science*, **287**, 2474–2476; Atkinson, M.P. *et al.* (2007) *Proceedings of the National Academy of Sciences of the United States of America*, **104**, 9540–9545.

2 Ivics, Z. *et al.* (2007) *Molecular Therapy*, **15**, 1137–1144.

3 Lipps, H.J. *et al.* (2003) *Gene*, **304**, 23–33; Duncan, A. and Hadlaczky, G. (2007) *Current Opinion in Biotechnology*, **18**, 420–424.

4 Zhang, Y.W. *et al.* (2006) *Stem Cells & Development*, **15**, 943–952; Keirstead, H. S. *et al.* (2005) *Journal of Neuroscience*, **25**, 4694–4705.

5 Lee, G. *et al.* (2009) *Nature*, **461**, 402–406.

6 Fenno, L.E. et al. (2008) *Current Opinion in Genetics & Development*, **18**, 324–329.

7 Gropp, M. *et al.* (2003) *Molecular Therapy*, **7**, 281–287.

8 Lombardo, A. *et al.* (2007) *Nature Biotechnology*, **25**, 1298–1306.

3) Perhaps sooner than expected a number of companies are already implimenting both systems biology and synthetic biology techniques: one example is Ambryx, Inc. (http://www.ambrx.com/wt/page/technology) uses the incorporation of novel non biological amino acids at precise location in proteins synthesized on modified ribosomes with modified tRNAs and tRNA synthetases (see for example [36]). The design of Zinc-finger nucleases to modify DNA during the induction of human stem cells is the novel technology basis for the Sangamo Biosciences Inc., Richmond, CA (http://www.sangamo.com).

4) As quoted by Paul Berg in his Nobel Prize lecture.

9 Zeng, J. *et al.* (2007) *Stem Cells*, **25**, 1055–1061.

10 Zwaka, T.P. and Thomson, J.A. (2003) *Nature Biotechnology*, **21**, 319–321.

11 Thyagarajan, B. *et al.* (2007) *Stem Cells*, Epub October 25.

12 Pushkarev, D., Neff, N.F., and Quake, S.R. (2009) Single molecule sequencing of an individual human genome. *Nature Biotechnology*, **27**, 847–850.

13 Eid, J. *et al.* (2009) *Science*, **323**, 133–138.

14 Clarke, J. *et al.* (2009) *Nature Nanotechnology*, Published online: 22 February 2009. doi: 10.1038/nnano.2009.12

15 Schoeberl, B. *et al.* (2002) *Nature Biotechnol*, **20**, 370–375.

16 Hoffmann, A. *et al.* (2002) *Science*, **298**, 1241–1245.

17 Cheong, R., Hoffmann, A., and Levchenko, A. (2008) *Molecular Systems Biology*, **4**, Article nr. 192.

18 von Meyenburg, H.K. (1973) Stable synchrony oscillations in continuous cultures of *Saccharomyces cerevisiae* under glucose limitation, in *Biological and Biochemical Oscillators* (eds B. Chance, E.K. Pye, T.K. Ghosh, and B. Hess), New York, Academic Press, pp. 411–417.

19 Borell, B. (2007) Selling systems biology [www.the-Scientist/article/print/53421/].

20 Ma, W. *et al.* (2009) *Cell*, **138**, 760–773.

21 Hendricks, B.S. *et al.* (2006) *IEE Proceedings Systems Biology*, **153**, 457–466.

22 Davies, K. (2009) Bio-IT World http://www.bio-itworld.com/archive/100902/firstbase.html.

23 Haseloff, J. and Ajioka, J. (2009) *Journal of the Royal Society Interface*, **6**, 389–391.

24 Rutz, B. (2009) EMBO reports 10 (Special issue) S14–S17.

25 Fraenkel-Conrat, H. and Williams, R.C. (1955) *Proceedings of the National Academy of Sciences of the United States of America*, **41**, 690–698.

26 Sanger, F. *et al.* (1977) *Nature*, **265** (5596), 687–689.

27 Smith, H.O. *et al.* (2003) *Proceedings of the National Academy of Sciences of the United States of America*, **100**, 15440–15445.

28 Lamb, R. A. and Jackson, D. (2005) Nature Medicine, **11**, 1154–56.

29 Lartigue, C. *et al.* (2007) *Science*, **317**, 632–638; Glass, J.I. *et al.* (2006) *Proceedings of the National Academy of Sciences of the United States of America*, **103**, 425–430; Gibson, D.G. *et al.* (2008) *Science*, **319**, 1215–1220.

30 Lartigue, C., Vashee, S. *et al.* (2009) Creating bacterial strains from genomes that have been cloned and engineered in yeast. *Science*, **325**, 1693–1696, Gibson, D. G. *et al.* (2010) Creation of a bacteria cell controlled by a chemically synthesized genome. *Science*, **329**, 52–56.

31 Wiltschi, B. and Budisa, N. (2007) *Applied Microbiology and Biotechnology*, **74**, 739–753.

32 Ferrer, M. *et al.* (2009) *Journal of Molecular Microbiology and Biotechnology*, **16**, 109–123.

33 Kiesling, J. (2008) Synthetic biology for synthetic chemistry. *ACS Chemical Biology*, **3**, 64–76.

34 Winkler, W.C. (2005) *Curr Op Chem Biol*, **9**, 594–602.

35 Win, M.N. and Smolke, C.D. (2007) *Proceedings of the National Academy of Sciences of the United States of America*, **104**, 14283–14288.

36 Barrett, O. P. T. and Chin, J. W. (2010) 'Evolved orthogonal ribosome purification for in vitro characterization' Nucleic Acids Research, **38**, 2682–2691.

14
Biotechnology and Intellectual Property

14.1
Introduction

Before the application of gene technology in the new Biotech 'start-ups' most biologists had worked in an academic environment with little pressure to consider possible applications of their work to practical or commercial problems. Scientists developed an appreciation of the necessity to protect their intellectual property in the early 1980s, ensuring the transfer of their work from the laboratory to an industrial setting. This later became a requirement for scientists receiving public funding in order to protect the intellectual property they generated.

This chapter discusses only the essentials of patent law as it relates to biotechnology and concentrates on recent developments, including such topics as the patenting of embryonic stem cells and transgenic animal and plant clones which have been the subject of much public debate. We recommend the reader to contact their research institute, company legal department, or a patent attorney before publishing or talking openly to anyone else about their results. None of the following statements should be considered as recommendations for action in a court of law. The authors take no responsibility for any consequences which may result from ignoring this warning.

14.2
Patents Ensure Growth and Rapid Dissemination of Knowledge

Patent literature represents the first publication of over 65% of original research, which, due to the requirement of demonstrating inventiveness, also contains excellent reviews of recent developments in that subject area. In spite of this patents are rarely cited in articles published in scientific journals.

During the infancy of the 'new biotech', life scientists were naïve and often completely ignorant of the whole procedure of patenting and the principles behind it. The basic idea of patenting which is derived from the Latin 'patere' (lie open), is to reveal, divulge or lay open for perusal. It is supported by government and is enforced

Concepts in Biotechnology: History, Science and Business. Klaus Buchholz and John Collins
Copyright © 2010 WILEY-VCH Verlag GmbH & Co. KGaA, Weinheim
ISBN: 978-3-527-31766-0

by national law to not only stimulate the economy but also to ensure that the invention enters the public domain. Stimulation of the economy is seen as the result of ensuring return on investment by issuing limited monopoly (the patent itself) in order to prevent (unfair) competition by those who did not shoulder the burden of the development costs. It was with this intention that patent and copyright was incorporated into the American Bill of Rights in 1791.[1]

14.3
Owning a Patent does not simply Mean that it can be Implemented: 'Freedom to Operate' (FTO)

One general misconception concerning patents is often the impression that the patent automatically allows the owner to practice the invention, that is, that he has a *freedom to operate* (FTO). This is not in fact the case. The patent solely entitles the owner to prevent others from using the invention unless they obtain a licence from the patent owner. Having the freedom to operate requires that an individual owns the intellectual property or obtains licences for all the steps involved in the operation, for example a production process. A rigorous check of this FTO status is usually required by investors before they commit funds for investment ('due diligence' requirement), for example before inaugurating a new company.

14.3.1
Patent Disputes Drive Licence Exchanges and/or Mergers

It was found in general that patent disputes were damaging to the public image of, and confidence in a particular company, resulting in depressed share values. The solution was then often a more altruistic approach to the problem, in the form of *exchange of technologies* (see the *Chugai/Medarex* cooperation in the use of transgenic mice for developing human antibodies); other solutions included splitting markets geographically or according to application area, or even amalgamation via one or other forms of buy-out.

14.3.2
Critical Timing for Biotech Cooperations

Some larger classical pharmaceutical industries seriously mistimed these opportunities, expecting that because of their large financial resources the opportunity to buy out the small high-tech biotech companies would present itself in good time for them

---

1) The idea that man has any *inalienable* rights, that is those which a government cannot remove goes back to *John Locke* (1632–1704) who said 'there are certain rights belonging to every single human being' (religious freedom, freedom of speech, freedom to acquire and possess property, freedom not to be punished on the basis of retroactive laws and of unfair criminal procedures).

to acquire the technology when necessary. For instance *Hoechst* did not buy *Genentech* when it was being offered for sale at some $5 million during financial difficulties in 1982. Only ten years later *Roche* (Basel) paid $2600 million for a 60% share in this pioneering 'new' biotech company and in 2008 they paid $90 billion for a 100% take-over. The 1992 buy-in set a precedent for other technology acquisitions of small biotech companies by their larger competitors.

14.3.3
Orphan Drugs: Special Status Under Patent Law

Orphan drugs are those used to treat a wide array of diseases with a relatively small prevalence that, when added up, affect a significant segment of the population. The number of orphan diseases has been estimated to be between 5000 and 8000 with 25 million people affected in the USA, another 25–30 million in Europe and untold numbers throughout the rest of the world. Orphan drug laws, enforced by the State encourage research, development, and marketing of products for rare diseases [1, 2].[2] The US Congress signed the Orphan Drug Act into law in 1983, recognizing that there are many life-threatening or very serious diseases and conditions that have not received attention because of their low prevalence [3]. The first product to be approved for the treatment of an orphan disease, claims market exclusivity in the USA for 7 years (a really exceptional patent privilege).

14.3.4
The Patentability of Inventions Relating to Plants and Animals (Europe)

The European Directive 98/44/EC distinguishes between plants and animals which are patentable and *varieties* of plants and animals which are not. Patent law thus endeavours to avoid issuing patents for a single variety of plant or animal, which is generally a product of a standard biological process such as a breeding programme. The generation of transgenic plants and animals involves well-defined entities such as genes which are introduced into the organism and their expression controlled by technical processes, for example genetic engineering.

Many patents have been granted in the field of biotechnology in Europe since the adoption of the Directive in July 1998. Some of these have provoked strong reactions from the public. One such case is a patent granted to the Seabright company relating to the creation of a transgenic fish. During discussions on this patent it was clarified that if the claimed invention involves a technical measure which is not confined to a particular animal variety it may be granted.

The European Directive expressly makes provision for the patentability of micro-biological processes and products directly obtained from such processes.

---

2) In the USA a disease qualifies as rare if the prevalence is less than 200 000 persons: Orphan Drug Act passed in the USA in 1983.

14.3.5
The Patentability of Inventions Relating to Elements Isolated from the Human Body

According to the law, the human body and the various stages of its formation and development which includes the embryo are not patentable. It follows that neither the discovery of the sequence of a gene, nor the crude data relating to the human genome constitute patentable inventions. However, inventions which *combine a natural element with a technical process*, enabling it to be isolated or produced *for an industrial application*, can be the subject of a patent application.

Two patents were granted to the company Myriad Genetics that were relevant to the above provisos and caused further debate. These are the patents for an invention which allows genetic screening for cancer of the breast and ovaries in women and are based on the DNA sequence for the BRCA1 gene. The questions focused principally on the dangers which granting the patents might pose for the freedom of research in the EU, in addition to the lack of accessibility to this technology for European patients, resulting from the rights inherent in the patents. Quoting from the yearly report of the European Commission '... the Commission stressed that Directive 98/44/EC was not intended to call into question the freedom of research in Europe. It reiterated that, if research results were commercialized and these results used a technique which has already been patented, a sub-licence should be obtained from the holder of the patent. In addition, most national legislation in the Member States had adopted the principle of exempting prior use, which allows anyone who has already used the invention in the European Community, or had made effective preparations for such use before the patent was filed, to continue such use or to use the invention as envisaged in the preparations'.

14.4
Life-forms as 'Novel Subject Matter' for Patents?

Pasteur patented several of his inventions including 'life-forms', for example his pure cultures of yeast which were used for brewing after inoculation on sterile media under sterile conditions. However, the debate on this subject was rekindled with the *Ananda Mohan Chakrabarty* case in 1980 in the USA.

Chakrabarty who worked for General Electric had developed bacterial strains which were able to break down oil and could therefore be used to deal with oil spills. However, his patent application for the strains and their commercial use were not allowed on the basis that it was not possible to have a patent on life-forms. Since patent law is based on precedence, the court decisions should not have ignored the precedent set by Pasteur's case. Finally the initial decision was overthrown at a subsequent handling of the case in the United States Court of Customs and Patent Appeals, where the Judge stated that 'the fact that micro-organisms are alive is without legal significance for purposes of the patent law'. After this latter decision had been challenged by *Sydney Diamond*, the Commissioner of Patents and Trademarks, it was reopened in the Supreme Court as '*Diamond vs. Chakrabarty*'. Chakrabarty won the case and his patent was granted. The Judges' statement contains the following: 'Judged in this light, the respondent's micro-organism plainly qualifies as patentable subject matter. His claim is...to a non-naturally occurring manufacture or composition of matter – a product of human ingenuity'.

14.4.1
General Considerations Relating to Biotechnology Patents

In terms of patenting life-forms, humans are generally excluded as the subject matter of claims. This is to prevent conflict with constitutional laws guaranteeing human freedom from slavery and is in general, in conformity with internationally ratified agreements on human rights. Almost any other life-form or tissue culture derived from it may be patented as long as:[3]

- there is a technical element involved in its production, or selection, for example a recombinant DNA technique
- and the immediate result is not identical with something that occurs in nature. The first generation recombinant DNA products were often identical to products that occur in nature, for example insulin and growth hormone, but they could now be produced in highly purified form, free of potential pathogenic virus contamination and in amounts often not available in nature (e.g. human blood-clotting factors, human DNAse). Since this includes natural substances as they occur in nature, claims should exclude this (these) substance(s) specifically if they could possibly be included in the general description of the invention. Methods of purifying such substances *can be* patented if they are novel, inventive and (commercially) advantageous. The involvement of an innovative technical step to achieve these goals was decisive.
- and it results in a commercial product.

---

3) The Biotechnology Patenting Directive (BPD; from the EPO) which has been in force since 6 July 1998 (EU directive 98/44) contains the following important statements:
(Art. 4)

- Inventions which concern plants or animals shall be patentable if the feasibility of the invention is not confined to a particular plant or variety,

and from Art. 5:

- The human body, at the various stages of its formation and development, and the simple discovery of one of its elements, including the sequence or partial sequence of a gene, cannot constitute a patentable invention.
- An element isolated from the human body or otherwise produced by means of a technical process, including the sequence or partial sequence of a gene, may constitute a patentable invention, even if the structure is identical to that of a natural element.

14.5
Technology 'State of the Art': Precedence/Directives, not Fixed by Law

Specific directives may be issued by patent offices to help their examiners deal with problems specific to a particular area of technology, but it is only recently that this has been complicated in the area of biotechnology by additional jurisdiction. This is discussed separately below, for instance with respect to the Biotechnology Patent Protection Act (BPPA), which is seen by many scholars as a retrograde act with respect to international efforts to unify patent law, and contrary to internationally ratified agreements of TRIPS (Trade Related Aspects of Intellectual Property Rights) which do not allow patent legislation relating to specific content from a particular technological area. The corollary is that patent jurisdiction can be, or should be, applicable in the same way to any type of technology be it related to physics, chemistry or biotechnology.

14.5.1
Conflict with Moral Codes

The content of a patent should not conflict with '*l'ordre publique*', i.e. *not in conflict with moral codes* [4].[4] Such a conflict has been seen in cases involving tissue of human origin (e.g. derived from a human cancer cell or *in vitro* fertilized egg, or stem cell) and in cases relating to embryonic stem cells as demonstrated by the laws forbidding the use of foetal tissue (to protect the embryo/unborn life).

The production of human iPS cells (see Section 13.3) without the involvement of embryos now provides patentable technology and products. This underlines previous warnings about making patent laws specific to certain technologies, that is stem cell technology, rather than relying on precedence cases. Making and repealing a law on a specific type of technology is expensive, slow and obstructive.

14.6
Who can make Decisions about Public Morality?

It is unclear whether an examiner, judge or small court is sufficiently well informed or representative of the general public's opinion with respect to moral correctness, to make such a decision, especially when detailed technical knowledge is also involved.

With respect to evaluation based on current public opinion or morals, it has been seen that attitudes change over time and that earlier decisions may therefore be overturned at later date. As long as civic law is not changed in order to block the implementation or approval of patents, precedence cases will normally regulate decision making in a fluid fashion. Once specific laws have been passed based on

---

4) In Europe Section 53a EPC.

public attitude to technical innovation, it is difficult to repeal them when in time public opinion changes, thus potentially blocking progress and transfer of technical innovation to the industrial sector. This is the basis of the international ruling that patent law should contain no *specifications relating only to a particular technology*. Law makers in the USA and to some extent in Europe have often conflicted with this guideline during the past 20 years in the aforementioned areas.

14.7
Biotechnology-orientated Directives Guide Patenting Decisions

The Biotechnology Patenting Directive (BPD) drawn up by the EPO and approved in 1999, included the following provisions on non-patentable subject matter on moral grounds:

- Processes for cloning human beings.
- Processes for germ-line modification of humans.
- The use of human embryos for industrial purposes.
- Processes for modifying the genetic identity of animals which are likely to cause them suffering without any substantial medical benefit to man or animal.

Debate on this latter point was intensified by the production of the 'OncoMouse®' at Harvard by *Phil Leder* and *Timothy A. Stewart*, in which tumours could easily be induced. This mouse line is used in research to develop and evaluate anti-cancer drugs at a preclinical stage. Eventually different forms of the patent were drawn up which were valid in Canada, Europe and the USA and were based on the original US 19840623774 application (priority 22.06.1984) with reduced claims that made it clear that human transgenics were excluded from the general method of producing transgenic animals. Further animal models have been patented for other diseases such as rheumatoid arthritis and Alzheimer's disease together with a CFTR knock-out mouse used in research into cystic fibrosis (CF).

References

1 Milne, C.-P. and Cabanilla, L.A. (2007) in *Comprehensive Medicinal Chemistry II*, vol. 1 (ed. P.D. Kennewell), Elsevier, Amsterdam, pp. 655–679.

2 Reichert, J.M. (2006) *Trends in Biotechnology*, **24**, 293–298.

3 Moos, W.H. (2007) *Comprehensive Medicinal Chemistry II*, vol. 7 (eds J.J. Plattner and M.C. Desai), Elsevier, Amsterdam, pp. 2–83.

4 Grubb, P.W. (1999) *Patents for Chemicals, Pharmaceuticals and Biotechnology*, Oxford University Press, London.

Part Three
Application

15
Bioprocess Engineering

15.1
Introduction

Applied microbiology, biochemistry and biochemical engineering have been established as subdisciplines of biotechnology (BT) since the 1960s. These areas of expertise explore microbial systems, enzymes and engineering tasks, taking into account the special requirements of BT, as well as the industrial scale-up and manufacture of microbial and enzymatic products. Textbooks on these subjects have been published since the 1960s and have now become classics in their field.[1] Biochemical engineering continues to represent the basic toolset for managing bioprocesses.

With the progress of molecular biology new tools were developed and their technologies are described as the 'omics': genomics, transcriptomics, proteomics, metabolomics, fluxomics and so on (see Section 15.6). Each of these involves a large database devoted to a particular aspect of the molecular biology of the cell, the organism or even interspecies comparisons, for example, genomics comprises DNA sequence comparisons, transcriptomics relates to mRNA transcripts and their splice products and proteomics to protein translation products. These specialities were first developed as independent entities, but mathematical tools which are referred to in sum as 'bioinformatics' have been developed in order to handle, process and correlate the huge amount of data generated by ever-faster analytical procedures most notably in the field of genomics. The obvious interdependence between the large volume of data and insight into their internal correlations, has led to an attempt to combine them in a 'holistic' approach known as biosystems engineering which is closely related to systems biology (see Section 13.7). Some aspects of applied microbiology,

---

1) Classical textbooks were published by Rehm [1], Pirt [2], Aiba *et al.* [3], and Bailey and Ollis [4]. Comprehensive texts have been collected together in encyclopaedias edited by Rehm and Reed (Biotechnology from 1981 onwards), Flickinger and Drew (Encyclopedia of Bioprocess Technology, 1999), and Flickinger (Encyclopedia of Industrial Biotechnology, 2009), and in the series 'Advances in Biochemical Engineering' (Fiechter, Scheper (eds)).

Concepts in Biotechnology: History, Science and Business. Klaus Buchholz and John Collins
Copyright © 2010 WILEY-VCH Verlag GmbH & Co. KGaA, Weinheim
ISBN: 978-3-527-31766-0

biocatalysis, selected aspects of biochemical engineering, and finally biosystems engineering, including the 'omics' tools, are briefly discussed in the following sections.

15.2
Aspects of Applied Microbiology

Applied microbiology has been at the core of bioprocessing for decades, notably since the demonstration of a spectacular increase in strain and process productivity in the large-scale production of penicillin during the 1940s and 1950s (see Sections 4.2.3 and 4.3.4). Today, managing strains and media still requires the traditional tools of applied microbiology; however, recombinant DNA techniques have taken the lead in innovation and optimization of bioprocesses. This is also true of the development of novel compounds and processes, where the emphasis has moved from screening large collections of microbial strains and their products towards techniques based on molecular biology. Micro-organisms, cells and enzymes represent the basic tools for biotransformations of substrates to yield valuable products. Several bacteria, yeasts and fungi are of technical relevance, as are certain plant and mammalian cell systems. Table 15.1 gives selected data relating to technically relevant micro-organisms (Figure 15.1). A comprehensive overview has been published by Sahm [5] and more recent short overviews by Heinzle *et al.* [6] and Hempel [7].

Bacteria are unicellular prokaryotes with a rigid cell wall. Media composition, temperature, gaseous environment, and pH which are controlled in pure culture are key determinants of bacterial growth. Aerobic bacteria require oxygen for growth, whereas anaerobes grow only in the absence of oxygen. Facultative anaerobic bacteria (and yeasts) are able to grow under both conditions. The optimal temperature ranges for growth is 30–40 °C for mesophiles, 45–60 °C for thermophiles, and 80–105 °C for extreme thermophiles. The optimum pH for growth of most bacteria is in the range of pH 6.5–7.5, although there are extremophiles that can tolerate pH values above or below that range. The sizes of the organisms are mostly in the range of 0.5 to a few µm, but in some cases can be up to 100 µm. Bacteria propagate by symmetric cell division.

Up until the 1990s random mutagenesis (using chemicals, or radiation) was used to modify micro-organisms with the aim of optimizing products and processes, including enhancing the product yield and extending the range of operation (e.g. low pH range). Genetic engineering techniques have largely replaced random mutagenesis as they provide a more specific and rational route to manipulate cellular metabolisms (see Section 13.6). Bacteria, as well as yeasts and fungi can be cultivated in large volumes with high productivity using inexpensive media. The wide variety of products that can be manufactured using such micro-organisms is summarised in exemplary form in Table 15.1 (see also Sections 16.4 and 17.4).

Fungi, or moulds, as well as yeasts, are eukaryotes in which the DNA (chromosomes) is enclosed in a membrane-bound nucleus and the mitochondrial DNA in the mitochondria within the cytoplasm. Fungi reproduce both sexually and asexually,

Table 15.1 Data for technically relevant micro-organisms. (A) Bacteria; (B) Fungi; (C) Yeasts [7].

Family	Description	Use in technical processes (examples)
(A) Bacteria		
Enterobacteriaceae	Gram negative, short rods	*Escherichia, Aerobacter*: different processes, including recombinant proteins
Bacillaceae	Gram positive, spore forming, rods, aerobes or anaerobes	*Bacillus* (aerobes): antibiotics, enzymes, *Chlostridium* (anaerobes): butanol, acetone
Corynebacteriaceae	Gram positive, rods, aerobes	Corynebacterium, Arthrobacter: amino acids
Pseudomonaceae	Gram negative, rods	*Pseudomonas*: hydrocarbon utilization, oxidation of steroids *Acetobacter*: oxidations, for example ethanol to acetic acid, sorbitol to sorbose
Lactobacteriaceae	Gram positive, *Streptococci*: rods	*Streptococcus*: lactic acid *Leuconostoc*: dextran synthesis, enzymes *Lactobacillus*: lactic acid, milk products
Streptomycetaceae	Gram positive, mycel formation, spore forming	*Actinomycetes* and *Streptomycetes*: many antibiotics, enzymes, vitamin B_{12}
Micrococcaceae	Gram positive, spherical	*Micrococcus*: oxidation of steroids *Methanococcus*: methane formation
(B) Fungi		
Aspergillaceae	Mycelia	*Aspergillus*: many organic acids, mainly citric acid, enzymes *Penicillium*: many antibiotics, organic acids, enzymes
Mucoraceae		*Mucor* and *Rhizopus*: organic acids
(C) Yeasts		
Saccharomycetaceae	Budding, also spore formation	*Saccharomyces*: beer, wine, industrial ethanol, bread
Cryptococcaceae	Budding, mycel formation possible	*Candida* and *Torulopsis*: proteins, feed, citric acid

mostly under aerobic conditions. They form dense filamentous mycelia in the form of cell aggregates or pellets which may lead to difficulties in mass transport and in which oxygen transport in particular may limit high productivity [8].

Yeasts usually reproduce by cell division or budding-off smaller cells. The yeast used most often is *Saccharomyces cerevisiae*, which is well characterized and grows quickly in inexpensive media. Traditionally it has been used to produce alcohol under anaerobic conditions in large industrial-scale reactors. Other yeasts, such as *Pichia*

(a) *E. coli*

(b) *Bacillus megaterium*

(c) *Bacillus subtilis*

(d) *Aspergillus niger*

Figure 15.1 Electron micrographs of micro-organisms used in technical processes: (a) *E. coli*; (b) *Bacillus megaterium*, grown in a shake flask (right) and a bioreactor (left, at high shear rate) (bar: 1 μm); (c) Bacillus subtilis (d) *Aspergillus niger*.

pastoris and *Hansenula polymorpha* have also been used, partially for the production of recombinant proteins where post-translational modification (glycosylation) may be important for protein function, solubility and/or stability.

Mammalian cell cultivation was established as early as 1907–1910 by Ross Granville Harrison at the Yale University, and advanced significantly in the 1940s and 1950s to support research in virology and for the manufacture of vaccines. It became of major importance in the 1980s for the production of recombinant human therapeutics. The majority of therapeutic proteins are produced by recombinant DNA mammalian cell cultivation (see for an example Section 17.3.2). Baby hamster kidney (BHK) cells and

Chinese hamster ovary (CHO) cells are most frequently used. Unlike most micro-organisms they produce correctly folded proteins and secrete them into the culture medium. In addition they can carry out post-translational modifications of proteins, for example glycosylation which is often essential for function, solubility and stability of proteins. Production titers have been optimized to yield up to 5 g/l. Mammalian cells have complex nutritional requirements, often requiring serum (e.g. foetal calf serum) which carries the risk of virus contamination. However, chemically-defined media have recently been developed. Furthermore mammalian cells grow slowly and are shear sensitive. Other systems that have been used less frequently for production include insect and plant cells. Although this system has not been widely applied, we note that insect cell systems can be extended simply by the use of viral vectors which results in contained sterile high-yield production in silk worm larvae without the need for sterile media and expensive apparatus in order to produce a high yield [9].

Nutrient requirements which are an essential part of process management, optimization and cost, are classified as follows: water (in some cases of special quality, for example highly purified water for pharmaceuticals production); macro-nutrients (concentrations $>10^{-4}$ M) and micronutrients. Macronutrients include carbon and energy sources, oxygen, nitrogen, phosphate, sulphate, and certain minerals, such as magnesium and potassium ions. The carbon and energy source is the dominant requirement. Heterotrophic organisms (most bacteria, fungi and yeast) require organic compounds for growth. These are often simple compounds such as glucose, sucrose or oil, whereas for technical processes, complex media are required, for example molasses, corn syrup and soybean oil. The latter are preferred for large-scale processes since they contain many other essential nutrients, and are rather cheap [10]. Autotrophic bacteria and algae can utilize carbon dioxide as a carbon source. Common sources of nitrogen are tryptone, ammonia, and ammonium nitrate. Soybean flour acts as a source of nitrogen and potassium and magnesium sulphate together with potassium phosphate are sources of sulphur, phosphorus, magnesium and potassium. The many other trace components required, such as metal ions (e.g. molybdenum and selenium) are usually provided by complex media such as molasses or corn steep liquor which are used in industrial fermentations [6, 7]. Most micro-organisms used in industrial fermentations require oxygen (being obligate aerobic organisms). Delivery of oxygen to the cells requires appropriate aeration systems (gas–liquid mixing) which in turn require considerable amounts of (electric) energy. A few anaerobic organisms that require oxygen-free media have been used for the production of organic solvents (acetone, butanol, see Section 4.3.2), for example *Clostridium* spp. Mixed anaerobic cultures are used on a large scale for the treatment of high load waste water, for example in the food industries. Facultative anaerobes can grow both with and without oxygen (slowly in the latter case). Yeasts used for the production of alcohol, beer, wine and bread are of great importance. They can be grown at an accelerated rate in the presence of oxygen to produce a high cell density, the cells then convert sugar into ethanol by fermentation under anaerobic conditions.

The basis for industrial fermentation is an appropriately high-producing strain which has been screened and selected for the specific purpose (normally in house, or in cooperation with commercial or academic partners). These strains are often the result of long-term development and optimization, the details of which are kept secret and are not made available to other parties (see e.g. [11]). The frozen starter culture is normally stored at −80 to −90 °C in sealed vials. It must be maintained at a high level of productivity by the common practice of strain maintenance and propagation. Where expression plasmids are involved, the strain may be regularly checked for genetic stability using DNA sequencing. This is a requirement in cases where proteins are being used for pharmaceutical use.

15.3
Biocatalysis

Biocatalysts (enzymes and enzyme systems) are nature's tools for catalysing all chemical reactions that are essential to living systems. Compared to non-specific chemical reactions, enzymes confer on biotransformations enantio-selectivity and specificity of the products and enable energy transfer from energy-rich compounds to be coupled to reversible endothermic reactions. Their application in industry includes the production of both bulk and high-added value products such as food (e.g. starch products, sweeteners, modified fats), fine chemicals (e.g. amino acids, vitamins) and pharmaceuticals (e.g. antibiotics, steroids). They are also used to produce washing and cleaning products (nearly all detergent formulations contain enzymes) and recyclable products which afford environmental protection (paper, textiles manufacture), in addition to their use for analytical and diagnostic purposes (Table 15.2) [12–14]. Biocatalysts comprise single enzymes, mixed enzyme systems and immobilized viable bacteria used for biotransformations. The large number of new enzyme processes (>100) that have been introduced in the last 30 years has been reviewed in detail [15–17]. The turnover of technical enzymes is estimated to be €2 billion/annum, and that of products made using biocatalysis to be €60 billion/annum [18].

The advantages of using biocatalysts in industrial processes are:

- Stereoselective production of chiral, enantiopure fine chemicals and therapeutics (chirality being a key requirement in drug manufacture)
- Mild temperatures (0–110 °C) required, low energy consumption
- Reactions in water (in general) at mild pH, 2–12
- Fewer by-products compared to most chemical processes
- Non-toxic when correctly used
- Can be reused (immobilized)
- Can be degraded biologically and in principle, can be produced in unlimited quantities

Insoluble, immobilized enzymes offer all the advantages of classical heterogeneous catalysis: convenient separation by filtration, centrifugation etc. for reuse after the reaction; application in continuous processes, fixed bed or fluidized bed and stirred tank reactors. The most obvious reason for immobilization is the need to reuse

Table 15.2 Products manufactured in quantities larger than 1000 t/a by different companies using biocatalysts.

Product	Enzymes	Free or immobilized enzyme	Companies
>10 000 000 t/a			
Glucose	Amylase	Free	Diverse
	Glucoamylase	Free	
Glucose-fructose syrup	Glucose isomerase	Immobilized	
>10 000 t/a			
Acrylamide	Nitrilase	Immobilized cells	Nitto, DSM
6-Aminopenicillanic acid (6-APA)	Penicillin amidase	Immobilized	Diverse
Cacao butter	Lipase	Immobilized	Fuji Oil, Unilever
Isomaltulose		Immobilized cells	Suedzucker
Galactose (Lactose free milk or whey)	β-Galactosidase	Free or immobilized	Diverse
>1000 t/a			
7-Aminocephalosporanic acid (7-ACA)	D-Amino acid oxidase	Immobilized	Diverse
	Glutaryl amidase	Immobilized	
L-Aspartic acid	Aspartase	Immobilized	Tanabe
Aspartame	Thermolysin	Immobilized	Tosoh, DSM
L-Methoxyisopropyl amine	Lipase	Immobilized	BASF
D-Pantothenic acid	Aldolactonase		Fuji Chem. Ind.
D-Phenylglycine	Hydantoinase, carbamoylase	Immobilized	diverse
L-Amino acids	Aminoacylase	Free	Evonik, Tanabe
>10 t/a			
Amoxicillin	Penicillin amidase	Immobilized	DSM
Cephalexin	Penicillin amidase	Immobilized	DSM
L-DOPA	β-Tyrosinase	Immobilized	Ajinomoto
Human insulin	Carboxypeptidase A	Free	Aventis
	Lysyl endopeptidase	Free	Diverse
	Trypsin	Free	
Sterically pure alcohols and amines	Lipase	Immobilized	BASF
D-Mandelic acid	Nitrilase	Immobilized	

Some other selected products made in lower quantities in recently developed processes are also included.

enzymes if they are expensive, in order to make their use in industrial processes economic. Today there are more than 15 processes of major importance on stream as well as over 100 further special applications (Table 15.2) [14, 15].

The impact of genetic engineering using recombinant techniques has played a major role in extending the fields of enzyme application. Thermal stability and selectivity can be improved by site-directed and random mutagenesis as well as by directed evolution. Amylases from hyperthermophilic micro-organisms are active even at temperatures of up to 130 °C and are used in industrial processes in a temperature range of 105–110 °C [14, p. 109–164]. The half lives of subtilisin (used in large quantities in surfactants) and a *Bacillus subtilis* lipase were increased 1500-fold and 300-fold, respectively, using a crystal structure to guide mutagenesis [20]. The (regio-) selectivity can be significantly improved in many cases and even the catalytic mechanism can be modified, for example by changing hydrolytic into synthetic catalysis and vice versa [21, 22].

A few innovative approaches using genetic engineering will be mentioned. Push–pull concepts extend the range of reactions with the yield limited by a non-favourable equilibrium, that is, thermodynamic constraints. They include coupled enzyme reactions such as oxidation and decarboxylation. Whole-cell biotransformations make use of engineered micro-organisms, for example *Pichia pastoris* which uses methanol as a co-substrate and produces CO_2 only as a by-product, has a high regeneration rate for NADH which then becomes available for dehydrogenase catalysis (2 moles of NADH per mole methanol); CO_2 as the by-product makes reduction reactions irreversible thereby driving the reaction towards the product [23]. By cascade reactions, C–C bond formation by aldolases has been coupled with an amino acid oxidase and an amino acid dehydrogenase to yield CO_2 as a by-product and shift (pull) the otherwise limited yield towards an α-hydroxyl acid [24]. Substrate and enzyme engineering can provide favourable thermodynamic conditions and improve reactivity, or extend the range of enzyme selectivity, for example for non-natural (activated) substrates, providing high bond energies that drive the reactions. Thus alternative glycosyl donors were synthesized including glycosyl fluorides and sucrose analogues. Cleavage of these glycosidic bonds is highly exothermic which provides favourable kinetics for synthetic reactions by glycosynthases or sucrases, both natural and modified, towards the formation of glycosidic bonds [25, 26].

The metagenome approach offers a considerable extension of the range of available enzymes. It provides access to large, hitherto unknown sequences of uncultivated, and uncultivable organisms, identifying, and making available (by DNA analysis e.g. of soil probes) new proteins, including enzymes of industrial interest [27, 28] (see Section 7.5).

As a bioinformatic tool for biocatalysis the BRENDA (BRaunschweig ENzyme DAtabase) (http://www.brenda-enzymes.org) provides a large freely available information system containing biochemical and molecular information on all classified enzymes. It contains information on classification and nomenclature, reaction and specificity, functional parameters, occurrence, enzyme structure and stability, mutants and enzyme engineering. It further presents software tools for querying the database and calculating molecular properties [29].

Future developments have been addressed by ESAB [30] including *in silico* design of fold and function, design of cascades for bioconversions, integrated process design, using microreactors and advanced downstream processing. Modelling tools to predict enzyme structure and function tend to be used routinely. However they require experimental proof of results. The use of various prediction algorithms and experimental approaches to check predictions is a recent example. Relevant structural details of a transmembrane protein or enzyme have been addressed, including reporter enzyme fusion protein analysis, site-directed mutagenesis and protein expression (Remminghorst *et al.*, 2008 [128]).

15.4
Biochemical Engineering

The aims of biochemical engineering are ([6, 7, 31])

- The quantitative investigation of biotransformations and modelling of the processes
- The development of bioreactors and downstream operations
- The transformation of laboratory results and theoretical approaches (models) into technical dimensions (scale up)

Fermentation using micro-organisms – bacteria, yeasts, fungi – has been well established for decades, and that using recombinant strains since the 1980s. Some microbial transformations produce yields close to the theoretical maximum, for example gluconic acid or alcohol fermentation. Others, notably those producing secondary metabolites operate at much lower yields – a strong motive for using metabolic engineering to achieve improvements ([32], [116]).

Stoichiometry is the basis for the quantitative analysis of chemical and biochemical reactions. Based on principles that had already been established in chemistry by Lavoisier it involves relating the quantities of the reactants (substrates) to the products that are formed. For single reactions (for many of the bulk products) the relationships are relatively simple in terms of their molar concentration or stoichiometry. An example for components A and B reacting to form product C is shown below (v_i is the stoichiometric coefficient for species i in the reaction, where the value of v is positive for the products and negative for the reactants) [6, p. 23–27]:

$$v_A \, A + v_B \, B = v_C \, C$$

Thermodynamics is important in calculating the heat of reactions (which may require appropriate cooling facilities in large fermenters) and thermodynamic equilibria and will not be discussed in further detail here (see [6, p. 28, 29], [33]).

The study of *kinetics* determines the time required for a desired conversion and thus the reactor size and associated investment cost. Only a few basic equations can be given here. Enzymatic conversion which in most cases uses a single enzyme follows relatively simple Michaelis–Menten-type kinetics, with first-order dependency on the substrate concentration in the lower and zero-order dependency in the higher concentration range (Equation 15.1). (In Equation 15.2 V_{max} denotes the maximal reaction rate at high substrate concentration as shown by $V_{max} = k_{cat}E$, where k_{cat} is a kinetic constant, E is the enzyme concentration, K_m is the Michaelis constant that equals the substrate concentration for half-maximal reaction rate and K_i

is the respective inhibition constant). In order to achieve maximum efficiency (including minimum reactor size and downstream cost), technical processes are mostly carried out using high concentrations of the reactants thus resulting in high levels of product. Therefore kinetics usually takes inhibition phenomena into account. For the common case of competitive product inhibition the reaction rate is given by Equation 15.2. For batch and continuous plug flow reactors the integrated equations are applied (Equation 15.3 for the most important case of competitive product inhibition).

From Equation 15.3 the amount of biocatalyst corresponding to V_{max} can be calculated for a given conversion $[S]_0 - [S]$. For (rather common) cases of product inhibition it is advisable to use a batch or a continuous plug flow reactor (such as a column with immobilized catalyst). For immobilized biocatalysts, mass transfer constraints occur which reduce efficiency under conditions of low substrate concentration. Further, enzyme inactivation kinetics need to be taken into account, notably for continuous processes that are typically run for several months. The inactivation of a biocatalyst can, in many cases, be described by a first-order process, where, in the exponential equation, k_i corresponds to the inactivation reaction rate and t to the reaction time (Equation 15.4) (see [14, p. 203–208, 333–367], [6, p. 29–32], [7]).

$$v = -\frac{d[S]}{dt} = \frac{k_{cat}[E][S]}{K_m + [S]} = \frac{V_{max}[S]}{K_m + [S]} \tag{15.1}$$

$$v = \frac{V_{max}[S]}{K_m + \frac{K_m}{K_i}[S]_0 + [S]\left(1 - \frac{K_m}{K_i}\right)} . \tag{15.2}$$

Integration gives:

$$V_{max} = \frac{([S]_0 - [S])\left(1 - \frac{K_m}{K_i}\right) + \left(K_m + \frac{K_m}{K_i}[S]_0\right)\ln([S]_0/[S])}{t} . \tag{15.3}$$

$$V_{max} = V_{max,0}\, e^{-k_i t} \tag{15.4}$$

For the growth of biomass the concentration of biomass x can, in many cases, accumulate at a rate μ which corresponds to exponential growth (Equation 15.5). Integration gives the concentration of biomass x at time t (where x_0 is the initial biomass concentration) and in many cases the lag phase must be taken into account (Equation 15.6). The growth rate μ depends on the limiting substrate concentration $[S]$ in a saturation-type curve, thus at low substrate concentration μ depends on $[S]$ according to Michaelis–Menten kinetics as shown by Equation 15.7 (μ_{max} is the maximum specific growth rate, K_S corresponds to the Michaelis constant in enzyme kinetics). From Equations 15.5 and 15.7 a function relating the growth of micro-organisms and the (limiting) substrate concentration (Equation 15.8) can be obtained using the rather simple equation derived by Monod, which assumes that there is one rate limiting step.

$$\mu = \frac{1}{x} \cdot \frac{dx}{dt} \tag{15.5}$$

$$x = x_0 \cdot e^{\mu \cdot t} \tag{15.6}$$

$$\mu = \mu_{max} \cdot \frac{[S]}{K_S + [S]} \tag{15.7}$$

$$\frac{dx}{dt} = \mu_{max} \cdot \frac{x \cdot [S]}{K_S + [S]} \sim \mu_{max} \cdot x, \ \text{if } S \gg K_S \tag{15.8}$$

Three typical kinetic patterns of growth and product formation are frequently observed. Most production processes are terminated when the growth phase becomes stationary (i.e. when the limiting substrate is depleted). However, in the case of secondary metabolites and also typically for heterologous proteins, production occurs only in the late phases of cultivation in which growth has slowed to a low rate [6, p. 29–32], [7]. The kinetics of product formation for different cases can be described by the Luedeking–Piret equation (Equation 15.9). The specific rate of product formation, q_P, is linked to growth by the parameter a. Non-growth-associated production is characterized by parameter b (more complex kinetics are common, for examples see [6]). The kinetics of growth and product formation determine the required reaction time and thus reactor size, the final product and by-product concentrations, as well as essential parameters for the design of downstream operations.

$$q_P = \frac{1}{x} \frac{dP}{dt} = a\mu + b \tag{15.9}$$

Technical fermentations, or biotransformations, typically comprise upstream processing, the bioreaction and downstream processing steps to isolate the product in appropriate purity and quality (see above) (for comprehensive overviews see [34, 35]). *Upstream processing* comprises media preparation (which must be appropriate for growth and product formation) and sterilization, either by filtration or by thermal – batch or continuous – inactivation of micro-organisms prior to inoculation. For aerobic processes sterile air must be supplied continuously using depth filtration. Aseptic, sterile operations throughout are mandatory [36, 37]. *Inoculation* starts first with liquid culture in test tubes, then in small (e.g. 100 ml) and subsequently in large (2–5 l) Erlenmeyer flasks or in small-scale fermenters (see [7]).

Appropriate *bioreactors* for production purposes must meet the requirements for mixing (agitation), aeration, corrosion resistance and aseptic operation (Figure 15.2) [7]. Bioreactors used in industry are typically stirred vessels and use

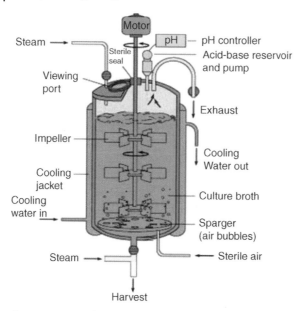

Figure 15.2   Typical bioreactor: stirred tank reactor (STR). (R. Stehr, with kind permission of Henkel KGa, Düsseldorf, Germany).

of the stirred tank reactor (STR) is standard. For scale up, the capacity of these bioreactors is in the range of for example 50, 100 or 300 l and 1 m³ and for production, 10–200 m³ or even 1000 m³ for very large volume processes such as glutamic acid and alcohol manufacture (Figure 15.3). In some cases other types of bioreactors are also used, for example surface culture for citric acid and vaccine production. Cleaning is often carried out 'in place' using CIP (clean in place) technology; sterilization prior to fermentation is essential and is commonly achieved using heat treatment at 120 °C for 20 min or sterile filtration. The standard operation is still the batch process. Mixing and oxygen transfer are essential parameters for growth and product formation. Mass transfer, the primary agitation criterion, has been investigated for all scales of reactor volume and substantially different levels of viscosity. Power input per volume is one of the lead criteria for scale-up ([38]; for an extended overview see [39]). However, since it is almost impossible to reproduce exactly the same fluid dynamic conditions as in small scale reactors, scale up remains one of the most complicated tasks of biochemical engineering. Thus it is rarely possible to simply transfer the optimal conditions identified at the pilot scale to large commercial plants. Mixing time is critical in large aerated stirred reactors and increases with fermenter volume, reaching about 100 s in a 100 m³ vessel. Thus, in a large scale process there is considerable heterogeneity within the fermentation broth, for example in relation to pH and oxygen concentration (Figure 15.3). The choice of an appropriate impeller type (e.g. multiple turbine impellers, typically with six blades) is therefore essential [6, 7, 38, p. 36–40], [37].

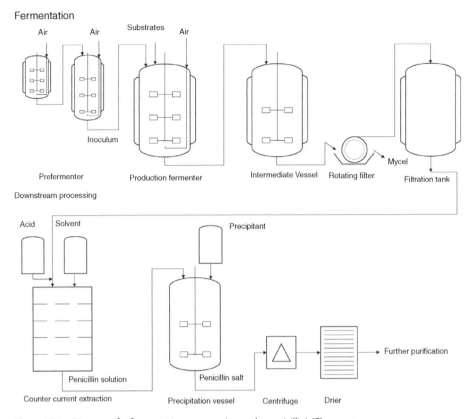

Figure 15.3 Diagram of a fermentation process (example: penicillin) [7].

Production is commonly carried out in batch mode; many processes are conducted as fed-batch fermentations, where nutrients, mainly the carbon source, but also precursors for biosynthesis of secondary metabolites, are re-supplied after consumption. This provides the advantages of avoiding substrate inhibition and significantly improving the concentration of the end-product. Stirred tank reactors may also be used for enzymatic processes; integrated membrane systems have been applied in cases where a soluble enzyme is to be recycled. Fixed bed reactors are the most common type of reactor used in processes employing immobilized biocatalysts [14, p. 369–417], [15].

In general *downstream processing* comprises a range of unit operations, typically including solid–liquid separation, isolation and purification of the product, and in the case of high value added products, polishing. All steps must be optimized with respect to the required purity of the product and maximum yield. A simplified scheme illustrating downstream processing is given in Figure 15.4 [7]. In-depth discussions of downstream operations have been given in Moo-Young (Ed., 1985) [41] and by Atkinson and Mavituna [34]. The essential steps in downstream processing are [6, 40–42, p. 40–50], [31]:

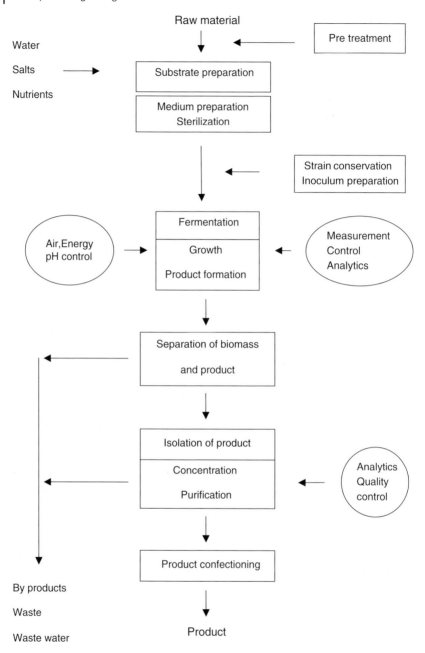

Figure 15.4 General scheme of fermentation and downstream processes [7].

- Cell disruption (to release intracellular or periplasmatic products) using homogenizers or controlled shear bead mills
- Solid–liquid separation by centrifugation or membrane filtration
- Membrane processes
  - Crossflow filtration (Solid–liquid separation)
  - Ultrafiltration (high–low Mw separation, concentration)
  - Reverse osmosis (salt removal, concentration)
- Product recovery and purification
  - Adsorption
  - Solvent or aqueous two-phase extraction
  - Distillation
  - Chromatography, including ion exchange, size exclusion, reversed phase or
  - hydrophobic, affinity chromatography, or continuous chromatography (SMB)[2]
- Concentration (ultrafiltration, evaporation, distillation, precipitation)
- Crystallization
- Finishing, polishing, including stabilization
- Waste treatment, reduction and recycling (including waste prevention)

The choice of downstream processing steps depends on the product quality required. There are significant differences in the requirements for different processes, thus for bulk products such as alcohols, organic acids and detergent enzymes, a high yield (typically > 90%) is essential and this can be achieved with only a few downstream operations. In contrast, the high purity required for fine chemicals and notably pharmaceuticals will involve several or many efficient purification steps including chromatography which will reduce the yield considerably (often by 45–65%). The type and sequence of purification steps should exploit the unique nature of the product, notably molecular weight, pI and polarity. The purification of pharma proteins may typically require more than 10 unit operations. A prerequisite for optimizing the number of operations is a detailed understanding of the thermodynamics and physics underlying the processes being utilized [44].

Whereas ion exchange chromatography is often employed as a first purification step, size exclusion may be chosen as a polishing step at the end of a purification sequence, to remove trace contaminants for example. Further techniques used are expanded bed adsorption, allowing for initial product recovery, high gradient magnetic separation, membrane based systems and extraction using aqueous two-phase systems. The product molecules can be captured directly by attaching specific tags to them using genetic engineering methods [45, 46]. The recovery of

---

2) SMB (simulated moving bed technology) may provide advantages such as reduced overall cost. Such technology has been introduced into the large industrial processes, for example chromatography of molasses to isolate sucrose, raffinose or amino acids in addition to chiral intermediates for pharmaceuticals, as carried out at Bayer Technology Services and Novartis [43].

lactoferrin from crude whey has been achieved using magnetic micro-ion exchangers in combination with high-gradient magnetic separation in an automated process [47]. Affinity chromatography may afford very high product enrichment and can minimize the number of actual chromatographic steps required, but in general this still remains a very expensive option [40].

Monitoring, measurement and *control* are essential tools that may include modelling of process steps or of the whole sequence. Measurement of temperature, pH and pO_2 are most essential steps and further analysis of cell density, foam, product concentration, for example enzyme activity, and many other process variables is common practice [48–50].

The *running costs* of a process are distributed as follows: about 22% for purchase of raw materials (bulk products), 33% for fermentation and 45% for downstream processing (without considering the cost of licensing patent technologies). Further ecologic and economic optimization comprising energy efficiency of the process and life cycle assessment have become common operations and tasks for development, including assessment of cost and effects of resources, production, product distribution, use and disposal (materials cycling, e.g. recycling organic solvents) (see Section 15.5) [51].

Integrated development of bioprocesses and process *integrated downstream operations* have received an increasing level of attention in recent years [6, 48, p. 50–59], [31]. Examples of process integrated product isolation include adsorption, membrane separation, crystallization, and so on [14, p. 412–417]. Nucleic acids can be isolated directly from crude sample materials such as blood, tissue homogenates, cultivation media and crude cell extracts using specifically functionalized magnetic particles. They can be removed easily and selectively even from viscous sample suspensions using small particles (diameter approx. 0.05–1 µm) in the presence of biological debris; the efficiency of magnetic separation is especially suited to large-scale purification [52]. The integrated production and purification of recombinant proteins deserves particular attention since the methods of manufacture of these high-added value products need to be improved in order to reduce the high manufacturing cost. Recent concepts for achieving this aim include integrating all the steps from genetic modification and process design to downstream processing in order to identify optimal process conditions using integrated process development, including systems biotechnology approaches (see Section 15.6).

15.5
Process Sustainability and Ecological Considerations

Sustainability or sustainable development has been defined as development that meets the needs of the present without compromising the ability of future generations to meet their own needs. Others define it as the optimal growth path that maintains economic development while protecting the environment and optimizing social conditions and relying on limited, exhaustible natural resources. Three dimensions should be taken into consideration: economic, environmental, and social development [6, p. 81–115].

Tools for *economic assessment*, a cost and profitability analysis, comprise the estimation of capital investment and operating cost and will be discussed only very briefly (for details see [6, 34, p. 81–115]). Capital investment should be based on a process flow diagram and a list of the essential equipment required for the process, including fermenter, media and product tanks, piping, downstream operation equipment, additional instrumentation and control facilities. A multiplier covers further costs, for example the planning cost. The operating costs include purchase of raw materials, energy, utilities and waste, and labour.

The *environmental or ecological dimension* relies on two aspects: the characteristics of the process and the properties of the components which together comprise the environmental indices (EI). The first includes the Mass Index (MI) which is calculated from the material balance of the process for all input and output components. In the case of the input materials the MI indicates how much of a component is consumed to produce a unit amount of the final product. In terms of the output components, the MI provides a measure of how much of a component is formed per unit final product, for example the volume of biomass produced in order to obtain a unit amount of purified enzyme. The sum of all input MIs (or output MIs) results in the Mass Index of the process which is an indicator of its material intensity. Not all components have the same environmental relevance. Therefore their impact is considered from the point of view of the properties of the components to give the second aspect, the environmental factors (EF). A range of (up to 15) impact categories (high, medium or low relevance) have been defined that represent the environmental relevance of a component, with respect to the environmental, health and safety aspects. These include availability of raw material, acute and chronic toxicity, ecotoxicity, contribution to global warming, acidification and eutrophication potential and so on. By introducing numerical values a weighing factor can be derived for every component. The environmental index (EI) is calculated by multiplying the MI and the EF, which leads to the identification of those components that are environmentally most relevant in the process (for the method of calculation see [6, p. 81–115]).

A range of these aspects has been introduced into the management of development work: minimization of material and energy intensity of products and services, minimization of toxic effects, increase in recycling potential of products, maximal use of renewable resources, and increase in the life cycle of products. BASF (Germany) even offers a service which analyses the eco-efficiency of products and processes [53].

Atom efficiency is basically a tool with which to evaluate technologies that reduce the generation of waste. It defines an E factor which is calculated from the quantity of waste generated per kg of product. This factor increases on a downstream basis from bulk to fine chemicals and specialities such as pharmaceuticals, where, however, the overall volume of product and thus waste, diminishes dramatically in general. The atom efficiency is a useful tool for the rapid evaluation of the quantity of waste generated by the routes used to synthesize a specific product. It is calculated by dividing the molecular weight of the desired product by the sum total of the molecular weight of all substances produced in the stoichiometric equation(s) involved. The theoretical E factor is readily derived from the atom efficiency (e.g. an atom efficiency of 40% corresponds to an E factor of 1.5). In practice, the E factor is much higher as the yield is not 100% and an excess of reagent(s) is often used, in addition solvent losses and salt generation need to be taken into account [54].

Indicators also play a key role in the *social assessment* of evolving technologies, however, they lack a general consensus. Such indicators have been developed by the Wuppertal Institute (justus.geibler@wupperist.org) for the biotechnology sector. The political relevance refers to political initiatives such as the sustainability strategies of a government. Further the relevance of stakeholders has been addressed with respect to different groups such as suppliers, customers, unions, industry, including competitors, financial, regulatory and legislative institutions, as well as NGOs. The entrepreneurial and product relevance has been determined by a survey of biotech companies, including information from rating agencies. Such a survey identifies relevant social aspects, as well as the possible contribution of BT products to the fulfilment of human needs and to the challenges and opportunities of the social field. Finally, eight significant aspects have been identified: health and safety; quality of working conditions; impact on employment policy; education and advanced training; knowledge management; innovative potential; customer acceptance; and societal dialogue [6, p. 81–115].

Manifold interactions exist between the three aspects of sustainability. Almost all environmental categories affect economic and social sustainability. The availability of raw materials takes into consideration the depletion of natural resources which may then lead to a steep increase in the price of the starting materials in the medium or long term. The plant capacity has a considerable impact on economic success, which is also influenced by competitors. Furthermore, government policies and legal constraints have a significant effect on a process; this is particularly true of pharmaceutical processes. A striking example of such interactions is the use of genetically modified (GM) crops in agriculture. The variation in attitude to the use of GM crops is illustrated by their rather high and low acceptance in America and the EU respectively. Thus all three dimensions of sustainability should be considered early in process development and there should be awareness of their possible interactions [6, p. 81–115].

15.6
Biosystems Engineering, including Omics Technologies

The aim of biosystems engineering or systems biotechnology is the integration of biology, mathematics, bioinformatics, and systems engineering to gain a holistic view of complex biological and biotechnological systems, including quantitative description and improvement of established or developing novel production processes.[3] When searching for a new product or process it has been proposed that the ideal strategy to follow is a rational procedure ranging from [55]: sequencing genomes of microbial species (bacterial, yeast, or fungal) of particular interest to the identification of host strains for use in metabolic engineering. In 2007 the genomes of 250 bacteria were available[4] (see Sections 7.5 and 13.6). The next step is annotation and reconstruction of the metabolism. Transcriptome, proteome and metabolome investigations are essential steps as the basis for rational optimization of a microorganism. Once the proteome has been elucidated the proteins synthesized by an organism can be identified and although this helps in the understanding of the cellular processes, the final synthesis of a model of an entire cell remains elusive.

The more detailed aim of biosystems engineering is to integrate data from the 'omics', notably from the transcriptome, proteome, fluxome, metabolome and environome, to mention only the most important. The prediction and optimization of biotechnological processes using modelling and experimental validation also fall under the remit of biosystems engineering. In order to realise even part of this highly complex goal, selective procedures for establishing models are applied, essentially selecting data and equations that are considered to be relevant. Since complete models are the most difficult to establish, partial models are used which are based for example on standardized fermentation conditions. In order to illustrate the problem, the numbers of variables (compared with the more simple task of modelling a bacterial system) are estimated to be around 5000 for the transcriptome, up to 500 for the proteome, some 100 for the fluxome, in the range of 20–100 for the metabolome, and about 10 for the environome. The analytical methods and tools used include DNA-arrays in transcriptomics (with up to 15 000 oligonucleotides, and producing quantitative information with an error of a factor of <2); 2D-gel electrophoresis for qualitative analysis, ELISA for obtaining quantitative data in proteomics; MFA (metabolic flux analysis via mass balances and including isotope materials), FBA (flux balance analysis via modelling) in fluxomics; GC-MS in metabolomics, and classical analysis of substrates, products, pH, pO_2, CO_2, temperature, time, power input and so on in environomics. The omics technologies should provide complete information about the state of cells and should characterize their phenotypes. Their integration using bioinformatic tools should enable the construction of networks and models using the genome and further experimental data obtained, both for production processes and target identification for drug design [56, 60, 61].

---

3) The highly sophisticated and complex nature of biological phenomena raises the question: Can we calculate life? Presently, any serious scientist will answer this question with a clear 'no' [56]. The vision is that a vast potential of multiple genes could be used to create entirely new pathways to produce a wide range of compounds from a diverse substrate potential [57]. The term 'biosystems engineering' has been introduced into the title of an established journal, which from 2001 onwards became known as *Bioprocess and Biosystems Engineering* and was intended to widen the scope and emphasize new approaches to the rational and evolutionary design of cellular systems by integrating recombinant technology and process design [58].

4) By April 2009 the genomes of over 1000 bacteria were completed and those of a further 928 were in progress; those of 22 eukaryotes and 10 fungi were also completed [59].

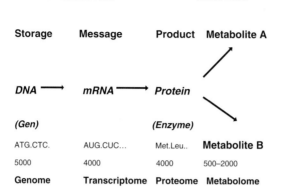

Figure 15.5 Basic molecular structure of biological processes. The information stored in DNA is transcribed into messenger molecules (m RNA) which in turn encode the synthesis of proteins on the ribosomes. Most of the proteins produced act as catalysts for the reactions in the metabolic network. The entirety of the DNA information is known as the genome, for microorganisms it typically contains up to 5000 identified genes. The proteome comprises about 4000 proteins and roughly 2000 metabolites can be identified as comprising the metabolome [56].

The basis of biosystems engineering resides in the molecular structure of biological processes and structures (Figure 15.5). The information is stored in the genes of an organism, the genome, and can be transcribed into messenger RNA (transcription), depending on physiological conditions and then translated into proteins (translation), depending in turn on regulation phenomena (transcriptomics). A typical scheme for protein synthesis, processing and secretion is shown in Figure 15.6. Proteins predominantly form enzymes. Gene expression is subject to regulatory control and genes with coherent functions are frequently organized into operons which are mutually controlled. Control may be by induction, repression or inhibition at different levels (transcription etc.) related to manifold stimuli, notably substrates (e.g. glucose, sucrose and lactose), but also toxic substances, and by virus and other microbial attack [62, 63]. Some more details on omics-technologies will be summarized subsequently.

For *eukaryotes* the picture has been made more complicated by the realization that small RNAs are also involved in regulating gene expression at many levels. Also long-term effects of the environment may feed back and influence gene expression through so-called 'epigenetic' effects where chromatin is specifically modified. Also conformational organization of chromosome segments in the nucleus influence gene expression in ways which are still in the process of being comprehended (see Section 11.3).

Genomics are discussed in detail in Section 7.5. Genomics encompasses fast automated sequencing, functional sequence analysis and genome annotation. It is the basis of functional genomics allowing identification of known and new genes (e.g. of enzymes) from genomes deciphered during recent years. Thus the automated

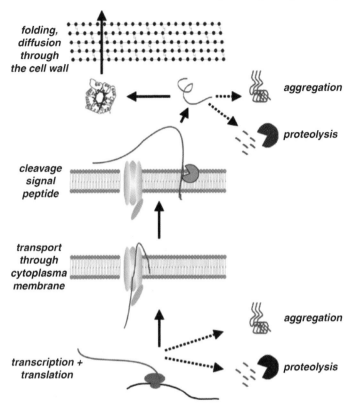

folding, diffusion through the cell wall

aggregation

proteolysis

cleavage signal peptide

transport through cytoplasma membrane

aggregation

transcription + translation

proteolysis

Figure 15.6 Schematic representation of the SEC pathway for protein secretion in *B. megaterium*. Also shown are points where unwanted aggregation and/or proteolytic degradation can occur during secretion [64].

annotation of the *B. megaterium* genome revealed about 5300 ORFs. The understanding of genome regulation has been advanced by comparative analysis of genomes. Thus sequencing of *Aspergillus nidulans* and comparative analysis with *A. fumigatus* and *A. oryzae* revealed 5000 actively conserved non-coding regions apart from open reading frames, in which potential functional elements were identified [65, 66].

A basic step in the post-genomic era is the evaluation of the functions of genes – *functional genomics* – and the elucidation of the relationships between the genome sequence information and non-linear cellular dynamics, initially at the protein expression level. Whereas the final objective – the dynamic simulation of the formation of the proteome *in vivo* – is still some way off, it is more realistic to envisage applications of *in vitro* protein biosynthesis, for example to identify and avoid bottlenecks in cell-free protein synthesis. The issues addressed are transcription, RNA degradation, translation and model validation [62].

The *metagenome* approach provides access to large, hitherto unknown sequences of uncultivated and uncultivable organisms which leads to the identification and availability of new proteins, including enzymes of industrial interest (see Section 7.5) [28].

Transcriptomics: the availability of complete genomes provides the basis for the utilization of the information contained in the sequence, both with respect to analysis and understanding and for technical purposes. Synthetic oligonucleotide (ON) DNA- and RNA-arrays are efficient tools that are used in transcriptomics. A major aim of transcriptomics is to optimize expression rates for protein production using different methods: enhancing transcription as well as translation rates and stabilizing mRNA. In addition, new or modified signal peptides have been introduced, including bioinformatic prediction of structure [61, 67].

In order to understand the regulation and dynamics of gene expression a dynamic model of prokaryotic gene expression was developed that makes substantial use of gene sequence information. Details of transcription kinetics as well as mRNA degradation kinetics were taken into account in this model. The model incorporates the mechanisms of initiation, elongation and termination of protein synthesis, taking into account the mechanistic roles of key translation factors and its emphasis on a systematic mechanistic model. Mutual interactions amongst ribosomes organized within a polysome structure are also taken into account. The application of the model was extended to cell-free protein synthesis dynamics. It provides, for both *in vivo* and *in vitro* protein synthesis, a comprehensive framework for a thorough analysis of sequence-related effects during mRNA synthesis and degradation, and ribosomal translation, as well as their non-linear interrelations [62].

Microarray-based gene expression profiling, protein expression profiling based on two-dimensional gel electrophoresis, and profiling analysis during recombinant protein production have been undertaken by Vijayendran and Flaschel [68].

Proteomics and notably functional proteomics, which utilize two-dimensional protein gels (2DE SDS-PAGE), analytical tandem machines (LC-MS, MS/MS etc.) and bioinformatic tools (Figure 15.7), have enabled the rapid identification of proteins and the correlation of their expression to specific metabolic states, for example stress response due to elevated temperature, low iron or phosphate concentrations and so on, gene overload of recombinant micro-organisms or cells, or shear stress due to reactor hydrodynamics. Using this technique, up to 1000 proteins can be separated and visualized as spots on the gel. Quantitative proteome analysis is also possible by estimation of the intensity of the spots. Genome expression or proteome regulation, respectively (e.g. as a function of temperature) can be investigated with an error rate of less than a factor of 1.5 (using Affymetrix systems). Analysis of the total proteome is currently not yet possible. However the reconstruction of parts of a metabolic network from proteome data has been successfully achieved, for example for the central carbon metabolism and biosynthesis of amino acids of *B. megaterium* (Figure 15.10) [56, 69]. Furthermore, functional proteomics can be used to relate protein expression to particular diseases, thus expanding the large field of medical research in correlation with

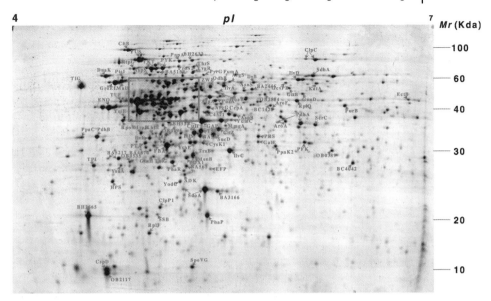

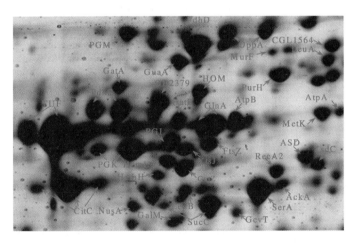

Figure 15.7 Two-dimensional gel electrophoretic separation (2DE IEF/SDS-PAGE) of intracellular proteins of *B. megaterium* MS 941. The proteins can be identified by excising the spots and analysing them using mass spectrometry (ESI-QqTOF MS/MS). The insert (red rectangle) in which every protein was identified is shown below at higher resolution [56].

the recent emergence of molecular medicine (see Sections 9.5 and 11.2). In order to keep up with the speed of gene sequencing and genomics, efforts were undertaken to establish high-throughput techniques for protein structure determination, 'protein factories' that work in terms of an *industrial type of research* [70, 71]. Commercial efforts continue to focus on providing information to pharmaceutical customers that relates to the suitability of specific proteins as drug targets [72]. However, after 8 years there has been no successful breakthrough.[5]

Glycomics, the analysis of structure and function of the glycosyl- (oligosaccharide-) moiety of glycoproteins and glycolipids, represents a recently emerged speciality aimed at understanding the communication processes of metabolites with cells or cell surfaces, cell–cell, virus–cell and micro-organism–cell communication. Mammalian cells are densely coated with a multiplicity of glycoconjugates, providing the so-called glycocalix. Expression of these molecules, especially their carbohydrate structures, is frequently specific to a cell or tissue type. Deciphering the information content of cellular glycomers is the current focus of the developing field of glycomics [79, 80]. Carbohydrate structures play a critical role in host–micro-organism interactions. Many host receptors for microbes are glycoconjugates; their structural diversity and selectivity of host tissue expression contribute significantly to the tropism of microbial infections. Thus, protein glycosylation often plays an essential role in viral and bacterial infection, cancer metastasis, inflammatory response, innate and adaptive immunity, and other receptor-mediated signalling processes [81, 82]. The best known interaction of glycans is the blood type classification of the ABO-system, based on the different glycostructures present on the red blood cells, the erythrocytes. These are highly relevant in blood transfusion. The carbohydrate motifs are *O*-linked glycans which differ in the terminal sugar unit (Figure 15.8a). A highly relevant and complex case is the development of tumour cells which are known to be highly glycosylated. Glycopeptides of tumours are promising candidates for use as antigens for immunotherapy. Synthetic carbohydrate antigens can be used as tumour markers and as vaccines for the induction of tumour antigen-specific antibody formation [83]. Several completely synthetic antigens, including combinations of several tumour antigens such as Globo-H and Tn, have been reported as multivalent vaccines for use in prostate cancer (Figure 15.8b) [84]. The analysis of the highly specific oligosaccharide–protein interactions represents a core problem, and new tools, including carbohydrate microarrays, to decipher the glyco code and carbohydrate-mediated recognition have been developed [26, 85, 86].

Metabolomics is an extension of proteomics and describes the analysis of the catalytic activities that make up the metabolic activities of a cell. Metabolomics is at the core of biosystems engineering and its purpose is to analyse the metabolome (all metabolites, roughly in the range of 500–2000) and their concentrations in the cell under given physiological conditions, as well as the dynamic response to changing

---

5) Instead the problems of structural proteomics, reasons for failure to obtain diffracting crystals, and lists of pitfalls, as well as experimental and theoretical approaches needed were discussed [73–75]. A number of consortia have formed worldwide to pursue the goals of structural genomics, that is, the annotation of DNA sequences to function, for example coding proteins. The Sanger Institute, Wellcome Trust, Cambridge, UK founded in 1993, is an outstanding example of industrially organized research [76]. The Berlin 'Protein Structure Factory' initiative focuses on human proteins, aiming at setting up an integrated local infrastructure for high-throughput protein structure analysis using X-ray crystallography and NMR spectroscopy [70, 71]. Other major activities, including commercial approaches, were initiated in North America and in Japan [72, 77, 78].

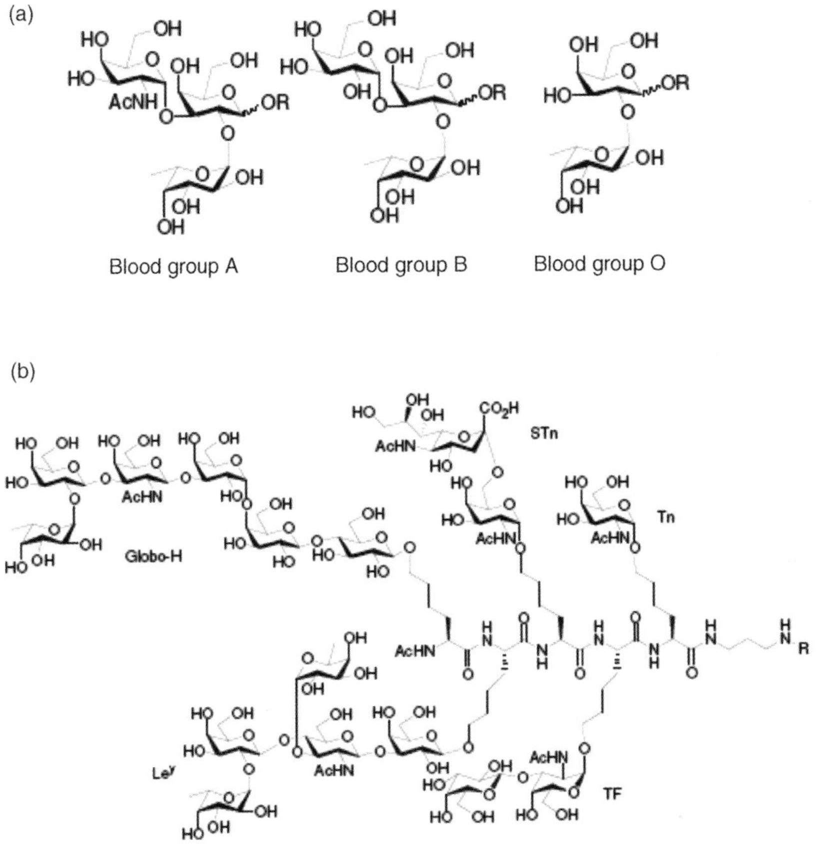

(a)

Blood group A   Blood group B   Blood group O

(b)

Figure 15.8 (a) Structure of the blood group ABO antigens. (b) Multivalent vaccine for use in prostate cancer [84].

environmental conditions (notably in a bioreactor). Its focus is on understanding the larger metabolic network in the cell and its utilization to produce various compounds more efficiently. This knowledge could be applied to the design of new biological pathways, systems and ultimately phenotypes through the use of recombinant DNA technology, and to the development of new products not presently available from wild-type micro-organisms or cells [57, 87, 88]. *Metabolic flux analysis and engineering* comprises the challenge of interactive analysis of metabolic networks and their directed modification. It relies on the quantitative analysis of metabolite formation and transformation (concentrations and reaction rates) by means of modern analytical tools such as liquid chromatography-mass spectrometry (LC-MS), tandem-MS and NMR [89–91]. It includes regulation phenomena and external influence (reactor environment) on the distribution of fluxes, as well as mathematical modelling. Its strength is seen to lie in going beyond more 'static' modelling by considering both the

GENE

Transcription

$$\frac{dmRNA_i}{dt} = \gamma_i D_i O_i \xi(T_F) \zeta_i (c_1, c_2, ...c_j...) - (\mu + k_i^d) \cdot mRNA_i$$

Translation

$$\frac{dE_i}{dt} = \lambda_i \cdot mRNA_i \cdot \phi_i (c_1, c_2, ...c_j...) - (\mu + k_i^s) E_i$$

Metabolic Fluxes

$$r_{ij} = \kappa_i E_i \psi(c_1, c_2, ...c_j, c_{j+1}...)$$

PRODUCT

$$\frac{dc_j}{dt} = \sum_i^n r_{ij} - \mu c_j$$

Figure 15.9 Differential equations for transcription, translation and metabolic fluxes of a cell. In simple cases such as tryptophan synthesis, the ratios of the metabolites '*j*' involved in '*i*' metabolic reactions can be formulated, which in turn are linked to the kinetics of transcription, translation, inhibition, repression, growth (dilution) and degradation (of enzymes and mRNA). Thus a (semi) quantitative description can be obtained [56].

regulatory and intracellular reaction networks.[6] Its applications range from industrial fermentation and its optimization to the recent advances in medical diagnosis. It may genetically widen or narrow the substrate spectrum and suppress disturbing side reactions that consume substrates.

In detail, the analysis is based on mass balances and stoichiometry, where manifold reactions with nodes and branches must be considered, in general under steady state conditions (mostly of continuous culture). Most relevant phenomena that should be considered are repressor levels, inhibition, notably feedback inhibition, and key metabolite concentrations, for example amino acids. Extracellular fluxes are taken into account: substrate consumption, oxygen uptake, biomass, product and by-product formation. However, the system of equations (Figure 15.9) used in metabolomics may appear to be ambiguous due to the occurrence of parallel and reversible metabolic pathways and futile cycles. Genetic approaches might suppress disturbing side reactions that consume substrates in order to improve yields. To analyse metabolic flux (*fluxome* analysis) ^{13}C (e.g. 1-^{13}C glucose) substrates are preferentially used as the analysis is based on the principle of pulse chase due to labelled pools from breakdown products, taking into account the distribution of the isotopomers of the metabolites [56, 92].

Amino acids (lysine), citric acid, lactic acid, propane diol, penicillin G, synthetic drug intermediates and therapeutic proteins are among the industrially relevant

---

6) Finding enzyme (gene) targets which have the most important influence on the formation rate of a product can be difficult because identification of a rate-limiting step is often not possible in metabolic networks, since the limitations may extend over many enzymatic steps [57].

products of fermentation and cell culture that have been targets for metabolic engineering. Some of this work has been adopted by industry (see Section 16.4.1). The major aim was to optimize the yields of industrial products. For example, after studying production of 1,2- and 1,3-propane diol by native organisms, specific enzymes have been transferred to *E. coli* to construct entirely new metabolic pathways that produce these compounds from sugar. Metabolic engineering and optimization of the pathways has significantly increased titers to the point where Dupont has commercialized the production of 1,3-propane diol via fermentation using corn starch [57]. Application of metabolic engineering to *Corynebacterium glutamicum* led to a significant increase in yields of amino acids, for example lysine and tryptophan [51, 89, 93, 94] (see Section 16.4.1). An example of significantly improved yield in glutamate production has been reported by Xiao *et al.* [95]. The method by which this was achieved is based on the analysis of the activities of two key enzymes involved in product and by-product formation, that is, glutamate and lactate dehydrogenases respectively, and the regulation of metabolic flux rates resulting in a 15% enhancement in glutamate concentration and improved productivity.

Bacillus megaterium and *Aspergillus niger* are further examples of cases where morphogenesis is relevant to productivity. These micro-organisms are being investigated as host organisms for the production of proteins including enzymes and variants of antibodies since their genomes are available [8, 96]. Data are integrated into models including the gene regulation network, stoichiometric analysis and metabolic network data (100 metabolites analysed). The tools used include DNA arrays (Affymetrix, providing access to 14 500 genes), 2D gels (580 proteins analysed), LC-MS/MS, construction of metabolic network using prediction of gene function, retro annotation and topologic network analysis [97–99]. Thus the reconstruction of an essential part of the metabolic network (central carbon metabolism and biosynthesis of amino acids) of *B. megaterium* from proteome data is shown in Figure 15.10 [56]. In an approach to the analysis of metabolic processes in *A. niger*, 120 reactions, 30 transport steps and 90 metabolites (81 internal, nine external) have been successfully modelled using a network analysis tool [100].

New medical applications may use metabolic engineering tools for clinical diagnosis and therapy, for example using isotopic tracers *in vivo* to establish a more quantitative analysis of the reaction networks that underlie physiology, and to create personalized medical treatment [57] (see Section 11.4). In terms of plant modification, the application of metabolic engineering to agriculture may reduce the costs of plant processing and milling, optimization of fermentation to increase yields and ultimately product purification.

The creation of intrinsically new methods has been at the centre of metabolic engineering. For example a new pathway has been created by combining enzymatic routes from different organisms for the production of vitamin C [101, 102]. An example of establishing new pathways to utilize methanol as an unconventional cheap substrate, as well as to synthesize new high added-value (chiral) products, for example drug intermediates, has been reported by Hartner [103] (Figure 15.11).

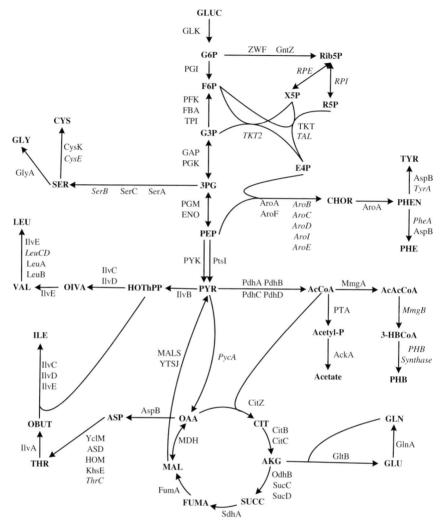

Figure 15.10 Reconstruction of the metabolic network *of B. megaterium* MS 941 from proteome data: central carbon metabolism and biosynthesis of amino acids (italic font: not yet identified proteins from proteome data) [56].

Environome: Metabolic flux in a bioreactor environment is correlated to various extracellular signals that may affect the cell machinery and influence the phenotype of the given genotype. The various effectors, as summarized in Figure 15.12, have a major influence on the functionality of the genome and the networks resulting therefrom, the nutrients being the most important stimuli for the metabolic network. Important cases for targeting the environome are processes controlled by the inhibition of excess substrate or product. In the first case specific feeding strategies

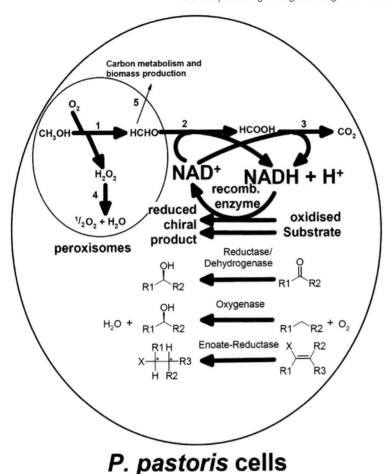

P. pastoris cells

Figure 15.11 New synthesis pathways: engineered whole cell biocatalyst of *Pichia pastoris*. Methanol serves as substrate for both carbon metabolism and redox equivalents for recombinant stereospecific redox enzymes from different sources to produce chiral intermediates [103].

and efficient mixing must be applied to adjust the substrate concentration to an appropriate level. In the second case process integrated separation is essential to avoid low reaction rates due to product inhibition. Biotechnological processes are essentially subject to dynamic changes. This is the general case for batch operation. Only simultaneous measurements of the proteome, the metabolome and the fluxome enable the development of suitable *in vivo* kinetics. Most important is the oxygen supply in aerobic fermentations, since in large-scale reactors oxygen concentration will vary over a large range as a function of time and space. Thus there will be fluxome changes at the transition from aerobic to anaerobic environments

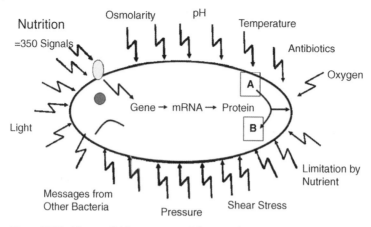

Figure 15.12 The manifold environmental factors influencing a cell [56].

within short time intervals. The problem must be approached by both computational fluid dynamics (CFD) and metabolic flux analysis, as will be discussed below (biosystems engineering). Furthermore shear stress due to high agitation rates must be considered, including the dependence of certain phenotypes on shear rates. Thus power input has a remarkable influence on the pellet density and diameter, and specific productivity in the case of *A. niger* [56, 104].

Bioinformatics aims to effectively deal with data pools from the omics to provide useful interpretations and models. It comprises understanding and modelling of genomes, proteomes, cell metabolism, whole-cell biotransformations and even incorporates the design of cell factories, in addition to modelling cells for new processes and the formation of new products.[3] Bioinformatics provides informational techniques specifically enabling access to and interpretation of large amounts of data generated in different fields of biosystems engineering. It is used to develop tools for annotating large genome sequences, analysing data produced by sequencing machines within a short time scale, interpreting data obtained from proteomics (two-dimensional gels and MS analysis), calculating mass balances in metabolic flux analysis and determining the kinetics and dynamics from the analysis of metabolic pathways. Bioinformatics thus represents a basic tool in biosystems engineering [99], (see also [97, 98, 100]).

Two databases are mentioned as examples that illustrate the potential of bioinformatics in the context of biosystems engineering. The first, MetaCyc (MetaCyc. org), is an universal database of metabolic pathways and enzymes, derived from more than 1000 metabolic pathways from a wide variety of organisms. Its goal is to serve as a metabolic encyclopaedia, containing a collection of non-redundant pathways central to small molecule metabolism. The pathways in MetaCyc are curated from the primary scientific literature, each reaction in a pathway being annotated with one or more well-characterized enzymes. The second, BioCyc

(BioCyc.org), is a collection of more than 370 organism-specific Pathway/Genome Databases[7] [105–107].

Biosystems engineering or *systems biotechnology*, has ambitious aims: to establish metabolic and regulatory networks of organisms by integrating data and information from the omics, and to use them to quantitatively and systematically comprehend, model and simulate the physiological performance of organisms. The approach includes using bioinformatic tools, analysis and data acquisition of the genome, different promotors, transcriptome, translation, codon usage, folding, processing, intracellular transport, membrane and cell wall proteome, excretion, and enviro-nomics. The first rational step of a systems biology approach is to construct networks from the omics-data. The proper information to achieve this is located in the genome. By decoding the genome using bioinformatic tools, the mechanisms of gene regulation can be elucidated, at least partially. A further approach is to construct the (potential) metabolic network *in silico* from the genome [56, 108]. A visualization of the integration of different tools has been presented by Papini *et al.* (2009) [115] (Figure 15.13).

A fundamental step in reconstructing a metabolic network is the identification of the metabolic enzymes (including the EC number) and the biochemical reactions catalysed by them. As an example an approach was undertaken for the construction of an *in silico* model for the central carbohydrate metabolism (CCM) of a living cell (of *Sulfolobus solfataricus*) that will enable computation of that metabolic network [109]. Another example used the genomic data of *A. niger* which recently became available. Figure 15.14 shows the metabolic network of *A. niger* in a modular structure. It contains enzymes with 434 unique EC numbers and 1172 metabolites. They correspond to the central C and N metabolism (specifically glycolysis, pyruvate metabolism, pentose phosphate pathway, citric acid cycle) while outliers with less branching represent pathways for the biosynthesis of lipids, amino acids and purines. The modular structure of the metabolic network obviously indicates that the reaction network of the catabolic processes is densely packed and exhibits numerous nodes and branch points. In contrast the anabolic reactions of purine and amino acid synthesis are located further out with fewer connections to other pathways. Whether or not a gene is switched on and mRNA produced can be directly determined by analysing the transcriptome [56].

A range of problems and challenges ranging from the control of gene expression, re-engineering translational control, codon optimization, mRNA structure and stability, pathway engineering including new tools such as synthetic DNA and

---

7) MetaCyc contains metabolic pathways, enzymatic reactions, enzymes, chemical compounds, genes and review-level comments. Enzyme information includes substrate specificity, kinetic properties, activators, inhibitors, cofactor requirements and links to sequence and structure databases. 'Superpathways' combine several more simple pathways for the synthesis of compounds, for example amino acids, including the genes, regulation, transport steps, enzymes and protein complexes involved. Each BioCyc Pathway/Genome Database contains the predicted metabolic network of one organism, that is metabolic pathways, enzymes, metabolites and reactions predicted by the Pathway Tools software, including predicted pathway hole fillers, using MetaCyc as a reference database [59, 105, 106].

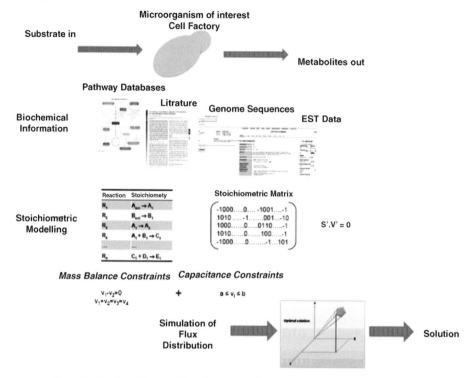

Figure 15.13 Simplified workflow diagram for the reconstruction of Genome-Scale Metabolic Models and their use in simulating cellular function through Flux Balance Analysis (Papini *et al.*, 2009) [115].

in silico approaches, has been addressed in a review by Stephanopoulos' group [63]. It includes goals such as the improvement of cellular properties, the intelligent design of biochemical pathways and beyond, the eventual design of new phenotypes and engineering microbial cell factories to produce fuels, chemicals, and pharmaceuticals.[8] Synthetic DNA and RNA would provide modified promoters, riboswitches, tunable intergenic regions, translation modulators, optimized mRNA secondary structures and RNA silencing to achieve the fine tuning of gene expression. In prokaryotes modifications of the start codon and its surrounding sites could drastically affect the rate of initiation of translation. The stability of mRNA transcripts is another determinant of translation efficiency. Incorporating selected stabilizing structural elements within the mRNA transcripts could, in theory, increase the lifetime of mRNA. RNA silencing, a well-established methodology for implementing

8) Thus, metabolic engineering should focus on the development of tools for finely tuning genetic expression to strike an optimal balance between pathway expression and cell viability, while avoiding deleterious effects such as impaired growth, metabolic imbalance and cytotoxicity due to metabolites [63].

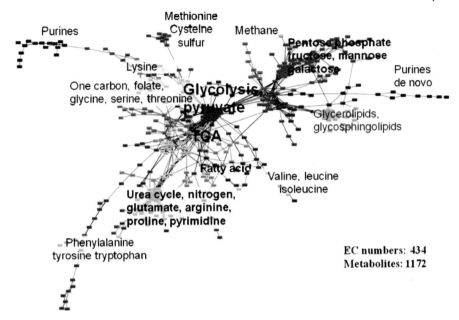

Figure 15.14 Modular structure of the potential metabolic network of A. *niger* inferred from its genome sequence. The network was built up by applying the program IdentiCS, and structured in modules. Densely connected regions (modules) with many nodes and branches can be seen [56].

genetic changes for metabolic engineering, can occur via a variety of mechanisms.[9] The ability to rigorously control gene expression via RNA silencing is particularly pertinent to metabolic engineering [63].

The next step in biosystems engineering must additionally take into account the environment of the cell in the bioreactor: concentrations (and their gradients) of substrates notably oxygen, products, temperature and pH, that is, the integration of *physiology* and *fluid dynamics* in bioprocesses, together with fluxomics and the environome, which in terms of modelling also comprise stress due to mixing and shear forces (Figure 15.12). This step may accelerate scale-up and optimize the overall process operation. This holistic approach incorporates the application of computational fluid dynamics (CFD) to three-dimensional, two-phase turbulence flow in stirred tank bioreactors (Figure 15.15). Examples of coupling momentum and material balance equations with unstructured kinetic models as well as a more complex approach based on structured metabolic models and the dynamic variations of the extracellular environment have been discussed by Schmalzriedt *et al.* [110]. More recently Lapin *et al.* [111] presented examples of the simulation of individual

9) All silencing pathways are triggered by 21–27 nucleotide-long RNA fragments collectively known as small RNA that include small interfering RNAs (siRNAs), repeat-associated small interfering RNAs (rasiRNAs) and micro RNAs (miRNAs). They are produced in response to regulatory processes as main agents of the gene silencing machinery [63].

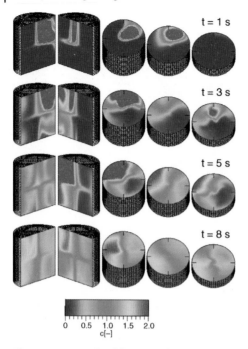

t = 1 s

t = 3 s

t = 5 s

t = 8 s

0   0.5   1.0   1.5   2.0
c[–]

Figure 15.15 Simulated dynamics of the tracer distribution at different times after a pulse onto the liquid surface in a stirred tank reactor (height/tank diameter ratio = 1.0, Rushton turbine, impeller/tank diameter ratio = 03, impeller clearance/height of liquid = 0.31) [110].

entities such as organisms or molecules within the cells, for assessing their effects on the dynamic behaviour of the system as a whole. One example is a structure-segregated model to study the lifelines of single cells in the environment of a three-dimensional turbulent field of a bioreactor, thus accounting for the heterogeneity in actual reactors, both in the fluid and cellular phases (Figure 15.15). As a representation of the simulation results at the cellular level, a schematic visualization framework has been implemented which shows all simulated components (Figure 15.16). The approach integrates computational fluid dynamics (CFD) and computational cell dynamics (CCD), including structured kinetic models. This approach allows the population behaviour resulting from the interaction between the intracellular state of its individual cells and the turbulent flow field in the bioreactor to be described, and permits the lifelines of individual cells to be analysed in space and time [111]. Theoretical analysis in biosystems engineering provides concepts for new experiments to optimize the models and thus to predict *experiments in silico* in order to select which experiments should be carried out [60].

A further step should address the *integration of all steps, from genetic engineering to downstream operations* right from the beginning of process development [48]. An example has been presented by Beshay *et al.* [112], in which the production and purification of a recombinant protein by affinity chromatography in *E. coli* has been

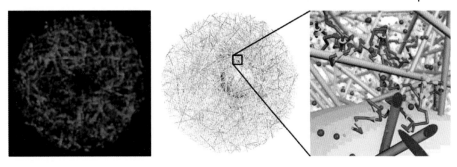

Figure 15.16 Visualization of a virtual cell. Left: fluorescence microscope image. Center and right: detailed images showing the cytoskeleton, nucleus, crowding molecules and signalling molecules, including the path of a signalling molecule (right: dark, with light arrow cones) [111].

integrated into the bioreactor design. A strongly expressed gene for a β-glucanase was fused with a metal-chelating affinity tag and a leader peptide in order to direct the fusion protein into the periplasmatic space. The bioreactor fitted with an external circuit produced an increased β-glucanase concentration during the process. Another approach involved the integration of an adsorption step into the bioreactor in order to isolate the product *in situ*, and the genetic modification of the enzyme (a glucosyl transferase) to achieve improved selectivity in the synthesis of an oligosaccharide (glucosylation of glucose with sucrose as the substrate to yield isomaltose). The first process variant avoids subsequent glucosylation of isomaltose reactions to higher oligosaccharides. Adsorption of the product onto a specific zeolite is achieved inside the enzyme reactor using a specially designed three-phase fluidized bed reactor with stationary immobilized biocatalyst and adsorbent which is removed continuously together with the adsorbed isomaltose. Both steps improved the yield, however improving the selectivity of the biocatalyst using genetic engineering (resulting in much reduced formation of higher oligosaccharides) proved to be the most efficient approach [113, 114].

A number of examples of biosystems engineering applied to industrial BT (or 'industrial systems biology'), notably concerning *Saccharomyces* and *Aspergillus* spp., has been summarized by Papini *et al.* [115] and Wittmann [116]. The examples include perspectives for commercially highly important products such as ethanol, butanol, polyhydroxyalkanoates, amino acids, polyketides, insulin and antibodies produced by *S. cerevisiae* (see also Sections 16.3.1 and 16.4.1).

15.7
Outlook and Perspectives

While progress in molecular biology over the last decade, including the elucidation of genome sequences and development of bioinformatic tools, led to considerable new knowledge and skills, further work will be required to transform these advances into technology and marketable products. It is hoped that the production of synthetic

Table 15.3 Perspectives in biochemical and biosystems engineering: selected topics [6, 19, p. 6, 7].

	Biochemical and biosystems engineering	Metabolic engineering	Genomics and bioinformatics
Type of research, need, (priorities, bottlenecks)	Alternative physiological concepts for continuous processes	Novel production hosts – investigation of microbial metabolism and regulation (stress) Dynamic models of metabolism Increased product concentration^a	Understanding of the regulatory networks under production conditions Gene annotation Genome/metabolome reconstruction software
	In situ removal of products, use of high capacity nanomaterials	Minimal genomes/ bacteria	Synthetic biology – the artificial strain
Tools	High-throughput development tools, microreactors	Comprehensive metabolome analysis Design of novel pathways *In silico* concepts of designed cells	Standardized bioinfomatics
Visionary projects	New products, high product concentration, proof on technical scale$^{a)}$	Sugar based production of acrylic acid, 1,3-propanediol, new materials, novel biopolymers Multivitamin cells Combinatorial synthesis Biochips for computers	Synthetic genes, genomes and pathways (the artificial strain vision)

a)   Example: ethanol concentration >20% (see Section 16.3.1) – this does not seem to be visionary. It
     may be the most difficult to achieve but is of most significant economic value.

genes, genomes and pathways, the 'minimal perfect host' that incorporates cascades of bioconversions, the *in silico* design of new catalytic activities, and even biochips for computers will be possible in the future (Table 15.3). The synthetic potential of genomics, metabolomics, bioinformatics and biosystems engineering is now described as *synthetic biology* (see Section 13.7). This term refers to the design and construction of new biological entities such as enzymes, genetic circuits and cells, or to the redesign of existing biological systems with the focus on the design and construction of engineered and integrated, biological systems that solve specific problems [117–121].

A new frontier that has been addressed is the creation of synthetic organisms, specifically user-defined biocatalysts for processes that are characterized by high productivities, yields and purity of products. However, the creation of multigenic traits such as tolerance to stress has still to reach the conceptual stage. The design and use of a *de novo* synthetic genome still remains a challenge [63, 119, 120]. However, much of this potential will depend both on the translation of the accumulated data into technical operations particularly on a large scale, and on public acceptance. Realization of this potential will in turn depend on a continuing dialogue based on sustainability assessment and a firm commitment to integrate the social, behavioural, cognitive and economic sciences into the research agenda [6, 19, p. 6, 7, 117].

An open question may be addressed here: despite their similarities in general structure and active centres, why are glycosidases hydrolytic enzymes but glycosyl-transferases in contrast, highly efficient synthetic enzymes? The mechanisms of these different reactions may in the first case, require highly efficient activation of their nucleophilic potential, either by water or by the acceptor (carbohydrate), to effect the transfer of a glycosyl group from the enzyme's active centre. Although the sophisticated mechanistic details of this hypothesis have been proposed, they have not as yet been verified. Answers to such mechanistic questions may catalyse further research into the design of new enzyme activities.

References

1 Rehm, H.J. (1967) *Industrielle Mikrobiologie*, Springer, Berlin.

2 Pirt, J. (1975) *Principles of Microbe and Cell Cultivation*, Blackwell Scientific Publications, Oxford.

3 Aiba, S., Humphrey, A.E., and Millis, N.F. (1965) *Biochemical Engineering*, Academic Press, New York.

4 Bailey, J.E. and Ollis, D.F. (1977) *Biochemical Engineering Fundamentals*, 1986 2nd edn, McGraw-Hill, New York.

5 Sahm, H. (1993) (ed.) Biological fundamentals, in *Biotechnology*, vol. 1 VCH, Weinheim, Germany.

6 Heinzle, E., Biwer, A.P., and Cooney, C.L. (2006) *Development of Sustainable Bioprocesses, Modelling and Assessment*, John Wiley & Sons, New York.

7 Hempel, D.C. (2007) Bioverfahrenstechnik, in *Dubbel Taschenbuch für den Maschinenbau*, 21st edn (eds K–.H., Grote and J. Feldhusen), Springer Verlag; Heidelberg, Berlin; New York, p. N34–N52 (English version in preparation).

8 Grimm, L.H., Kelly, S., Krull, R., and Hempel, D. (2005) Morphology and productivity of filamentous fungi. *Applied Microbiology and Biotechnology*, **49**, 375–384.

9 Reis, U., Blum, B., von Specht, B.-U., Domdey, H., and Collins, J. (1992) Antibody production in silkworm cells and silkworm larvae infected with a dual recombinant Bombyx mori nuclear polyhedrosis virus. *BioTechnology*, **10**, 910–912.

10 Stoppok, E. and Buchholz, K. (1996) Sugar-based raw materials for fermentation applications, in *Biotechnology: Products of Primary Metabolism*, vol. 6 (eds H.-J. Rehm and G. Reed), VCH, Weinheim u.a., p. S5–S29.

11 Minas, W. (2005) Production of erythromycin with saccharapolyspora erythrea, in *Microbial Processes and Products* (ed. J-.L. Barredo), Humana Press, Totowa, New Jersey, USA.

12 Aehle, W., Perham, R.N., Michal, G. *et al.* (2003) Enzymes, in *Ullmann's Encyclopedia of Industrial Chemistry*, Wiley-VCH, Weinheim, New York, www. mrw.interscience.wiley.com/emrw/.

13 Bommarius, A.S. and Riebel, B.R. (2004) *Biocatalysis*, Wiley-VCH, Weinheim.

14 Buchholz, K., Kasche, V., and Bornscheuer, U. (2005) *Biocatalysts and Enzyme Technology*, Wiley-VCH, Weinheim, p. 412–417.

15 Liese, A., Seelbach, K., and Wandrey, C. (2006) *Industrial Biotransformations*, Wiley-VCH, Weinheim.

16 Herbots, I., Kottwitz, B., Reilly, P.J. *et al.* (2007) Enzymes, non-food application, in *Ullmann's Encyclopedia of Industrial Chemistry*, Wiley-VCH, Weinheim, New York, www.mrw.interscience.wiley.com/emrw/.

17 Schäfer, T., Borchert, T.V., Skovgard Nielsen, V. *et al.* (2007) Industrial enzymes. *Advances in Biochemical Engineering/Biotechnology*, **105**, 59–131.

18 BIOCOM (2007) *Biotechnica Journal*, 22.

19 EU (2005) Workshop Strategic Research, Brussels.

20 Bommarius, A., Broering, J.M., Chapparo-Riggers, J.F., and Polizzi, K.M. (2006) High-throughput screening for enhanced protein stability. *Current Opinion in Biotechnology*, **17**, 606–610.

21 Beier, L., Svendsen, A., Andersen, C., Frandsen, T.P., Borchert, T.V., and Cherry, J.R. (2000) Conversion of the maltogenic α-amylase Novamyl into a CGTase. *Protein Engineering*, **13**, 509–513.

22 Kelly, R.M., Leemhuis, H., and Dijkhuizen, L. (2007) Conversion of a cyclodextringlucano-transferase into an α-amylase: assessment of directed evolution strategies. *Biochemistry*, **46**, 11216–11222.

23 Jahic, M., Rotticci-Mulder, J., Martinelle, M., Hult, K., and Enfors, S.-O. (2002) Modelling of growth and energy metabolism of *Pichia pastoris* producing a fusion protein. *Bioprocess and Biosystems Engineering*, **24**, 385–393.

24 Vedha-Peters, K., Gunawardana, M., Rozzell, D., and Novick, S. (2006) Creation of a broad-range and highly stereoselective d-amino acid dehydrogenase for the one-step synthesis of d-amino acids. *Journal of the American Chemical Society*, **128**, 10923–10929.

25 Jahn, M. and Withers, S.G. (2003) New approaches to enzymatic oligosaccharide synthesis: Glycosynthases and thioglycoligases. *Biocat Biotrans*, **21**, 159–166.

26 Seibel, J., Hellmuth, H., Hofer, B., and Schmalbruch, B. (2006) Identification of new donor specifities of glycosyl-transferases R via the aid of substrate microarrays. *Chembiochem*, **7**, 310–320.

27 Lorenz, P., Liebeton, K., Niehaus, F., and Eck, J. (2002) Screening for novel enzymes for biocatalytic processes: accessing the metagenome as a resource of novel functional sequence space. *Current Opinion in Biotechnology*, **13**, 572–577.

28 Lorenz, P. and Eck, J. (2005) Metagenomics and industrial applications. *Nature Reviews Microbiology*, **3**, 510–516.

29 Chang, A., Scheer, M., Grote, A., Schomburg, I., and Schomburg, D. (2009) BRENDA, AMENDA and FRENDA the enzyme information system: New content and tools in 2009. *Nucleic Acids Research*, **37** (SUPPL 1), D588–D592.

30 ESAB (2005) (European Section of Applied Biocatalysis) and EuropaBio, 2005: Workshop Strategic Research Agenda Industrial Biotechnology, 2005. Working document, Brussels, march 17–18th 2005, M. Franssen, W. Soetaert coordinators.

31 Weuster-Botz, D., Hekmat, D., Puskeiler, R., and Franco-Lara, E. (2007) Enabling technologies: fermentation and downstream processing. *Advances in Biochemical Engineering/Biotechnology*, **105**, 205–247.

32 Buckland, B.C. and Lilly, M. (1993) Fermentation: an overview, in *Biotechnology*, vol. 3 (ed. G. Stephanopoulos), VCH, Weinheim, Germany, p. 7–22.

33 Heijnen, J.J. (2010) Impact of Thermodynamic Principles in Systems Biology. *Advances in Biochemical Engineering/Biotechnology*, **121**, 139–162.

34 Atkinson, B. and Mavituna, F. (1991) *Biochemical Engineering and Biotechnology Handbook*, 2nd edn, MacMillan Publisher Ltd., New York.

35 Stephanopoulos, G. (ed.) (1993) Bioprocessing, in *Biotechnology*, vol. 3, VCH, Weinheim, Germany.

36 Raju, G.K. and Cooney, C.L. (1993) Media and air sterilization, in *Biotechnology*, vol. 3 (ed. G. Stephanopoulos), VCH, Weinheim, Germany, p. 157–184.

37 Krahe, M. (2003) Biochemical engineering, in *Ullmann's Encyclopedia of Industrial Chemistry*, Wiley-VCH, Weinheim, New York, www.mrw. interscience.wiley.com.

38 Benz, G.T. (2008) Piloting Bioreactors for Scale-Up. *Chemical Engineering Progress*, February, 32–34.

39 Hempel, D.C. (1988) Fundamentals of scale up for biotechnological processes in stirred fermenters, in *Biotechnology Focus* (eds R.K. Finn *et al*..), Hanser Publishers, Munich.

40 Spears, R. (1993) Overview of downstream processing, in *Biotechnology*, vol. 3 (ed. G. Stephanopoulos), VCH, Weinheim, Germany, p. 39–55.

41 Moo-Young, M. (ed.) (1985) *Comprehensive Biotechnology*, vol. 1 and 2, Pergamon Press, Oxford.

42 Becker, T., Breithaupt, D., and Doelle, H.W. (2007) *Biotechnology. Ullmann's Encyclopedia of Industrial Chemistry*, Wiley-VCH, Weinheim, New York, www.mrw.interscience.wiley.com.

43 Schmidt, T. (2009) SMB-Chromatographie für die Prozessentwicklung. *CHEManager*, **3**, 10.

44 Hubbuch, J. and Kula, M.-R. (2007) Isolation and purification of biotechnological products. *Journal of Non-Equilibrium Thermodynamics*, **32**, 99–127.

45 Schügerl, K. and Hubbuch, J. (2005) Integrated bioprocesses. *Current Opinion in Biotechnology*, **8**, 294–300.

46 Hubbuch, J., Matthiesen, D.B., Hobley, T.J., and Thomas, R.T. (2001) High gradient magnetic separation versus expanded bed adsorption: a first principle comparison. *Bioseparation*, **10**, 99–112.

47 Meyer, A., Berensmeier, S., and Franzreb, M. (2007) Direct capture of lactoferrin from whey using magnetic micro-ion exchangers in combination with high-gradient magnetic separation. *Reactive and Functional Polymers*, **67**, 1577–1588.

48 Hempel, D.C. (2006) Development of biotechnological processes by integrating genetic and engineering metods. *Engineering in Life Sciences*, **6**, 443–447.

49 Becker, T., Hitzmann, B., Muffler, K., Poertner, R. *et al.* (2007) Future aspects of bioprocess monitoring. *Advances in Biochemical Engineering/Biotechnology*, **105**, 249–293.

50 Frerick, C., Kreis, P., and Górak, A. (2008) Optimierung von Proteinaufreinigungsprozessen basierend auf einem generischen Prozessmodell. *Chemical Engineering & Technology*, **80**, 97–106.

51 Ostermann, E. (2008) Lecture Vom Gen zum Produkt oder aus der Uni in die Praxis (Evonik Degussa) SFB Kolloquium 9.6.08 Tech. Univ. Braunschweig.

52 Berensmeier, S. (2006) Magnetic particles for the separation and purification of nucleic acids. *Applied Microbiology and Biotechnology*, **73**, 495–504.

53 Brinkmann, T. (2009) Nachhaltigkeit von biotechnologischen Produkten. *CHEManager*, **9**, 32.

54 Sheldon, R.A. (2000) Atom efficiency and catalysis in organic synthesis. *Pure and Applied Chemistry*, **72**, 1233–1246.

55 Pühler, A. (2007) Genomforschung – Motor zur Entwicklung von mikrobiellen Produktionsorganismen. *Laborwelt*, **8** (2), 4–5.

56 Deckwer, W.-D., Jahn, D., Hempel, D., and Zeng, A.-D. (2006) Systems Biology Approaches to Bioprocess Development. *Engineering in Life Sciences*, **6**, 455–469.

57 Raab, R.M., Tyo, K., and Stephanopoulos, G. (2005) Metabolic engineering. *Advances in Biochemical Engineering/Biotechnology*, **100**, 1–17.

58 Reuss, M. (2001) *Bioprocess and Biosystems Engineering*, editorial, 24, No 1, January.

59 Caspi, R. (2009) The MetaCyc Database and the BioCyc collection of pathway/genome databases. Lecture, SFB-meeting, Biocenter, Tech. Univ. Braunschweig, April 30.

60 Melzer, G., Eslahpazir, M., Franco-Lara, E., Wittmann, C. (2009) Flux Design: In silico design of cell factories based on correlation of pathway fluxes to desired properties. *BMC Systems Biology*, **3**, art. no. 120.

61 Gamer, M., Fröde, D., Biedendieck, R., Stammen, S., Jahn, D. (2009) A T7 RNA polymerase-dependant gene expression system for *Bacillus megaterium*. *Applied Microbiology and Biotechnology*, **82**, 957–967, 1195–1203.

62 Arnold, S., Siemann-Herzberg, M., Schmid, J., and Reuss, M. (2005) Model-based inference of gene expression dynamics from sequence information. *Advances in Biochemical Engineering/Biotechnology*, **100**, 89–179.

63 Klein-Marcuschamer, D., Yadav, V.G., Ghaderi, A., and Stephanopoulos, G.N. (2010) De Novo Metabolic Engineering and the Promise of Synthetic DNA. *Advances in Biochemical Engineering/Biotechnology*, **120**, 101–131.

64 Hollmann, R., Malten, M., Biedendieck, R., Yang, Y., Wang, W., Jahn, D., and Deckwer, W.-D. (2006) Bacillus *megaterium* as a host for recombinant protein production. *Engineering in Life Sciences*, **6**, 470–474.

65 Bunk, B. (2009) Methoden der Genomsequenz- und –annotation. Lecture, SFB-meeting, St. Andreasberg/Braunschweig, Germany, 16.3.09.

66 Galagan, J.E., Calvo, S.E., Cuomo, C. *et al.* (2005) Sequencing of Aspergillus nidulans and comparative analysis with A. fumigatus and A. oryzae. *Nature*, **438**, 1105–1115.

67 Stammen, S., Müller, B.K., Korneli, C., *et al.* (2010) High-yield intra- and extracellular protein production using *Bacillus megaterium*. *Applied and Environmental Microbiology*, **76**, 4037–4046.

68 Vijayendran, C. and Flaschel, E. (2010) Impact of Profiling Technologies in the Understanding of Recombinant Protein Production. *Advances in Biochemical Engineering/Biotechnology*, **121**, 45–70.

69 Jänsch, L. (2009) Quantitative Proteomanalyse in der Systembiotechnologie. Lecture, SFB-meeting, St. Andreasberg/Braunschweig, Germany, 16.3.09.

70 Heinemann, U. (2000) Structural geneomics in Europe: slow start, strong finish? *Nature Structural Biology*, **7**, 940–942.

71 Heinemann, U., Illing, G., and Oschkinat, H. (2001) High-throughput three-dimensional protein structure determination. *Current Opinion in Biotechnology*, **12**, 348–354.

72 Dry, S., McCarthy, S., and Harris, T. (2000) Structural genomics in the biotechnology sector. *Nature Struct Biol Structural Genomics Supplement*, 946–949.

73 Banci, L., Baumeister, W., Enfedaque, J. *et al.* (2007) Structural proteomics: from the molecule to the system. *Nature Structural & Molecular Biology*, **14**, 3–4.

74 Niesen, F.H., Koch, A., Lenski, U. *et al.* (2008) An approach to quality management in structural biology: biophysical selection of proteins for successful crystallization. *Journal of Structural Biology*, **162**, 451–459.

75 Structural Genomics Consortium *et al.* (2008) Protein production and purification. *Nature Methods*, **5**, 135–146.

76 Sanger (2002/2009) The Wellcome Trust Sanger Institute, www.sanger.ac.uk/.

77 Terwilliger, T.C. (2000) Structural genomics in North America. *Nature Structural Biology*, **7**, 935–939.

78 Yokoyama, S., Hirota, H., Kigawa, T. *et al.* (2000) Structural genomics projects in Japan. *Nature Structural Biology*, **7**, 943–945.

79 Pratt, M.R. and Bertozzi, C.R. (2005) Synthetic glycopeptides and glycoproteins as tools for biology. *Chemical Society Reviews*, **34**, 58.

80 Wong, C.-H. (2005) Protein glycosylation: new challenges and opportunities. *The Journal of Organic Chemistry*, **70**, 4219–4225.

81 Varki, A. (1993) Biological roles of oligosaccharides: all of the theories are correct. *Glycobiology*, **3** (2), 97–130.

82 Varki, A. (2006) Nothing in Glycobiology Makes Sense, except in the Light of Evolution. *Cell*, **126**, 841–845.

83 Freire, T., Bay, S., Vichier-Guerre, S., Lo-Man, R., and Leclerc, C. (2006) Carbohydrate antigens: synthesis aspects and immunological applications in cancer. *Mini-Reviews in Medicinal Chemistry*, **6**, 1357–1373.

84 Keding, S.J. and Danishefsky, S.J. (2004) Prospects for total synthesis: a vision for a totally synthetic vaccine taqrgetting epithelial tumors. *Proceedings of the National Academy of Sciences of the United States of America*, **101**, 11937–11942.

85 Wang, D., Liu, S., Trummer, B.J., Deng, Ch., and Wang, A. (2006) Carbohydrate microarrays for the recognition of cross-

reactive molecular markers of microbes and host cells. *Nature Biotechnology*, **20**, 275–281.

86 Feizi, T. and Chai, W. (2004) Oligosaccharide microarrays to decipher the glyco code. *Nature Reviews. Molecular Cell Biology*, **5**, 582–588.

87 Baily, J.E. (1996) Metabolic engineering. *Advances in Molecular and Cell Biology*, **15**, 289–296.

88 Sinskey, A.J. (1999) Foreword. *Metabolic Engineering*, **1** (3), iii–iii.

89 Marx, A., Eikmanns, B.J., Sahm, H., de Graaf, A.A., and Eggeling, L. (1999) Response of the central metabolism in *Corynebacterium glutamicum* to the use of an NADH-dependent glutamate dehydrogenase. *Metabolic Engineering*, **1**, 35–48.

90 van Gulic, W.M., de Laat, W.T.A.M., Vinke, J.L., and Heijnen, J.J. (2000) Application of metabolic flux analysis for the identification of metabolic bottlenecks in the biosynthesis of penicillin-G. *Biotechnology and Bioengineering*, **68**, 602–618.

91 Börner, J., Buchinger, S., and Schomburg, D. (2007) A high-throughput method for microbial metabolome analysis using gas chromatography/mass spectrometry. *Analytical Biochemistry*, **367**, 143–151.

92 Xiu, Z.-L., Zeng, A.-P., and Deckwer, W.-D. (1997) Model analysis concerning the effects of growth rate and intracellular tryptophan level on the stability and dynamics of tryptophan biosynthesis in bacteria. *Journal of Biotechnology*, **58**, 125–140.

93 Sahm, H., Eggeling, L., and de Graaf, A. (2000) Pathway analysis and metabolic engineering in *Corynebacterium glutamicum*. *Biological Chemistry*, **381**, 899–910.

94 Pfefferle, W., Möckel, B., Bathe, B., and Marx, A. (2003) Biotechnological manufacture of lysine. *Advances in Biochemical Engineering/Biotechnology*, **79**, 59–112.

95 Xiao, J., Shi, Z., Gao, P. *et al.* (2006) On-line optimization of glutamate production based on balanced metablolic control by RQ. *Bioprocess and Biosystems Engineering*, **29**, 109–117.

96 Kelly, S., Grimm, L.H., Jonas, R., Hempel, D.C., and Krull, R. (2006) investigations of the morphogenesis of filamentos microorganisms. *Engineering in Life Sciences*, **6**, 475–480.

97 Choi, C., Muench, R., Bunk, B. *et al.* (2006) SYSTOMONAS – an integrated database for systems biology analysis of *Pseudomonas*. *Nucleic Acids Research*, **35**, 533–537.

98 Hiller, K., Grote, A., Manek, M., Muench, R., and Jahn, D. (2006) JVirGel 2.0: computational prediction of proteomes separated via two dimensional gel electrophoresis under consideration of membrane and secreted proteins. *Bioinformatics*, **22**, 2441–2443.

99 Muench, R. and Schomburg, D. (2008) Integrierte Datenbanken, bioinformatische Werkzeuge, Analyse und Modellierung für die Systembiologie mit *B. megaterium* und *A. niger*. Project B9, 397–418, Collaborative Research Center SFB 578 "From Gene to Product", Technical University, Braunschweig.

100 Melzer, G., Eslahpazir, M., Franco-Lara, E., and Wittmann, C. (2009) Flux Design: In silico design of cell factories based on correlation of pathway fluxes todesired properties. *BMC Systems Biology*, **3**, art. no. 120 .

101 Anderson, S., Berman Marks, C., Lazarus, R., Miller, J., Stafford, K., Seymour, J., Light, D., Rastetter, W., and Estell, D. (1985) Production of 2-Keto-L-Gulonate, an intermediate in L-ascorbate synthesis by a genetically modified. *Erwinia Herbicola Science*, **230**, 144–149.

102 Grindley, J.F., Payton, M.A., van den Pol, H., and Hardy, G. (1988) Conversion of Glucose to 2-Keto-L-Gulonate, an Intermediate in L-Ascorbate Synthesis, by a Recombinant Strain of *Erwinia citreus*. *Applied and Environmental Microbiology*, **54**, 1770–1775.

103 Hartner, F.S. (2007) Engineering *Pichia pastoris* for whole cell biotransformations. PhD thesis, Graz University of Technology.

104 Krull, R., Cordes, C., Horn, H., Kampen, I., Kwade, A., Neu, T. R., and Nörtemann, B. (2010) Morphology of Filamentous Fungi: Linking Cellular Biology to

Process Engineering Using Aspergillus niger. *Advances in Biochemical Engineering/Biotechnology*, **121**, 1–21.

105 Caspi, R., Foerster, H., Fulcher, C.A., Hopkinson, R., Ingraham, J., Kaipa, P., Krummenacker, M., and Karp, P.D. (2006) MetaCyc: a multiorganism database of metabolic pathways and enzymes. *Nucleic Acids Research*, **34** (Database issue), D511–D516.

106 Caspi, R., Foerster, H., Fulcher, C.A., Kaipa, P., Krummenacker, M., Latendresse, M., Paley, S., Karp, P.D. *et al.* (2008) The MetaCyc Database of metabolic pathways and enzymes and the BioCyc collection of pathway/genome databases. *Nucleic Acids Research*, **36** (Suppl 1), D623–D631, 25.

107 Caspi, R. and Karp, P.D. (2007) Using the MetaCyc pathway database and the BioCyc database collection, Chapter 1, in *Current Protocols in Bioinformatics* (ed. A.D. Baxevanis), p. Unit1.17 0.

108 Sun, J. and Zeng, A.-P. (2004) IdentiCS – identification of coding sequence and *in silico* reconstruction of the metabolic network directly from unannotated low-coverage bacterial genome sequence. *BMC Bioinformatics*, **5**, 112.

109 Albers, S.V., Birkeland, N.K., Driessen, A.J., Gertig, S., Haferkamp, P., Klenk, H.P., Kouril, T., Zaparty, M. *et al.* (2009) SulfoSYS (Sulfolobus Systems Biology): towards a silicon cell model for the central carbohydrate metabolism of the archaeon Sulfolobus solfataricus under temperature variation. *Biochemical Society Transactions*, **37** (Pt 1), 58–64.

110 Schmalzriedt, S., Jenne, M., Mauch, K., and Reuss, M. (2003) Integration of physiology and fluid dynamics. *Advances in Biochemical Engineering/Biotechnology*, **80**, 21–68.

111 Lapin, A., Klann, M., and Reuss, M. (2010) Multi-Scale Spatio-Temporal Modeling: Lifelines of Microorganisms in Bioreactors and Tracking Molecules in Cells. *Advances in Biochemical Engineering/Biotechnology*, **121**, 23–43.

112 Beshay, U., Miksch, G., Friehs, K., and Flaschel, E. (2009) Integrated bioprocess for the production and purification of recombinant proteins by affinity chromatography in Escherichia coli. *Bioprocess and Biosystems Engineering*, **32**, 149–158.

113 Ergezinger, M., Bohnet, M., Berensmeier, S., and Bucholz, K. (2006) Integrated enzymatic synthesis and adsorption of Isomaltose in a multiphase fluidized bed reactor. *Engineering in Life Sciences*, **6** (5), 1–8.

114 Seibel, J., Jördening, H-J., and Buchholz, K. (2010) Extending Synthetic Routes for Oligosaccharides by Enzyme, Substrate and Reaction Engineering. *Advances in Biochemical Engineering/Biotechnology*, **120**, 163–193.

115 Papini, M, Salazar, M, Nielsen, J. (2010) Systems biology of industrial microorganisms. *Advances in Biochemical Engineering/Biotechnology*, **120**, 51–99.

116 Wittmann, C. (2010) Analysis and engineering of metabolic pathway fluxes in *Corynebacterium glutamicum* for amino acid production. *Advances in Biochemical Engineering/Biotechnology*, **120**, 21–49.

117 Erickson, B.E. (2009) Synthetic biology. *Chemical and Engineering News*, 23–25.

118 Carothers, J.M., Goler, J.A., and Keasling, J.D. (2009) Chemical synthesis using synthetic biology, *Current Opinion in Biotechnology*, **20**, 498–503.

119 Gibson, D.G, Glass, J.I, Lartigue, C. *et al.* (2010) Creation of a bacterial cell controlled by a chemically synthesized genome. *Science*, **329**, 52–56.

120 Bornscheuer, U.T. (2010) The first artificial cell – a revolutionary step in synthetic biology? *Angewandte Chemie-International Edition*, **49**, 5228–5230.

121 Ferrer, M. *et al.* (2009) *Journal of Molecular Microbiology and Biotechnology*, **16**, 109–123.

16
Industrial Biotechnology

16.1
Introduction

Industrial biotechnology (BT) – often termed White BT, evokes the following *visions*:

- Large processes, industrial products and business, providing
- commodities, including ethylene, organic acids, plastics,
- fine chemicals – chiral synthons and products for pharmaceuticals, agricultural products, food, enzymes and so on, based on
- renewable resources with the challenge of replacing oil-based chemicals and fuel

For hundreds of years industrial BT has been a large, traditional technology concerned with producing beer and wine but has nevertheless been a dynamic field which makes use of the most advanced tools such as recombinant DNA (rDNA) technologies and systems biotechnology. Only a short overview will be given here. However, a few examples of processes will be discussed in some detail in order to illustrate the technology and its complexity and also the associated problems, solutions, economic aspects, challenges and trends, for example in the cases of ethanol, starch processing, and fluidized bed technology in modern waste water treatment. The chapter is subdivided into sections according to the type of products and the specific problems associated with each type, for example the scale of the production process (ranging from a few m^3 to some $10^3\ m^3$ reactor volume, or some 100 kg to several millions of tons of product) as well as the price and economics.

The National Research Council of the USA has set ambitious targets to be achieved by the year 2020 for biobased product industries to 'eventually satisfy over 90% of US organic chemical consumption and up to 50% of liquid fuel needs . . .' and even 'a biobased industrial future . . .' [1, p. 4,11], [2, p. 11]. A 2007 energy law requires the production of 136 mn (million) m^3 of biofuels by 2022, including 60 mn m^3 of cellulosic ethanol and 19 mn m^3 of advanced biofuels and biodiesel. According to this law a biofuel must achieve at least a 20% reduction in greenhouse gas emissions (as compared to petroleum-based fuel) based on life-cycle analyses [3]. A report by SusChem [4], European Technology Platform for Sustainable Chemistry, offers a certain degree of vision and data despite being verbose and imprecise: the

Concepts in Biotechnology: History, Science and Business. Klaus Buchholz and John Collins
Copyright © 2010 WILEY-VCH Verlag GmbH & Co. KGaA, Weinheim
ISBN: 978-3-527-31766-0

eco-efficient use of renewable resources in an environmentally sustainable manner and the concept of biorefinery; the target in the EU is to supply between 27 and 48% of the road transport fuel needs by 2030 [5].

A short general definition of industrial BT has been given by the United Nations Convention on Biological Diversity (http://www.cbd.int./): 'any technological application that uses biological systems, living organisms, or derivatives thereof, to make or modify products or processes for specific use' [6]. It comprises a considerable range of current industrial activities and includes not only synthesis and manufacture of products, but also services including environmental processes (sometimes referred to as Grey BT). Both fermentation processes which use some 100 strains in about 130 industrial processes, and biocatalysis, are applied in industrial BT with whole-cell biocatalysts and enzymes used in processes on a large scale as well as in sophisticated systems [4, 7, p. 43, 85]. Relevant arguments in favour of biobased industries are more biodegradable products, less harmful environmental impacts and mitigation of the projected global climate change [1, p. 3]. A wide field of diverse applications and tools of modern bioprocess engineering are thus assembled under the heading of White BT (see Chapter 15) [4, 9, 10].

The *business* potential is considerable or even vast, depending on the source cited. In 2005 turnover was estimated to be in the range of €34 billion (bn)/annum ($50 bn/a) worldwide (about 2% of that of chemicals with a turnover of €1700 bn/a) including biofuels (€18 bn/a), pharmaceuticals (€8 bn/a), bulk chemicals (€7 bn/a) and others [11–13]. Recently world sales have been estimated at €55 bn (about US$77 bn) or even at €77 bn according to McKinsey [14, 15].

The classes of products manufactured using White BT, in increasing added value and (usually) decreasing volume (Table 16.1, see also Table 16.3) are as follows:

- Fuel, e.g. ethanol and biodiesel, and energy, e.g. biogas
- Commodities: organic acids, acrylamide, detergents, biopolymers, for example biodegradable polylactate (PLA) and polyhydroxybutyrate (PHB)
- Food and feed ingredients, textile and paper, for example amino acids, starch derivatives and sweeteners
- Fine and speciality chemicals, for example antibiotics, chiral intermediates for pharmaceuticals, agrochemicals, sugars and derivatives such as sorbitol, enzymes, vitamins, dyes, fragrances, cosmetics and polysaccharides.

Environmental processes play a major role in industrial BT, enabling clean and *sustainable production* and furthermore providing a *safe environment* most notably in crowded areas, big cities and industrial areas. These technologies represent big business. They comprise waste water and exhaust air purification as well as soil remediation. Data on most relevant current products can be found in Dechema [9]. Many details concerning the prospects of future products have been compiled by the US Department of Energy [2]. Table 16.1 shows a small selection of products only, many more are actually marketed. It also shows the economic relevance of bulk production at low prices as well as low volume speciality production at higher prices. Several classical products in the food sector that are manufactured by fermentation are still produced in a higher total volume and are of greater value, for example beer, wine and yeast which are produced in quantities of 138, 27 and 1.8 million t/a, respectively.

Table 16.1 Selected products made by fermentation (approximate data, worldwide, 2003–2005) ([9, 10], other see notes below). The inverse relation between production volume and price is clear.

Product/Process	Production (t/a)	Price (€/kg)	Market value (million €)	Company
Ethanol	37 500 000[a)]	0.4	15 000	Diverse
Starch products	>10 000 000			Diverse
L-glutamate	1 500 000	1.20	1800	Ajinomoto, Tanabe Seiyaku (Japan)
Citric acid	1 100 000[b)]	0.80	880	Diverse
Enzymes			1 830	Novozymes (DK), Genencor (USA)
HFCS	8 000 000	0.80	6400	ADS, A.E. Staley, Cargill, CPC (USA), a.o.
Isomalt	>100 000			Suedzucker, Cerestar (D)
Xanthan	40 000	8.40	336	Evonic (D)
Penicillins	45 000[c)]		Total antibiotics: 19 000	DSM (NL), Bayer (D), Kaneka (Japan), a.o.
Cephalosporins	30 000[c)]			
Riboflavin (B$_2$)	30 000			BASF(D), DSM (NL)
Vitamin C	80 000	8	640	Roche (CH), BASF (D), Takeda (Japan)
Ephedrine	1500	60–90	110	BASF (D)
Cyclodextrins	5000	10	50	Wacker (D)

a) [16].
b) [17].
c) [18, 19].
a.o., and others.
ADS, Archer Daniels Midland.
CPC, CPC International.
HFCS, High Fructose Corn Syrup, Glucose/Fructose syrup (sweetener).
Isomalt, hydrogenated isomaltulose (sweetener) (production volume: [20, 23]).

The considerable range of major companies that are active in the field comprises chemical, oil and biofuel, agro, food and pharmaceuticals business, as well as start-up and medium-sized companies (see e.g. Table 16.1). The essential role of small and medium-sized companies should be emphasized with respect to innovation.

16.2
General Aspects

16.2.1
Raw Materials, Resources

The major source of *biomass* is produced by photosynthesis with an annual total of about 170 billion tons, of which carbohydrates are the most important constituents accounting for some 75%. 'Annually, plant biomass produces eight times as much energy as we consume globally. Therefore, if we could harness 12% of plant biomass for the production of energy and industrial materials, in place of fossil fuels, we could establish a sustainable world' [21]. Among the industrially used raw materials cellulosic matter is the most abundant carbohydrate produced with approximately 2 billion tons being used annually for bioenergy and bioproducts (wood is composed of ca. 40–50% cellulose), followed by starch, sucrose and inulin. These carbohydrates are used as renewable resources by processing biomass in vast quantities [20, 22–24]. However, it must be borne in mind that cellulose is used mainly in the classical sectors of wood, paper and pulp manufacture and that starch and sugar, both of which are used in the food sector, must take priority. Resources for industrial biotechnology mainly comprise classical agro products: starch, sucrose, molasses (containing ~50% sucrose) and cellulose which currently accounts for only a minor fraction of the raw materials, in addition to plant oils (palm and rape seed oil used for the production of biodiesel) (Table 16.2).

The main sources of *industrial starch* which is produced in quantities of 48.5 mn t/a, are given in Table 16.2 as is their origin; it is noteworthy that America is the dominant supplier. It is highly important to be aware of the size of the input of plant material required to manufacture products from agricultural sources: 44 mn t of maize are required to produce 27.6 mn t of starch, 5.3 mn t of wheat to produce 2.9 mn t of starch, thus leaving 16.4 and 2.4 mn t of by-products and residues respectively, to be disposed of, an important part of which is used as feed material. The by-products represent an essential part of the overall process economy [22]. The major proportion (55%) of industrial starch is used in the food and beverages sectors for the manufacture of food and food additives (see Section 16.5). Of the total industrial starch produced 45% is employed in the manufacture of non-food products with about 27% being used in paper and pulp manufacture [25]. In the USA a major

Table 16.2 Production of renewable resources worldwide; (rough estimates for 2004) [25–27].

Type	Million (t/a)	%	Origin	Sources
Cellulose	150	29.6		Wood, straw
Lignin	50	9.8		Wood, straw
Sucrose	160	31		Sugar cane, sugar beet
Molasses[a]	20	3.9		
Industrial Starch[b]	48.5	9.5	America, 51% (in USA mainly corn)	Maize (corn), 74%
			Asia, 26%	Tapioca, 10%
			Europe, 20% (mainly wheat)	Wheat, 8%
				Potato
Fats and oil	81	15.9		
Total	509.5	100		

a) Containing about 50% sucrose.
b) Another source reports about 37 million t/a [22].

proportion of the starch produced is used as the raw material for ethanol fermentation to provide fuel. The price of corn starch as a fermentation substrate is in the range of $0.25–0.6/kg hexose and as the starting material for the production of glucose syrups the price range is $0.3–0.6/kg hexose [24].

Sucrose is used in the food sector mainly as a sweetener. Sucrose and molasses serve as major feedstocks for the fermentation of ethanol, citric acid and oligo- and polysaccharides used in the food industry and together with other applications this accounts for over 13 million t/a. The price for sucrose as a fermentation substrate is in the range of 0.2–0.3 $/kg hexose [24]. In Brazil sucrose is used in ethanol fermentation for the large-scale production of ethanol. Over 80% of molasses is used as a feed component and a further important percentage serves as substrate for fermentation purposes, for example in the production of organic acids (citric and gluconic acids etc.) at a cost of $0.15-0.25/kg hexose [24, 27]. Several other raw materials play a significant role in the manufacture of speciality products such as inulin (production is about 300 000 t/a, [28]), tapioca, additional plant oils, soya (protein) products and so on.

Future resources may essentially rely on wood and agricultural residues such as corn stover, straw, bagasse and sugar beet pulp which are composed of 95% lignocellulosics and are available in large quantities. The US National Research Council estimates that 'enough waste biomass is generated each year – approximately 280 million tons – to supply domestic consumption of all industrial chemicals . . . and also contribute to the nation's liquid transportation fuel needs.' Assuming that additional land, including some 35 million of marginal cropland, can be made available and yields can be increased fourfold (up to $25\,t\,ha^{-1}$, for example with switch grass), approximately 46 mn (million) t of additional biomass feedstock would become available in the USA [1, 2, p.11]. It has been estimated that in the EU there is the potential to produce 300 mn t of straw from cereals, 60% of which would in principle be available for economic utilization, a figure that is greatly in excess of that required to meet the target of reducing the use of fossil fuels by 10% [15, 29]. According to Novozymes, a new cellulase system will cost about 13 cents per L, and China is expected to produce ethanol from agricultural waste at a cost of $2.5/gallon ($0.66/L) by 2010 and $1.5/gallon ($0.40/L) by 2015 [30].

Success in achieving these objectives depends on the development of new and efficient technologies. Such advancements will require integrated processing and utilization as projected in biorefineries (see below). The sustainable production and eco-efficient utilization of renewable raw materials must be the focus of these objectives [4, 17]. The huge reservoir of available cellulosic materials is currently receiving attention from several major companies, for example Dow, Dupont, and Cargill. Enzyme suppliers such as Novozymes, Genencor, Diversa and BRAIN, partly funded by the US Department of Energy, are developing technologies to substantially reduce the cost of enzyme production and are engaged in producing significantly cheaper cellulase systems by the end of 2010 [31–33].

The prospects for modifying raw materials by genetic engineering including the modification of plant metabolism to increase the range of varieties and the yields, are already under discussion [1, 34] (see also Sections 18.3.2 and 18.3.3). An effective approach to enhancing the yield of carbohydrates in plants using genetic engineering to overcome the sugar-yield ceiling in plants, has recently been published by Wu and Birch [35]. The authors introduced an appropriate sucrose isomerase gene into sugar cane which resulted in the accumulation of isomaltulose in the storage tissues without a concomitant decrease in the concentration of the stored sucrose, thus effectively doubling the sugar concentration of the harvested juice.

16.2.2
Biorefineries

The concept of integrated biorefineries producing commodities from renewable resources has been envisioned. This involves processing various feedstocks derived from a range of different biomolecules including lingocellulosics such as wood straw and whole crops (e.g. maize) into a variety of useful products. The manufacture of such products which include fuel, energy and materials, building blocks for chemical synthesis and chemicals, would make use of the entire biomass including the by-products [9, p. 16], [2, p. 13], [36, 37]. A federal programme launched in the USA in 2009 provides nearly $800 million to support biofuels R&D and to encourage the development of commercial-scale biorefineries which can meet the projected target of producing 60 million m^3 of cellulosic ethanol by 2022 [3]. The objectives of these projects still need to be tested for success in commercial plants (for bioethanol pilot plants see Section 16.3.1). Further, competition by burning the material to generate electricity and process steam (as is common practice with bagasse, and has been successfully tested for sugar beet pulp) must be considered.

The concepts behind integrated biorefineries are [38–40]

- a holistic approach, use of various complex feedstocks and handling of multiple intermediates and products
- fermentation using new feedstocks
- logistics and management of biorefineries
- feedstock engineering
- link to chemicals production, including bulk and higher added value products
- minimization of energy consumption, optimization of processes, flowsheeting and reduction of environmental impact, minimization of waste.

In view of the raw materials available, a priority will be the utilization of (ligno-) cellulose as it accounts for over 90% of the bulk of renewable raw materials outside the food sector. The materials, intermediates and products should be used for both energy (heat and power) and the production of various fuels and chemicals [38]. Whole corn (maize)-based refineries using raw materials including starch, ligno-cellulosics and oil as the main components, will yield alcohol for fuel, protein for animal feed, oil, starch and fibre products ([1], see also the concept of a whole-crop biorefinery [37]).

A concept for an integrated approach for the utilization of lignocellulosics '... needs a myriad of logistical and technical issues to be solved.' [33]. This includes research concerning the bottlenecks of logistics, physical and/or physicochemical pre-treatment, for example using low Mw alcohols, as well as economic and ecological assessment. Unsolved problems resulting from the proposed use of lignocellulosics such as straw, as alternative feedstocks arise from logistics as well as from pre-treatment procedures, due to the highly complex structure of the material and the fact that it is highly resistant to enzymatic attack. Regarding the availability of straw, it has been calculated that 1 t of cereals (wheat) gives approximately 0.8 t of straw. The logistic problems result from the low density (180–220 kg/m³) of the material and the vast quantities that need to be transported.[1] The cost of straw as a raw material was estimated to be in the range of 30 to 90 €/t including collecting, drying, pressing, storage and transport. If the cost of conversion into sugar is taken into account the price of the ethanol produced would be much higher than the actual fuel prices; thus there are currently no plans to establish a large commercial plant [41].

The highly complex raw material comprises different sugars as basic building blocks, for example glucose, fructose, xylose and arabinose (pentoses from hemi-celluloses) which, in addition to lignin, will be obtained as primary hydrolysis products. Glycerol and various fatty acids are the primary hydrolysis products from oil. Their biotransformation by chemical routes, enzymes and/or fermentation with bacteria, fungi or yeasts, yields a wide variety of products and building blocks such as methanol, ethanol and other alcohols, acids, for example propionic, succinic, itaconic acids and also lignin and/or its constituents (Figure 16.1). In part, these are important commercial products but may also serve as intermediates in many different subsequent process steps in the manufacture of chemicals and commodities such as surfactants. Secondary chemicals include ethers and esters, acrylates, polyesters polyamides and so on. The flow scheme for biorefinery production will be highly complex as shown in Figure 16.1 due to the many components and constituents of the biomass and the resulting conversion into intermediates and products (see [2, p. 18], [29]). Whilst many of these products are available from alternative sources, others are currently not commercially available and must be introduced into the market in competition with established products, not an easy task. Thus, despite considerable efforts in the production, including the establishment of pilot plants, and utilization of furfurals and derivatives during the 1980s, commercial success was not achieved and was in fact 'hampered by their prohibitive selling price' [43–45]. Integrated biorefinery concepts are under discussion at several chemical and food companies; however the target date for realisation of these aims has been shifted beyond 2020 to 2050 ([36, 46, 47, 54]).

1) To illustrate the logistical problems, the proposed scale of fuel production (>200 000 t/a) should be taken into account. To produce 1 t of ethanol it is assumed that 5–10 t of straw will be required; thus about 8000 t of straw must be transported by 500 trucks per day and assembled in an area of up to 100 km in radius from the factory [41]. Similar problems are associated with using wood as a raw material since the transport costs are estimated to be over €60/t [42]. Physical pre-treatment which is essential for optimizing enzyme hydrolysis represents another bottleneck. Steam explosion which was developed several decades earlier, has not been applied on a large scale over extended periods of time. As alternatives, thermal and chemical processing is being investigated and is under development at the pilot scale.

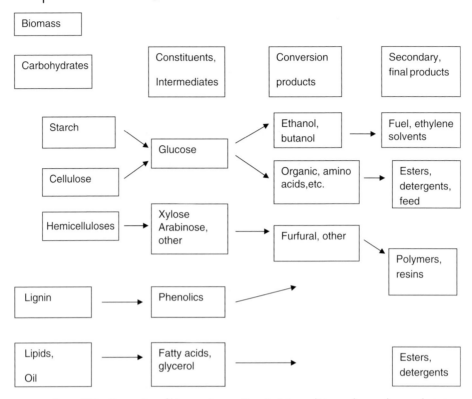

Figure 16.1 Conversion of biomass to constituents, intermediates and secondary products in a biorefinery (selection from a wide range of further constituents and potential products).

Genomic strategies are expected to provide clues to unravelling cellulosic biomass (see Section 18.3.2). The plan is to identify plants that are more easily digested and might only require hot-water pre-treatment rather than treatment with chemicals. A further goal is a one-pot-process which involves engineering a multi-talented microbe that can disassemble the plant cell wall and ferment the resulting sugars into biofuel in one go, for example using *Clostridium thermocellum* as a model organism [48].

16.3
Commodities

16.3.1
Biofuels

Although the production of biofuels is of considerable economic relevance it is also characterized by heated political, ecological and social debate. Recombinant technologies and second generation fuels are expected to contribute to a reduction in the dependence on fossil resources and to significantly reduce greenhouse gas emissions – a calculation of emissions gave 94 for gasoline, 77 for currently available bioethanol, and 11 for cellulosics-based ethanol (measured in kg of CO_2 equivalent per MJ of fuel production and burning) [49].

Ethanol is both currently and traditionally one of the most important BT products to emerge since the nineteenth century although its production has been variable, depending on the availability of competing fuels and wartime requirements (see Section 4.3.3). A total of 46 mn (million) m^3 of ethanol was produced worldwide in 2006 and currently 47 and 28 mn m^3 being produced in Brazil and the USA, respectively. Among the largest producers of bioethanol are Archer Daniel Midland and Shell in the USA, Petrobas in Brazil, and Suedzucker in the EU [16, 23, 50–53]. Currently ethanol can be blended with standard fuel up to 10% in the USA and up to 5% in Europe, whereas in Brazil all gasoline contains 20–25% ethanol. Other major ethanol-producing countries include France and Spain. It is essential to recognize that government incentives and large-scale economies are crucial for fuel production, ethanol being a very cheap product (depending on government funding and/or tax reduction or legislation in relation to blending gasoline with bioethanol).

Problems in ethanol production occurred in 2007 due to the high prices of the raw materials, notably wheat in Europe. Several plants were closed down [54]. Major *crises* occurred in 2007 and most notably in mid-2008, at which time food prices increased dramatically. The growing use of cereals for biofuels was thought to be in part responsible for this price increase. It was mainly poor populations in third world countries who suffered; there were protests and even riots in at least 24 countries around the world, for example in Haiti and Bangladesh [55, p. 19], [56, p.1, 4]. An OECD expert estimated that one-third of the rise in prices were due to biofuel production. Even in late 2008 when food prices began to fall, critics argued that the food crisis was still not over [57].

The most common *sources* of raw materials for ethanol manufacture by fermentation are sucrose from sugar cane (Brazil) or sugar beet (EU), glucose from corn (maize) (USA), and wheat and barley (EU) (sugar cane is the cheapest raw material and as such gives the lowest production costs and the optimum energy balance [58]. Fermentation of sucrose is a straight forward process whereas fermentation of cereal crops is more complex. Two principally different production methods are used, the whole grain process and the wheat starch process based on the fermentation of lower grade starches and glucose syrup. In the latter process starches of lower quality – the so-called C-starches – are used as feedstocks. Raw material and yield followed by energy consumption have the highest impact on overall cost [20, 23].

16.3.1.1 Processing

Some details of bioethanol production processes, as an industrially and economically most important topic, deserve to be mentioned (taken from [20, 59]; for starch processing see also [7, p. 218, 219], [51, 60]). Whereas most fermentation processes work in batch mode, continuous systems with cell recycling are currently preferred

Whole-Grain Wheat Fermentation Process

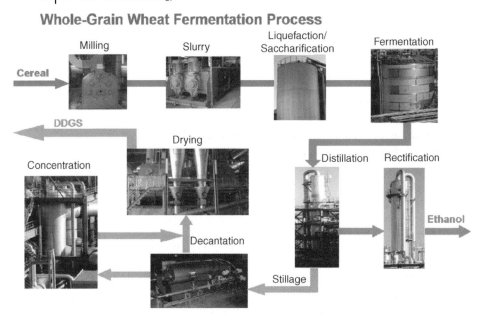

Figure 16.2 Whole grain process for bioethanol manufacture [20].

since productivity (in the range of $30\text{--}50\,\mathrm{g\,L^{-1}\,h^{-1}}$) is higher by an order of magnitude using this method as compared to batch fermentation (in the range of $2\text{--}2.5\,\mathrm{g\,L^{-1}\,h^{-1}}$). The biomass from the fermentation broth is separated by centrifugation and returned to the reactor(s). The whole-grain process is becoming the work-horse of bioethanol production both in the USA and Europe. In this process ground kernels (corn or wheat) containing all the starch and non-starch components such as pentosans and other hemicelluloses, β-glucans, proteins (soluble and insoluble) and fats are introduced into the process and thus influence the fermentation and subsequent steps. The three main stages are upstream processing, fermentation and downstream processing (Figure 16.2) [20, 23, 51, 59].

In *upstream processing* the major steps are: milling; starch hydrolysis to sugars and enzymatic degradation of hemicelluloses; sterilization. Starch is converted by the synergistic action of thermal denaturation and hydrolysis with thermostable α-amylases first at about $110\,^\circ\mathrm{C}$, then at $90\,^\circ\mathrm{C}$ into dextrins followed by conversion with amyloglucosidase (glucoamylase) at $60\,^\circ\mathrm{C}$ into glucose [7, p. 218, 219]. Other enzymes added include proteases, β-glucanases to convert β-glucans to glucose, arabinoxylanases, xylanases, other hemicellulases and pectinases in order to partially hydrolyse the corresponding polysaccharides and thus reduce the viscosity of the solution.

As a substrate for *fermentation* a 22% glucose solution is fed into the reactor to yield 12–13% (V/V) ethanol when over 90% of the starch has been converted to ethanol (about 48 h residence time at $32\,^\circ\mathrm{C}$) (for principles of fermentation see Section 15.4). Propagated yeast is used as inoculum for the fermentation. The capacity of a fermenter can be up to 1000 m³. Recycling of the thin stillage (see Figure 16.2) and the reduction of freshwater consumption are crucial for reducing energy consumption – in addition to other measures such as multistage distillation and rectification. Batch, fed batch and continuous processing are all possible.[2] Fermentation in a commercial plant is run continuously for 2 to 3 weeks, with inoculation taking place in the first fermenter. Virtually all sugars or starches should be converted during fermentation [20, 23, 59].

In *downstream operations* the mash is freed from alcohol by distillation mostly in a two-step procedure, the aqueous alcohol is further rectified and dried (also using molecular sieves). The liquid is partly recycled in order to reduce the requirement for freshwater. Contamination (formation of acids by lactic acid bacteria or glycerol by *Saccharomyces* under stress) both in upstream processing and fermentation easily leads to relatively high ethanol losses. Specific energy saving techniques such as multi effect, high pressure evaporation in cascades and enzyme technology have greatly contributed to the reduction in overall energy consumption [20, 51, 59].

Large-scale processes offer significant economic advantages; a capacity of at least 200 000 m³ per year has been considered advantageous (Figure 16.3); a large production facility is shown in Figure 16.4. Economies-of-scale imply reduced fixed cost for large plants (the investment for a large plant will be over 100 million €) (for cost analysis see [61]). Plants should be able to work with different substrates such as sugar beet and cereals [20, 23]. By-product utilization plays a crucial role both due to high volume and economic significance. This is obvious from the ratio of wheat input to ethanol output of about 3.4 (t/t). Per t of ethanol produced about 1.4 t of solid by-products (plus 0.96 t CO_2) have to be marketed or disposed of. The main

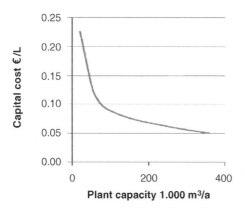

Figure 16.3 Economies of scale for ethanol production (capital cost, € per l ethanol as a function of plant size, in 1000 m³/a) (Kunz, 2006, personal communication).

2) The main advantages of batch fermentations are: higher yield, lower risk of contamination, and lower risk of by-product formation; disadvantages are: higher investment costs, more sophisticated instrumentation and control measures, and the by-product glycerol is more deleterious in continuous compared to batch fermentation [20, 23].

Figure 16.4 Production facility for bioethanol manufacture (with kind permission from Suedzucker, 2008). The figure shows ethanol tanks (left), milling, saccharification (at the rear), fermentation station (background, left) and grain silos (background, middle).

by-products are bran, husk, sugars from hemicelluloses and protein (mainly in the stillage), all of which are largely sold as feed material.

16.3.1.2 Future Developments and Challenges
To improve the existing technology the challenges are [20, 23, 62]:

- Optimized *Saccharomyces cerevisiae* strains having higher osmotic and alcohol tolerance and sufficiently high fermentation rates (yeast being very sensitive to alcohol concentration of $100 \, \mathrm{g \, l^{-1}}$).
- The hydrolytic enzymes for hydrolysis of the raw material must be improved to reduce the influence of the minor components of the cereals, especially β-glucans, pentosans, and proteins which all impact the viscosity of the thin stillage, in order to minimize the overall energy consumption in ethanol processing.
- Purification of the product using advanced membrane technologies.

Metabolic engineering of *Saccharomyces cerevisae* is a promising tool for *strain optimization*. The complete genome sequence has been known since 1996. Strategies and concepts for improvements are based on metabolic flux analysis and genetic engineering of the cells. In particular the goals include extension of the substrate range to include for example pentoses, improvement of productivity and yield, elimination of by-product formation and improvement of cellular properties.

New bioengineering concepts for ethanol fermentation are being undertaken, thus for example a continuous two-stage bioreactor system with cell recycle, the first stage dedicated to yeast cell growth, the second to ethanol production, including an ultrafiltration module for high cell density processing, resulting in ethanol productivity of up to $41\,\mathrm{g\,l^{-1}\,h^{-1}}$ [63]. A further example is the development of a very high gravity ethanol fermentation with simultaneous saccharification and fermentation [64]. Such approaches might be successful particularly in combination with progress in engineered yeast which exhibit enhanced ethanol tolerance.

Second generation biofuels based on non-food feedstocks such as woodchips and straw have been discussed in Section 16.2.2 (for policy goals see Section 16.1) and will not be widely available before about 2015. Production of corn-based ethanol in the USA will level off at 56 to 68 million $\mathrm{m^3}$ (15–18 billion gal) per year. That volume would consume around half the current US corn crop [50]. The US 'energy bill' of 2007 requires that by 2022 15% of the annual fuel consumption must be substituted by biofuels (137 mn $\mathrm{m^3}$) and in addition 60% of biofuels are to be derived from lignocellulosics [3, 15]. However, the differing boundary conditions and assumptions have been shown to result in largely contradictory estimates of the potential of biofuels [65]. A federal programme launched in 2009 in the USA provides nearly 800 million $ to support biofuels R&D and to encourage commercial-scale biorefinery demonstration projects [3]. USDA (US Department of Agriculture) funding for cellulosic ethanol would be 400 million $ over 10 years. The DOE (US Department of Energy) contributed 250 million $ to cellulosic research centres for research into new technologies for breaking down cellulose. A biorefinery funded by the DEO is proposing to expand its plant in Iowa to increase annual production from 190 000 to 470 000 $\mathrm{m^3}$; 110 000 $\mathrm{m^3}$ of this will be produced from corn stover [3, 50]. Shell (GB/NL) was one of the first energy companies to invest in second generation biofuels with a demonstration plant. In a joint venture, DuPont (USA) and Genencor (Danisco, DK) are in the process of building a production unit based on switchgrass and corn stover with an investment of 140 million $, scheduled for the end of 2009 [53, 66], [67]. Canadian enzyme producer Iogen operates the world's only large-scale demonstration facility in which cellulosics-based ethanol is made from agricultural residues, using up to 33 t of feedstock per day [68]. A considerable range of unsolved problems for using cellulosic biomass must still be taken into account, from transport and logistics to physical pre-treatment, raw materials at reasonable cost, government support and finally the attraction of sufficient private investment. Thus, a specialist's question might be, 'how are you going to get enough stover off the fields, move it, store it, and process it in order to supply a 375 thousand $\mathrm{m^3}$ ethanol plant?' [50, 68].

An alternative potential source for biofuels is engineered algae which use sunlight as an energy source to convert waste CO_2 into sugars and ethanol which is then excreted. Dow Chemical (USA) is planning a pilot project with a biofuel start-up Algenol Biofuels [69]. Alternatives to ethanol might be butanol, with higher energy density and suitability for use in unmodified combustion engines. BP and DuPont plan to build a research and demonstration facility to produce biobutanol [70]. In order to simplify fermentation it could be produced by engineered *E. coli* instead of by

Clostridium acetobutylicum [71]. Biodiesel based on the Jatropha plant is considered as still another fuel alternative, the plant grows on soil that is not used for food production and thus does not compete with food production [72]. Microalgae, which use sunlight to convert CO_2, are considered to be an efficient source of biodiesel all-year-round. Prospects and current limitations have been discussed by Chisti [73] and Schenk *et al.* [74] and aspects of bioreactor design, performance, light and mass transfer and physiology of microalgae have been summarized by Posten [75].

16.3.2
Biogas, Energy

With respect to energy, biomass was the source of 10–15% of the world's energy needs in 2005; the expectation is that by 2030 it will provide about 33% [76]. Biogas could, in principle, provide the solution to energy production from any complex biomass, including the huge reservoir of available residues and wastes, estimated at a potential of some 100 million t/a (oil equivalent) [29]. Methanogenesis was detected by Volta in 1776 but it was over 100 years later that this biological process was first exploited technologically [77].

Biogas, the product of methanogenesis by complex consortia of (obligate and facultative) anaerobic and facultative aerobic bacteria, is essentially a mixture of mainly methane and carbon dioxide (in the range of 40–75% methane, 25–60% carbon dioxide, 0–1% hydrogen, 0–0.5% hydrogen sulphide). Biogas is currently used for heating, drying, and production of electricity. It has two major favourable characteristics: it can be produced from nearly every biological material, easily and quickly from sugars, from factory waste water and highly complex biomass including slowly degradable lignocellulosics, however all these reactions occur with vastly differing kinetics [78, 79]. The second positive aspect is that methane is nearly insoluble in water and therefore readily separates from aqueous solution (e.g. waste water). It does not require any energy input for this step in contrast to most other energy sources such as ethanol. Further favourable characteristics are that biogas production requires little process energy, far less than its output, and that it produces very little surplus biomass to be disposed of. On the other hand there are major constraints: its formation requires highly complex microbial consortia (archeae) depending on their synergistic action, that grow slowly and under certain conditions only, requiring strict control of pH and temperature, absence of oxygen (in general provided by facultative aerobe bacteria) (Figure 16.14) [7 p. 309–317] [77]. In reality major facilities require reactor volumes of several thousand or tens of thousands of m^3 for the production of relevant biogas volumes (see also Section 16.6.2). Anaerobic technology producing biogas has traditionally been used on a very broad basis for decades in waste management and waste water treatment for industries such as food, textiles, pulp and paper mills. Sewage sludge from municipal aerobic waste water treatment plants is treated routinely using anaerobic digestion (about 27 400 plants in the EU), as is animal waste, mainly manure, or cattle dung. More than 3 million digesters are in operation in China and over 1 million in India [77].

Biogas can also be used for the production of electricity by modified diesel motors for example, or can potentially be used in fuel cells with significantly higher efficiency. The biggest biogas unit in Europe went on stream in 2008 in south Germany, producing 16 mn m^3 gas annually, based on 80 000 t maize, grass and intermediate crops [80]. The potential of electricity production in Germany by biogas is estimated to be 5% [81].

Future options for sources of biogas may include energy crops, whole plant material grown primarily as an energy source, including corn (maize), special grasses, water hyacinths, algae, or microalgae which provide high biomass yields, and others [77, 82]. Thus corn residues are available in quantities of about 100 million metric tons annually in the USA, and much more in total, approximately 280 million tons (dry mass) [1]. It is essential to note that biogas production on a large scale depends on government funding in order to be competitive. The production of hydrogen from biomass may be a further option. The selection of micro-organisms which best fit this difficult task as well as sophisticated reactor design, for example for two-stage processes, will require much more work [83].

Microbial fuel cells (MFC) represent a most promising concept that makes use of microbial catabolic activity to directly generate electricity from the degradation of organic matter providing access to cheap and environmentally friendly energy sources. A MFC is an electrochemical device in which microbially-produced reduction equivalents are utilized to deliver electrons to a fuel cell anode. Substrates may be various types of organic material in solution, such as methanol, ethanol, glucose and other sugars (e.g. hydrolysate of hemicellulosics), and, most important, waste water constituents including acetate, lactate and other organic acids. Energy conversion can be achieved using MFCs in which anaerobic micro-organisms serve as biocatalysts and from which electrons are diverted and transferred to an electrode to generate electricity. Finding an efficient way to 'wire' the microbial activity is the key to the success of the concept [84, 85]. Different concepts and mechanisms for electron transfer from the biocatalyst to the anode of the fuel cell have been proposed. Thus conductive bacterial pili ('nanowires') have been observed. Further, membrane-bound or excreted bacterial redox mediators are assumed to enhance the electron transport to the anode and between multiple cell layers (Figure 16.5) [86, 87]. Connecting several microbial fuel cell units in series or parallel can increase voltage and current. Six units in a stacked configuration produced a maximum power output of 258 W m^{-3} h^{-1} [88]. Still another approach uses electrocatalytic anode catalysts based on platinum–polyaniline and platinum–poly(tetrafluoroaniline) sandwich electrodes or on tungsten carbide which allows the efficient *in situ* oxidation of microbial hydrogen and a number of organic acids. Current densities in the range of 1.5–3 mA cm^{-2} have been achieved using a large variety of fermentative, photo-fermentative and even photosynthetic microbial activities [89].

Microbial fuel cells can be operated discontinuously or continuously – the continuous mode being the most appealing mode (for a recent overview see [90]). The most promising systems are mixed bacterial biofilms that grow on and adhere to anode surfaces and do not require sterile conditions, which is essential for the utilization of waste materials. An important fact is that they assemble spontaneously

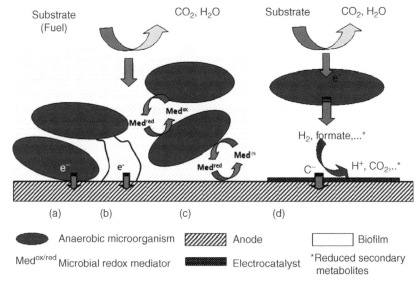

Figure 16.5 Identified electron transfer mechanisms in MFCs. Electron transfer via (a) cell-membrane bound cytochromes, (b) electrically conductive pili (nanowires), (c) microbial redox mediators, and (d) oxidation of reduced secondary metabolites [89].

under appropriate conditions and that inocula from waste water systems and even soil, both of which are rich in all types of micro-organisms can be used. Fermentative bacteria convert different types of substrate (carbohydrates, organic acids etc.) into hydrogen, formate or acetate; these are further oxidized by electrochemically-active bacteria to yield H_2O and CO_2 and electrons which are then collected by the anode (Figure 16.5d). Electrotransfer in the biofilm is assumed to proceed by cell–cell transfer by bacterial redox mediators (e.g. outer membrane redox proteins) and/or by conductive pili. Thus *Geobacter* sp. which is known to have evolved electro-conducting pili has been identified in such systems. Remarkably such systems have been found to operate even under conditions of stress, for example in the presence of toxic substances or at non-optimal temperature, the heterogeneous biofilm clearly stabilizes the whole system. Currently a power output of $1 \, \mathrm{kW \, m^{-3} \, h^{-1}}$ has been achieved [90, 91]. The Integration of MFCs into existing water or process lines has been discussed and a pilot test system has been run at a brewery in Brisbane (Australia) [92].

16.4
Chemicals

16.4.1
Bulk Products

There is a considerable number of traditionally established bulk chemicals produced by bioprocesses, mostly by fermentation (Table 16.3), in quantities of typically over

Table 16.3 Bulk products of bioprocesses (worldwide) [9, 10, 95, 96].

Product	Production (t/a)	Price (€/kg)	Market value (10⁶ €)	Main application
Glucose[a]	ca. 40×10^6	ca. 0.3	12 000	Fermentation substrate, sweetener
Isoglucose[b]	10×10^6	0.80	8000	Sweetener
Baker's yeast	1.8×10^6		2300	Food, beverages
Amino acids:				
L-Glutamic acid	1 500 000	1.20	1800	Flavour enhancer
L-Lysine	850 000	2.00	1400	Animal feed additive
L-Threonine	30 000	6	180	Animal feed additive
L-Phenylalanine	10 000	10	100	Aspartame manufacture
L-Tryptophan	1200	20	24	Animal feed, nutrition
L-Arginine	1000	20	20	Medicine, cosmetics
L-Methionine	400	20	8	Animal feed additive
L-*tert.* Leucine	10	500	5	
Organic acids:				
Citric acid	1 000 000	0.80	800	Food, medicine, detergents
Lactic acid	150.000	1,80	270	Food, leather, textiles, PLA[c]
Gluconic acid	100 000	1.50	150	Food, textiles, metal
Biopolymers:				
PLA[c]	140 000	2.52	315	Packaging etc.
Xanthan	40 000	8.40	336	Food, oil production
Vitamins:				
Vitamin C	80 000	8	640	Food, animal feed
Riboflavine (B₂)	30 000			Medicine, animal feed
Enzymes:				
Total			1830	
Detergents			580	Detergent ingredients
Starch processing			500	Starch hydrolysis, dextrins, glucose production
Textile, leather			250	Tanning
Antibiotics:				
Penicillins	45 000	300	13 500	Medicine, animal feed
6-APA[d]	10 000			Semi-synthetic antibiotics
Cephalosporins	30 000			Medicine, animal feed
Acrylamide[e]	100 000	1.40	140	Polymer synthesis

a) Enzymatic process; including glucose syrup for fermentation to ethanol.
b) Enzymatic process; glucose/fructose syrup (HFCS).
c) PLA, Polylactic acid.
d) Enzymatic process; 6-APA, 6-amino penicillanic acid, intermediate for semi-synthetic antibiotics.
e) Enzymatic process.

10 000 t/a. Among traditional bulk products is glucose for fermentation to ethanol, amino acids, organic acids, and so on, and as a sweetener, manufactured on a scale of some 40 million t/a. Amino acids and organic acids have been produced in large quantities for many decades, mainly by fermentation but also by enzymatic processes, in the range of over 1 million t/a. Over the last four decades there has been a steadily increasing interest in and market for technical enzymes, the current revenue from which stands at about 2.3 billion €, due to improved production methods using recombinant bacterial strains and a significantly extended range of applications (see Section 15.3). Antibiotics such as penicillin and cephalosporin and their derivatives (e.g. aminopenicillanic and aminocephalosporanic acid) are considered here since the quantities produced are high and the prices rather low, in contrast to most pharmaceuticals. New products entering the market during recent years include lactate and 1,3-propanediol as building blocks for polymer synthesis (see Section 16.4.2). The production of 100 000 ton per annum of acrylamide from acrylonitrile was an important economic breakthrough. The clearly inverse relationship of production scale to price level can be seen from Table 16.3. Companies which are active in the field are listed in Table 16.1.

The cost of the manufacturing process for products made in large quantities must be very low, particularly in the case of bulk commodities or feed. The product yield to substrate ratio must be high and should not be reduced to any considerable extent by by-product formation. Genetic engineering and metabolic analysis are key tools in the improvement of micro-organisms and biocatalysts and in the analysis of why yields are less than that calculated from the stoichiometric correlation of the reaction sequences involved; this is not an easy task when the many, sometimes unexpected and hidden metabolic pathways are taken into account (see Section 15.6). Solutions to some of the problems were found, mostly by the cooperation between major academic research groups, research institutes and industry under economic pressure to improve biocatalysts and product yields in highly competitive markets. More problem-solving activity is required in the future in order to meet the substantial expectations of the energy, fuel and chemical industries. Improving yields that are in general high (90–95%), by one or a few percent may be essential for competitive production and may produce an increase in yield of several thousand tons or even more than 100 000 t of the final product. This will be illustrated by the examples of L-lysine and glucose (see below and Section 16.5). Future development will include biosystems engineering both as a challenge and a tool, notably including metabolic engineering. This is the key to producing large amounts of product at low prices [93, 94]. From the wide range of products and processes, some selected details considered to be typical of and relevant to this topic will be discussed here.

Examples of progress facilitated by the development of new tools will be mentioned, including new processes, some process details, economic data and possible new routes using engineered cells.

16.4.1.1 **Amino Acids**
Over 20 amino acids are produced industrially, two (L-glutamic acid and L-lysine) in very large amounts and several in quantities of over 1000 t/a (Table 16.3). Stereo-

selective production yielding either the L- or the D-enantiomer is mandatory. Amino acids are used in many different fields, such as human nutrition: L-glutamic acid or monosodium glutamate are used as taste enhancers for food flavourings and both L-aspartic acid and L-phenylalanine are used as starting materials for the synthesis of aspartame (a peptide sweetener); animal nutrition: L-lysine, D-, L-methionine, L-threonine and L-tryptophan which constitute the largest market share (56%); the pharmaceutical and cosmetic industries and for the synthesis of chiral ingredients of pharmaceuticals, cosmetics and agrochemicals. The total amino acid market in 2004 was estimated to be worth about 3.2 billion € (4.5 billion US$) [97]. Fermentation and enzymatic routes are dominant production processes, yielding enantiopure L- or D-amino acids. Since the 1950s fermentation processes using *Corynebacterium* species have been refined to the highest level of efficiency and economic value and are used to produce L-glutamic acid and L-lysine as bulk products. Furthermore, L-valine, L-isoleucine, L-threonine and other amino acids are also being produced by this species [98].

In fact high performance microbial strains (*Corynebacterium glutamicum, E. coli*) are used in fermentation processes where the carbon source is provided by sucrose, molasses, or glucose. Classical genetics and mutagenesis as well as recombinant DNA techniques are generally employed to modify the biosynthetic pathway and optimize the bacteria to overproduce the corresponding metabolites which results in a striking increase in, for example lysine formation [97, 99]. The identification of key enzymes in various syntheses together with pathway modelling including metabolic balancing using data obtained from ^{13}C-isotope experiments, has revealed a coordinated flux through the pentose phosphate cycle and the futile cycling of C3 and C4 compounds produced during glycolysis and by the tricarboxylic acid cycle, respectively. Thus a unified comprehensive analysis of the pathways together with computational approaches and directed strain engineering can be achieved [100–103].

The availability of the *Corynebacterium glutamicum* genome in 2003 had and continues to have a major impact on the improvement of the manufacture of L-lysine, L-glutamate and other amino acids. At least three different biotechnological companies (BASF, Degussa and Kyowa Hakko) were involved in sequencing, annotation and establishing a network. As a result, all the genes responsible for the carbon flow from sugar uptake to lysine secretion have been identified and are accessible to manipulation. Thus, 3002 protein-coding genes have been identified and functions were assigned to 2489 of these by homologies to known proteins. For biotechnological application the complete genome sequence was used to reconstruct the metabolic flow of carbon into a number of industrially important products derived from L-aspartate as an intermediate metabolite. Preventing the feedback inhibition of aspartate kinase which controls the pathway flux, was one of the most important aims and the creation of feedback-resistant enzyme variants, as well as over-expression of further limiting enzymes resulted in increased lysine production [101, 104–108].

16.4.1.2 Organic Acids

Microbial oxidation and reduction are used in several processes including citric and gluconic acid manufacture, both of which are major products of high economic

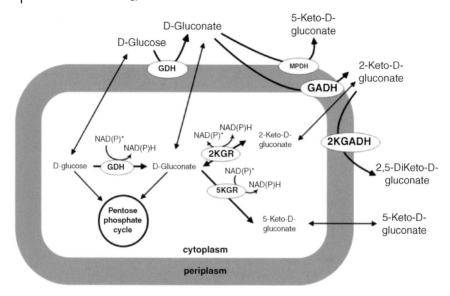

Figure 16.6 Pathways of glucose oxidation in *G. oxydans*. GDH, glucose dehydrogenase; GADH, gluconate dehydrogenase; 2KGADH, 2-keto-D-gluconate dehydrogenase; 2KGR, 2-keto-D-gluconate reductase; 5KGR, 5-keto-D-gluconate reductase; MPDH, major polyol dehydrogenase [110].

significance (Table 16.3), and also in the production of other acids and keto acids such as 2,5-diketogluconic acid and 5-ketogluconic acid. The major organisms used are *Aspergillus* sp. (industrial strains include *A. niger, A. oryzae* and *A. terreus*) and *Gluconobacter oxydans* [96, 109, 110]. They are amongst the most frequently used micro-organisms in industrial processes. Thus citric acid is produced with high efficiency from glucose using *A. niger* (>200 g/l, that is over 95% of the theoretical yield – 1 mole of glucose giving 1 mole of citric acid), as well as gluconic acid by *Gluconobacter oxydans*, also in high yield [111]. From the pathways of glucose oxidation in *G. oxydans* the formation of gluconate, 5-keto-D-gluconate, 2-keto-D-gluconate and 2,5-keto-D-gluconate can readily be deduced (Figure 16.6). The glucose catabolism of *Gluconobacter oxydans* has recently been analysed quantitatively and a metabolic network model has been established which can be used to model all relevant fluxes, based on metabolic flux analysis. Most of the glucose in the medium is oxidized in the periplasm and excreted mainly as gluconate (82%) and to a lesser extent as keto-gluconates; the relevant enzyme (mostly membrane bound) activities have been measured quantitatively [112].

The genomes of four species including the industrial strains *A. niger, A. oryzae* and *Gluconobacter oxydans*, have been sequenced. From this information and knowledge of the metabolic pathways, remarkable insights have been gained: the genomes of the industrial strains are considerably larger than the wild-type strains and contain more metabolic genes which probably provide enhanced metabolic flexibility. This notably influences secondary metabolism, transport and secretion capabilities. The genomes of *A. niger* and *A. oryzae* contain 14 097 and 14 063 predictable genes, respectively. Analysis of the genome shows that this capability is paralleled by a multiplication of several of the genes involved in citric acid metabolism (citrate synthase, aconitase, malate dehydrogenase). There are unanswered questions which relate to the futile cycle of citrate in mitochondria, the role of which is unknown. The ability of efficient gluconate formation is characterized by four genes for glucose oxidases, four for gluconolactonases and 11 for catalases, the latter providing efficient protection against attack by the hydrogen peroxide formed during glucose oxidation. The production of extracellular enzymes and particularly starch hydrolyzing enzymes, for example glucoamylase, is of considerable industrial relevance. Further a large number of genes coding for carbohydrate active enzymes (CAZY) have been identified in the bacterial genomes: 171 in that of *A. niger* and 217 in that of *A. oryzae*. Current genetic information makes optimization more feasible and offers new opportunities for the production of speciality chemicals, novel primary and secondary metabolites and enzymes [113, 114]. Thus pyranose oxidases may provide access to various keto-sugars and keto-sugar acids which may serve as chiral intermediates [115, 116].

16.4.1.3 Vitamins

For many decades vitamin C (ascorbic acid) has been a major product, however, it was mainly produced using the Reichstein process which is a complex production process that gives an overall yield of only 60% using glucose as the starting material. The process includes one regioselective biocatalytic step, the oxidation of D-sorbitol to L-sorbose, and several chemical steps. This method is still used on a large scale by Roche and Takeda and takes advantage of the economy of scale [117]. It is only recently that a new process which was developed in China has been used on a large industrial scale. This new process is based on a two-stage fermentation procedure and replaces the Reichstein process. It comprises the conversion of sorbitol to L-sorbose and subsequently to 2-keto-L-gulonic acid which is readily transformed into ascorbic acid. The process uses a mixed culture of *Gluconobacter oxydans* and *Bacillus thuringiensis* and produces a yield of 85%. The fixed and capital costs as well as production costs of this process are lower than those of the Reichstein process. The Chinese process has been licensed to some Western producers including Roche, BASF and others, and is used by all Chinese companies to give an estimated production of 43 000 t/a [117–120]. A one-step conversion was developed using a recombinant strain of *Erwinia* sp. which oxidizes D-glucose to 2,5-diketo-D-gluconic acid. Furthermore, this strain encodes a recombinant reductase from *Corynebacterium* sp. producing 2-keto-L-gulonic acid which is easily rearranged to produce ascorbic acid (Figure 16.7) [121]. Other vitamins produced by fermentation include vitamins B_1, B_6, and B_{12} [122].

16.4.2
Biopolymers

Polymers synthesized from renewable resources comprise less than 0.1% of the total output. However, McKinsey has emphasized the significant potential of biopolymers

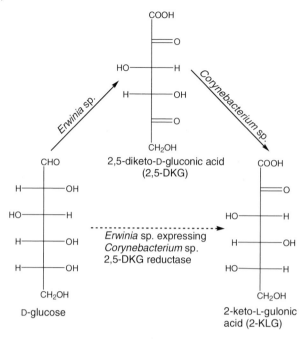

Figure 16.7 Whole-cell biotransformation of D-glucose to 2,5-diketo-D-gluconic acid and 2-keto-L-gulonic acid using a recombinant strain of *Erwinia* sp. (glucose and sugar acids are present in solution in equilibrium as cyclic pyranose and lactones, respectively) [121].

and has predicted that they will account for 6–12% of the market in polymers in 2010 [123]. Polyhydroxybutyric acid (PHB) a biodegrodable polyester produced by bacteria was formerly a major issue [124]. Despite major technological development work by ICI, technical problems and unfavourable economics made these investments a failure and production units were closed down in 2004. Nevertheless R&D in polyhydroxyalkanoates is continuing and engineering of plant species is of current interest (see Section 18.3.4) [125, 126]. On the large scale polylactic acid (PLA) which is based on lactic acid as the monomer, has been established on the market; Nature-Works LLC (Cargill-Dow LLC) produces 140 000 t/a of PLA at its plant in Nebraska, USA. Lactic acid (or its cyclic dimer) is produced by fermentation of glucose and is further processed by catalysis into the polymer. The polymer is fully biodegradable and is used in the textile industries and as packaging material [123, 127].

The production of several monomer building blocks for synthesizing polymers is being undertaken or is at least in the planning stages at various companies. In 2006 Dupont opened a new plant in London manufacturing 1,3-propanediol (PDO) from glucose using an engineered strain of *E. coli* containing genes from yeast and *Klebsiella pneumonia*. PDO is used for the manufacture of polytrimethylene terephthalate (PTT) fibre as part of a drive for 'converting all sales to biobased materials' [128–130]. Other monomers that are either being produced or are of interest include propylene glycol (ADM and Cargill), succinic acid, 1,4-butanediol

(DSM and Mitsubishi Chemical, Roquette), and even methacrylate monomers (Rohm & Haas and Ceres) [123, 131].

A considerable variety of biopolymers, such as polysaccharides, polyesters, and polyamides are naturally produced by micro-organisms. Their use in the food and cosmetics industries is well established due to their favourable properties: lack of toxicity, provision of solution viscosity of various rheological characteristics, ability to form gels of flexible strength and their protective effects for skin and hair. Furthermore, they offer considerable potential for expanded application in the fields of pharmaceuticals and medicine. Commercial products include xanthan, alginate, dextran, (bacterial) cellulose, hyaluronic acid and polyhydroxyalkanoates as mentioned previously [125, 126]. Xanthan is produced by bacteria of the genus *Xanthomonas* as an exopolysaccharide. Because of its high viscosity in solution, it is widely used as a thickener, emulsifier or stabilizer by both food and non-food industries. Dextran has mainly been used as a carrier material in chromatography (e.g. Sephadex®). Hyaluronic acid has found valuable applications in the pharmaceutical and cosmetic industries [125, 126].

Sophisticated biohybrid materials that comprise inorganic crystalline components and organic matrices have become of interest and both classical and biological polymers, for example peptides or agarose, have been investigated in this regard [132, 133]. Their properties of compatibility with cells and bones, cell adhesion and so on, make them interesting for biological and medical purposes. Nanoparticles, both organic and inorganic, can be synthesized by a wide range of micro-organisms. They may be used in different areas, including medical and pharmaceutical applications, such as test systems, biomarkers or drug carriers [134, 135].

16.4.3
Fine Chemicals – Biotransformations and Chiral Synthons

Fine chemicals made using BT comprise a wide field of products, including traditional products such as antibiotics; the more recent innovations such as vitamin B_{12}, and the expanding field of chiral organic building blocks for pharmaceuticals. The assignment to either bulk or fine chemicals is difficult and changes with time, as exemplified by antibiotics where the classical penicillins and cephalosporins are considered to be bulk products and other antibiotics are specialities. One criterion may be the amount produced: over 10 000 t/a for bulk chemicals and less than 1000 t/a for specialities. The price range for bulk chemicals is typically in the range below 20 €/kg, whereas fine chemicals mostly lie in the range above 100 €/kg; both criteria have an obvious intermediate range. Further typical differences are related to engineering: bulk chemicals are produced in large fermenters, in general devoted to one product and fine chemicals are produced in more flexible bioreactors and multipurpose systems that may serve for different products, each for a limited period. Low price products are associated with a lower level of purity whereas sophisticated purification is necessary for high price products. Still another relevant aspect worth consideration is that large companies are often not interested in the development of products with a small market (below some 10 million €); therefore such products are typically more the focus of small companies. Also classification into groups is difficult; this may be related to application, such as chiral synthons for synthesis, food ingredients, cosmetics, biocatalysts including enzymes and cells, or to the biocatalytic system involved, or may follow a chemical classification. Overviews can be found under different headings: products of secondary metabolism [136], biotransformations (typically one or two biocatalytic steps) [137–140], or building blocks [141]. Here a pragmatic principle will be applied with either type of classification. Only examples selected from the broad field of compounds produced are reviewed here in some detail. Pharmaceuticals are discussed in Chapter 17, since they represent a class of products with specific standards, for example of purity, testing and regulatory (approval) requirements.

Since the mid-1970s, a new approach and impetus to systematic studies aimed at application of biocatalysis and biotransformation in organic synthesis was undertaken, including semi-synthetic penicillins, esters and glycerides [142] as well as optically active compounds [143]. It was recognized that biocatalysts could be used for synthesis in organic chemistry, and notably for pharmaceuticals and agrochemicals incorporating chirality, that is, the production of pure compounds consisting of one isomer only as the key requirement.[3] Economic considerations led to straight enzymatic synthesis providing elegant and favourable solutions. It has been estimated that about 5% of fine chemicals are currently produced by biological or hybrid chemo-enzymatic routes and are worth 25 billion US-$ [144, 145]. An important economic breakthrough was the enzymatic synthesis of chiral aspartame, an intense sweetener, the sales of which are in the region of 850 million $ (Figure 16.8) [146].

Several reasons can be given for the development and application of biocatalysts in organic syntheses [147]:

- Organic chemists tend to accept the use of biocatalysts
- Biocatalysis may save additional reaction steps compared to organic synthesis thus reducing production costs
- Enzymes often exhibit excellent enantio- and stereospecificity, satisfying the demand for optically pure (chiral) compounds
- Biocatalysis often results in sustainable technology
- Many enzymes have become commercially available and cheaper due to recombinant technology.

3) Chiral compounds are isomers that contain at least one asymmetric carbon atom, and as such represent mirror images of each other; they are chemically essentially identical, but optically active and polarize light differently. They behave differently in biological systems, where e.g. receptors recognize only one of the isomers, thus only one of two (or more isomer) compounds may be an active drug. A major problem arises when the other exhibits side effects that may be detrimental. One of the worst cases was thalidomide, one component of two isomers being an efficient sleeping pill, the other, however, caused dramatic side effects, embryopathy (malformation of embryos). Therefore pure compounds containing only one isomer are essential in pharmaceuticals, food, feed, agrochemicals etc.

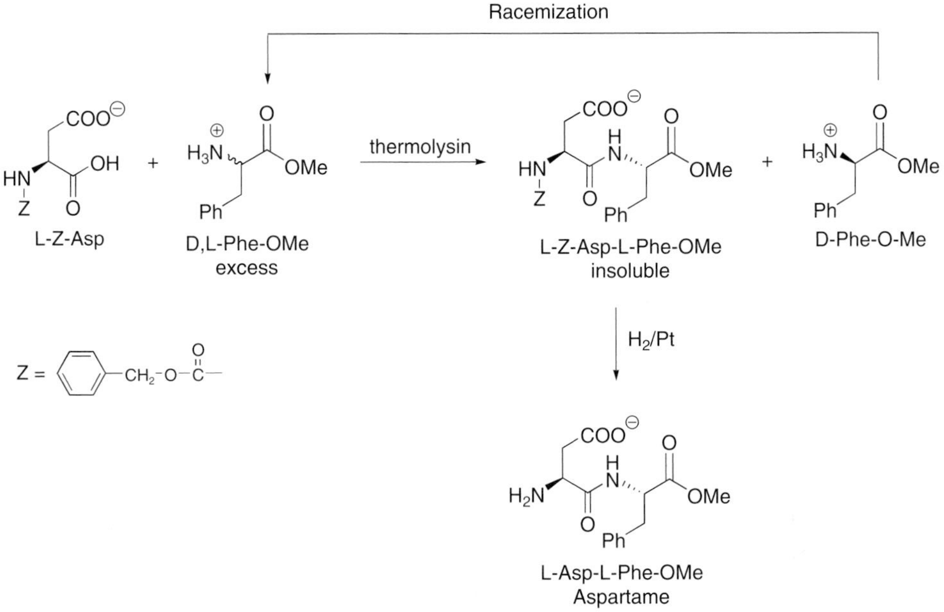

Racemization

Figure 16.8 Synthesis of aspartame using immobilized thermolysine in an organic medium [146].

The types of reactions and classes of enzymes used comprise oxidoreductases (monooxygenases, dehydrogenases (DH)), transferases (aminotransferases, glycosyltransferases), hydrolases (amidases, esterases, glycanases, lipases, epoxide hydrolases), lyases and synthases, isomerases, and ligases (bond formation reactions, hydratases).

The trend continues with considerable intensity, as is evident from the eminent number of papers published since 2000 [148–150]. The number of industrial processes has increased substantially to more than 140 [141, 144, 151, 152]. Liese *et al.* [8] and Straathof *et al.* [140] analysed a substantial number of industrial processes and developed criteria for their technical application. Some examples of biotransformations from a wide range of applications are given in Table 16.4. These examples illustrate the diverse range of biotransformations elegantly used in the synthesis of chiral building blocks.

16.4.3.1 Oxidoreductases

Monooxygenases, including P450-monooxygenases (also termed cytochrome P450 enzymes, E.C. 1.14.x.y) are widespread in nature and play a key role in various steps of primary and secondary metabolism as well as in the detoxification of xenobiotic compounds. A range of reactions are catalysed by these enzymes (Figure 16.9) which all include the transfer of molecular oxygen to non-activated aliphatic or aromatic X-H bonds (X: –C, –N, –S) [149]. An example of reactions catalysed by oxygenases is (*R*)-2-(4′ hydroxy phenoxy) propionic acid, produced at BASF by the oxidation of (*R*)-2-phenoxypropionic acid.

Table 16.4 Selected biotransformations used for the manufacture of chiral products [141, 149, 152].

Biocatalyst	Substrate	Product	Purpose/Application
Oxidoreductases			
Gluconobacter oxydans	Aminosorbitol	1-Deoxynojirimycin	α-Glucosidase inhibitor[a]
Monooxygenases			
Nocardia sp.	Terminal alkenes (*R*)-2-Phenoxy-propionic acid	Epoxides (*R*)-2-(4' Hydroxy phenoxy) propionic acid	
Dehydrogenases[b] (from *Candida* sp.)	α-Keto isocaproate[b]	L-Leucine	Food, pharmaceuticals
		L-*tert*-Leucine Pseudoephedrine α-Halo alcohols, *S*-chlorohydrin	(Fuji Chem. Ind.) Synthesis of cholesterol antagonist (BMS)
D-Lactate DH (*Leuconostoc*)		Fluorophenyl-2-hydroxy propionic acid	Building block for Rupintrivir[c]
Keto-reductase		4-Chloro-3-hydroxybutyrate[d]	Building block for atorvastin
Transferases			
Aminotransferase and reductase	L-Lysine	L-Piperidine-2-carboxylic acid	Synthesis of local anaesthetic Bupivacaine
L-Amino acid deaminases and D-amino acid aminotransferases		D-Amino acids (alanine, leucine etc.)	Synthesis of pharmaceuticals
Glucosyltransferases	Sucrose, maltose	Isomaltooligosaccharides	Food, prebiotics
Fructosyltransferases	Sucrose	Fructooligosaccharides	Food, prebiotics
Cyclodextrin glycosyltransferases	Dextrins	Cyclodextrins	Food, pharmaceuticals, textile
Hydrolases			
Esterases	Acetylated D/L-amino acid mixtures	L-Amino acids[e]	Food, feed, pharmaceuticals
Amidases	6-APA, 7-ACA[f]	Semi-synthetic β-lactam antibiotics	Pharmaceuticals
Nitrile hydratase and amidase	Racemic 2-phenylpropionitrile	(*S*)-2-Phenylpropionic acid	
Lipases	Amines	(*R*)-Amines via (*R,S*)-acetyl amines[g]	Synthons for pharmaceuticals, agrochemicals (BASF)
Epoxide Hydrolases	Racemic indene oxide	(1*S*,2*R*)-Indene oxide	Precursor for a HIV protease inhibitor

Table 16.4   (*Continued*)

Biocatalyst	Substrate	Product	Purpose/Application
Lyases and synthetases			
Tryptophan synthase	Indole and serine	Tryptophan	Pharmaceuticals
Hydroxynitrile Lyases	o-Chlorobenzalde-hyde, HCN	o-Chloromandelic acid	Synthon for antide-pressant (Clopidogrel).
Hydratase:			
Maleate hydratase (*Pseudomonas* sp.)	Maleic anhydride	D-Malic acid	(DSM, NL)
Nitrile hydratase (*Rhodococcus*)	3-Cyano pyridine	Nicotinamide	Vitamin for feed (Lonza, CH)
Three-step biotransformation	2-Propionic acid	β-Hydroxy-isobutyric acid	Synthon for captotril[h] (Kaneka, J)

a)   [164].
b)   Coenzyme and co-substrate required for cofactor regeneration, see text.
c)   Rhinovirus protease inhibitor.
d)   Introduction of a cyano group by exchange of Cl with a halohydrin dehalogenase.
e)   [165].
f)   6-APA, 6-aminopenicillanic acid, 7-ACA: 7-aminocephalosporanic acid.
g)   Stereoselective hydrolysis of the (*R,S*)-acetyl amines and physico-chemical separation.
h)   ACE (angiotensin converting enzyme) inhibitor.

Dehydrogenases (*DH*) require coenzyme regeneration with co-substrates either donating or accepting hydrogen and involving a cycle with a second enzyme. An example is the production of L-leucine from α-keto isocaproate with leuDH, formiate DH and a polymer-bound coenzyme in a membrane reactor (Figure 16.10). Formic acid or its salts is a suitable co-substrate for the regeneration of NAD^+ since it is cheap and the required enzyme, formiate DH (FDH) from *Candida boidinii* is readily available. In addition, the CO_2 produced can be easily separated from the reaction solution thus shifting the equilibrium towards the product, L-leucine, which is produced by Degussa (Germany) [153]. The same technology is used on a ton scale for the production of L-*tert*-leucine, a non proteinogenic amino acid used as a building block for a range of pharmaceuticals. L-and D-specific hydroxy isocarproate de-hydrogenases were made available through recombinant DNA technology [238].

Certain *chiral alcohols* are used as intermediates in the synthesis of acetylcholine esterase inhibitors, as well as in the synthesis of Trusopt, a glaucoma remedy produced on a ton scale, and in the production of C_3 and C_4 building blocks for the synthesis of pharmaceuticals [152]. One-pot chemical/enzymatic syntheses have been developed to facilitate production routes, for example by the combination of a Wittig reaction and an enzymatic ketone reduction or catalytic and biotransformation steps [154, 155].

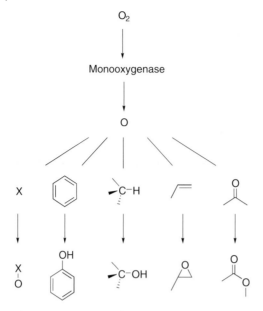

Figure 16.9 Overview of reactions catalysed by P450-monooxygenases [149].

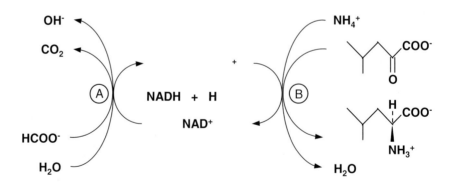

(A) Formiate-Dehydrogenase (FDH)

(B) Leucine-Dehydrogenase (Leu-DH)

Figure 16.10 Reaction scheme for continuous production of L-leucine from α-keto isocaproate using leuDH (B), formiate DH (A) and polymer-bound coenzyme [153].

16.4.3.2 **Transferases**

Both L-amino acid deaminases and D-amino acid *aminotransferases* are non-specific, enabling the production of D-alanine, D-leucine, D-glutamic acid and D-tyrosine from the corresponding L-amino acids in a two-enzyme sequence. First an L-amino acid deaminase generates the α-keto acid. Subsequently the amino group of an inexpensive D-amino acid such as asparagine is transferred to yield the D-amino acid required.

Cyclodextrins represent a remarkable example of fine chemicals with an extended application range, including such diverse areas as food (flavour protection), pharmaceuticals (drug stabilization, solubilization and slow release) and commodities (odour adsorption or release in textiles) [156]. The enzymes used for their manufacture from starch or dextrins are *cyclodextrin glycosyltransferases*; several have been crystallized and their structure and reaction mechanisms elucidated. Enzyme production, including improved selectivity, has been optimized with the aid of genetic tools, in order to increase the yield of specific cyclodextrins, α-, or β-, or γ-, which best correspond to particular applications [156–161]. *Oligosaccharides* (OS) and *polysaccharides* (PS), manufactured using *glycosyltransferases*, are of considerable interest and have manifold applications in the fields of food and food ingredients, pharmaceuticals and cosmetics as a result of their various specific properties (see Section 16.5) [162, 163].

16.4.3.3 **Hydrolases**

Glycosidases have traditionally been used in starch processing on a very large scale for the manufacture of the substrate for ethanol fermentation and the production of starch derivatives and sweeteners (see Table 16.3; see also Section 16.5). It is mainly *Bacillus*, but also *Aspergillus* sp. that is used for production of amylases, glucoamylases, glycosyltransferases and so on. Systems biology tools have been used to improve and modify enzymes as well as their production, including excretion and purification [166]. *Amidases* are used for the enzymatic synthesis of a range of semisynthetic β-lactam antibiotics (see Section 17.4.1).

A sequence of two enzymes, nitrile hydratase and amidase (in a whole-cell processes) catalyse the formation of (S)-2-phenylpropionic acid from racemic 2-phenylpropionitrile [144]. A further step is integrated in a process in which the nitrile group of 2-cyanopyrazine is hydrolysed to pyrazine carboxylic acid followed by the regioselective hydroxylation to 5-hydroxypyrazine carboxylic acid by the same bacterial cells, a process applied by Lonza [151].

Lipases (triacylglycerol hydrolases) are probably the most frequently used hydrolases in organic synthesis for the production of optically active compounds [149, 150, 167, 168]. They are used for a range of industrial processes due to their high activity in a range of non-aqueous solvents, often excellent stereoselectivity and acceptance of nucleophiles other than water (i.e. alcohols, amines), properties which are necessary for the synthesis reactions. Kinetic resolution has shown that lipases are extremely efficient in the synthesis of optically pure *amines* which are versatile synthons for a wide variety of products. An example is the highly enantioselective acylation of (R,S)-1-phenethyl amine in a process established at BASF. The (R)-amide is separated from the (S)-amine by distillation or extraction and the free (R)-amine is released by basic hydrolysis. Since the lipase from *Burkholderia* sp. shows broad substrate specificity, a wide variety of further amines and amino alcohols can be resolved, in some cases on the multi-ton scale. Undesirable enantiomers can be racemized. BASF produces more than 2500 t/a using this process. Further industrial processes based on lipase catalysis have been established in the past few years, for example at the DSM company for the production of a Captopril intermediate [8, 152].

Researchers at Merck identified some enantioselective *epoxide hydrolases* capable of resolving indene oxide which is a precursor of the side chain of a HIV protease inhibitor. These enzymes exclusively produce the desired (1*S*,2*R*)-indene oxide in a yield of 14% [169].

16.4.3.4 Lyases and Synthetases, Hydratases

Two applications based on *nitrile hydratase* from *Rhodococcus rhodochrous* have been commercialized: the large-scale production of acrylamide by Nitto Chemical (Yokohama, Japan) (Table 16.3) and the conversion of 3-cyanopyridine to nicotinamide (a vitamin in animal feed supplementation); the second process has been industrialized by Lonza (CH) on a scale of >3000 t/a.

Important growing *market sectors* include cosmetics, care chemicals, health and nutrition which incorporate products that are based largely on renewable resources, thus serving both wellness and sustainability requirements and bestowing competitive advantages on the bio-label. Procter & Gamble (USA), Cognis and Henkel (Germany), L'Oréal (F) and Nestlé (CH) are the leading companies in the market of consumer products and are active in this field [170]. Oligosaccharides, lipids, polyalcohols and polyacids, flavours and fragrances, alkaloids and vitamins are among the products manufactured in such diverse fields [136, 138, 162].

16.4.3.5 Outlook

Extending the *application range of biocatalysis* as well as the search for new solutions in synthesis, notably of chiral compounds, has been a continuing challenge. More recently, new strategies have been developed to include the plethora of 'non-culturable' biodiversity in biocatalysis: (i) the metagenome approach and (ii) sequence-based discovery [171] (see Section 7.6.7). The major aims of biocatalyst development have been altered specific activity and selectivity. In principle, two major strategies for the improvement of an enzyme on the amino acid sequence level can be utilized: (i) rational protein design, which requires the availability of the three-dimensional structure (or a homology model) necessary to identify type and position for the introduction of appropriate amino acid changes by site-directed mutagenesis or (ii) directed evolution[4] (see Chapter 9) [149, 173, 175]. The potential for extended

---

4) Directed evolution essentially comprises two steps: first, the random mutagenesis of the gene(s) encoding the enzyme(s) and second, the identification of the desired biocatalyst variants within these mutant libraries by screening or selection (protocols are available from [172, 173]). The major challenge in directed evolution is the identification of desired variants within large mutant libraries, preferably using high-throughput robot-assistance which enables several tens of thousands of variants to be screened within a reasonable time [174, 175].

and new synthetic pathways is exemplified by the reactions which alter the synthetic (cyclization) activity of a cyclodextrin transferase (CGTase) into hydrolytic activity and the engineering of a hydrolase (an α-amylase) to function as a CGTase, that is, a synthetic enzyme [176, 177].

Biotransformation by designed recombinant *whole-cell systems* represents a field of high current interest, and a range of examples has been successfully developed for several multi-step reactions. Metabolic engineering offers a key tool for improving yields by optimizing metabolic fluxes (see Section 15.6). As an example steroid pathway intermediates formed during sterol catabolism have been recognized as important precursors for drug synthesis. Different approaches have been developed to overcome difficulties due to the limited capabilities of micro-organisms in such pathways. One strategy has focused on the bioengineering of molecularly defined mutant strains of *Rhodococcus* species blocked at the level of steroid delta 1-dehydrogenation and steroid 9-alpha-hydroxylation. A molecular toolbox has been developed including characterization of the steroid catabolic pathway, cloning and expression vectors for use in the construction of genomic libraries, which enabled the rational construction of *Rhodococcus* strains with optimized properties for sterol/steroid bioconversions [178].

Marine prokaryotes and *fungi* represent unconventional sources of new metabolites and enzymes. They include bioactive metabolites with cytotoxic, antibacterial and antifungal properties. Screening barophilic strains originating from deep-sea or from hydrothermal vents poses a particular challenge [179].

A remarkable perspective has been developed by Griengl [180]; he has proposed that any chemical reaction can in principle, be catalysed enzymatically. Genetics and molecular biology may provide the evidence for this within the next 10 years. Griengl thus found that a cyanhydrin lyase can transfer a nitro group instead of a cyano group onto an aldehyde, a reaction that is not found in nature [181].

16.5
Food Processing and Products

Biotechnology has traditionally played a major role in the manufacture of food, cheese and other milk products and beverages. Thus beer and wine were produced at an output of 170 million t/a worth 200 billion €, cheese at an output of 13 million t/a worth 100 billion € and bakers yeast at a production rate of 1.8 million t/a, in addition to products such as the Asian specialties, sake and soya sauce (data for 2003; [7, p. 4]). These traditional products will not be discussed here despite the fact that new developments have taken place (e.g. improved yeast strains for brewing, wine making and baking [182]); however, selected examples will be reviewed in some detail in order to illustrate recent progress in fundamental insight and technology.

Starch processing is one of the most important technologies and therefore will be discussed in more detail. The industrial production of starch is about 45 million t/a (in the EU ca. 7.5 million t/a). The most important starch sources are corn (maize), wheat and potato. Enzymes used for starch processing, including α-amylase, glucoamylase, pullulanase, isoamylase and glucose isomerase, comprise about 30% of the world's industrial enzyme production. Starch is processed (hydrolysed) in very large quantities (over 10 million t/a are used in the manufacture of dextrins) for applications in the food industries to produce glucose syrups, crystalline glucose and glucose–fructose syrups as sweeteners. Furthermore, over 25 million t/a are hydrolysed to yield the substrate for ethanol, amino acid, organic acid and vitamin fermentation (for details see [60, 162]).

A number of starch-converting enzymes belong to a single family: the α-amylases or family 13 glycosyl hydrolases [183]. Three-dimensional structures, mechanistic principles deduced from structure–function relationships and properties such as kinetics, selectivity and stability have been investigated, reported and summarized in several reviews [158, 159, 184]. Major research efforts have had considerable success in the investigation of thermostable α-amylases and have revealed the structural determinants responsible for the high thermostability of *Bacillus* enzymes [185] which can be used in processing at temperatures of 105–115 °C.

Glucose manufacture is achieved using glucoamylase (and additional enzymes) [7 p. 212–220, 186]. Glucoamylase (GA, known industrially as amyloglucosidase) is an exo-acting inverting glycoside hydrolase (α-(1 → 4)-D-glucan glucohydrolase) that catalyses the release of β-D-glucose. Glucoamylase is mainly produced by *Aspergillus* sp., in a greater tonnage than almost any other industrial enzyme and thus is available at a low price. Some structural details will be discussed subsequently in order to illustrate the relevance of basic knowledge to the improvement of the technical process. The catalytic domain of GA consists of a $(\alpha/\alpha)_6$ barrel with six interior α-helices surrounded by six exterior α-helices (Figure 16.11). The helices form a bed supporting a network of loops, in the centre of which is the active site. This consists of a well that is 10 Å deep and 15 Å wide at its mouth. After deep penetration

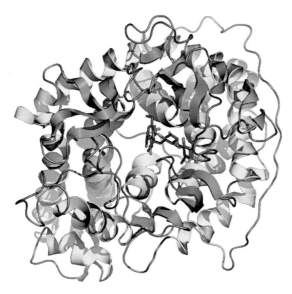

Figure 16.11 The acarbose complex with glucoamylase [187, 188].

of the well the substrate undergoes cleavage. Subsequently the remaining chain leaves the well followed by the liberated glucose before the next reaction can occur. This explains the exo-acting nature of GA [158, 159, 186, 187].

Two major problems were motives for extensive research into the genetic modification of glucoamylase. First, at a high dissolved solid concentration glucoamylase condenses some of the glucose formed to di-,tri- and tetrasaccharides, the most important being isomaltose. Combining favourable mutations was successful in decreasing the kinetics of the enzyme towards the formation of isomaltose, increasing its glucose yield from 96 to 97.5% (equivalent to 200 thousand extra tons on the actual production scale). Second, glucoamylase is not as stable as α-amylase and glucose isomerase and is used at 60 °C in starch processing. Stiffening the α-helices by Gly → Ala and Ser → Pro mutations as well as creating disulfide bonds across two loops contributed to a fourfold increase in the thermostability of A. niger glucoamylase [189, 190]. (For further details of the mechanism and structure see van der Maarel et al., 2003; [158, 159, 184, 187]).

Some details of starch processing will be given subsequently in order to illustrate the technology (for more details see [186, 191], [7, p. 212–220]). In an initial step gelatinization of suspended ground starch is carried out by thermal treatment (105–110 °C) in a tubular reactor in order to produce polysaccharide particles which are then accessible to α-amylase. Some of the (hyper-) thermostable bacterial α-amylase is added in this step (at T ≥ 105 °C) making use of the synergistic effect. The second part of the α-amylase is added in the second step (partial hydrolysis) at 85–95 °C to yield dextrins (Figure 16.12) [191]. The hydrolysis step to produce glucose syrup is carried out subsequently at a lower temperature and pH in large stirred tanks with a high residence time. The yields of glucose syrups are in the range of 96–97% and that of the by-products, di- and oligosaccharides, in the range of 3–4%.

Glucose isomerization by glucose isomerase (GI) is the basis for the largest process undertaken using immobilized enzymes (about 1500 t of biocatalyst) to produce a glucose–fructose syrup (HFCS, high fructose corn syrup) at an output of

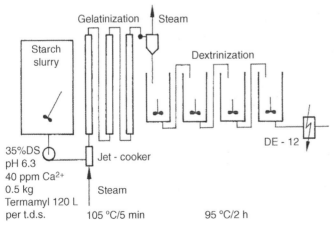

Figure 16.12 Reactor configuration for the starch liquefaction process. (Termamyl is a bacterial thermophilic α-amylase) [191].

more than 10 million t/a of product [192, 193]. Glucose is isomerized in part to fructose which is approximately twice as sweet as glucose. Many GIs (which are basically D-xylose ketol-isomerases, EC 5.3.1.5) have been characterized and their tertiary structures are now known [183, 186, 194]. GI has been successfully genetically engineered with respect to thermal stability, with reference to substrate binding and the tight binding of Mg^{2+} required at the active centre of the enzyme [195]. GI is immobilized in order to keep process costs low. The productivity of the biocatalyst is in the range of 12 to 20 t (dry substance) per kg of biocatalyst with a half-life of 80 to 150 days; various commercial biocatalysts are available. The dimensions of the reactor are typically 1.5 m diameter and 5 m height for fixed bed bioreactors operating at 58–60 °C [186, 196].

Isomaltulose (Palatinose) manufacture is another large process that utilizes the glucosyl transfer activity of a sucrose mutase from *Protominobacter rubrum*. Sucrose is the substrate in which the α-1,2′-β-glycosidic bond is rearranged to an α-1,6′ glycosidic bond as a result of an intramolecular rearrangement, giving a yield of between 84 and 85% isomaltulose and a production rate of about 100 000 t/a. The catalyst consists of immobilized cells of *Protominobacter rubrum* (the cells are non-viable but the sucrose mutase remains active) in alginate. At high initial sucrose concentration (550 g/l) the biocatalyst exhibits high stability (half-life 120–140 days, utilization over a period of 180–200 days) and is used in large fixed bed reactors (Figure 16.13). Isomaltulose has been used in the food sector in Japan and more recently in Europe in sports drinks as an energy carrier with retarded metabolism. The major proportion of the isomaltulose produced is hydrogenated using classical Raney nickel catalysts to give isomalt, a mixture of two sugar-alcohol isomers, glucosyl-sorbitol and glucosyl-mannitol, which is used worldwide in the food sector as an alternative sweetener that has non-cariogenic properties and is suitable for use by diabetics. It is largely used in caramels, chewing gum, tablets and so on ([197, 239]).

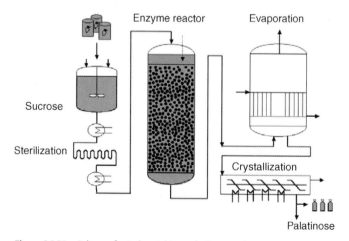

Figure 16.13 Scheme for industrial isomaltulose production [239].

In a considerable number of further processes immobilized glycosyltransferases are used for converting sugars and polysaccharides into new products and for the synthesis of oligosaccharides (OS) and derivatives such as malto- and isomalto-oligosaccharides, galacto- and fructo-oligosaccharides. These are mostly used as sweeteners, food additives and functional food due to their properties such as flavour, sweetness and 'body' which enhances other tastes, they are also used in other products relevant to their properties such as prebiotics, skin protection, and even in products aimed at stimulating the immune system. They are produced by immobilized glycosyltransferases in quantities in the range of 3000 to 7000 t/a [162, 198–200].

Enzymatic processes fully exhibit their potential in the carbohydrate field as they are capable of regio- as well as stereospecificity which make them superior to chemical and classical catalytic reactions. The potential to create different linkages via α- and β- bonds with different regio-positions, that is, positions 1 to 6 in hexose sugars, makes chemical synthesis of oligosaccharides dramatically complex; in general enzymes easily form one single bond or one main product in high yield (mostly >80%) with low by-product formation of isomers.

Enzyme and substrate engineering are among the new approaches to extending the range of OS synthesized by glycosyltransferases, using for example glycosyl fluorides and the development of glycosynthases including thioglycoside synthesis [201, 202]. The development of sucrose analogues has been used for the synthesis of new OS via both wild-type and modified fructosyltransferases, for example levansucrase enzymes to synthesize unique oligo- and polyfructans, such as 1-kestose- and 1-nystose-analogues terminally linked with monosaccharides other than glucose, for example galactose or xylose [163, 203–205].

16.6
Environmental Processes

Environmental processes play a major role in industrial BT, both enabling clean and sustainable production in industry and providing a sound environment. They comprise waste water and exhaust air purification as well as soil remediation, and they represent big business. ([206, 210], [7, p. 309–321]). Thus environmental BT is of major importance considering that all cities worldwide and every town in western countries are obliged to undertake biological waste water treatment, and that most industries including not only food, pharma and chemistry but also car factories have a requirement for biological waste water treatment and in many cases for biological exhaust air treatment systems as well. In 1988 there were about 27 400 aerobic municipal wastewater treatment plants in the EU [77]. The cost of waste water treatment worldwide has been estimated to be 30 billion $ in 1980, and 68 billion $ in 1990, with an annual growth rate of 9% [207]. Waste water treatment is an economically important factor for governments and authorities as they draw up the legislation detailing the standards that must be met, while aware of the consequences of ignoring them. Waste water treatment is also of great economic relevance to those industries involved in the building of treatment plants and those that sell the equipment used in the treatment processes, the success of both these industries depends on substantial financial investment. Thus in China the central government, local governments and banks will provide as much as 58 billion $ in funding for new facilities through 2011 to build water treatments plants or provide water suitable for drinking [208]. Dow Chemicals expect sales of the company's waste water treatment technology to be more than 360 million $ by 2010 [209].

Comprehensive overviews of waste water treatment have been compiled by Winter [240] and Jördening and Winter [210]. Three topics will be discussed here in further detail: aerobic waste water treatment, advanced anaerobic immobilized systems and exhaust air purification; soil remediation, composting and biological processes in landfills will also be addressed. Biochemical engineering is the major tool of process development and optimization. It must be emphasized that non-sterile operation using mixed cultures is standard, in contrast to production pro-cesses which use pure cultures of high performance strains under sterile conditions. Environmental processes are open systems which work continuously to treat the inflow of large volumes of waste water or contaminated air, which makes it neither possible nor advantageous to maintain sterile conditions since many microbial species are required to degrade the waste materials.

Biological degradation of biomass and organic chemicals during solid waste or waste water treatment proceeds either in the presence of molecular oxygen by respiration (aerobic conditions) or under anoxic conditions by denitrification or under anaerobic conditions by methanogenesis or sulfidogenesis. Conversion of soluble organic compounds or extracellularly solubilized biopolymers such as carbohydrates, proteins, fats or lipids in activated sludge systems leads to the formation of CO_2, water and a significant quantity of surplus sludge. Under strictly anaerobic conditions, soluble carbon compounds produced from wastes and waste water are degraded in a stepwise manner to methane, CO_2, NH_3, and H_2S via a synthrophic interaction of fermentative and acetogenic bacteria containing metha-nogens or sulphate reducers (for fundamentals see [211, 212]). If a producer of waste water has to decide whether to install an aerobic or an anaerobic waste or waste water treatment system, several points will need to be taken into consideration:

- In general anaerobic treatment does not lead to low pollution standards of BOD_5 or COD that can be met with aerobic systems and which are required by environmental laws.
- Waste water with high organ content should in general be treated anaerobically in a first step because of the possibility of energy recovery in biogas and the much lower amounts of surplus sludge to be disposed of.
- Wastewater with a low concentration of organic pollutants should be treated aerobically due to its higher process stability at low pollutant concentrations [211, 212].

The criteria for processing as well as for the final concentrations required by the regulatory authorities, are the biochemical oxygen demand (BOD_5, or BOD_{20}, incubation period of 5 or 20 days), the chemical oxygen demand (COD), or total organic carbon.

The dissolved oxygen concentration (DO), the pH and the toxic substances contained in the waste water are considered to be environmental factors which are of major importance for the biological reactions. Problems may arise due to insufficient sludge recycling, high hydraulic load and content of the solids. Biodegradable toxic organic substances such as for example phenols and cyanides can be almost completely removed (both in waste water and in exhaust gas) if they are continuously present in the waste water so that the micro-organisms can adapt and accumulate the appropriate enzymes, and if their concentration in the mixed liquor (waste water containing suspended bacteria) is kept as low as possible at all times; shock loads, therefore, should be avoided.

Since many anaerobic bacteria grow slowly *immobilized systems* have been developed. Thus a large number of fluidized bed reactors with immobilized viable anaerobic mixed cultures, either on a carrier or as granular bacterial aggregates, are in operation in the food industry for the purification of high load waste water. Trickle bed systems are used for denitrification and exhaust gas purification systems in a large number of production facilities. The advantage of systems that use immobilized micro-organisms are the high density of organisms and thus the high biocatalyst activity when compared to freely suspended micro-organisms, thus leading to high-capacity reactors which therefore are of much smaller volume and are less costly [213, 214].

Production integration is a trend that aims to avoid or to reduce the volume of waste water. The main principle is that avoidance should have priority over utilization, and utilization over disposal. Measures include:

- Careful treatment of (agricultural) raw materials (short storage, careful handling)
- Changes in the transport facilities (avoiding damage to raw materials)
- Changes in the production methods (reduction of water demand, improvements in the organization of production)
- Identification and avoidance of production losses
- Production circuits and multiple use of water (reverse flow cleaning, cooling water recirculation)
- Utilization of production residues as raw materials (protein coagulation, valuable substrate recycling, forage production) [215].

16.6.1
Aerobic Waste Water Treatment

Most municipal and industrial waste waters are mixtures of soluble and particulate organic matter in addition to salts. The microbial degradation of organic carbon requires certain amounts of nitrogen, phosphorus, calcium, sodium, magnesium, iron and other essential trace elements to build up biomass. At industrial waste water treatment plants the missing elements need to be added. Since domestic waste water contains an excess of all the necessary elements it is sensible to treat special industrial waste water together with municipal waste water. Typical types of reactors and tanks for mixing and aeration are used. The mixing tanks for denitrification may be square, rectangular or circular and fitted with either mixers in the centre or with propellers. In closed loop tanks either propellers or vertical shaft impellers maintain the circulation of flow [216].

Aeration equipment (blowers) as well as appropriate mixers must be installed. Mixing may also be achieved using rotating bridges on which the diffusers are mounted. In practice simultaneous nitrification and denitrification is mainly undertaken in closed loop tanks equipped with horizontal axis surface aerators, for example mammoth-rotors or vertical shaft surface aerators as in the carousel process. Simultaneous denitrification may also be carried out if air diffusers are arranged in 'fields' in closed loop tanks [216, 217].

Nitrogen elimination from waste water is a standard requirement since the limiting values set by the authorities for total nitrogen containing compounds, and notably ammonia, require this step. Nitrogen elimination is a subsequent process step following either aerobic or anaerobic degradation of organic compounds. During anaerobic degradation of nitrogen-containing compounds most of the nitrogen is converted to ammonia. Subsequent nitrification and denitrification converts it to nitrogen (N_2) gas and to a lesser degree to N_2O, which should be avoided as far as possible (for details of this and phosphorus removal see [216]). In the case of nitrifying bacteria which grow slowly, immobilization has been used, for example in trickle bed systems (for the elimination of nitrogen-containing compounds).

16.6.2
Anaerobic Waste Water Treatment

Anaerobic waste water treatment offers considerable advantages as compared to aerobic treatment. Whereas a high level of energy is needed for aeration in aerobic waste water treatment plants, the anaerobic treatment produces energy in the form of useable biogas. On the other hand this also means that the growth of anaerobic bacteria is slow and for this reason in particular the start-up of an anaerobic plant is slow, and the mixed culture needs tight process control, notably of pH. Furthermore investment costs are higher than for aerobic systems. Figure 16.14 presents an overview of the most important reaction pathways of anaerobic degradation of biomass. The anaerobic treatment of waste water includes the conversion of (poly-) saccharides, proteins and fat by acidifying bacteria mainly into fatty acids (in general a rather fast reaction, except for cellulosics), subsequent reactions by acetogenic bacteria to yield primarily acetate and formiate, which are finally used by methanogenic bacteria to form biogas, CH_4, CO_2 and H_2 [218]. The microbial consortia differ significantly in terms of nutritional needs, growth kinetics and sensitivity concerning the environmental conditions, notably pH [211, 212, 219]. Therefore these two steps are preferably physically separated. Thus two-stage systems may be preferred since the performance in terms of stability and space–time–yield will be superior to one-stage systems.

One very broad and most successful application of high-performance anaerobic treatment in fixed and fluidized bed reactors relates to waste water from industries based on agricultural and forestry products with typically high concentrations of organic substrates which are readily degraded by anaerobic bacteria. This type of

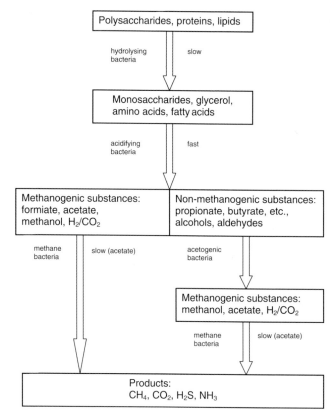

Figure 16.14 Scheme of anaerobic degradation reactions.

treatment offers the advantage of high-load systems which require much lower volume and less space, and hence less investment as compared to conventional systems. A large number of different types of plants using immobilized anaerobic bacteria are currently on stream, among these over 50 are fixed bed and about 20 are fluidized bed reactors that employ carriers; the dominant type of plant is the expanded granular sludge bed (EGSB) of which there are about 200, in addition there are about 700 upflow anaerobic sludge blanket (UASB) type systems which operate without carrier, using granular bacterial aggregates instead [220, 221, 241]. An example of a fluidized-bed reactor is shown in Figure 16.15. It is used in the large scale (up to $500 \, m^3$) treatment of anaerobic waste water from sugar factories; the system has a turnover of 18 to 20 t (CSB)/day under continuous operating conditions [222].

16.6.3
Exhaust Gas Purification and Further Techniques

Exhaust gas purification is a prerequisite of many industrial fields notably the food industries, since a wide variety of organic substances and even some inorganic compounds responsible for odour can be eliminated from the waste gas [224–226].

Figure 16.15 Industrial fluidized bed reactor for anaerobic waste water treatment; working volume 500 m^3, height 30 m (with kind permission of Nordzucker AG, Braunschweig; [223]).

The technology has found successful applications in the food industry for the removal of animal, meat, fish and yeast aromas from exhaust gas; other diverse examples include the degradation of both aliphatic and aromatic hydrocarbons, phenols, sulfides, amines, amides and even halogenated compounds in exhaust gases from foundries, and from spray cabins used in the automobile industry. The advantages of the biological treatment of exhaust gas are operation at low temperature and the efficient elimination of low concentrations of toxic and odorous substances in the waste gas stream at a rather low cost. The microbiological and engineering fundamentals of this technology have been summarized by Klein and Winter [206].

The biofilter is the most commonly used reactor found in many hundreds of installations. These are fixed bed reactors where the waste gas flows through a bed of carrier material (mostly natural fibre material) with mixed bacterial cultures adhering to the surface. Trickling filters are another standard method and are capable of degrading more contaminants per unit volume than biofilters, and can be used to treat exhaust gas at a rate of up to $100\,000\,\mathrm{m^3\,h^{-1}}$. In a trickle bed reactor, an aqueous phase is continuously circulated through a bed of inert material (Raschig rings, saddles etc.) which carries the biofilm. The waste air flows through the packing co- or counter-currently to a re-circulated liquid. Further bioscrubbers and membrane bioreactors are in operation. The mode of operation of all these reactors is similar. Air containing volatile compounds is passed through the bioreactor where the volatile compounds are transferred from the gas phase into the liquid phase. The micro-organisms responsible for the biodegradation, a mixture of different bacteria, fungi and protozoa, usually grow as a mixed culture of organisms utilizing the compounds transferred as substrates. They are generally organized into thin layers known as biofilms. The pollutants in the air (such as ethanol, carboxylic acids, esters, aldehydes, methane, dichloromethane, toluene etc.) usually act as a source of carbon and energy for the growth and maintenance of the micro-organisms [225–227].

In addition, *soil remediation* should be mentioned which comprises aerobic processes for the efficient biological treatment of contaminated soils. It is important to optimize the environmental conditions for the micro-organisms (oxygen supply, water content, pH value, etc.) in order to enhance the degradation processes and the degree of degradation. Inoculation with micro-organisms, addition of limiting nutrients and substrates, aeration and moisture are all relevant parameters [228].

Other relevant topics in environmental biotechnology can only be mentioned here: *composting* and *biological processes in landfills*. Composting is the biological decomposition of organic compounds from wastes under controlled aerobic conditions. The objectives of composting are stabilization, volume and mass reduction, elimination of undesired seeds, and sanitation. The end-product of composting can be used as soil conditioner or fertilizer [229]. In landfills, biodegradation is subject to a variety of hydrological, physical, chemical and biological processes that include convection, diffusion and hydrodynamic dispersion, and microbial degradation reactions. Carbon dioxide and methane are usually the major gaseous products and the reactions producing them continue for at least 15 years. The majority of degradable waste will decompose during this period [230].

A general *overview* shows the rather highly developed status of environmental technology in developed countries but the urgent need to improve the situation in newly industrialized countries such as China and India, and more so in developing countries. Fortunately environmental activists are becoming more influential, for example Ma Jun in China who has been named as one of the 100 most influential people in the world [231]. It is the pivotal role of governments and authorities to fix standards as well as to control adherence to them.

One example for future progress using recombinant tools may be the deciphering of the genome sequence of the cosmopolitan marine bacterium *Alcanivorax borkumensis* which uses oil hydrocarbons as its exclusive source of carbon and energy. This might provide optimized strategies for the bioremediation of the several million tons of oil which are discharged into the oceans every year [232].

16.7
Summary, Trends and Perspectives

Ambitious aims and optimistic views characterize the prospects for the future – mostly in the form of the topics that have been discussed in the preceding chapters:

- Meeting the challenge of identifying renewable resources for fuels and chemicals with respect to the priority of food production, making accessible non-food resources, notably lignocellulosics, and exploiting the potential of genetically engineered plants (Section 16.3.1); developing biorefineries to meet these goals (see Section 16.2.2)
- Widen the scope and development of new bulk chemicals and commodities manufacture, including new biopolymers and building blocks for polymers for the vast plastics market; exploitation of unconvential properties such as the greater-than-steel strength of amyloid fibrils with their flexibility and ability to self-assemble [233]
- The development of modified and improved biocatalysts in order to manufacture new products or to make processes easier and ecologically more favourable with the use of new strategies including the metagenome approach and sequence-based discovery (see Section 7.6.7)
- Making use of advanced and sophisticated tools for efficient production: metabolic engineering and biosystems ingineering; designed cells for synthesis (15.6); visionary projects including the multivitamin/vitamin E/vitamin C bug; novel antibiotics/pharmaceuticals from pathway shuffling/combinatorial synthesis; synthetic genes, pathways, and genomes; synthetic organisms (see Section 13.5) [234]
- Satisfying the requirements of environmentally friendly, lean production, and significantly reduced greenhouse gas emissions. According to McKinsay carbon dioxide emissions could be reduced by as much as 180 million t/a worldwide by the application of BT processes [235]
- Introduce environmental processes for the protection of water, air and soil, in countries like China and India, and in third world countries that do not meet or are a long way from attaining proper standards (Section 16.6)
- Exploit new opportunities such as mechanical systems – nanomotors – biomolecular robots using motor proteins capable of converting chemical energy into directed motion and transport, for example transport of DNA molecules by cargo-carrying microtubules based on polysaccharides [236–237].

Identification of specific targets is essential to ensure progress but these targets must be considered carefully and critically, bearing in mind the failures of the past. Proposals for furthering progress should take into account the critical discussions related to ecological aspects and the long-term needs of agriculture and ecology in addition to the ever-expanding manufacture of products based on renewable biomass [242]. Continued discourse with the public will be mandatory for the realization of these far-reaching and optimistic aims.

References

1 National Academies (2000) *Committee on Biobased Industrial Products, National Research Council, Biobased Industrial Products: Research and Commercialization Priorities*, National Academy Press, Washington, www.nap.edu/catalog/5295.html.

2 US Department of Energy (2004) *Energy Efficiency and Renewable Energy, Biomass: Top Value Added Chemicals from Biomass* (eds T. Werpy and G. Petersen), US Department of Commerce, Springfield, VA, USA.

3 Johnson, J. (2009) Supporting biofuels. *Chemical & Engineering News*, May 11, p. 8.

4 SusChem (2005) p. 43–59, http://www.suschem.org.

5 European Commission (2006) Biofuels, A vision for 2030 and beyond http://ec.europa.eu/research/energy.

6 Taussig, M.J. (2008) New biotechnology and the European Federation of Biotechnology. *New Biotechnology*, **25**, 1–2.

7 Buchholz, K., Kasche, V., and Bornscheuer, U. (2005) *Biocatalysts and Enzyme Technology*, Wiley-VCH, Weinheim.

8 Liese, A., Seelbach, K., and Wandrey, C. (2006) *Industrial Biotransformations*, Wiley-VCH, Weinheim.

9 Dechema, Q. (2004) Weisse Biotechnologie, Positionspapier, Dechema, Frankfurt.

10 Soetaert, W. and Vandamme E. (Eds.) (2010) *Industrial Biotechnology*. Wiley-VCH, Weinheim, Germany.

11 Hambrecht, J. (2005) Weisse biotechnologie. *Bioforum*, (5), 14–16.

12 Kircher, M. (2006) White Biotechnology: Ready to partner and invest in. *Biotechnology Journal*, **1** (7/8), 787–794.

13 Zinke, H. (2007) Nachhaltiger industrieller Trend oder nur ein Strohfeuer? *Transkript*, **13** (Sonderheft), 86–88.

14 Kirchner, M. (2008) Industrielle biotechnologie. *CHEManager*, 5/2008, p. 25.

15 Koltermann, A. (2009) Von der petrochemie zur biorafinerie. *CHEManager*, 9/2009, p. 8.

16 RFA (Renewable Fuels Association) (2005) http://www.ethanolrfa.org/industry/locations/.

17 Heinzle, E., Biwer, A.P., and Cooney, C.L. (2006) Development of Sustainable Bioprocesses. Wiley, New York.

18 Bruggink, A. (2001) *Synthesis of β-lactam Antibiotics*, Kluwer Academic Publ., Dordrecht.

19 Elander, R.P. (2003) Industrial production of β-lactam antibiotics. *Applied Microbiology and Biotechnology*, **61**, 385–392.

20 Kunz, M. (2007) Bioethanol - experiences from running plants, optimization and prospects. *Biocat Biotrans*, **26**, 128–132.

21 Shinmyo, A. (2008) Plant biotechnology: a key technology in the 21st century. *Plant Biotechnology (Sheffield, England)*, **25**, 211–212.

22 Kraessig, H., Schurz, J., Steadman, R.G. et al. (2004) Cellulose, in *Ullmann's Encyclopedia of Industrial Chemistry*, Wiley-VCH, Weinheim, New York, www.mrw.interscience.wiley.com/emrw/.

23 Kunz, M. (2007) Lecture, 7th Carbohydrate Bioengineering Meeting, April 2007, Braunschweig, Germany.

24 Peters, D. (2007) Raw materials. *Advances in Biochemical Engineering/Biotechnology*, **105**, 1–30.

25 Vorwerg, W., Radosta, S., and Dijksterhuis, J. (2005) Technologie der Kohlenhydrate – 2: Stärke, in *Winnacker-Küchler. Chemische Technik: Prozesse und Produkte*, vol. 8 (eds R. Dittmeyer, W., Keim, G., Kreysa and A. Oberholz), Wiley-VCH, p. 355–406.

26 Buchholz, K. and Ekelhof, B. (2005) Technologie der Kohlenhydrate – 1: Zucker, in *Winnacker-Küchler. Chemische Technik: Prozesse und Produkte*, vol. 8 (eds R. Dittmeyer, W., Keim, G., Kreysa and A. Oberholz), Wiley-VCH, p. 315–354.

27 Stoppok, E. and Buchholz, K. (1996) Sugar-based raw materials for fermentation applications, in *Biotechnology: Products of primary metabolism*, vol. 6 (eds H.-J. Rehm and G. Reed), VCH, Weinheim u.a, p. 5–29.

28 Walter, M. (2005) Technologie der Kohlenhydrate – 3: Inulin, in *Winnacker-*

Küchler. *Chemische Technik: Prozesse und Produkte*, vol. 8 (eds R. Dittmeyer, W., Keim, G., Kreysa and A. Oberholz), Wiley-VCH, p. 407–412.

29 EU (2005) http://ec.europa.eu/research/ energy/index_eu.htm.

30 *Chemical & Engineering News*, 4/2009, April 27, p.14; 2/2010, February 22, p. 20.

31 Mitchinson, C. (2004) in Int. Congr. Biocat. Hamburg (Germany).

32 Vlasenko, E. (2005) in: 6th Carbohydrate Bioengineering Meeting, Barcelona, Spain.

33 Novozymes (2009) www.biotimes.com/ en/Articles/2009.

34 EU (2007) EU 25 biomass for energy production, European Environmental Agency; European Commission: Biofuels, 2006 http://ec.europa.eu/ research/energy.

35 Wu, L. and Birch, R.G. (2007) Doubled sugar content in sugarcane plants modified to produce a sugar isomer. *Plant Biotechnology (Sheffield, England)*, **5**, 109–117.

36 O'Driscoll, C. (2006) An insatiable hunger. *Chemistry and Industry (London)*, 24–27.

37 Kamm, B. and Kamm, M. (2007) Biorefineries – Multi product processes. *Advances in Biochemical Engineering/ Biotechnology*, **105**, 175–204.

38 Busch, R., Hirth, T., and Liese, A. *et al.* (2006) The utilization of renewable resources in German industrial production. *Journal of Biotechnology*, **1**, 770–776.

39 Kamm, B. (2006) Biorefineries, in *Industrial Processes and Products*, Wiley-VCH, Weinheim.

40 Kamm, B., Gruber, P.R., and Kamm, M. (2007) Biorefineries, in *Ullmann's Encyclopedia of Industrial Chemistry*, Wiley-VCH, Weinheim, New York, www. mrw.interscience.wiley.com/emrw/.

41 Nolte, B. (2008) Logistik für agrarrohstoffe an den beispielen stroh, zuckerrüben und weizen. Lecture, Dechema conference Industrielle Nutzung nachwachsender Rohstoffe, February, 18/19, 2008, Frankfurt, Germany.

42 Schweinle, J. and Fröhling, M. (2008) Rohstofflogistik für Bioraffinerien – das

Bespiel Holzlogistik. Lecture, Dechema conference Industrielle Nutzung nachwachsender Rohstoffe, February, 18/19, 2008, Frankfurt, Germany.

43 Schiweck, H., Munir, M., Rapp, K.M., Scheider, B., and Vogel, M. (1991) New developments in the use of sucrose as an industrial bulk chemical. *Lichtenthaler*, **1991**, 57–94.

44 Lichtenthaler, F., Cuny, E., and Martin, D. *et al.* (1991) Practical routes from mono- and disaccharides to building blocks with industrial application profiles, in *Carbohydrates as Organic Raw Materials* (ed. F.W. Lichtenthaler), VCH Weinheim, New York, p. 207–246.

45 Lichtenthaler, F., Immel, S., Martin, D., and Müller, V. (1993) Some disaccharide-derived building blocks of potential industrial utility, in *Carbohydrates as Organic Raw Materials* (ed. G. Descotes), VCH Weinheim, New York, p. 59–98.

46 Sanders, J., Scott, E., Weusthuiz, R., and Mooiebroek, H. (2007) Bio-refinery as the bio-inspired process to bulk chemicals. *Macromolecular Bioscience*, **7**, 105–117.

47 Scott, E., Peter, F., and Sanders, J. (2007) Biomass in the manufacture of industrial products – the use of proteins and amino acids. *Applied Microbiology and Biotechnology*, **75**, 751–762.

48 Ritter, S.K. (2008) Genes to gasoline. *Chemical & Engineering News*, December 8, p. 10–17.

49 *Chemical & Engineering News*, 12/2007, The costs of biofuels. December 17, p. 12–16 (emissions adapted form Science, 2006, 311, 506).

50 Johnson, J. (2007) Corn based ethanol. *Chemical & Engineering News*, May 7, p. 50–52.

51 Ingledew, W., M. (2010) Ethanol fuel production: Yeast processes. *Encyclopedia Industrial Biotechnol*. New York, John Wiley & Sons, Inc.; http://tinyurl.com/ wileyEIB.

52 Ritter, S. (2007) Biofuel nations. *Chemical & Engineering News*, June 4, www.CEN-online.org.

53 Shell (2007) www.shell.com/biofuels.

54 Wach, W. (2007) personal communication.

55 SZ (2008) Sueddeutsche Zeitung, June 6.

56 NY Times (2008) The New York Times, June 23.

57 von Braun, J. (2008) The food crisis isn't over. *Nature*, **456**, 701.

58 CHEManager 21/2007, p. 24.

59 Kosaric, N., Duvnjak, Z., Sahm, H., and Bringer-Meyer, S. (2002) Fermentation, in *Ullmann's Encyclopedia of Industrial Chemistry*, Wiley-VCH, Weinheim, New York, www.mrw.interscience.wiley.com/emrw/.

60 Schäfer, T., Borchert, T.V., and Skovgard Nielsen, V. *et al.* (2007) Industrial enzymes. *Advances in Biochemical Engineering/Biotechnology*, **105**, 59–131.

61 Goebel, O. (2002) Comparison of process economics for synthetic and fermentation ethanol, in *Ullmann's Encyclopedia of Industrial Chemistry*, Wiley-VCH, Weinheim, New York, www.mrw.interscience.wiley.com/emrw/.

62 Sanchez-Gonzales, Y., Cameleyre, X., Molina-Jouve, C., Goma, G., and Alfenore, S. (2009) Dynamic microbial response under ethanol stress to monitor *Saccharomyces cerevisiae* activity in different initial physiological states. *Bioprocess and Biosystems Engineering*, **32**, 459–466.

63 Chaabane, B., Aldiguier, A.S., and Alfenore, S. *et al.* (2006) Very high ethanol productivity in an innovative continuous two-stage bioreactor with cell recycle. *Bioprocess and Biosystems Engineering*, **29**, 49–57.

64 Devantier, R., Scheithauer, B., Villas-Bôas, S.G., Pedersen, S., and Olsson, L. (2005) Metabolite profiling for analysis of yeast stress response during very high gravity ethanol fermentations. *Biotechnology and Bioengineering*, **90**, 703–714.

65 Moore, A. (2008) Biofuels are dead: long live biofuels. *New Biotechnology*, **25**, 6–12.

66 *Chemical & Engineering News*, 7/2008, July 28, p. 27.

67 *CHEManager*, 14/2008, p. 5.

68 Voith, M. (2009) Cellulosic scale-up. *Chemical & Engineering News*, April 27, p. 10–13.

69 *Chemical & Engineering News*, 7/2009, July 6, p. 10.

70 *Chemical & Engineering News*, 7/2007, July 7, p. 8.

71 Keasling, J.D. and Chou, H. (2008) Metabolic engineering delivers next-generation biofuels. *Nature Biotechnology*, **26**, 298–299.

72 *CHEManager*, 7/2008, p. 9.

73 Chisti, Y. (2007) Biodiesel from microalgae. *Biotechnology Advances*, **25**, 294–306.

74 Schenk, P.M., Thomas-Hall, S.R., and Stephens, E. *et al.* (2008) Second generation biofuels: High efficiency microalgae for biodiesel production. *Bioenergy Research*, **1** (No. 1) 20–43.

75 Posten, C. (2009) Design principles of photo-bioreactors for cultivation of microalgae. *Engineering in Life Sciences*, **9**, 167–177.

76 Noweck, K. (2007) Biokraftstoffe heute und morgen. *CHEManager*, 2/2007, p. 11.

77 Gallert, C. and Winter, J. (2005) Bacterial metabolism in wastewater treatment systems. In: Jördening, H.-J. and Winter, J., *Environmental Biotechnology*, pp. 1–48, Wiley-VCH, Weinheim.

78 Buchholz, K., Stoppok, E., Emmerich, R., and Bartz, U. (1987) Zur anaeroben Fermentation von Rübenschnitzeln und Rübenabfall – Reaktionstechnik und scale up. *Zuckerind*, **112**, 605–610.

79 Arntz, H.-J. and Buchholz, K. (1988) Production of extracellular enzymes by anaerobic mixed cultures, in *Enzyme Engineering IX*, New York Academy of Sciences, p. 126–134.

80 Bettzieche, J. (2008) Saubere Arbeit. Sueddeutsche Zeitung, November 12, p. V2/3.

81 *CHEManager*, 6/2007, p. 5.

82 Posten, C., Schaub, G., Lehr, F. et al. (2008) *Chemical Engineering & Technology*, **80**, 1371.

83 Wukovits, W., Friedl, A., Schumacher, M. *et al.* (2007) Identifizierung von geigneten Routen zur nicht thermischen Produktion von Biowasserstoff. *Chemical Engineering & Technology*, **79**, 1335.

84 Schröder, U., Niessen, J., and Scholz, F. (2003) A generation of microbial fuel cells with current outputs boosted by more than one order of magnitude. *Angewandte Chemie*, **115**, 2986–2989.

85 Schröder, U. (2007) Anodic electron transfer mechanisms in microbial fuel cells and their energy efficiency. *Physical Chemistry Chemical Physics*, **9**, 2619–2629.

86 Rabaey, K., Boon, N., Siciliano, S.D., Verhaege, M., and Verstraete, W. (2004) Biofuel cells select for microbial consortia that self-mediate electron transfer. *Applied and Environmental Microbiology*, **70**, 5373–5382.

87 Rabaey, K., Boon, N., Verstraete, W., and Höfte, M. (2005) Microbial phenazine production enhances electron transfer in biofuel cells. *Environmental Science & Technology*, **39**, 3401–3408.

88 Aelterman, P., Rabaey, K., Phan, H.T., Boon, N., and Verstraete W. (2006) Continuous electricity generation at high voltages and currents using stacked microbial fuel cells. *Environmental Science & Technology*, **40** 3388–3394.

89 Rosenbaum, M., Zhao, F., Schröder, U., and Scholz, F. (2006) Interfacing electrocatalysis and Biocatalysis with tungsten Carbide: A high performance, noble-metal-free microbial fuel cell. *Angewandte Chemie*, **118**, 6810–6813.

90 Rabaey, K., Angenent, L., Schroeder, U. and Keller F J. (eds) (2009) *Bioelectrochemical Systems*, IWA Publishing, London.

91 Schröder, U. (2009) personal communication.

92 Rabaey, K., Keller, J., Angenent, L., Lens, P., and Schroeder, U. (2009) Introduction – MFCs and BESs in the context of environmental and industrial biotechnology In: Rabaey *et al.*, [90].

93 Petersen, S., Mack, C., and de Graaf, A. *et al.* (2001) Metabolic consequences of altered phosphoenolpyruvate carboxykinase activity in *Corynebacterium glutamicum* reveal anaplerotic regulation mechanisms *in vivo*. *Metabolic Engineering*, **3**, 344–361.

94 Papini, M., Salazar, M., and Nielsen, J. (2009) Systems biology of industrial microorganisms. *Advances in Biochemical Engineering/Biotechnology*, **120**, 51–99.

95 Flaschel, E. and Sell, D. (2005) Charme und chancen der weissen biotechnologie. *Chemie Ingenieur Technik*, **77**, 1298–1310.

96 Becker, T., Breithaupt, D., and Doelle, H.W. (2007) Biotechnology, in *Ullmann's Encyclopedia of Industrial Chemistry*, Wiley-VCH, Weinheim, New York, www.mrw.interscience.wiley.com.

97 Leuchtenberger, W., Huthmacher, K., and Drauz, K. (2005) Biotechnological production of amino acids and derivatives: current status and prospects. *Applied Microbiology and Biotechnology*, **69** (1), 1–8.

98 Hermann, T. (2003) Industrial production of amino acids by coryneform bacteria. *Journal of Biotechnology*, **104**, 155–172.

99 Drauz, K., Grayson, I., Kleemann, A. *et al.* (2007) Amino acids, in *Ullmann's Encyclopedia of Industrial Chemistry*, Wiley-VCH, Weinheim, New York, www.mrw.interscience.wiley.com/emrw/.

100 Marx, A., Eikmanns, B.J., Sahm, H., de Graaf, A.A., and Eggeling, L. (1999) Response of the central metabolism in *Corynebacterium glutamicum* to the use of an NADH-dependent glutamate dehydrogenase. *Metabolic Engineering*, **1**, 35–48.

101 Sahm, H., Eggeling, L., and de Graaf, A. (2000) Pathway analysis and metabolic engineering in *Corynebacterium glutamicum*. *Biological Chemistry*, **381**, 899–910.

102 Pfefferle, W., Möckel, B., Bathe, B., and Marx, A. (2003) Biotechnological manufacture of lysine. *Advances in Biochemical Engineering/Biotechnology*, **79**, 59–112.

103 Wendisch, V.F., Bott, M., and Eikmanns, B.J. (2006) Metabolic engineering of *Escherichia coli* and *Corynebacterium glutamicum* for biotechnological production of organic acids and amino acids. *Current Opinion in Microbiology*, **9**, 268–274.

104 Eggeling, L., Oberle, S., and Sahm, H. (1998) Improved L-lysine yield with *Corynebacterium glutamicum*: use of *dapA* resulting in increased flux combined with growth limitation. *Applied Microbiology and Biotechnology*, **49**, 24–30.

105 Kalinowski, J., Bathe, B., and Bartels, D. *et al.* (2003) The complete *Corynebacterium glutamicum* ATCC 13032 genome sequence and its impact on the production of L-aspartate-derived

amino acids and vitamins. *Journal of Biotechnology*, **104**, 5–25.

106 Kelle, R., Hermann, T., and Bathe, B. (2005) L-Lysine production, in *Handbook of Cornyebacterium glutamicum* (eds L. Eggeling and M. Bott), CRC Press, Boca Raton, p. 465.

107 Ohnishi, J., Katahira, R., Mitsuhashi, S., Kakita, S., and Ikeda, M. (2005) A novel gnd mutation leading to increased l-lysine production in Corynebacterium glutamicum. *FEMS Microbiology Letters*, **242**, 265–274.

108 Wittmann, C. (2009) Analysis and engineering of metabolic pathway fluxes in *Corynebacterium glutamicum* for amino acid production. *Advances in Biochemical Engineering/Biotechnology*, **120**, 21–49.

109 Roehr, M. and Kubicek, C.P. (1992) Industrial acids and other small molecules. *Biotechnology*, **23**, 91–131.

110 De Muynck, C., Pereira, C., Naessens, M., Parmentier, S., Soetaert, W., and Vandamme, E.J. (2007) The genus *Gluconobacter Oxydans*: Comprehensive overview of biochemistry and biotechnological applications. *Critical Reviews in Biotechnology*, **27**, 147–171.

111 Hustede, H. (2000) Gluconic acid, *Ullmann's Encyclopedia of Industrial Chemistry*, Wiley-VCH, Weinheim, New York, www.mrw.interscience.wiley.com.

112 Deppenmeier, U. and Ehrenreich, A. (2009) Physiology of acetic acid bacteria in light of the genome sequence of Gluconobacter oxydans. *Journal of Molecular Microbiology and Biotechnology*, **16**, 69–80.

113 Cullen, D. (2007) The genome of an industrial workhorse. *Nature Biotechnology*, **25**, 189–190.

114 Karaffa, L. and Kubicek, C.P. (2003) Aspergillus niger citric acid accumulation: do we understand this well working black box? *Applied Microbiology and Biotechnology*, **61**, 189–196.

115 Röper, H. (1990) Selective oxidation of D-Glucose: Chiral Intermediates for Industrial Utilization. *Starch/Stärke*, **42**, 342–349.

116 Giffhorn, F., Köpper, A., Huwig, A., and Freimund, S. (2000) Rare sugars and sugar-based synthons by chemo-enzymatic synthesis. *Enzyme and Microbial Technology*, **27**, 734–742.

117 BASF AG and Takeda Chemical Industries Ltd (1999) 2001: A report on the acquisition by BASF AG of certain assets of Takeda Chemical Industries Ltd; http://www.w3c.org/TR/1999/REC-html401-19991224/loose.dtd; http://www.competition-commission.org.uk/rep_pub/reports/2001/456basftake.htm.

118 Rueckel, M. (2000) EP 0972843 Hoffmann-Laroche AG, Schweiz.

119 Troostemberg, J.-C., Debonne, I.A., Obyn, W.R., and Peuzet, C. (2003) WO 03016508 Cerestar Holding BV, Netherlands.

120 Troostemberg, J.-C., Debonne, I.A., Obyn, W.R., and Peuzet, C. (2003) WO 03016508 Cerestar Holding BV, Netherlands.

121 Grindley, J.F., Payton, M.A., van den Pol, H., and Hardy, K.G. (1988) Conversion of Glucose to 2-Keto-L-Gulonate, an Intermediate in L-Ascorbate Synthesis, by a Recombinant Strain of *Erwinia citreus*. *Applied and Environmental Microbiology*, 1770–1775.

122 Eggersdorfer, M., Adam, G., John, M. et al. (2000) Vitamins, in *Ullmann's Encyclopedia of Industrial Chemistry*, Wiley-VCH, Weinheim, New York, www.mrw.interscience.wiley.com/emrw/.

123 CHEManager 6/2007.

124 King, P.P. (1982) Biotechnology, an industrial view. *Journal of Chemical Technology and Biotechnology*, **32**, 2–8.

125 Steinbüchel, A. (2002) *Biopolymers, Polysaccharides I: Polysaccharides from Prokaryotes; Biopolymers. 4: Polyesters; 3. Applications and Commercial Products*, Wiley-VCH, Weinheim, Germany.

126 Rehm, B. (ed.) (2009) *Microbial Production of Biopolymers and Polymer Precursors: Applications and Perspectives*, Caister Academic Press.

127 Horn, S., Bader, H.J., and Buchholz, K. (2003) Kunststoffe aus nachwachsenden Rohstoffen, *Green Chemistry-Nachhaltigkeit in der Chemie*, GDCh Hrg, Wiley-VCH, p. 55–74.

128 Tullo, A. (2007) Chemicals from renewables. *Chemical & Engineering News*, May 7, **2007**, p. 14.

129 Tullo, A. (2007) A living plant. *Chemical & Engineering News*, June 25, p. 36.

130 Tullo, A.H. (2007) Soy rebounds. *Chemical & Engineering News*, August 20, p. 36.

131 *Chemical & Engineering News*, 8/2007, July 7, p. 25.

132 McGrath, K.M. (2001) Probing material formation in the presence of organic and biological molecules. *Advanced Materials*, **13**, 989–992.

133 Halford, B. (2009) Calcite close up. *Chemical & Engineering News*, November 30, p. 7.

134 Chan, W.C.W., Maxwell, D.J., and Gao, X.H. *et al.* (2002) Luminescent quantum dots for multiplexed biological detection and imaging. *Current Opinion in Biotechnology*, **13**, 40–46.

135 Perner-Nochta, J., Krumov, N., Oder, S. *et al.* (2009) Biopartikel: Eine Alternative zur Produktion nanoskaliger anorganischer Partikel. *Chemical Engineering & Technology*, **81**, 685–697.

136 Kleinkauf, H. and von Döhren, H. (eds) (1997) Biotechnology, in *Products of Secondary Metabolism*, vol. 7, VCH, Weinheim, Germany.

137 Kelly, D.R. (1998/2000) Biotechnolgy, in *Biotransformations*, vol. 8a, b, Wiley-VCH, Weinheim, Germany.

138 Ghisalba, O., Meyer, H-P., and Wohlgemuth, R. (2010) Industrial Biotransformation. In: *Encyclopedia of Industrial Biotechnology*. New York, John Wiley & Sons, Inc.; http://tinyurl.com/wileyEIB.

139 Faber, K. (2004) *Biotransformations in Organic Chemistry*, 5th edn, Springer, Berlin.

140 Straathof, A.J.J., Panke, S., and Schmid, A. (2002) The production of fine chemicals by biotransformations. *Current Opinion in Biotechnology*, **13**, 548–556.

141 Hilterhaus, L. and Liese, A. (2007) Building blocks. *Advances in Biochemical Engineering/Biotechnology*, **105**, 133–173.

142 Andersen, O. and Poulsen, E.B. (1983) Application of enzymes to organic synthesis, *Enzyme Technology* (ed. R.M. Lafferty), Springer, Berlin, p. 179–188.

143 Klibanov, A. and Cambou, B. (1987) Enzymatic production of optically active compounds in biphasic aqueous-organic systems, in *Methods in Enzymology*, vol. 136 (ed. K. Mosbach), Academic Press, Orlando, p. 117–137.

144 Schmid, A., Hollmann, F., Park, J.B., and Bühler, B. (2002) The use of enzymes in the chemical industry in Europe. *Current Opinion in Biotechnology*, **13**, 359–366.

145 McKinsey Company (2004) *Transkript*, vol. 10, p. 59.

146 Cheetham, P. (2000) Case studies in the application of biocatalysts, in *Applied Biocalysis* (eds A.J.J. Straathof and P. Adlercreutz), Harwood Academic Publishers, Amsterdam, p. 93–152.

147 Bornscheuer, U. and Buchholz, K. (2005) Highlights in biocatalysis – historical landmarks and current trends. *Engineering in Life Sciences*, **5**, 309–323.

148 Bommarius, A.S. and Riebel, B.R. (2004) *Biocatalysis*, Wiley-VCH, Weinheim.

149 Bornscheuer (2005) Enzymes in Organic Chemistry, in: *Biocatalysts and Enzyme Technology* (eds K. Buchholz, V., Kasche and U. Bornscheuer), Wiley-VCH, Weinheim, p. 109–170.

150 Bornscheuer, U. and Kazlauskas, R.J. (2006) *Hydrolases in Organic Synthesis - Regio- and Stereoselective Biotransformations*, Wiley-VCH, Weinheim.

151 Schmid, A., Dordick, J.S., Hauer, B., Kiener, A., Wubbolts, M., and Witholt, B. (2001) Industrial biocatalysis today and tomorrow. *Nature*, **409**, 258–267.

152 Breuer, M., Ditrich, K., Habicher, T., Hauer, B., Keßeler, M., Stürmer, R., and Zelinski, T. (2004) Industrial methods for the production of optically active intermediates. *Angewandte Chemie (International Edition in English)*, **43**, 788–824.

153 Kragl, U., Kruse, W., Hummel, W., and Wandrey, C. (1996) Enzyme engineering aspects of biocatalysis: Cofactor regeneration as examples. *Biotechnology and Bioengineering*, **52**, 309–319.

154 Kraußer, M., Hummel, W., and Gröger, H. (2007) Enantioselective one-pot two-step synthesis of hydrophobic allylic alcohols in aqueous medium through the combination of a Wittig reaction and an enzymatic ketone reduction. *European Journal of Organic Chemistry*, **2007**, 5175–5179.

155 Burda, E., Hummel, W., and Gröger, H. (2008) Modulare chemoenzymatische Eintopfsynthesen im wässrigen Medium: Kombination einer Palladiumkatalysierten Kreuzkupplung mit einer asymmetrischen Biotransformation. *Angewandte Chemie*, **120**, 9693–9696.

156 Wimmer, T. (2003) Cyclodextrins, in *Ullmann's Encyclopedia of Industrial Chemistry*, Wiley-VCH, Weinheim, New York, www.mrw.interscience.wiley.com/emrw/.

157 Dijkhuizen, L. and van der Veen, B. (2003) Cyclodextrin glycosyltransferase, in *Handbook of Food Enzymology* (eds J.R. Whitaker, A.G.J., Voragen and D.W.S. Wong), M. Dekker, New York, p. 615–627.

158 Kelly, R.M., Dijkhuizen, L., and Leemhuis, H. (2009) The evolution of cyclodextrin glucanotransferase product specificity. *Applied Microbiology and Biotechnology*, **84**, 119–133.

159 Kelly, R.M., Dijkhuizen, L., and Leemhuis, H. (2009) Starch and α-glucan acting enzymes, modulating their properties by directed evolution. *Journal of Biotechnology*, **140**, 184–193.

160 Leemhuis, H., Kelly, R.M., and Dijkhuizen, L. (2009) Engineering of cyclodextrin glucanotransferases and the impact for biotechnological applications. *Applied Microbiology and Biotechnology*. doi: 10.1007/s00253-009-2221-3.

161 Wimmer, T. (1999) München, Wacker.-Chemie, personal communication.

162 Buchholz, K. and Seibel, J. (2008) Industrial carbohydrate biotransformations. *Carbohydrate Research*, **343**, 1966–1979.

163 Seibel, J. and Buchholz, K. (2010) Tools in oligosaccharide synthesis: current research and application. *Advances in Carbohydrate Chemistry and Biochemistry*, **63**, 101–138.

164 Schedel, M. (2000) Regioselective oxidation of aminosorbitol with *Gluconobacter oxydans*, a key reaction in the industrial 1-deoxynojirimycin synthesis, in *Biotechnology*, vol. 8b, Wiley-VCH, p. 295–307.

165 Chibata, I., Tosa, T., and Sato, T. (1987) Application of immobilized biocatalysts in pharmaceutical and chemical industries, in *Biotechnology*, vol. 7a (eds H.J. Rehm and G.H. Reed), Verlag Chemie, Weinheim, p. S653–S684.

166 Bunk, B., Biedendieck, R., Jahn, D., and Vary, P. (2010) Industrial production by

Bacillus megaterium and other Bacilli, in *Encyclopedia of Industrial Biotechnology* (ed. M.C. Flickinger), John Wiley & Sons, Inc., Hoboken, NJ, USA., http://tinyurl.com/WileyEIB.

167 Schmid, R.D. and Verger, R. (1998) Lipases - interfacial enzymes with attractive applications. *Angewandte Chemie (International Edition in English)*, **37**, 1608–1633.

168 Kazlauskas, R.J. and Bornscheuer, U.T. (1998) Biotransformations with lipases, in *Biotechnology*, vol. 8a (eds H.J. Rehm, G., Reed, A., Pühler, P.J.W., Stadler and D.R. Kelly), Wiley-VCH, Weinheim, p. 37–191.

169 Zhang, J., Reddy, J., Roberge, C., Senanayake, C., Greasham, R., and Chartrain, M. (1995) Chiral bio-resolution of racemic indene oxide by fungal epoxide hydrolase. *Journal of Fermentation and Bioengineering*, **80**, 244–246.

170 Trius, A. (2007) *CHEManager*, 18/2007, p. 1, 6, see also www.cognis.de.

171 Lorenz, P. and Eck, J. (2005) Metagenomics and industrial applications. *Nature Reviews Microbiology*, **3**, 510–515.

172 Bornscheuer, U. and Pohl, M. (2001) Improved biocatalysts by directed evolution and rational protein design. *Current Opinion in Chemical Biology*, **5**, 137–142.

173 Reetz, M.T. (2004) Controlling the enantioselectivity of enzymes by directed evolution: practical and theoretical ramifications. *Proceedings of the National Academy of Sciences of the United States of America*, **101**, 5716–5722.

174 Bornscheuer, U. (2001) Directed evolution of enzymes for biocatalytic applications. *Biocatalytic Biotransformation*, **19**, 84–96.

175 Jaeger, K.E., Eggert, T., Eipper, A., and Reetz, M.T. (2004) Changing the enantioselectivity of enzymes by directed evolution. *Methods in Enzymology*, **388**, 238–256.

176 Kelly, R.M., Leemhuis, H., and Dijkhuizen, L. (2007) Conversion of a cyclodextrin glucanotransferase into an α-amylase: assessment of directed evolution strategies. *Biochemistry*, **46**, 11216–11222.

177 Beier, L., Svendsen, A., Andersen, C., Frandsen, T.P., Borchert, T.V., and Cherry, J.R. (2000) Conversion of the maltogenic α-amylase Novamyl into a CGTase. *Protein Engineering*, **13**, 509–513.

178 van der Geize, R., Hessels, G.I., and Dijkhuizen, L. (2002) Molecular and functional characterization of the kstD2 gene of *Rhodococcus erythropolis* SQ1 encoding a second 3-ketosteroid δ¹-dehydrogenase isoenzyme. *Microbiology*, **148**, 3285–3292.

179 Lang, S., Hüners, M., and Lurtz, V. (2005) Bioprocess engineering data on the cultivation of marine prokariotes and fungi. *Advances in Biochemical Engineering/Biotechnology*, **97**, 29–62.

180 Griengl, H. (2006) Personal communication.

181 Purkarthofer, T., Gruber, K., Gruber-Khadjawi, M., Waich, K., Skranc, W., Mink, D., and Griengl, H. (2006) A biocatalytic Henry reaction--the hydroxynitrile lyase from *Hevea brasiliensis* also catalyzes nitroaldol reactions. *Angewandte Chemie (International Edition in English)*, **45**, 3454–3456.

182 Donalies, U., Nguyen, H., Stahl, U., and Nevoigt, E. (2008) Improvement of *Saccharomyces* yeast strains used in brewing, wine making and baking. *Advances in Biochemical Engineering/Biotechnology*, **111**, 67–98.

183 CAZy, 21.10.2006.

184 van der Maarel, M.J., van der Veen, B., Uitdehaag, J.C., Leemhuis, H., and Dijkhuizen, L. (2002) Properties and applications of starch–converting enzymes of the α-amylase family. *Journal of Biotechnology*, **94**, 137–155.

185 Declerck, N., Machius, M., Wiegand, G., Huber, R., and Gaillardin, C. (2000) Probing structural determinants specifying high thermostability in *Bacillus licheniformis* alpha-amylase. *Journal of Molecular Biology*, **301**, 1041–1057.

186 Reilly, P. and Antrim, R.L. (2003) Enzymes in grain wet milling, in *Ullmann's Encyclopedia of Industrial Chemistry*, Wiley-VCH, Weinheim, New York, www.mrw.interscience.wiley.com/emrw/.

187 Reilly, P. (2003) Glucoamylase, in *Handbook of Food Enzymology* (eds J.R. Whitaker, A.G.J., Voragen and D.W.S. Wong), Marcel Dekker, New York, p. 727–738.

188 Coutinho, P.M. and Reilly, P.J. (1997) Glucoamylase structural, functional, and evolutionary relationships. *Proteins*, **29**, 334–347.

189 Reilly, P.J. (1999) Protein engineering of glucoamylase to improve industrial performance- a review. *Starch/Stärke*, **51**, 269–274.

190 Suvd, D., Fujimoto, Z., Takase, K., Matsumura, M., and Mizuno, H. (2001) Crystal structure of *Bacillus stearothermophilus* alpha-amylase: possible factors of the thermostability. *Journal of Biochemistry*, **129**, 461–468.

191 Olsen, H.S. (1995) Use of enzymes in food processing, in *Biotechnology*, vol. 9 (eds G. Reed and T.W. Nagodawithana), VCH, Weinheim, p. 663–736.

192 Antrim, R.L. and Auterinen, A.L. (1986) A new regenerable immobilized glucose isomerase. *Stärke*, **38**, 132–137.

193 Misset, O. (2003) Xylose (glucose) isomerase, in *Handbook of Food Enzymology* (eds J.R. Whitaker, A.G.J., Voragen and D.W.S. Wong), Marcel Dekker, New York, p. 1057–1077.

194 Lavie, A., Allen, K.N., Petsko, G.A., and Ringe, D. (1994) X-ray crystallographic structures of D-Xylose isomerase-substrate complexes position the substrate and provide evidence for metal movement during catalysis. *Biochemistry*, **33**, 5469–5480.

195 Genencor International (1990) Glucose isomerases having altered substrate specificity EP 1 264 883.

196 Pedersen, S. and Christensen, M.W. (2000) Immobilized biocatalysts, in *Applied Biocatalysis* (eds A.J.J. Straathof and P. Adlercreutz), Harwood Academic Publishers, Amsterdam, p. 213–228.

197 Schiweck, H., Bär, A., Vogel, R., Schwarz, E., and Kunz, M. (2000) *Sugar Alcohols*, Ullmann's Encyclopedia of Industrial Chemistry, http://mrw.interscience.wiley.com.

198 Dols-Lafargue, M., Willemot, R.-M., Monsan, P., and Remaud-Simeon, M.

(2001) *Biotechnology and Bioengineering*, **75**, 276–284.

199 Fuji, S. and Komoto, M. (1991) Novel carbohydrate sweeteners in Japan. *Zuckerindustrie*, **116**, 197–200.

200 Nakakuki, T. (2005) Present status and future prospects of functional oligosaccharide development in Japan. *Journal of Applied Glycoscience*, **52**, 267–271.

201 Hancock, S.M., Vaughan, M.D., and Withers, S.G. (2006) Engineering of glycosidases and glycosyltransferases. *Current Opinion in Chemical Biology*, **10**, 509–519.

202 Jahn, M., Marles, J., Warren, R.A., and Withers, S.G. (2003) Thioglycoligases: mutant glycosidases for thioglycoside synthesis. *Angewandte Chemie (International Edition in English)*, **42**, 352–354.

203 Seibel, J., Moraru, R., Gotze, S., Buchholz, K., Na'amnieh, S., Pawlowski, A., and Hecht, H.J. (2006) Sucrose analogues synthesis with the fructosyl transferase of *B. subtilis* NCIMB 11871 provides mechanistic insights. *Carbohydrate Research*, **341**, 2335–2349.

204 Seibel, J., Beine, R., Moraru, R., and Buchholz, B.C.K. (2006) A new pathway for the synthesis of oligosaccharides by the use of non-leloir glycosyltransferases. *Biocatalysis and Biotransformation*, **24**, 157–165.

205 Zucarra, A., Götze, S., Kneip, S., Dersch, P., and Seibel, J. (2008) Tailor-made fructooligosaccharides by a combination of substrate and genetic engineering. *Chembiochem*, **9**, 143–149.

206 Klein, J. and Winter, J. (eds) (2000) Biotechnology, in *Environmental Processes III*, vol. 11c, Wiley-VCH, Weinheim.

207 Adler, I. (1986) Biotechnologie ist mehr als Gentechnologie. *Chemische Rundschau*, September 19.

208 Trembley, J.-F. (2009) Cash flows from China' s water. *Chemical & Engineering News*, May 11, p. 18–21.

209 Voith, M. (2008) The other scarce resource. *Chemical & Engineering News*, October 6, p. 12–19.

210 Jördening, H.-J. and Winter, J. (2005) *Environmental Biotechnology*, Wiley-VCH, Weinheim.

211 Gallert, C. and Winter, J. (2005) Bacterial metabolism in wastewater treatment systems, in *Environmental Biotechnology* (eds H-.J. Jördening and J. Winter), Wiley-VCH, Weinheim, p. 1–48.

212 Gallert, C. and Winter, J. (2005) Perspectives of wastewater, waste, off-gas and soil treatment, in *Environmental Biotechnology* (eds H-.J. Jördening and J. Winter), Wiley-VCH, Weinheim, p. 439–451.

213 Jördening, H.-J. and Buchholz, K. (1999) Fixed film stationary-bed and fluidized-bed reactors, *Biotechnology*, vol. 11a (ed. J. Winter) Wiley-VCH, Weinheim, p. 493–515.

214 Jördening, H.-J. and Buchholz, K. (2005) High rate anaerobic wastewater treatment, in *Environmental Biotechnology* (eds H-.J. Jördening and J. Winter), Wiley-VCH, Weinheim, p. 135–162.

215 Rosenwinkel, K.-H., Austermann-Haun, U., and Meyer, H. (2005) Industrial wastewater sources and treatment strategies, in *Environmental Biotechnology* (eds H-.J. Jördening and J. Winter), Wiley-VCH, Weinheim, p. 49–77.

216 Kayser, R. (2005) Activated sludge process, in *Environmental Biotechnology* (eds H-.J. Jördening and J. Winter), Wiley-VCH, Weinheim, p. 79–119.

217 Kayser, R. (2000) Activated sludge process, *Environmental Processes III*. *Biotechnology*, vol. 11c (eds J. Klein and J. Winter), Wiley-VCH, Weinheim.

218 McInerny, M.J. (1999) Anaerobic metabolism and its regulation, in *Biotechnology*, vol. 11a (ed. J. Winter), Wiley-VCH, p. 455–478.

219 Demirel, B. and Yenigün, O. (2002) Two-phase anaerobic digestion processes: a review. *Journal of Chemical Technology and Biotechnology (Oxford, Oxfordshire: 1986)*, **77**, 743–755.

220 Lettinga, G., Hulshof Pol, L.W., van Lier, J.B., and Zeemann, G. (1999) Possibilities and potential of anaerobic waste water treatment using Anaerobic Sludge Bed (ASB) reactors, in *Biotechnology*, vol. 11a (ed. J. Winter) Wiley-VCH, p. 517–526.

221 Frankin, R.J. (2001) Full-scale experiences with anaerobic treatment of

industrial wastewater. *Water Science and Technology*, **44**, 1–6.

222 Jördening, H.-J. (2009) personal communication.

223 Jördening, H.-J. (1996) Scaling-up and operation of anaerobic fluidized bed reactors. *Zuckerindustrie*, **121**, 847–854.

224 VDI (1991) VDI-Richtlinien 3477, Biological Waste Gas/Waste Air Purification. Biofilters (German, English).

225 VDI (1996) VDI-Richtlinien 3478, Biological Waste Gas Purification. Bioscrubbers and Trickle Bed Reactors (German, English).

226 VDI (2002) VDI-Richtlinien 3477, Biologische Abgasreinigung. Biofilter (German).

227 Waweru, M., Herrygers, V., Van Langenhove, H., and Verstraete, W. (2000) Process engineering of biological waste gas purification, in *Environmental Processes III. Biotechnology*, vol. 11c (eds J. Klein and J. Winter), Wiley-VCH, Weinheim, p. 259–273.

228 Konig, M., Hupe, K., and Stegmann, R. (2005) Soil remediation and disposal, in *Environmental Biotechnology* (eds H.-J. Jördening and J. Winter), Wiley-VCH, Weinheim, p. 259–274.

229 Schuchardt, F. (2005) Composting of organic waste, in *Environmental Biotechnology* (eds H.-J. Jördening and J. Winter), Wiley-VCH, Weinheim, p. 333–354.

230 Haarstrick, A. and Völkerding, I.I. (2007) Biodegradation modelling - an overview, in *Landfill Modelling* (eds A. Haarstrick and T. Reichel), CISA, Publisher, Padua (Italy).

231 Trembley, J.-F. (2007) *Chemical & Engineering News*, October 1, p. 26.

232 Schneiker, S., Martins dos Santos, V., and Bartels, D. *et al.* (2006) Genome sequence of the ubiquitous hydrocarbon degrading marine bacterium *Alcanivorax borkumensis*. *Nature Biotechnology*, **24**, 997–1004.

233 Arnold, C. (2008) From diseases to devices. *Chemical & Engineering News*, July 21, p. 48–50.

234 ESAB (2005) (European Section of Applied Biocatalysis) and EuropaBio, 2005.

235 GDCh (2008) Gesellschaft Deutscher Chemiker, Wissenschaftlicher Pressedienst Nr. 08/08, Frankfurt (Germany).

236 Dietz, S., Reuther, C., Dinu, C. *et al.* (2003) Stretching and transporting DNA molecules using motor proteins. *Nano Letters*, **3**, 1251–1254.

237 Tsuchiya, T., Komori, T., Hirano, M., and Shinkai, S. *et al.* (2009) A Polysaccharide-based container transportation system powered by molecular motors. *Angewandte Chemie (International Edition in English)*, **122**, 736–739.

238 Lerch, H-P., Blöcker, H., Kalwas: H., Hoppe, J., Tsai, H., and Collins, J. (1989) Cloning, sequencing and expression in *E. coli* of the D-hydroxycaproic acid dehydrogenase gene of *Lactobacillus casei*. *Gene*, **78**, 47–57; Lerch, H-P, Frank, R., and Collins, J. (1989) Cloning, sequencing and expression of the L-2-hydroxyisocaproate dehydrogenase-encoding gene of *Lactobacillu confusus* in *Escherichia coli*. *Gene*, **83**, 263–270.

239 Rose, T. and Kunz, M. (2002) In: *Landbauforschung; Bundesanstalt für Landwirtschaft: Braunschweig, Germany*, Vol. 241, p 75–80.

240 Winter, J. (ed.) (1999) *Environmental Processes*, in: Biotechnology, vol. 11a, Wiley-VCH, Weinheim.

241 Macarie, H. (2001) Overview of the application of anaerobic treatment to chemical and petrochemical wastewaters. *Water Science and Technology*, **44**, 201–214.

242 Osseweijer, P., Amman, K., and Kinderlerer, J. (2010) Societal issues in Industrial Biotechnology. In: Soetaert, W., Vandamme E. (Eds.), *Industrial Biotechnology*, pp. 457–483, Wiley-VCH, Weinheim, Germany

17
Pharmaceutical Biotechnology

17.1
Introduction

'What exactly is a biopharmaceutical? The term has now become an accepted one,... but a clear concise definition is absent.... A general consensus evolved that it represents a class of therapeutic products produced by modern biotechnological techniques' [1]. Another definition comprises 'applications in medicine or in the pharmaceutical industry... and contributions to improving human health and life expectancy' ([2], p. 6). In contrast to bulk products, as well as fine chemicals, pharmaceuticals and the processes for their production are subject to higher levels of scrutiny and control, including essential nontechnical aspects. They must fulfil specific requirements to a high level with respect to:

- Purity and quality
- Chirality
- Posttranslational modification
- Biological activity
- Absence of carcinogenic, mutagenic, teratogenic, toxic side effects (or at least a highly beneficial ratio of positive and negative effects)
- Strict safety, testing in a range of preclinical and clinical phases
- Approval through authorities such as the Food and Drug Administration (FDA, USA), or the European authority for approval of pharmaceuticals (EMEA).

Technologies and business involved in health and disease are inevitably affected by *politics*, and by *ethics* (for a more general discussion see Chapter 8). Health innovations depend on scientific and technological advances that improve human health by reducing the incidence of disease and improving the quality of life. They include better pharmaceuticals and diagnostic tools, as well as improved data management for health research and healthcare. Ensuring high standards is primarily a task for the government of a country. Another task is protection of inventions by the granting of patents (see Chapter 14). First, basic research depends on public research institutions and universities, and the funds they have available – a highly political issue with respect to the health needs of the population. The role of governments is to determine

Concepts in Biotechnology: History, Science and Business. Klaus Buchholz and John Collins
Copyright © 2010 WILEY-VCH Verlag GmbH & Co. KGaA, Weinheim
ISBN: 978-3-527-31766-0

research politics, spur innovation, and includes big projects such as the Human Genome Project. In 2003, the government of the USA allocated 31.5 billion $ to health and environment programs, in the EU 15 governments allocated a total of 10 billion $ [3]. Application depends chiefly on industrial research and development but approval of a newly developed drug is – necessarily – under control of government institutions that are responsible for safety and its control (see Section 17.2.2).

Orphan drugs also represent a political issue. They relate to a wide array of diseases with a relatively small prevalence that, when added up, affect a significant segment of the population (see also Section 14.3.3). The number of orphan diseases has been estimated to be between 5000 and 8000 with 25 million people affected in the USA, another 25–30 million in Europe, and untold numbers throughout the rest of the world. Orphan drug laws, by gaining political attention and public perception through the actions of patient organizations, were designed to encourage research, development, and marketing of products for rare diseases. (In the USA a disease qualifies as rare if the prevalence is less than 200 000 persons) [4]. Thus the US Congress first signed the Orphan Drug Act into law in 1983, recognizing that there are many life-threatening or very serious diseases and conditions that have not recieved attention because of their small market size [5]. Amongst these are 210 cancer related, 50 infectious diseases, 44 autoimmune disorders, and many others [6]. In 2007, the FDA created an incentive, the Priority Review Vouchers (PRVs), essentially to shorten the review line for neglected diseases to within four to 12 months [7]). Recent initiatives include public/private partnership [8].

Traditionally, low molecular weight (M_w) biopharmaceuticals became of importance when penicillin first became available as a drug during the mid 1940s. It was of outstanding success on a large scale particularly for the treatment of infections of combat wounds (see Section 4.3.4). Penicillin was rapidly followed by a large range of further antibiotics. Natural products such as pig or sheep insulin, and human growth hormone, extracted from tissue and blood coagulation factors, isolated from blood were being introduced as pharmaceutical products. During the 1980s and 1990s recombinant DNA technology provided clean reliable sources for these biopharmaceuticals and many more. This provided much needed therapies for various lifethreatening diseases including diabetes (insulin), many forms of cancer (monoclonal antibodies and interferons), end-stage renal disease (erythropoietin), hepatitis (interferon), clotting disorders (Factor VII, VIII, IX), and others [9].

A historical landmark was the approval of recombinant human insulin in 1982, produced in *E. coli* and developed by Genentech in cooperation with Eli Lilly in the late 1970s. By 2006, some 165 biopharmaceuticals had gained approval for human use in the EU and/or the USA ([10, 11], pp. 8–11). These drugs include hormones, soluble hormone receptors (such as hormone antagonists), blood factors, thrombolytics, interferons, monoclonal antibodies, vaccines, and therapeutic enzymes, most protein-based drugs. Epothilone represents an example of a recently developed low molecular weight chiral biopharmaceutical compound for tumour therapy. Approximately one in four of all genuinely new drugs currently coming on to the market is a biopharmaceutical. This sector doubled in value between 1999 and 2005, being estimated to be worth in excess of 40 billion $ in 2006 [6, 10]. Over 400 biopharmaceuticals were in various stages of clinical evaluation in 2008 (for more details see Section 9.7.2). Chirality is a key requirement in drug manufacture: In 2006, 80% of small-molecule drugs approved by FDA were chiral, and 75% were single enantiomers (for definition see Chapter 16.4.3, note 3) [12].

17.2
Drug Targeting, Discovery Strategies and Development

There are multiple hit identification strategies, and new drugs come from many sources, both rational and brute force.[1] These sources include focused synthesis of new chemical structures, modification of old drugs (new analogs), new uses for old drugs, endogenous substances found in humans, natural products from nonhuman sources, and random screening. Selecting the right target is more an art than a science. Rothberg *et al.* [13] claim to have identified 6273 potential drug targets defining for the first time a complete Pharmaceutically Tractable Genome (PTG), using both laboratory and computational strategies. The major drug discovery target classes are: G protein coupled receptors (GPCRs), kinases and phosphatases, ion channels, nuclear hormone receptors, nucleic acids, and others. Amongst these, a major target represent kinases that are implicated directly or indirectly in more than 400 human diseases, notably with inflammatory or proliferative responses, including arthritis, asthma, cancer, cardiovascular disorders, and neurological disorders. Drug research has yielded many antibodies and some small molecule drugs on the market or in development (see Section 10.8) [5].

Many big pharmaceutical companies abandoned natural product discovery in the mid-to-late 1990s because known natural products were being rediscovered with increasing frequency, and have changed their strategy to rely on combinatorial chemistry since the end of the 1990s. But few hits resulted from combinatorial chemistry and HTS (high throughput screening) targeting at small molecules. Therefore, currently, orientated combinatorial synthesis tends to avoid proceeding blindly in favour of a more balanced approach, e.g. using building blocks more likely to yield biologically active compounds [5, 15].

17.2.1
Discovery Strategies

Structure-based strategies have experienced significant improvements with respect to speed and efficacy, and the development of future drugs will use a combination of methods that will contain a major component of structure-based design [5, 15, 16] (for a more detailed discussion see Sections 9.5, 9.6 and 13.7).

---

1) For selecting drugs in development projects Lipinski's remarkably simple 'rule of five' has been adopted throughout the industry a decade after publication in 1997, a useful but overemphasized guideline for small-molecule drug design, referring to molar mass, the numbers of hydrogen-bond donors, and N and O atoms, a.o. [14, 15].

Furthermore, studies have shown that much structural novelty remains undiscovered and that organisms from underexplored environments (e.g. marine actinomycetes) produce novel bioactive metabolites, some of which have recently proceeded into clinical trials. Thus, *myxobacteria* turned out to be a very good source of new and active compounds, and new niches provided interesting new metabolites [17]. Almost 30% of NCEs (new chemical entities) approved by the FDA between 1981 and 2002 were natural products or derivatives [5].

It is also expected that genome sequences and the metagenome approach, together with mechanistic insight into antibiotic resistance, might provide access to new, secondary metabolites, for example polyketide synthase gene clusters from the genomes of *Myxococcus xanthus* and *Sorangium cellulosum* (Gehrt and Müller, 2005) (see Section 7.5). Many antibiotics target ribosomes, and high-resolution structures of ribosomal complexes justify expectations for structural-based improvement of properties of existing compounds as well as for the development of novel drugs (Nobel Prize 2008 for A. Yonath and T.A. Steitz) [18, 19]. However, this remains largely conjectural at the time of writing.

Synthetic strategies compete with biotechnological as sources for new drugs. Modern research groups are often large and interdisciplinary so that both approaches can be used in parallel. Currently *chemical synthesis* of natural products and their analogues continues to represent a valid strategy. This may include multiple and/or parallel syntheses, providing families and/or variants of structures. Thus a structurally focussed library with typically 1000 up to 10 000 compounds may offer a reasonable chance of finding a drug [17].

Chemical synthesis may proceed with 3 different approaches:

- *de novo* synthesis,
- synthesis orientated towards natural structures,
- chemical derivatisation of natural products.

The second and third approaches correspond to about 50% of the strategies applied. Chemical synthesis of natural products may also create analogues and/or offer access to variants that cannot be obtained by derivatisation of the natural compound (see for example epothilone; Section 17.2.3). Further, it may provide a better alternative for existing patented biosynthesis (see also chapter 14) [17].

Chemical biology represents a promising current strategy to identify new antiinfective substances. It includes structure-based drug design and peptide combinatorial synthesis and screening [20]. Structure-based drug design is based on identification of the structure of the target protein, and identification of a lead compound that binds to the target with good affinity, that is with a dissociation constant in the micromolar range or better, and a synthetic route to produce the designed compounds. Targets comprise small molecules that bind specifically to cellular biomolecules such as proteins, DNA and RNA. Further drug design considerations take account of drug metabolism in the body or in the cell, absorption or uptake of the drug, distribution in the tissues, toxicity, and so on. These comprise the 'ADMET factors' which are discussed further in Section 9.7.3 [15, 21, 22]. The role of protein glycosylation in the development, regulation and progression of disease has recently come under increased scrutiny. Cell lines have been developed for engineering species and tissue-specific glycosylation. Products produced in these engineered strains have been used in the development of novel diagnostic and therapeutic interventions [23, 24] (see Sections 9.3, 9.4).

Strategies for the development of *anticancer drugs* are either based on small molecules, for example blocking MET, a tyrosine-kinase receptor implied in many cancers, or antibodies [14, 25–30]. The first drug that came out of the concept of targeted therapies and the monoclonal antibody (mAB) concept was Herceptin® (trastuzumab) of Genentech/Roche, approved in 1998 in the USA and in 2000 in the EU, for treatment of breast cancer of woman [32]. It is based on an elevated level of the HER cell-surface antigen on particularly aggressive mammo-carcinomas. The successful targeting of a cancer-specific kinase, which culminated in the development of Glivec (Gleevec), FDA approved in 2001 for chronic myeloid leukaemia (CML), is considered as the first validation of the concept of drug development based entirely on the known structure of a defined target (see Section 9.6).

17.2.2
Approval Procedure

Following on the discovery of lead compounds and active derivatives the further steps in drug development comprise a long, tedious pathway through the testing and approval process, as well as the parallel development of a technical production and purification process to make it available in the required amounts (Table 17.1) ([5]; for details see [33, 34]). In many countries national authorities regulate the approval process. Testing addresses the properties required of a useful drug, principally

Table 17.1 Pharmaceutical R&D and approval process, timeline, and costs [5].

	Preclinical	Clinical[a]		
	Discovery and early development	Phase I	Phase II	Phase III
Years[a]	6.5	1.5	2	3.5
Test subjects	*In vitro, in vivo,* animal studies	Healthy volunteers 20–100 subjects	Patient volunteers 100–500 patients	Patient volunteers 1000–5000 patients
Purpose	Bioactivity, safety, pharmaceutical properties	Safety and dose	Efficacy and side effects	Efficacy, adverse reactions, longer term use
Success rate	>5000 compounds evaluated	5 drugs enter clinical trials		After phase III: 1 drug approved
Cost	335 million US $			467 million US $ (All Clinical phases)

a) Mean clinical and approval times for new biopharmaceuticals in the USA increased from 7 years for the period 1969–2000 to 8.5 years for the period 2001–2005 [4].

bioavailability and lack of any toxicity, carcinogenicity and/or teratogenicity of the compound or its metabolites. Furthermore, shelflife and *in vivo* stabilities are measured. Of high significance are the issues of reproducibility of process parameters and of achievable product quantity and quality [21]. The most important regulatory authorities are the FDA, the Food and Drug Administration in the USA, and EMEA, the European authority for approval of pharmaceuticals. Low rates of success in obtaining approval and the very high cost of the approval process for a new drug have resulted in the pharmaceutical industry concentrating on the development of drugs for common diseases that may have a large market with high sales. The consequence of this is that the development of pharmaceuticals for rarer diseases has been neglected (for 'orphan drug' legislation to counteract this tendency see Section 17.1). Average capitalized costs of a new drug R&D rose dramatically from 138 in the 1970s to over 318 million $ in the 1980s to about 1.2 billion $ per drug in 2008. Submission of total drugs peaked with 160 per annum, in 1998, and decreased to some 115 in 2002. Submission of new biologics steadily decreased from about 70 in 1993, over some 45 (1995), 30 (1998), to some 20 in 2000 and 2002 [5, 35, 36]. This is even though, during the last decade, the total budgets of both NIH and US pharmaceutical R&D spending have more than doubled [5]. FDA approvals (new molecular entities and biologics) dropped from about 38 (1999) to 28 (2003), and 19 (2007) ([37], see also [4, 38]). Of biotech drug candidates, in 2004 only 6 were approved by the FDA, well below the average of 12 approvals annually for the first half of this decade and of 20 in the second half of the 1990s [39, 40]. After a review process and approval by the FDA a postmarketing phase follows with surveillance, where valuable data accumulates on up to millions of patients, as well as other studies perhaps demanded by the FDA. The latter may involve additional costs in the range of 95 million US $. The results of such further studies can lead to withdrawal or suspension of the approval on the basis of severe side-effects. Rare immune-mediated reactions (e.g. 1 in 10 000 patient-years) will only become apparent through robust post-marketing surveillance [9].

17.2.3
The Case of Epothilone – The Story of a Hit

The *case of Epothilone* illustrates the story of a hit, but also portrays how the entire process can be 'a protracted and often interrupted development' (there have been more cases like this one) [17, 41]. Epothilones (EP) are secondary metabolites of the myxobacterium *Sorangium cellulosum*. They are macrolactones of a novel structure type. Epothilone was discovered by the group of Reichenbach at GBF (Gesellschaft für Biotechnologische Forschung, today HZI, Helmholtz-Institute for Infection Research, Braunschweig, Germany) in 1986 because of its antifungal activity. One EP variant of Bristol-Myers Squibb (BMS) finally was approved in 2007 for treatment of multi-drug resistant (MDR) human adenocarcinoma and is considered by many as the next generation equivalent of the anti-cancer blockbuster drug Taxol, exhibiting less side effects.

At the GBF Hans Reichenbach and his group, in 1975, began to isolate strains of myxobacteria from soil samples collected all over the world. In 1978 G. Höfle joined

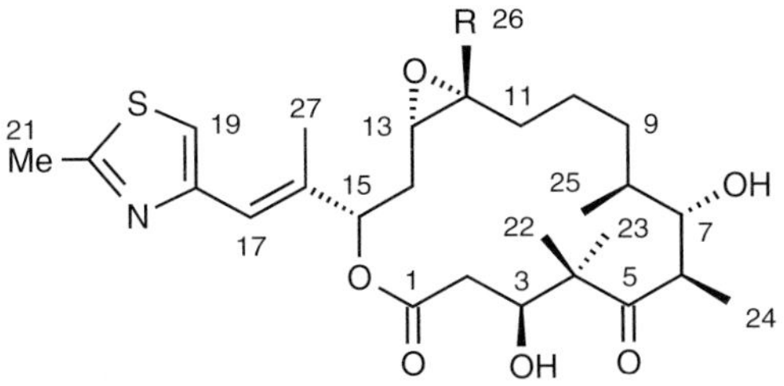

Epothilone A   R = H   (1)
          B   R = Me   (2)

Figure 17.1   Structures of Epothilones A and B; the 7 stereocenters are obvious from the figure [43].

them purifying active compounds and defining chemical structures. In 1987 two closely related antifungal compounds were isolated, epothilone A and B, a difficult task at the low concentration of about $1–2\,mg\,L^{-1}$. In 1987 an X-ray structural analysis of crystalline epothilone B confirmed the structure which had been initially proposed as well as the configuration of the seven stereocenters [41]. The structure was first disclosed by Höfle in 1991 (Figure 17.1). (The B variant exhibits an extra methyl group at the epoxide and is more active by a factor of 10) [42–44]. Screening had first been performed for several targets, including cytostatic activity, but plant protection showed the first hits in 1987. In 1990 Ciba Geigy's plant protection department noticed a selective activity of epothilones against oomycetes, a crop pest. Since this seemed highly promising as a fungicide with optimal effectiveness a German patent was filed in 1991 (granted in 1994), and an international one in 1992. These include application for plant protection or as a component of pharmaceuticals on the basis of its cytotoxic activity. The impressive antitumor activity was not, as yet, recognized. Due to limited interest in purely cytotoxic compounds, investigations in this direction were discontinued, partly because of teratogenic side effects, and the patents were not maintained (part of a cost-reduction campaign of the BMFT in 1994) [17]. At this time, 1993 at the latest, a mouse xenograft study would have been appropriate to discover the anticancer activity *in vivo* – it can only be speculated why this was not done [17, 41].

When scientists at Merck, Sharp and Dohme (MSD) screened extracts from myxobacteria (*Sorangium* strains, obtained from J. E. Peterson) for Taxol mimics, they rediscovered EP. Following the publication of a Merck article with exciting results from a collection of 7000 extracts only, from plants, microorganisms (including 300 strains of myxobacteria) and other sources, with one confirmed hit, in 1995 the EP story took a new turn. (The Upjohn company had screened 67 000 synthetic compounds and 70 000 extracts from natural sources for taxol mimics with no direct hit). The article by Merck immediately triggered a variety of activities in pharmaceutical companies and in academia [17].

Chemical synthesis was performed in 1996/7 by the groups of Danishewky (New York), Nicolau (La Jolla, CA, USA) and Schinzer (Braunschweig, now Halle, Germany), a worthwhile strategy in order to have access to derivatives. The first two groups produced an abundance of hundreds, if not thousands, of structural analogues, of which a large portion was synthesized in a combinatorial approach. This was the basis for a broad variation of the natural structures, independent of biological sources, for the investigation of the structure–activity relationship and *in vivo* studies. EP was also chemically synthesized by Schering (today Bayer Pharma), which filed patents [41].

The group of Höfle and Reichenbach came to a material-, know-how- and technology transfer and joint development agreement with Bristol-Myers Squib (BMS) in 1997 and started to work on the scale-up of production. BMS relied on semisynthesis to improve the pharmacological properties of epothilone B. This was successful through a change to the corresponding lactam with improved metabolic stability. A further modification resulted in higher water solubility [41, 44]. More than 30 000 mutants of the producer strain were made via classical mutation in 1996 and screened by the group of Höfle and Reichenbach, resulting in an epothilone titre improved by more than a hundred fold. Genetic engineering was not available for Myxobacteria at that time although the cloning of the genomes had been initiated in 1980 (J. Collins, personal communication). The complete biosynthesis gene, including a clustered a 56-kilo basepair stretch coding for over 30 enzymatic steps, was cloned by Novartis and Kosan Biosciences, but mutants obtained did not improve yields [43]. Several companies, including Schering, Novartis and BMS (with the lactam analogue), initiated clinical trials with epothilones and analogues, in phases I to III. Thus 'a protracted and often interrupted development' culminated in a success story, when finally a BMS variant was approved in 2007 [44].

17.3
Pharmaceuticals Production

Biopharmaceuticals are produced in general by two types of platforms, microbial or mammalian cell systems, depending on the complexity of the human protein, for example the absence or presence of post-translational modification, particularly glycosylation. Further production hosts are insect cell systems, or plant systems (tobacco, moss, carrot), or animals [9]. An important driver for intentionally introduced structural changes is the pharmacokinetic properties (see e.g. insulin and analogs) [9]. The most common microbial strains used are *E. coli*, still accounting for around 40% of all protein biopharmaceuticals approved before 2006; it displays several advantages as a production system. Its molecular genetics are well established, it grows easily and rapidly on inexpensive media, and high product expression levels are generally achieved; however, in many cases, products accumulate as inclusion bodies that require solubilisation and renaturation. Other strains used are *Pseudomonas fluorescens*, *Streptomyces sp.*, (of particular interest for the production of e.g. polyketides with anticancer activity, antibacterials), *Saccharomyces sp.*, *Pichia pastoris* and *Hansenula polymorpha*. They serve – mostly as recombinant strains – for the manufacture of both low molecular weight products, such as antibiotics, anti-cancer drugs, modified steroids and so on, and recombinant proteins. The glycosylation machinery, that is essential for glycoprotein production, is absent in *E. coli*. In *Saccharomyces cerevisiae* it is present, being of a simpler structure than that of mammalian cells. Glycosylation is highly conserved and complex amongst mammalian cells. This topic is dealt with in detail in Section 9.4.1. Mammalian cells used most frequently for recombinant protein production are baby hamster kidney (BHK) and chinese hamster ovary (CHO) cells. Although more technically complex and expensive, they are capable of carrying out post-translational modifications that may be necessary for biological activity; examples include EPO, many interferons, blood factor VIII and others [1, 9, 45–47].

The bulk of first-generation (early approved) biopharmaceuticals were unaltered mABs or replacement proteins such as insulin and blood factor VII. An increasing number of modern biopharmaceuticals, however, have been engineered in order to tailor their therapeutic properties, in most cases by altered amino acid sequence for improved pharmacokinetic characteristics in general, alteration of the biological half-life of the protein, reduction or elimination of product immunogenicity, generation of fast- or slow-acting products (e.g. variants of insulin), and generation of novel, hybrid protein therapeutics. Thus, domain-deleted engineered tissue plasminogen activator (tPA) variants, display much longer half-lives, in the region of 15–20 min, as compared to unmodified tPA, with a half life of some 3 min after administration. The majority of antibody-based biopharmaceuticals (mAbs) gaining approval in recent years are engineered murine-human hybrid antibodies in order to reduce or effectively eliminate severe immunogenic properties [1, 9] (see Sections 10.3, 10.4, 10.6, 10.8).

Engineering by post-translational modification has been developed and applied more recently (see Section 9.4). The changes introduced may entail the covalent attachment of a chemical group to the protein, or the alteration of the glycocomponent of glycoproteins. Engineering of the glycocomponent can extend the products serum half-life significantly, for example in the case of EPO, but may also promote targeted delivery of the biopharmaceutical to the cell type most affected. Covalent attachment of a chemical group in most cases comprised PEGylation (attachment of polyethylene glycol), that generally increases the plasma half-life of a protein, but also attachment of fatty acid groups and glycosylation [1, 9].

17.3.1
Microbial Fermentation

Fermentation with pure strain cultures of microorganisms on a large scale – bacteria, yeasts, fungi – has been well established since the 1940s, and with genetically modified organisms (recombinant DNA strains) since the 1980s. Fermentation starts with an appropriate strain, purchased from a culture collection, or in industrial processes with a high producing strain screened and selected for the specific purpose (normally in house, or from commercial or academic cooperation partners). Reactors, stirred tanks, used in industry for production are in the range 10–200 m^3 (see

Section 15.4, Figures 15.2 and 15.3). Sterile operation is essential. The standard operation is still the batch process (for a more detailed treatment see Section 15.4). Downstream processing requires many steps in addition to convential operations in order to ensure high product purity and quality, including polishing (see Figure 17.5 below). All steps must be optimized with respect to the product purity required and high yield [2, 48]. Downstream processing typical for therapeutic proteins is discussed subsequently.

Examples for the microbial production of recombinant proteins are insulin (Novo, Sanofi-Aventis), recombinant hepatitis B surface antigen (SmithKline Beecham), tissue plasminogen activator (rPA, Reteplase®,/Rapilysin®; Roche), interferon (Roferon®, Pegasys®: Roche), as well as recombinant anticoagulant hirudin. Several products are formed as inclusion bodies and require renaturation technologies to give the active proteins [1].

Yeast (*Saccharomyces cerevisiae*) was established as a successful host for producing recombinant proteins with the production of recombinant human insulin. This involved expression and secretion of a precursor protein that subsequently could be converted to human insulin by the action of the specific proteases.[2] An interesting insulin analogue with a substitution in B28 (proline to aspartic acid) resulted in a fast acting insulin (Aspart) [49]. Further investigations comprised signal peptides, modified leader peptides, modified spacer peptides, resulting in increased yield, structural mutations, and *in vitro* folding stability. It was found that the genetic stability of *S. cerevisiae* for extended periods of time when grown on the large scale was a great advantage for industrial production. Control of fermentation, based on carbohydrates, is achieved by monitoring the oxygen consumption and carbon dioxide production rates. These variables are used as input for control algorithms to adjust the fermentation process parameters. The brew leaving the fermenter is clarified by centrifugation or filtration, followed by product purification on chromatography columns. Conversion of the insulin precursor to human insulin is elegantly achieved through the use of enzymatic processes based on serine endopeptidases, trypsin and carbopeptidase B [49].

Studying the expression of insulin in *S. cerevisiae* has highlighted the importance of addressing every event leading to secretion of the insulin precursor from the initial translocation, its folding in the endoplasmatic reticulum (ER), transport through the Golgi apparatus, to maturation and ultimately secretion via relocalization to secretory vesicles. The demand for improved pharmacokinetics in the patient prompted the design and development of new fast- and slow-acting insulin analogues produced by recombinant techniques [49]. Also monoclonal antibodies (mAbs), for example humanized IgGs with specific human N-glycan structures (fragments), have been produced in yeast by glycoingeneered *Pichia pastoris* but not commercially applied [50].

---

2) After the discovery by Banting and Best in 1921/22, insulin extracted from bovine and porcine pancreas was the drug available for some 60 years to treat type 1 diabetes. With the rapid increase in the incidence of diabetes the requirement for insulin (estimated to be 15–20 t/a in 2005) could no longer be satisfied from animal sources, furthermore it is associated with allergic problems due to slight differences from human insulin [49].

Mammalian glycosyltransferases have been successfully cloned and expressed in other cell lines where they allow the production of glycoproteins with correspondingly more complex glycosylation side-chains. Even in insect cells, which can scavenge sialic acids from glycoproteins added to the medium, such modifications have been successfully carried out. It would appear that in the near future there should be no limitation to finding appropriate host cells for production of glycoproteins with any desired modification. Further alternatives being explored include plant cell culture [51, 52].

17.3.2
Mammalian Cell Culture – Principles

The clinical and commercial success that recombinant proteins have had since the mid-1990s has clearly stimulated development of mammalian cell culture technology. Antibody or antibody-like proteins given to patients in rather large doses (hundreds of milligrams to grams per patient) require large facilities for manufacturing. New medical indications and off-label use increase the demand even further. The success of several antibody and antibody-fusion products, for the treatment of cancer and diseases such as rheumatoid arthritis, has driven huge investment in order to assure market supply.[3] Therapeutic antibodies are expected to reach sales of 20 billion $ in the USA by 2010 [53] (see Section 10.8).

The first recombinant therapeutic protein made in cultivated mammalian cells obtained market approval in 1986, making Chinese hamster ovary (CHO) cells cultivated in large-scale bioreactors known to a wider public. They are now the dominating host system for manufacturing more than 60% of all new target proteins in the clinical pipelines. The basis for the success of this and other systems was the rapid and often dramatic yield improvements with such processes, bringing the volumetric productivity of mammalian cells from a titre range of $5–50\,mg\,L^{-1}$ in the early 1980s to $0.5–5\,g\,L^{-1}$ in 2002–2004, an increase by two orders of magnitude. This made it possible to produce complex recombinant proteins for clinical application in kilogram quantities, or even a ton per year. The products exhibit all the necessary secondary modifications that only a higher eukaryote can execute, and that are often essential for biological activity and pharmacokinetics: proper folding, disulfide bridge formation, oligomerization, proteolytic processing, phosphorylation and glycosylation. Moreover, all proteins developed from gene constructs allow secretion of the desired protein into the culture medium, thus substantially facilitating product isolation and purification [53, 54]. Approval depends on tight constraints with respect to the operational procedure in the manufacturing process. Thus due diligence and protocolled process supervision for quality assurance are mandatory [55].

Reactor technology remains diversified, with reactor types ranging from roller bottles to stacked plates, hollow fibres, stirred tank reactors or disposables such as

---

3) All mammalian cells used for large scale production of recombinant proteins are considered 'immortalized', as they can be grown continuously for an indefinite period. Controversial discussions about risks initiated more than two decades earlier have been settled meanwhile [53].

single use bioreactors, notably for low volume and speciality applications, such as the production of viral vaccines [56]. Both CHO and NSO (Mouse myeloma derived) cells grow well in suspension with high cell densities. In early EPO production, where only small doses are required, adherent cells have been cultivated in *roller bottles,* on the inner surface of cylindrical bottles having a volume of 1–3 L, a standard 2 L bottle containing about 300 mL of medium and providing an inner surface of 850 cm^2. This process can easily be scaled up with a large number of roller bottles handled in parallel. 1000 roller bottles will provide from 300 to 600 L of supernatant. Over a one-year period such a process will deliver 15 000–30 000 L product solution, providing protein in the kg range. The current Epogen® process is a robot-based manufacturing procedure where all the critical handling steps – including seeding, filling with media and harvesting of cell culture fluids – are executed within air-filtered environments and without human interaction [53]. In contrast, modern EPO production systems, for example at Roche, use stirred tank systems [57]. *Single use bioreactors* (SUBs) are used in the 15–125 L range for seeding larger stainless steel reactors, particularly for mammalian cell culture. Such systems include disposable sensors for dissolved oygen, pH and temperature. Also, new microreactor systems for development purposes have been introduced, available in 24 well or 96 well disposable format, that yield results compatible with standard shake flasks [58, 59].

The standard batch *stirred-tank bioreactor* has emerged as the industry's technology of choice. The issues of adapting cells to suspension culture, shear sensivity and oxygen supply have been largely resolved. The scale-up to very large volumes now occurs rather rapidly, usually by diluting the entire volume of one bioreactor into 5–20 volumes of fresh medium in a larger reactor. Within 10–15 days, a suspension culture at the 50 L scale can be used as an inoculum. A typical series of bioreactors for scale up comprises sizes of 80, 400, 2000 and 10 000 L working volume. The medium is needed initially to support rapid growth to achieve the highest possible cell density. In the last phase it provides the nutritional basis to maintain viability and productivity for extended periods of production (6–14 days). At a titre of, for example, 4 g L^{-1} and a volume of 10 000 L a single production run can produce some 30 kg of purified product (assuming a recovery yield of about 70%). Large bioreactor systems with 20 000 L volume are in operation at Lonza (CH) and Genentech (USA) [53, 54, 57].

Adherent cell culture and perfusion technology: Several processes have been developed in the human and animal vaccine industry, as well as for tissue engineering, using the microcarrier concept with cells that have a high anchorage dependence (adherent cells, to be grown in stirred tank bioreactors). The system allows easier scale-up and increased homogeneity in the supply of nutrients in media, oxygen and carbon dioxide exchange. The cells grow on the microcarrier beads. Widely used micro-carriers are based on a crosslinked dextran [53]. In vaccine production, the cells serve as substrates for the multiplication of viruses such as measles, polio or mumps. CHO cells are used for the production of several human recombinant proteins on microcarriers in stirred bioreactors (most notably at Merck-Serono). The continuous perfused production process aims at the highest cell concentration possible. Perfused cultures can be maintained for many weeks or months, whereby product harvesting is carried out repeatedly throughout that period.[3)] The antihemophilic protein factor VIII is reliably being manufactured using perfusion technology with BHK cells.

Other products from CHO cells have also been approved using this technology ([2], Section 9.1, [53]).

Two examples will be given here for production of Herceptin, and a scFv, a monoclonal antibody fragment for diagnostic purposes.

Herceptin, with Trastuzumab as the active protein, approved in 2002, was developed by Genentech, now part of Roche. Roche invested 290 million € in a new production unit (Biologics IV) that was completed in September 2007 for enlargement of the production capacity. It will serve the production of this *humanized antibody*, for the treatment of breast cancer (mammocarcinoma) [60]. The new production unit comprises: a preculture fermentation unit; a fermentation unit with 3 fermenters of 13 m^3 capacity (working volume) each (Figure 17.2); a cell harvest facility and a downstream processing line (Figure 17.3). A genetically modified CHO cell line is used in production, which is inoculated from a working cell bank. Cultivation takes place in volumes of up to 3 L for two weeks. The fermentation media are prepared in a central technical unit, passed through sterile filtration and transferred by pipelines to the bioreactors. Further precultures are performed in fermenters of 20 and 2500 L for three days each. They are then transferred into the production bioreactors, where the cultivation takes 14 days. In order to provide high standards of purity and sterility, cleaning and sterilization is performed by an automated system *in situ* (CIP and SIP). All processes, including recipes, are computerized (Figure 17.4). Cell growth and product formation are analysed and documented [61]. Downstream processing starts with centrifugation of the cells. The therapeutic protein is secreted into the medium which is stored in a tank. Several steps provide for the separation of the product from other components in solution: Purification is afforded by four specific chromatographic steps and a

Figure 17.2 Antibody production fermenter (with kind permission by Roche Diagnostics GmbH, 2007).

Figure 17.3 Antibody production unit (with kind permission by Roche Diagnostics GmbH, 2007).

subsequent ultra-/diafiltration unit, giving an aqueous solution with 99.9% purity that is transferred to vessels, frozen and stored [60].

A second process concerns the production and purification of a *recombinant antibody* at Bayer Schering Pharma (Germany) [62]. This human scFv- antibody for clinical *in vivo* tumour diagnostics targets, with high affinity, ED-B Fibronectin, a

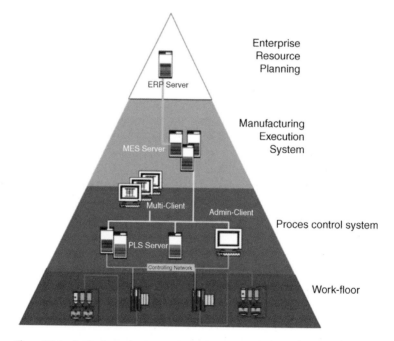

Figure 17.4 Antibody production automation system, with lowest work-floor machine control facilities, process control system, manufacturing execution system and enterprise resource planning [61] (with kind permission by Roche Diagnostics GmbH, 2007).

specific marker of tumour angiogenesis. Fermentation of recombinant *Pichia pastoris* resulted, after optimization, in a product concentration of 400 mg L^{-1}. The challenge of downstream processing was solved by combining affinity chromatography with protein A, reduction (to maintain terminal cystein-SH, avoiding covalent dimer formation, and to facilitate later conjugation with the radioisotope), size exclusion chromatography (SEC) and desalting. A highly efficient process has been developed with high product purity and a final yield of 60–70 mg L^{-1}. A long shelf-life is required and achieved with an appropriate storage buffer. The final cysteine-functionalized product is labelled subsequently by conjugation with the radioisotope ^{99}Technetium for detection of the tumour via positron emission tomography (PET). A team of about 30 scientists was involved in the development, including major efforts required for the development of different analytical methods; speed of development was a critical parameter [62].

Downstream processing for most biopharmaceuticals generally follows a sequence of unit operations which had altered little during the last 20 to 30 years (see e.g. [63]) (Figure 17.5). However, large scale fermentation technologies with high titre cell cultivation and dramatically increased yields focussed developments in downstream processing towards innovative techniques for protein purification [57]. These include membrane and aqueous two-phase systems, preparative chromatography and crystallization. Some principles developed more recently include inclusion body refolding and affinity-based procedures particularly for factor VIII and antibody-based products. Further techniques that are currently being developed use functionalized magnetic nanoparticles with ligands, such as highly specific recombinant antibodies, for on-line analytics as well as for selective protein isolation from complex mixtures, for example fermentation fluids, followed by separation in a magnetic field and desorption [64, 65]. Safety aspects are important and must ensure removal of any virus particles or nucleic acid from the product. The final product may be formulated in liquid or freeze-dried form and sterilized by filtration followed by aseptic processing and addition of stabilizers for example human serum albumin, sorbitol, or polysaccharides [1, 54, 57, 61, 62]. Of high significance for the evaluation of drugs and therapeutic proteins by regulatory agencies are *quality aspects* including the issues of reproducibility of process parameters and of achievable product quantity and quality. Recently the US FDA (Food and Drug Administration) introduced a PAT (Process Analytical Technology) initiative, aiming particularly at a holistic process orientated approach including QM (Quality Management) and validation of production processes [66, 67].

17.4
Products, Pharmaceuticals Made by Biotechnology

Two main developments may be discerned as being the roots of the present formidable success of biopharmaceuticals, both with respect to new therapies and their application, and to economic growth: the discovery and manufacturing of penicillin (and other antibiotics), and the "New Biotechnology" based on recombinant

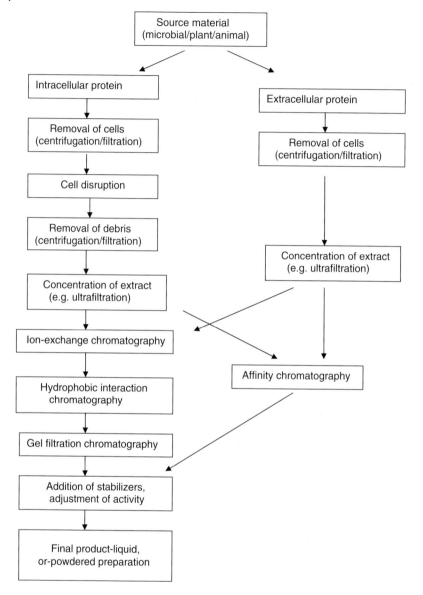

Figure 17.5 Overview of downstream processing steps applied to the production of therapeutic proteins. Remarks: protein refolding is often required following microbial fermentation; affinity chromatography may save several other chromatographic steps as indicated; additional viral inactivation steps and final product sterilization by filtration are often included; automation systems may be used to insure product quality and economic processing (modified from [1, 61, 62]).

Table 17.2 Main areas of application of approved biopharmaceuticals (2005) ([6]; www.Pharma. org).

Area of application	Approved Biopharmaceuticals[a]
Cancer related	16
Infectious diseases, with special emphasis on AIDS and hepatitis	26
Autoimmune disorders, with special emphasis on diabetes; neurological disorders, respiratory disorders	12
Cardiovascular diseases, myocardial infarction	7
Blood disorders, haemophilia	7
Growth disorders, human growth hormone	10

a) A total of about 165 biopharmaceuticals have been approved up to 2006 ([10], [11], pp. 8–11).

technologies. These events created new techniques to make drugs, and prompted innovative research including the new disciplines of microbiology and genetics. They finally led to important changes in the pharmaceutical industries [68–70]. (See Sections 4.3.4, 5.3.2, 6 and 7.2).

Infectious diseases top the list of *areas of application* of biopharmaceuticals. Mortality due to infectious disease has decreased steadily over the past 50 years in the industrialized world; however, for the rest of the world's population the opposite is true. More than 17 million people die from infectious diseases each year. 42 million people are infected by AIDS, 29 million in Africa. For malaria up to 500 million new cases are reported each year. Some 2 billion people are infected with hepatitis B, 350 million individuals suffer from lifelong chronic infections, and more than 1 million infected patients die each year from liver cirrhosis and/or liver cancer due to the infection. Based on such figures the global market for anti-infectives is estimated at over $30 billion. An overview of the most relevant areas of application is presented in Table 17.2. Further applications address skin disorders, eye conditions and genetic disorders [6].

17.4.1
Low Molecular Weight Agents: Antibiotics, Steroid Biotransformation

17.4.1.1 Antibiotics
Antibiotics are still the most important strategy against bacterial infections. The world production of antibiotics is estimated at over 60 000 t/a, valued at more than 30 billion $/a [71, 72]. They comprise a diverse group of potent low molecular weight (M_w) drugs, with diverse structures and modes of action. They have been discovered amongst the products (secondary metabolites) of many different bacteria, actinomycetes and fungi. They target not only a wide range of bacteria, but also parasites and even some types of cancer. A phenomenon that has limited the effective application of antibiotics is the rapid development of bacterial or viral resistance.

Despite measures aimed at curbing resistance, a study with data spanning nearly 70 years suggests that it might still be on the rise [73]. This represents a permanent challenge to biology, to screen for new antibiotics, and to chemistry to develop a large number of derivatives. This is an expensive and time-consuming undertaking where there have been many failures, but also some successes.

Classic antibiotics such as penicillin and cephalosporin are usually considered as low price products manufactured by fermentation. They are sold as bulk products on the world market, often by Asian companies which do not necessarily make the final drug confectioning themselves. Hence, the basic fermentation products are considered as part of industrial biotechnology (Section 16.4.1). High value products are derivatives of the raw material, modified by subsequent chemical or enzymatic process steps, for example ampicillin and amoxicillin. The properties of derivatives were optimized for stability (including low pH-resistant products for oral application), pharmacokinetics, activity against resistant strains, and lower and fewer side effects [74].

The action of antibiotics can be interpreted as a direct or indirect inhibition of certain enzyme systems. Thus β-lactam antibiotics act as a practically irreversible inhibitor for an enzyme that catalyzes the formation of the glycoprotein cell wall of gram-positive bacteria. (Similar enzymes do not exist in mammals, which explains the fewer side effects observed in their use compared to other antibiotics) [72]. The choice of a drug depends primarily on the (selective) sensitivity of the microorganism to the drug. The clinical basis of effective therapy is that the antimicrobial agent will target the disease-causing pathogen with minimal side effects in the mammalian host. Various undesirable side effects are known and must be considered [74, 75]. A rational approach to classifying antimicrobial agents derives from consideration of the target site of action that underlies their activity. A convenient survey, following the basic structure of types of antibiotics, referring to the target in microorganisms and main indications, is presented in Table 17.3.

Penicillins and cephalosporins, both of which possess a β-lactam ring, are designated β-lactam antibiotics, produced for example by *Penicillium* sp. Extensive attempts to improve their antibacterial spectra through chemical modifications led to the development of many kinds of semisynthetic penicillins and cephalosporins [75]. They are the main antibiotics for human use, with a market share of about 45–65% [71]. The β-lactam precursors of all penicillins and cephalosporins are produced by fermentation in large fermentors of up to 1000 m^3 (see chapter 15.4). They have emerged as drugs of choice to treat many infections for the following reasons: they cover a broad antibacterial spectrum, they are safe, and are bactericidal, and some representatives can be administered orally. However, the tremendous therapeutic potential has been threatened by the emergence of increasingly resistant bacterial strains as a natural consequence of their use. In clinical settings, more than 50% of *Escherichia coli* isolates and more than 90% of *Staphylococcus aureus* isolates are ampicillin resistant. Factors that exacerbate this phenomenon are misuse and overuse, and the widespread use of antibiotics in aquariums, in agriculture and animal husbandry [72].

Table 17.3 Survey of major antibiotics classes, selected examples, targets and main indications [72, 74, 79–81].

Type of antibiotic	Commercial preparations (selected)	Target in microorganisms	Main indications (selected)
β-Lactam antibiotics Penicillins	Penicillin G Penicillin V[a] Isocillin	Cell wall synthesis	Broad antibacterial spectrum; Angina, bronchitis, scarlatina, gonorrhoe, rheumatic fever, meningitis, urinary infections
	Ampicillin Amoxicillin		Hospitalism infections (e.g. pneumonia) Wound and tissue infections
Cephalosporins Cefazolin Cefuroxim	Gramaxin® Zinacef®	Cell wall synthesis	Broad antibacterial spectrum
Tetracyclines	Aureomycin, oxytetracycline	Protein synthesis (peptidyl transferase)	Broad spectrum of Gram-negative and Gram-positive bacteria, mycoplasmas, parasites
	Minocycline, doxycycline		Activity against resistant strains, acne vulgaris, prophylactic against malaria
Macrolides	Erythromycin A	Protein synthesis (23 S ribosomal peptide transferase)	Respiratory tract, pneumonia, bronchitis, skin, soft tissue, and urogenital infections
Aminoglycosides	Streptomycin, Kanamycin, Gentamicin	Protein synthesis (23 S ribosomal peptide transferase)	Tuberculosis (*Mycobacterium tuberculosis, E. coli, Klebsiella pneumoniae, Enterobacter spp., Pseudomonas aeruginosa*)
Peptide antibiotics	Actinomycin Gramicidin		Antibacterial, antifungal, antitumoral

a) Oral application.

For the synthesis of semisynthetic penicillins and cephalosporins the fermentation products penicillin G and cephalosporin, respectively, are hydrolysed by immobilized enzymes to yield the acid form with the intact lactam ring as the active principle (Figure 17.6): 6-aminopenicillanic acid (6-APA), 7-aminocephalosporanic (7-ACA), or 7-aminoacetoxy cephalosporanic acid (7-ADCA) (for comparison the older chemical procedure that required several toxic chemicals is outlined). Organon developed an interesting variant immobilizing urease together with the penicillin amidase allowing pH control at the site of the reaction by addition of urea to the substrate flowing through the column. This lengthened the half-life of the amidase several-fold. The production is in the range of over 10 000 (6-APA) and over 1000 t/a (7-ACA), respectively. Common products include ampicillin, amoxicillin, cephalexin and claforan (Figure 17.7). The market value of the products is in the range 20–30 billion €. Also, the synthetic step by enzymatic catalysis has been introduced on an industrial scale ([76], Kasche, in [77], 381–392).

Aminoglycosides, including streptomycin, kanamycin and gentamicin, are most often used as empiric therapy to treat serious infections caused by aerobic Gram-negative and Gram-positive bacilli, and as a specific medicine for tuberculosis (Table 17.3) [79]. Streptomycin, discovered by Waksman in 1944, is produced by *Streptomyces griseus*. Kanamycin, discovered by Umezawa in 1957, is especially effective against resistant bacteria, and serves as a second-line antituberculosis drug. These antibiotics are called aminoglycosides because their structural units are

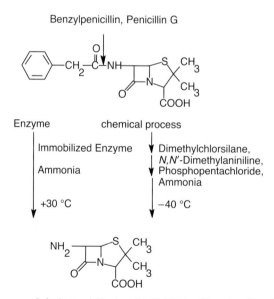

Figure 17.6 Hydrolysis of penicillin G by penicillin amidase (comparing of the chemical, and the enzyme process). The product 6-aminopenicillanic acid (6-APA) is used for the synthesis of semisynthetic penicillins with other side chains than phenylacetic acid (from [78]).

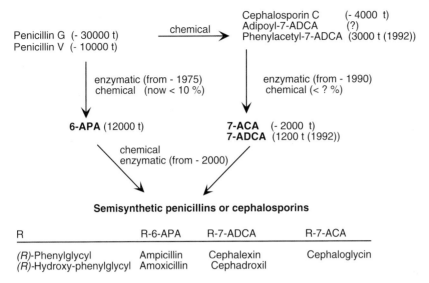

R	R-6-APA	R-7-ADCA	R-7-ACA
(R)-Phenylglycyl	Ampicillin	Cephalexin	Cephaloglycin
(R)-Hydroxy-phenylglycyl	Amoxicillin	Cephadroxil	

Figure 17.7 The enzymatic and chemical production of semisynthetic penicillins and cephalosporins from the hydrolysis products (6-APA, 7-ACA, 7-ADCA) of the fermentation products. The chemical conversion of penicillin G to Cephalosporin C is by ring expansion. The by-products phenylacetate and adipate can be recycled in the fermentations. The amounts produced are estimated from literature data (Kasche, in [77], 381–392, [71, 76]).

aminosugars, sugars and amino acids. The mechanism of resistance to aminoglycosides has been closely investigated and derivatives for use active against resistant bacteria have been developed [75]. Combinations of aminoglycosides plus a penicillin or cephalosporin are synergistic against a variety of bacterial pathogens, for example penicillin-resistant pneumococci [79]. Actinomycin is a *peptide antibiotic*, produced by *Streptomycetes*. It belongs to one of the major antibiotics groups, includes actinomycins, gramicidins and others, that possess diverse activities, that is antibacterial, antifungal and antitumor activities. Their structures are varied with often unusual amino acids as synthons [75].

Tetracyclines are a family of polyketide natural products that exhibit extraordinary activity against a broad spectrum of bacteria. Their structure is based on the naphthacene fused tetracyclic ring sytem; they have been isolated for example from *actinomycetes* and include aureomycin, first isolated in 1948. They are produced by aerobic fermentation of, for example *Streptomyces aureofacines* and *S. rimosus*. They are used for treatment of a wide range of pathogens amongst the Gram-negative and Gram-positive bacteria, but also atypical intracellular pathogens, such as parasites, rickettsias, and so on. They are active against infections caused by mycoplasmas and protozoas. Semisynthetic second-generation derivatives, including minocycline and doxycycline, and the third generation brought increases in potency, spectrum, oral availability, safety, and activity against resistant strains. Doxycycline has been used clinically for malaria prophylaxis [80].

Macrolide antibiotics, represented by erythromycin A, the most widely used, are large polyketide natural products and their semisynthetic derivatives. Erythromycin A was discovered in 1952 and was quickly accepted for medical use due to its safety and spectrum of activity. It continues to be used in the treatment of respiratory tract, skin, soft tissue, and urogenital infections. The naturally occurring macrolide antibiotics are mostly produced by the actinomycete species with structures characterized by a 12–16-membered lactone ring and by the presence of at least one sugar moiety attached to the macrolactone ring. Macrolide antibiotics exert their antibacterial activity by inhibiting protein synthesis, however, protein synthesis in eukaryotic cells is not inhibited. In addition, their relative safety makes macrolides the drugs of choice for many respiratory infections of paediatric patients. Their first and foremost use is in the outpatient treatment of community-acquired respiratory infections, including pneumonia, bronchitis, associated with the principle pathogens *Streptococcus pneumoniae* and *S. pyrogenes*. Second-generation macrolides, amongst them clarithromycin and azithromycin, with improved pharmacokinetic properties and increased spectra of activity, are at present the commercially most important [81].

Certain antibiotics are effective for clinical *treatment of cancer*, and occupy an important position amongst the various agents for cancer chemotherapy. At present, doxorubicin, daunorubicin, mitomycin C, and several others are used clinically. They interact with DNA to inhibit polymerases of DNA and RNA, or to cause DNA strand breakage. The first *nucleoside antibiotic* was isolated in 1950. Since then various others have been isolated and are important as antibacterials, antineoplastics and agricultural chemicals [75].

The use of *antibiotics in agriculture*, notably for the treatment of animal and plant diseases, and to stimulate animal growth, is still a most critical point that should be controlled by authorities, in order to avoid, or reduce, the danger of pathogens developing resistance. Microbial resistance can be acquired through a spontaneous or induced mutation in any of a number of promiscuous microorganisms that can transfer it rapidly to others (e.g. within the group of the Enterobacteriaciae which includes many pathogens such as some *E. coli* and *Salmonella spp.*). Therefore, worldwide efforts are directed at avoiding the indiscriminate or over-use of therapeutically important antibiotics, particularly the penicillins and tetracyclines, as feed additives [75]. An account of the dramatic consequences of uncontrolled use has been given by Bud [68].

A few *other inhibitors* are produced by recombinant technologies. In order to deal with influenza A (H_1N_1) virus infections, notably the 'swine influenza' that developed into a threat worldwide early in 2009, the FDA authorized the emergency use of Tamiflu. This led to a jump in the shares of Roche and Gilead Sciences which market the approved drug. The active drug component oseltamivir, an *inhibitor of neuraminidases*, active as an antiviral agent against influenza A (H_1N_1), was discovered only in 1995 by Gilead Sciences (USA) and marketed as early as 1999 by Hoffmann-La Roche AG (CH) [82]. A fermentation process has been developed by the group of Frost [83]. Further examples include a potent *α-glucosidase inhibitor* with antiviral action, 1-deoxynojirimycin, which is manufactured by regioselective oxidation of aminosorbitol with *Gluconobacter oxydans*, [84].

17.4.1.2 Steroid Biotransformation

Steroids became a topic of high public interest when the anti-baby pill became available in 1958. Since 1960 it represents the most frequently used means for birth control. It is based on the female steroid hormones estrogens and gestagens (or variants) [85]. It not only was a new drug, but caused a social tremble, a 'moral Waterloo' for conservative patriarchal societies. Moreover, steroids became fashion pharmaceuticals in sports and body building, with doping as a major problem, anabolics (testosterone derivatives, male steroid hormones) representing the group of most frequently used drugs. Thus at the Olympic games in Montreal in 1976 doping controls were introduced for the first time. Additional problems came to be known, including side effects such as cardiac attacks, also as a result of high blood viscosity where cyclists were using EPO (erythropoietin) ([85], www.Wissen.Spiegel.de).

Microbial conversion of steroids became a relevant topic for pharmaceuticals production, based on the classical discovery of Peterson *et al.*, in 1952, that *Rizopus arrhizus* converts progesterone and related steroids to the corresponding 11α-hydroxy derivatives [86]. Progesterone, a female sex hormone, is produced industrially in large amounts by oxidative degradation of stigmasterol, available from soja, or by conversion of steroid precursors from other plant sources. For most of the relevant hydroxylations and oxidations, regio- and stereo-selective microbial transformation reactions have been found (for chemical reactions a highly tedious task), including positions 3, 5, 11, and 17 of the basic steroid structure, and α- or ß-stereospecificity. Microbial oxidation of the C-11 position allows the synthesis of analogues of corticosteroids (relevant for control of metabolic activities). Favourite objects of studies, in addition to oestrogen, remain the group of androgens, that is male sex hormones, androstan, testosterone, and the pregnane series (gestagen group, including progesterone), and cholesterol. The development aimed at the production of different steroid hormones to provide treatment of various diseases [86, 87].

17.4.2
Recombinant Proteins

Protein-based drugs constitute about a quarter of new approvals with a majority being glycoproteins. Overall, some 165 biopharmaceutical products have gained approval up to 2006, with a market size estimated at some $ 35 billion in 2006, and projected to reach $ 50 billion by the end of the decade (see Section 9.7). The average annual growth rate has been 20% ([11], pp. 8–11, [10, 52, 88]) (for overviews of approved recombinant biopharmaceuticals see: [1] [13], see also www.centerwatch.com/patient/drugs).

Approved recombinant proteins comprise a number of different biopharmaceuticals: blood factors, thrombolytics, insulins, other hormones, hematopoietic growth factors, bone morphogenetic proteins, interferons (IFNs) and interleukins (ILs), monoclonal and engineered antibodies (mABs), vaccines, therapeutic enzymes and enzyme inhibitors [1] (Table 17.4). Most of the drugs listed in the table are *blockbuster* with sales of 1 billion US $ per year or more. They result from concentration of

Table 17.4 Major classes of biopharmaceuticals, selected drugs (mostly blockbuster), indication, manufacturers and sales [6] (data for 2001), [1, 10, 11, 25–31, 91, 92]); the data can only taken as rough information, since data from different sources are not always consistent, and change rapidly with time.

Class	Indication/ Therapeutic area	Market size 2001 (Billion $)	Market size, world, 2006 (Billion $)
Hormones/Growth factors			
Erythropoietin	Anaemia	6.8	17.9 (2006/08)
Epogen (Amgen)			~4[b),f)
Aranesp (Amgen)			5.1[a)
Neupogen, Neulasta (Amgen)			~3 (Neulasta)[c)
Procrit (Johnson & Johnson)[c)			4[f)
Human growth hormone	Growth disorders	1.7	(14.1 USA)[c)
NeoRecormon/Epogin (Roche)			1.8 (2008)[h)
Human Insulin	Diabetes	4.0	>4.5 (2005)[g)
			(3.8, USA)[c)
Humulin, Humalog (Eli Lilly)			(38%)[i)
Lantus (Sanofi-Aventis)			(37%)[i)
Novolin, Novorapid (Novo Nordisk)			(26%)[i)
Blood clotting factors	Haemophilia	2.6	(0.78, total blood factors, USA)
Factor VIII, factor VIIa, Factor IX			
Cytokines			(3.4, USA)[c)
Interferon beta	Multiple sclerosis, Hepatitis	2.1	(2.0, USA)[c)
Interferon alpha	Cancer, hepatitis	1.8	
Pegasys (Peginteferon α-2a) (Roche)	Hepatitis C		1.6 (2008)[h)
Monoclonal antibodies	*Cancer*	1.75	(5, USA)[c),i)
MabThera/Rituxan (Roche Genentech)	(Non-Hodgkin)		5.9[h)
Neupogen, Neulasta (Amgen)	(white blood cell)		~ 1 (both)[b)
Avastin (Roche)	(colon, lung)		5.2 (2008)[h)
Herceptin (Roche/Genentech)	(breast)		5.1 (2008)[h)
Erbitux (Merck, Germany)	(colorectal)		0.66[c)
Rituxan (Genentech)	(Non-Hodgkin lymphoma)		1.5[f)
Monoclonal antibodies	*Various (autoimmune, inflammation etc.)*	1.15	(6.3, USA)[c),i)
Embrel (recombinant fusion Protein[l)) (Amgen)	Anti-TNF (Autoimmune treatment, Rheumoid arthritis and other)		4.4[d),k)

Table 17.4 (Continued)

Class	Indication/ Therapeutic area	Market size 2001 (Billion $)	Market size, world, 2006 (Billion $)
Humira (Abbot Lab.)	As before		$2.0^{d)}$
Remicade (Johnson & Johnson, Centocor)$^{m)}$	As before, Crohn's disease		$3.6^{d)}$
Truvada (Gilead)$^{d)}$	HIV		$3.9^{d)}$
Vaccines			
Gardasil (Merck, USA)$^{c)}$	Cervical cancer		$(1.2, \text{USA})^{c),n)}$
Recombivax (Merck, USA)$^{g)}$	Hepatitis B		
Comvax (Merck, USA)$^{g)}$	Hepatitis B$^{o)}$		

a) [91].
b) [26].
c) [10].
d) [92].
e) [94].
f) [11], p. 8.
g) [1].
h) [95] (figures using a rate of approximately 1:1 for $ and CHF, October 2009).
i) Insulin market share of companies [10].
j) Total mAB sales in the USA: 11.3 billion $; for oncology ~ 5; for autoimmune and inflammation diseases: 4.5 billion $ [10]; for antibodies against TNF-α sales of 13.5 billion $ in 2008 have reported (Pappas, D.A., Bachon J.M., 2009, Nature Rev. Drug. Discov. **8**, 695–6).
k) Sales of TNF inhibitors: 6.7 billion $, second to EPOs as the largest selling molecules [10].
l) Enbrel: recombinant TNF receptor fused to IgG fragment [10].
m) Chimeric mABs.
n) Estimated from sales of 598 million $, first half year, 2007 [10].
o) Combination vaccine including Hepatitis B.

research and development on diseases of high importance, affecting major parts of the population. Almost two thirds of sales and growth is driven by three categories: growth factors, mAbs and hormones (notably insulin) [10, 89]. Many of these biopharmaceuticals are glycoproteins where the glycan moiety may have no known function or one of the following various functions. It may contribute to correct protein folding, quaternary interactions, stability, hydrophobicity and/or exhibit specific functions such as being a biochemical signal in lectin-mediated information transfer, intracellular targeting to the lysosome, cell–cell communication, and, notably, in infection and immune regulation. Therefore, recognition of sugar-based interactions holds great promise for deciphering disease mechanisms and for the development of novel diagnostic and therapeutic interventions [23, 24]. Most of the newly approved medicines have been developed to address indications which are the major causes of disease and mortality in the industrialized world, notably infections, with Hepatitis B and C as the most targeted indications, diabetes, haemophilia, myocardial infarction and various cancers [6]. These categories have become major markets because of the existence of a large unmet need and constant innovation from biotech companies. Some of these biotech drugs, especially anti-TNFs, anti-cancer mABs and insulin analogues, are positioned in the treatment regimen where traditional small molecules have already maximized their benefit, have failed, or cannot be used [10]. The most important types of medical products will be treated briefly with some information on therapeutic indication, production and the companies involved.

17.4.2.1 Blood Factors and Thrombolytics

Blood factors aim to treat haemophilia, and all (about 7) are produced in enginerered animal cell lines in order to faciliate product glycosylation (for a detailed discussion see [11], pp. 329–353). Blood factor VIII-based products are indicated for treatment and prophylaxis of patients with haemophilia A, a genetic disease characterized by the total lack or low levels of blood clotting factor VIII. Similarly, blood factor IX is indicated for patients with haemophilia B, caused by a deficiency of coagulation factor IX. Tissue plaminogen activator- (tPA-) based thrombolytics, as well as urokinase, help in accelerated removal of blood clots that form under inappropriate conditions, especially in cases of occlusion of brain arteries (ischaemic stroke). A recombinant form of native human tPA was first marketed by Genentech in 1987. Currently, different host systems are used, including mammalian cells; most tPA-based products (about 5) are engineered in some way in order to extend their plasma half-lives [1, 11, 57]. An interesting variant and highly potent plasminogen activator was found in 1980 in the saliva of the vampire bat *Desmodus rotundus*. The corresponding gene was cloned, patented and produced as an alternative product to tPA, but so far (end of 2009) has not shown efficacy in clinical trials for ischaemic stroke. This is one of many hopeful developments which failed, dashing the hopes of investors and start-up biotechs, contributing to the over-cautious attitude of the pharmaceutical industry. Recently, the FDA has allowed the first human therapy by a product from a transgenic animal, a goat. Human antithrombin protein (ATryn) is extracted from the goat's milk. Since the product is in short supply, the approval was fast-tracked by the FDA as an orphan drug: as an anticoagulant also for replacement therapy of patients lacking this factor [90].

Hirudin, a polypeptide of molecular mass 7000 Da, is a highly potent anticoagulant, originally of the medicinal leach (*Hirudus medicinalis*), inhibiting thrombin. Two recombinant products have been approved, the first in 1997, both expressed and secreted by *S. cerevisiae* ([11], pp. 342–344). It should be remembered that long term or repeated injection of proteins of non-human origin can be associated with strong immune response, in the extreme culminating in anaphylactic shock and death. This usually limits application of such products to external use.

17.4.2.2 Cytokines, Interferons (IFNs) and interleukins (ILs)

Cytokines, including IFNs, ILs, and tumour necrosis factor (TNF)-α, are produced in the response of mammals to challenge by a pathogen, bacterial products, viruses (α- and β-interferons), and agents that cause physical damage (for details see Section 10.1) ([11], pp. 205–263). They are small proteins (many glycosylated) released by various cells in the body. They, in turn, diffuse to adjacent cells in tissue, or are distributed in the circulation, inducing responses on binding to specific cell surface receptors. IFNs are produced by virus-infected cells and serve to inhibit viral replication and to activate natural killer (NK) cells.

With new efficient technologies for the large scale production of recombinant proteins, it became possible to produce such cytokines which had previously only been available in trace amounts and barely characterized. This facilitated the development of several potent drugs for the treatment of cancer, viral infections and arthritis. Cytokines are the fourth largest category of recombinant drugs in the US biotech sector, with sales of $ 3.4 billion and a growth rate of 10% in 2006 [10]. The first recombinant interferon, Hoffmann La-Roche's Roferon A (IFN-α), developed in cooperation with Genentech, was approved as early as 1986, and has 'reigned supreme for decades as biotech's blockbuster antiviral' [35, 36]. Following this start-up, quite a number of recombinant IFN-and IL-based products have gained approval over the last 12 years. Currently 19 different brands of cytokines are on the US market, including 16 interferon-based biopharmaceuticals, with four major types of molecules – IFN-α, IFN-β and IFN-gamma, and the interleukins. An early breakthrough constituted a recombinant interferon β-1a for the treatment of multiple sclerosis, 'a stunning commercial success' by meeting the demands of a needful drug with appropriate indications. IFN-β brands indicated for multiple sclerosis are the highest sellers with almost 60% of the cytokine market ([6, 10], [11], pp. 9, 224).

Eight interferon-based products and eight prophylactic vaccines against hepatitis B were approved during 2001–2004 (three brands from Schering AG, and two from Roche: interferon-based vaccine Roferon; and Pegasys, a peggylated Roferon) [10, 57]. The amount of activity which has been focused on the development of treatments for this viral infection reflects its global significance. Some 2 billion people are infected with hepatitis B per annum [6]. The α-IFNs, including PEGylated forms with extended plasma half-life, have found application mainly in the treatment of certain viral diseases and cancer types. ILs have been applied for example for treatment of renal cell carcinoma, and, in addition, an IL-1 receptor antagonist has gained approval for the treatment of rheumatoid arthritis [1]. TNF-α is an important cytokine that triggers local containment of infection [11].

17.4.2.3 Insulin

Insulin, a peptide hormone synthesised in the pancreas, is essential for carbohydrate metabolism and regulates the blood glucose level. The demand for insulin had steadily increased since Banting and Best, after isolating insulin in 1920, first described its use in the treatment of diabetes, a life-saving therapy for many. The development of recombinant human insulin in 1978, the first recombinant thera-peutic protein, provided the basis for one of the most successful biotech companies, Genentech (since 2009 part of Roche). In cooperation with Ely Lilly approval by the FDA was accomplished in 1982. Eli Lilly's Humulin® received the status of a 'blockbuster' product with annual sales of one billion $ in 2002. This success had the greatest impact on the development of recombinant medicines (Section 5.3.2). The modern insulin industry is dominated by Lilly, Novo, and Sanofi/Aventis. The

WHO estimates that some 170 million people suffer from diabetes, a figure likely to double by 2030. Although a minority of these require daily insulin injection the market for human insulin has been valued at over $ 4.5 billion in 2005. Both first-generation human insulin and engineered second-generation insulins are marketed (see Section 9.3). 'Fast-acting' insulin has an amino acid sequence that leads to conformational changes, rapid deoligomerization upon injection and thus an increased rate of uptake in the blood stream. The 'slow-acting' product provides an extended base level of insulin due to the attachment of a C14 fatty acid at a lysine residue that promotes binding of the insulin analogue to albumin and a constant and prolonged release of free insulin into the blood [1, 6].

17.4.2.4 Additional Recombinant Hormones

Nineteen additional recombinant hormones have been approved. These include erythropoietins and a number of recombinant versions of human growth hormone (hGH), including a PEGylated analogue, various gonadotropins, glucagon, parathyroid hormone and calcitonin. Genentech was the first company that produced and obtained an approval for the use of hGH for treatment of hGH deficiency in children [1].

Growth factors represent the most lucrative fraction of biopharmaceuticals, with sales of $14.1 billion in 2006 in the USA, and an annual growth rate of over 10% since 2001. Of a total of 11 brands most fall into the category of erythropoietins (EPOs) and colony stimulating factors (CSFs). This market was established with Amgen's launch of Epogen for anaemia approved in 1989 and Neupogen (a CSF) for chemotherapy-induced neutropenia in 1991. EPO is a *hematopoietic (glycoprotein) growth factor*, stimulating the formation of erythrocytes. It has been developed by Amgen for the treatment of anaemia associated with various medical conditions, notably dialysis patients with chronic renal failure, who lose their own EPO during dialysis. This represents tens of thousands of patients each requiring a very small dose. It is sold under the brand names Epogen and Aranesp (Amgen), and Procrit (Johnson & Johnson) in the USA, as Recormon, later Neorecormon (Boehringer Mannheim, now Roche) in Germany and Epogin (Chugai) in Japan. It has been the best selling, single type of molecule in the biotech drug market, experiencing a doubling of US sales from $ 5 billion (2001) to $ 10 billion (2006), recently, however, with a decline in sales, due to risks and warnings by the FDA. Nespo (Aranesp in the USA) is an engineered EPO analogue, with five instead of three carbohydrate side chains, produced in a CHO cell line. It has an extended serum half life [1, 10, 53, 57]. In different disciplines of sports, such as cycle racing, EPO is (mis-)used for doping. Carl Djerassi, the 'father of the pill', ironically proposed 'doping, but correct', a test run to prove the effect, and the risks, of the drug[4] [96].

---

4) Litigation has occurred between Amgen and (i) Genetics Institute, (ii) Johnson & Johnson, and (iii) Transkaryotic Technologies, with the outcomes mostly favoring Amgen; further, there is an ongoing patent dispute with former Boehringer Mannheim GmbH, now Roche. The original patent for EPO expired in 2004 in Europe, but other patents are valid until 2012–2015 [10].

Another CFS blockbuster in Amgen's haematology franchise is Neulasta (sales of $ 3 billion in 2006). Its main indication is for use in stimulating the production of neutrophils and leukocytes before chemotherapy and during bone marrow transplants. A further six products in this class have been approved and marketed up to 2005 [1, 10]. New classes of biopharmaceuticals approved during recent years include various bone morphogenetic proteins (BMPs), for example BMP-7 and BMP-2, that can accelerate bone fusion in slow-healing fractures [6].

17.4.2.5 Monoclonal Antibody (mAb)-Based Products

Antibodies are immunoglobulins (glycoproteins) produced by B-lymphocytes and plasma cells as part of the immune reaction, namely the humoral response, against an antigen, such as a virus (see Sections 10.3 and 10.4; for industrial production see Section 17.3.2). Glycosylation plays a major role in recognition of antigen by immunoglobulins, influencing its solubility, its half-life in serum, and in interaction with attacking cells. Antibody-based products represent the single largest category of biopharmaceuticals approved for general medical use, and, in the biotech drug market in the USA, the second largest category, with sales of $ 11.3 billion in 2006 and an average yearly growth rate of 35% since 2001 (for mAbs approved and in clinical trial see Section 10.8) ([1, 10], [11]). mABs are effective for the treatment of a wide range of diseases. Most of the approved mABs are used to either detect or treat various forms of cancer, and in treating autoimmune and inflammation disorders, for example Crohn's disease and rheumatoid arthritis (see Table 17.4). Further applications are prophylaxis or reversal of acute kidney transplant rejection, prevention of blood clots, or as selective diagnostic tools. Detection (diagnosis) of solid tumors *in situ* is facilitated by the conjugation of a radioactive tag to an appropriate antibody in combination with tomography. Examples of such products have been approved. The use of antibody conjugates with highly fluorescent fluorophores is a growing diagnostic market for biopsies and blood samples as well as use *in vivo* for detecting pathogens [1].

The category of *fusion proteins* includes only four drugs, but their sales put them amongst the most lucrative biotech sectors in the US market, with Enbrel® (Amgen) accounting for $ 4 (USA: 3) billion in sales in 2006. It comprises a recombinant TNF receptor fused to the Fc region of the IgG fragment, and it is thought to act by neutralizing the pro-inflammatory cytokine TNF with the Fc portion of the molecule, thereby facilitating aggregation and uptake by phagocytes. It is used to treat rheumatoid arthritis [6, 10].

17.4.2.6 Vaccines

The manufacture of recombinant vaccines by microbial fermentation was introduced successfully in the 1980s. Merck Research Laboratories Inc. introduced a vaccine for meningitis made by chemically conjugating polysaccharides from *Haemophilus influenzae* type b with an outer membrane protein complex from *Neisseria meningitis*. It is a member of a new class of 'conjugate vaccines'. A vaccine for hepatitis B was introduced by Merck & Co., Inc. using *Saccharomyces cerevisiae* as the host for production of the active ingredient, the rHbS viral capsid protein antigen [97]. Eight prophylactic vaccines against hepatitis B were approved during 2001–2004. The activity which has been focused on this viral infection reflects its global significance. More recently, approved vaccines also contain an identical S. cerevisiae-derived hepatitis B surface antigen (rHbsAg). More generally, it represents one component of multicomponent vaccine preparations [1, 6]. Combination vaccines are for example recombinant hepatitis A and B surface antigens for immunizing adults against hepatitis A and B and GSKs Pediatrix® that vaccinates children for five diseases (diphtheria, tetanus, acellular pertussis, hepatitis B and polio) [10]. Novartis is developing the first flu vaccine to be produced in mammalian cells, rather than in chicken eggs [25]. Expectations for a vaccine against malaria are related to the synthesis of an oligosaccharide that was identified on the surface of the pathogen *Plasmodium falciparum* [98]. Recently, the group of Danishefsky published the synthesis of a vaccine against ovarian cancer. In this strategy the carbohydrate moiety is coupled to an immunogenic aglycon such as BSA (bovine serum albumin) or KLH (keyhole limpet haemocyanin), leading to a concerted immune response in mice. The carbohydrate units have also been coupled to T- and B-cell oligopeptide epitopes successfully [99].

Several veterinary biopharmaceuticals have also gained approval, such as bovine somatotropin (recombinant bovine growth hotmone), used to boost the milk yields of dairy cattle. The majority, however, are engineered vaccines for vaccination to prevent rapid spread of disease through high density populations characteristic of modern agricultural practice [1].

17.4.3
Further Biopharmaceuticals

Therapeutic enzymes include recombinant enzymes used for the treatment of rare genetic disorders, which allows these therapies to be launched in part with orphan drug status and often with a high price tag. They are treated in Section 7.5.6, and summarized in Figure 7.1. The first successful application of *enzyme replacement therapy* for an inherited disease was the treatment for adenosine desaminase (ADA) deficiency using the modified enzyme [57].

Roche is producing Tamiflu as a therapeutic *enzyme inhibitor* for influenza A (H_1N_1) virus (with oseltamivir, a neuraminidase inhibitor, as the active ingredient), with sales of 2.1 billion $ in 2006, high due to the avian (bird) flu, significantly less, about 610 million, in 2008; they are expected to be high again in 2009 due to the swine flu [57].

The first of a novel class of antiretroviral drugs is *Enfuvirtide* (INN), a HIV *fusion inhibitor*, used in combination therapy for the treatment of HIV-1 infection. It is marketed under the trade name Fuzeon® (Roche) (www.fuzeon.com/). Enfuvirtide works by disrupting the HIV-1 molecular machinery at the final stage of fusion with the target cell, preventing uninfected cells from becoming infected. As a biomimetic peptide, enfuvirtide was rationally designed to mimic components of the HIV-1 fusion machinery and displace them, preventing normal fusion [100]. This synthetic peptide, also called T20, is produced on a very large scale (see also Section 7.4).

In the field of *nucleic acids, or oligonucleotides*, despite some 15 years of research, only two products have gained approval in the USA or the EU. Isis Pharmaceuticals (CA, USA) Vitravene® (fomivirsen) is used for cytomegalovirus infection of the eye; and Eyetech's (Pfizer) Macugen® (pegaptanib, the first aptamer approved) indicated AMD, cancer and diabetic retinopathy. Both of them are dosed intravitreally (eye injection). In 2003 China approved a further one – Shenzhen Gentzech SiBiono's Gendicine® – for head and neck cancer. The product is a replication-incompetent human serotype-5 adenovirus engineered to contain the human wild-type *TP53* tumour-suppressor gene, expression leading to cell cycle arrest and apoptosis [10, 88]. The emerging field of RNA silencing research has shown that it is a common phenomenon that small RNAs (siRNAs, miRNAs) are regulating gene expression in cells. This knowledge is expected to radically alter the potential of RNAs as drugs, in spite of difficulties remaining (see Section 13.7). *Viral vectors* provide a convenient means to deliver vaccine antigens to selected target cells or tissue. A broad spectrum of replicating and non-replicating vectors is available [101]. For a status report and opinion on the future of gene therapy see Section 13.3.

17.5
Medicinal Techniques, Diagnostics

17.5.1
Data Banks, Collections

The aim of data banks, or biobanks, is to collect and provide access to biomedically relevant samples and data, to enhance therapy of and prevent common and rare diseases. An initiative for collecting a standardized bank of blood and cell probes was launched in 2008: the Biobanking and Biomolecular Resources Research Infrastructure (BBMRI), which claimed to be the worlds biggest. It will establish catalogues, quality-assessed samples, standardized working protocols and analytical procedures that enable the comparison of probes and data (with links to related clinical and epidemiological information). It will comprise around 12.5 million probes from blood, tissues and body fluids and make them available for academic as well as industrial research, in order to identify biomarkers, targets, diagnostics, as well as provide additional resources for cDNA- and antibody collections, cell cultures and so on. Synergism is intended to be achieved by interlinking, standardising and harmonising – sometimes even just cross-referencing – a large variety of well-qualified, up-to date, existing and *de novo* national resources [102, 103]. (See Section 7.6.10 for sequences available through GenBank).

As discussed in Section 13.4, a collection of some 50 induced pluripotent stem cell lines are being established and stored using the Yamanaka factors which should be available for over 86% of the japanese tissue types and applicable for regenerative medicine [156]. This is representative of similar initiatives in other countries.

17.5.2
Diagnostic Enzymes and Antibodies

The primary goal of diagnostic testing has long been the detection and quantification of disease-specific analytes ranging from simple species, such as ions, through complex biomolecules, such as drugs, hormones and proteins, to complex analytes, such as cells and viruses. The recent shift in emphasis towards detecting genetic predispositions and prophylactic testing is dealt with in Section 11.4 on personalized medicine. As most analytes occur at low concentrations in complex biological samples from the bodily fluids: blood, plasma, sweat, urine, faeces, or from tissue biopsy, a high analytical sensitivity and specificity is required. Both antibodies and enzymes can be applied in suitable detection systems. The antibodies exhibit very high selective affinity for their antigens and the enzymes a very high substrate specificity (see Section 10.3). Coupled to highly fluorescent tags antibodies can be detected at very low concentrations, or in combination with radionuclides emitting positrons can be used in PET tomography for high resolution detection of solid tumours *in vivo*. If the analyte is a substrate for a specific enzyme, an enzymatic assay can be applied to determine its concentration. A great number (several hundred) such test systems have been developed to a high level of sophistication.They can be implimented simply and with high accuracy, using standardized methods, for a wide range of analytes, metabolites and drugs. They may even be applied by the patient on himself in a few examples such as glucose determination, a standard test for diabetic patients ([104, 105], p. LXXXI; for a comprehensive overview on diagnostic enzymes see [106]). The leading company for diagnostics, Hofmann La-Roche AG, division diagnostics (Basel, CH) had sales of some 9.7 billion $ in 2008 [95].

Chip technologies and antibody *microarray analysis* have been established as a high-performaqnce technology platform. They are treated in detail in Section 11.4.1. Microarray-based expression profiling for studying biological mechanisms is used to target complex proteomes, as well as for developing clinically valuable predictive classifiers. The first chips for clinical application were approved for use in the predictive diagnostics of drug susceptibility [107, 108]. Carbohydrate microarrays – *glycochips* –address carbohydrate-mediated recognition and responses to infection, for example, to detect and type the various strains of the influenza virus [109–111]. Commercial products for medical diagnosis of cancer and autoimmune diseases, as well as blood tests for the prediction of MS are mentioned as examples now available [112].

17.5.3
Pharmacokinetics, Controlled Release and Drug Targeting

Pharmacokinetics is a key topic in pharmaceuticals development, including absorption, distribution, metabolism in, and excretion of drugs from the body (see

ADMET factors, Section 9.7.3) ([11], pp. 74–80). A pharmacokinetic study comprises dosing and sampling of probes from animals and subjects, bioanalysis of the biological samples (e.g. blood, plasma, tissues) and analysis of the resulting blood, plasma or serum concentration versus time. In drug discovery research pharmacokinetic studies are most commonly conducted in rodent species, then subsequently in larger animal species such as dog or monkey to support toxicology studies. The latter generate data more likely to be useful in predicting human pharmaocokinetics [113, 114]. See Section 10.7 for recent problems that have been encountered in this area.

Controlled release represents a challenge. Tissue engineering has been applied to provide a constant supply of a potent hormone or an enzyme for replacement therapy. Cell biocompatibility and biodegradability of the drug itself are important aspects to be considered. Proteins and peptides, potent and specific therapeutic agents, have intrinsic limitations ascribed to low stability *in vivo*, a short residence time in the body, and potential immunogenicity, or even allergenicity, especially in the application of non-human proteins. In the case of low-M_w drugs, a short *in vivo* half life (due to rapid clearance by the kidney or enzymatic degradation in the liver) is often observed. Sometimes there may be physicochemical drawbacks, such as low solubility or instability. Several drug delivery systems have been developed in the last few years to improve the pharmacokinetic profiles of many drugs, in addition to addressing drug targeting to a specific site. These approaches can be based on tailor-made formulation of the drug, such as liposomal preparation, controlled release formulations, or via a modification of the drug molecule itself (see examples of insulins, Section 17.4.2.3), or by covalent attachment to a polymer. When modified in this way, the drug can often achieve a prolonged residence time in the serum, an increased resistance to degradation, increased water solubility, enhanced bioavailability, protection from degrading enzymes and decreased side effects (e.g. reduction of immunogenicity and antigenicity). In sum this is generally correlated with improved patient compliance. Polyethylene glycol (PEG) emerged as the best candidate for protein modification (PEGylation). Some of those products have become blockbusters, for example Pegasys® (Roche) for therapy of Hepatitis B + C. Other polymer systems are under investigation, including specific spacers that may control the release rate, or target the release to certain cellular compartments ([115], see also [88], [11], pp. 74–80). Specific drug delivery systems can make use of, for example, tertratricopeptide repeat proteins (TPRs) that recognize cell receptors, such as the HER2 growth factor receptors on cancer cells. Thus designed they would or should be internalized only by cancer cells [116].

Other routes of administration are also available, for example through the mucous membranes or subcutaneous micro-capillary pumps. Delivery to the lung depends on producing the correct particle size for the droplets in the aerosol to ensure lung penetration and high residence time and adsorption, representing an effective method for uptake in an acid-free environment. Acceptance is highly variable from country to country, being hardly accepted at all in the USA.

17.6
Business: Companies and Economic Aspects

17.6.1
General Overview

Remarkable changes in the structure of the pharmaceutical industry occurred during the last century, first the decline of the 'world pharmacy', that of two German chemical companies, second the trend towards big pharma, mainly in the USA and Britain, by mergers and acquisitions (globalisation) in the late 1990s and subsequent years, and third the emergence of new biotech companies (see Sections 6, 17.6.2 and 17.6.3). Currently, the world market for pharmaceuticals is over 600 billion US $ (2007), the largest being that of the USA (219), followed by Japan (57), Germany (30), France (28), and the UK (17, all billion US $), respectively [5, 118, 119]. For several years the pharmaceutical market exhibited growth rates of 7.5–10%. Ten companies account for 36% of the market. Several blockbuster drugs per year – termed a 'blockbuster mania' – would be required to achieve the expected return on investment – EBIT margins of over 20% – of their stock-holders. However, an extensive series of mergers amongst the giants, as well as their acquisitions of smaller biopharmaceutical companies did not solve the problem, apparently due to strategic management problems arising after a merger which often resulted in the newly formed company being less innovative [5, 120]. Dramatic reductions in staff of pharmaceutical companies were under way in 2007, in the range of 5–10%, top management departures, and downsizing and shrinking the number of production sites (e.g. at Pfizer, BMS, and Merck) and other cost cutting regimes [26–28, 89, 121–123]. In addition, as Malthus might have pointed out, increase in spending, public and private, in health care cannot continue at such rates indefinetely.

The *economic problems* of the pharmaceuticals industries – which may even be described as a crisis – were obvious from 'the lower number of significant new drugs coming to market, further involving a high attrition rate during drug development (the average clinical success rate was only 11% from 1991–2000).... This was accompanied by a fall from grace in the public eye as a result of concerns over the safety and price of drugs...' [22, 123]. Further problems were combating generic competition to their products, due to patent expiration (patent protection generally holds for 20 years), 'in doing damage control following unexpected drug withdrawals, and maintaining a well-oiled R&D engine – all while trying to meet the lofty sales and profit expectations of shareholder and financial analysts'. When Pfizer announced abandoning Exubera, its inhaled insulin product, the company wound up incurring a 'breathtaking $2.8 billion write-off asset' for the debacle. Also Eli Lilly and Novo Nordisk undertook the write-off of large amounts of investment in similar developments [26–28, 124, 125]. The financial crisis that started about late 2008 or early in 2009 led to low growth rates of the global pharmaceutical market. Nevertheless, sales and profits of the pharma industries are expected to grow at a rate of 5–7% in the near future [119, 126]. An ongoing debate questions the legitimation, in terms of ethical acceptance of unusually high profits in the health and pharmaceuticals sector. Also, the largest biotech companies, such as Amgen and Biogen Idec are having a difficult time reaching the double-digit sales and profit growth expected. Genentech (that has recently been integrated into Hoffmann La-Roche AG) appears to be the most innovative amongst the larger established 'new biotechs' [31].

The role of small- and medium-sized enterprises has been acknowledged, as well as the role of certain productive clusters, including those in Boston, San Diego and the 'Bay area' in and around San Francisco, California (USA), Cambridge and Oxford (UK), Munich (Germany), Leiden (Netherlands) and Paris (France), as well as the 'Medicon Valley' on the canal border between Denmark and Sweden [127].

17.6.2
Companies

Traditional *pharmaceutical companies* make the larger part of their sales with drugs made by chemical synthesis; they are, nevertheless, increasing their activity in developing and marketing biopharmaceuticals. The largest pharmaceutical companies are compiled in Table 17.5. Also given are the largest biotech companies which have their core activity in development, production and marketing of biopharmaceuticals by biotechnology. Biotech drug sales were, in the USA, 40 billion $ in 2006, showing an average annual growth rate of some 20%. R&D investment by US pharmaceutical companies was 14% of sales in 2008.

The strong performance of the core *biotech firms* – small- or medium-sized enterprises – in drug discovery has been appreciated [127]. A boost in the foundation of new companies followed during the 1980s and 1990s, first in the USA, then later in the 1990s in Europe (see Chapter 6 and Section 7.5.7). Start-ups were founded often on a key technology platform, for example based on peptide display and novel combinatory DNA methods (Section 10.4). In 2000 there were around 1.273 new biotech companies with 162 000 employees and 21 billion $ turn over in USA, 1570 new biotech companies with 61 000 employes and 7.7 billion $ turn over in Europe (data from and analysis of companies with a maximum of 500 employees in 2000) [128].

Economic data for the biotech firms (not including major pharmaceutical companies (Table 17.5) have been collected by Lähteenmäki and Lawrence [39, 40]. Biotech companies, whose shares are traded publicly on stock exchanges – 309 firms – generated almost 47 billion $ in revenue in 2004, 404 firms generated more than 63 billion $ in 2005 in the USA, accounting for 30% of the revenue, and 20% of R&D spending, an increase by 11% over 2004. Venture capital played a significant role [133]. Initial public offering (IPOs) showed a remarkable shift in so far as, in 2005, only about one-third were companies based in the USA, whereas, in 2004, it was six out of ten.

Data from Ernst &Young indicate that over the past few years biotech firms have become the primary source of new medical entities (NME) approvals. In 2005 major pharmaceutical companies garnered only 11 NME approvals whereas biotechs had 18, with a research budget only about one-quarter the size of that of the pharmaceutical industry. Another representation would be that the small innovative

Table 17.5 (a) The largest pharmaceutical companies, with sales in 2008 and 2007 [89]; (b) the largest biotech companies with sales [89, 129].

(a) Pharmaceutical companies	Sales (billion US $) 2008 (2007)	(b) Biotech companies	Sales (billion US $) 2008 (2006)
Johson & Johnson (USA)	63.7 (61.1)	Amgen Inc. (USA)	14.7 (12.0)
Pfizer (USA)[a)]	48.3 (48.4)	Genentech (USA)[c)]	10.5 (7.6)
GlaxoSmithKline (GB)	45.2 (42.2)	Novo (DK)	(3.7)
Wyeth (USA)[a)]	44.8 (22.4)	Gilead Sciences Inc.	4.7[f)] (3.0)
Roche (CH)	42.2 (42.6)	Genzyme Corp. (USA)	4.2 (2.9)
Novartis (CH)	41.5 (38.1)	Serono[d)] (CH)	(2.8)
Sanofi-Aventis (F)	40.6 (41.3)	CSL Ltd.	(2.2)
Astra Zeneca (GB/S)	31.6 (29.6)	Chiron[e)] (USA)	(1.9)
Abbot Laboratories (USA)	29.5 (25.9)	Biogen Idec Inc. (USA)	2.8 (1.8)
Merck & Co. (USA)[b)]	24.9 (24.2)	Cephalon Inc.	(2.1)[g)] (1.7)
Eli Lilly (USA)	20.4 (18.6)	MedImmune Inc.	(1.3)
Bristol-Myers Squibb (USA)	20.1 (19.3)	Celgene Corp.	(2.5)[g)] (~1.2)

a) Wyeth has been acquired by Pfizer in 2009 for 68 bn $; the combined sales are expected to be 75 bn $; 10% of the jobs will be cut [130].
b) Merck & Co. will take over Schering-Plough (USA) for 41 bn $. In 2008 Merck & Co. had sales of 24.9, Schering-Plough 18.5 bn $, so that both together will total some 43 bn $ [131].
c) Taken over by Hoffman La-Roche AG, CH, in 1990, with a majority share of 60%, with full control and integrated into the company in 2009.
d) Taken over by Merck KGA, Germany, in 2006.
e) Taken over by Novartis, CH, in 2005/2006.
f) Company report.
g) 2009, estd., [132], see also Company report.

biotech was some sixfold more efficient than the established industry [39, 40]. A most important market is that of *diagnostics*, with sales of about 35 billion $, dominated by Roche with 9.2 billion $ in 2008, followed by Abbot, Johnson & Johnson and Bayer [95, 134]. For biopharmaceuticals in the pipeline and in clinical trials see Section 9.7. Interesting trends in big companies concerning the development of new pharmaceuticals with respect to licencing have been identified (Table 17.6) [157].

The marketing of *generics*, (biosimilars, or follow-on, or generic medicinal products), that are copies of the original biopharmaceuticals after patent protection has expired, represents a major economic relief for patients and their health insurance companies, but a threat to big pharma. Some ten companies world wide are developing and marketing biosimilars for prices that are typically lower by around 30%. These are (with approximate sales, in billion $, 2007): Teva (Israel) (4.3), Sandoz/Novartis (CH) (including Hexal) (3.3), Mylan (USA) (2.2), Ratiopharm

Table 17.6 Projects under development, in house project versus licenced products [157].

Company	Total (numbers)	In house (numbers) (%)		Licenced (%)
Roche[a]	53	32	60	40
Aventis[b]	45	16	36	64
Glaxo Smith Kline	45	10	22	78
Johnson & Johnson	29	21	72	28
Pfizer	27	3	11	89
Bristol-Meyers Squibb				79
Merck AG (Germany)				86

a) Roche in total sells 14 biopharmaceuticals and owns one of the most promising biotechnology pipelines [60].
b) Now part of Sanofi-Aventis.

(Germany), Barr (USA), Watson (USA), Actavis (Germany) and Stada (Germany) (all with sales of 1 to 1.2 billion $) [135]. The FDA currently has no clear framework to approve generics. The EU, in contrast, has put forward a legislation that allowed the launch of a few biosimilars. EMEA requires for approval a considerable body of material concerning comparable clinical behaviour, with associated cost up to 200 million $. It had accepted 10 applications for approval and about 5 were meanwhile approved. The first approval in 2006 by EMEA was a biosimilar of EPO (Omnitrope of Sandoz), followed by several others in 2007, the price being about 75% of the originals [7, 134]. The sheer size of the EPO market has led to the largest number of patent disputes in the history of the US biotech market.[4] It is estimated that by 2010 biosimilars will have a market share of some 20% of biopharmaceuticals [10, 57, 134, 136].

17.6.3
Mergers, Acquisitions, Alliances and Cooperation

Vary large-scale mergers and acquisitions changed the situation of pharmaceutical companies during the 1990s a great deal. In many significant examples a schism occurred in the chemical-pharmaceutical industry between 'life-sciences' and 'chemistry' [137]. One significant event is the decline, even *eclipse* of German pharmaceutical companies, earlier termed 'the pharmacy of the world'. During most of the twentieth century, Hoechst AG and Bayer AG were the largest pharmaceutical manufacturers, at least till the early 1980s, ending as minor players (Hoechst, after merger into Aventis, in position 8, Bayer in position 16 in 2001/02) [138, 139]. One reason seems to be the hubris of the management relying on chemical synthesis, allowing a nonchalant neglect of biotechnological research, development and production. Another factor was the emergence of giant specialized pharmaceutical companies by the mergers and acquisitions mentioned. Further factors are related to an explosive increase in the cost of research, development, clinical testing, approval and marketing of new drugs. The blockbuster-mania then led to high profits, the

Table 17.7 Mergers and acquisitions of large pharmaceutical companies during the 1990s (data in billion (bn) $).

Bristol-Myers and Squibb merge to form Bristol-Myers Squibb (BMS) Company (USA) in 1989 [140]

Roche (CH) takeover of majority of Genetech in 1990

Novartis formed out of life-sciences activities of Ciba-Geigy ad Sandoz all CH) in 1996 [139, 141].

Roche (CH) acquired Boehringer Mannheim (Germany), in 1997, to become the world's largest diagnostics manufacturer [139]

BMS acquired DuPont Pharma (both USA), in 1997, for 7.8 bn $ [141, 142]

AstraZeneca (UK/S) formed by a merger of Astra and Zeneca, announced in 1998, with expected sales of 6.3 bn $ [142]

Pfizer (USA) aquired Werner–Lambert in 2000, and Pharmacia (Sweden), in 2003 for an estd. 50 bn $ [141, 142].

Aventis (F) was formed, in 1999, by the merger of Hoechst AG (Germany) and Rhône-Poulenc (F), with sales epected to reach 15.4 bn $ [143]

Upjohn (USA) and Pharmacia (Sweden) merged in 1999 [141, 142]

Pfizer acquired Warner-Lambert (USA), in 1999/2000, with expected combined sales of 28 bn $ [144, 145]

GlaxoSmithKline plc (UK) formed by the merger of GlaxoWellcome and SmithKline Beecham at the end of 2000 – 'The worlds biggest ever corporate merger . . .'. Sales of 9.1 bn $ per year were expected [146–148].

source of ongoing acquisitions. Important mergers are compiled in Table 17.7. Remarkable successes were achieved by the Swiss chemical and pharmaceutical companies in building two world leaders in pharmaceuticals, Novartis and Hoffman-La Roche. Ciba and Geigy merged in 1970, Ciba-Geigy then merged with Sandoz to form Novartis in 1996. Roche paid 2.1 bn $ in 1990 for its majority stake in Genentech, and later acquired Boehringer Mannheim, building up a highly successful centre for protein biopharmaceuticals in Penzberg (Germany).[5] The sales of Swiss companies in pharmaceuticals went up from some 3 billion in 1980 to about 39 billion $ in 2006 [139].

Currently, most large companies try to overcome the difficulties mentioned by continuing acquisitions, alliances or cooperation. Major examples of *acquisitions*, ranging from some 10 to nearly 70 billion $, should be mentioned: the Bayer takeover of Schering (both Germany), in 2006, for 24.4 bn $; Schering-Plough (USA) take over of Akzo Nobel's (Netherlands) pharma division in 2007, for 16.3 bn $; Merck KGaA (Germany) acquisition of Europes biggest biotech company Serono (CH), in 2007, for 15.7 bn $; Roche's takeover/full control of Genentech, in

---

5) Genetech's sales mushroomed from about 400 mn (million) $ in 1990 to 11.7 bn $ in 2007; Roche initially had agreed to let the biotech company operate independently to maintain its enterprising spirit; the company had been formed over beers at a pub in 1976 by Herbert Boyer of the University of California and venture capitalist Robert A Swanson [124, 125].

2008 for 90 bn $; Pfizer's announcement of the take over of Wyeth in 2009 for 68 bn $; Merck & Co.'s takeover of Schering-Plough in 2009 for 41.1 bn $; considerably more deals took place also in Japan and India [39, 40, 127, 129, 134, 149–151]. Trends are obvious from recent aquisitions of SIRNA by Merck & Co. (for 1.1 billion US$) and Alnylam Europe by Roche (800 million US$) showing the current interest of big pharma in oligonucleotide drugs (including RNA interference (RNAi) technology) [27, 152]. Generics company Teva (Israel) acquired Ratiopharm (Germany) for nearly 5 bn $ in 2010, extending its sales to 16,2 bn $ [158].

Amongst *alliances, and cooperations*, major deals have been: a cooperation on antibodies between Pfizer and Medarex (both USA) in 2004 ($ 510 million), and between Pfizer and Incyte (USA) on an antagonist for rheumatoid arthritis ($ 803 million). Novartis has formed a 10-year alliance with MorphoSys (Munich, Germany) on antibody technology, which could be worth more than 1 billion $ [122, 153].

China and India recently received much attention with respect to cooperation and alliances, a few examples should illustrate the trend. Eli-Lilly & Co. inaugurated a research partnership with Chinese Hutchinson MediPharma, aiming not at outsourcing, 'but rather about sharing the risks and rewards of pharmaceutical research'. Chinese Hutchinson MediPharma signed discovery partnerships with Merck KgaA (Germany) and Procter and Gamble in 2006. In India Eli-Lilly & Co is collaborating in drug discovery with Nicholas Piramal [154]. Merck & Co. provided two targets for new cancer drugs to be developed by Nicholas Piramal India LTD, who will earn as much as 175 mn $ in milestone payments for each target, as well as potential future royalties [155]. Such collaborations take account of the lower salaries of a highly skilled labour-force in China and India.

17.6.4
Conclusions and Future Prospects

Conclusions and future prospects are discussed in detail in Chapter 13. Here two aspects should be mentioned in addition.

Recent developments by a Japanese group suggest that biocontainer transport might be used to deliver drugs, or deliver gene therapies directly to their target (Figure 17.8) [117]. A molecular motor (myosin) would effect transport along actin filaments – cellular highways inside cells – via a polysaccharide biocontainer – schizophyllan – that can form complexes with various cargoes, for example a carbon nanotube to include drugs, or DNA (Figure 17.8) [117].

The Human Genome Project (HGP) was achieved in 2000 with great effort and money, by a major international consortium (cost more than 400 million $), and by Craig Venters group, and it was communicated in the presence of the President of the United States. After 10 years expectations, e.g. with respect to drug targeting, have been commented, saying that "... a transformational technology will always have its immediate consequences overestimated and its long -term consequences underestimated", and ..."you may just start to imagine all the projects that will spin-off ..." [159].

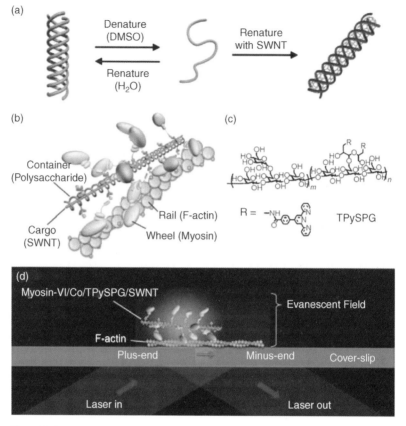

(a)

Denature
(DMSO)

Renature
(H₂O)

Renature
with SWNT

(b)

Container
(Polysaccharide)

Cargo
(SWNT)

Rail (F-actin)

Wheel (Myosin)

(c)

R = TPySPG

(d)

Myosin-VI/Co/TPySPG/SWNT

Evanescent Field

F-actin

Plus-end   Minus-end   Cover-slip

Laser in   Laser out

Figure 17.8 Biocontainer transport with a molecular motor (myosin) along actin filaments, via a polysaccharide biocontainer – schizophyllan – that can form complexes with various cargoes, for example a carbon nanotube to include drugs, or DNA [117].

References

1 Walsh, G. (2005) Current status of biopharmaceuticals: approved products and trends in approvals, in *Modern Biopharmaceuticals*, vol. 1 (ed. J. Knaeblein), Wiley-VCH Verlag GmbH, Weinheim, Germany, pp. 1–34.

2 Becker, T., Breithaupt, D., Doelle, H.W. *et al.* (2007) Biotechnology, in *Ullmann's Encyclopedia of Industrial Chemistry*, Wiley-VCH Verlag GmbH, Weinheim, New York, www.mrw.interscience.wiley.com.

3 Ozawa, J.K. and Franco, Q.C. (2007) The role of government in Health research, in *Comprehensive Medicinal Chemistry II*, vol. 1 (ed. P.D. Kennewell), Elsevier, Amsterdam, pp. 725–752.

4 Reichert, J.M. (2006) Trends in US-approvals: new biopharmaceuticals and vaccines. *Trends in Biotechnology*, **24**, 293–298.

5 Moos, W.H. (2007) The intersection of strategy and drug research, in

Comprehensive Medicinal Chemistry II, vol. 7 (eds J.J. Plattner and M.C. Desai), Elsevier, Amsterdam, pp. 2–83.

6 Melmer, G. (2005) Biopharmaceuticals and the industrial environment, in: *Production of Recombinant Proteins. Novel Microbial and Eukariotic Expression Systems* (ed. G. Gellissen), Wiley-VCH Verlag GmbH, Weinheim, pp. 361–383.

7 Jarvis, L.M. (2009) Biosimilars bet. *Chemical & Engineering News*, January 12, p. 28–29.

8 Mullin, R. (2009) Paying attention to neglected diseases. *Chemical & Engineering News*, April 20, p. 21–24.

9 Dingermann, T. (2008) Recombinant therapeutic proteins. *Biotechnology Journal*, **3**, 60–97.

10 Aggarwal, S. (2007) What's fueling the biotech engine? *Nature Biotechnology*, **25**, 1097–1104.

11 Walsh, G. (2007) *Pharmaceutical Biotechnology*, John Wiley & Sons Ltd., Chichester, England.

12 Thayer, A.M. (2007) Centering on chirality. *Chemical & Engineering News*, August 6, p. 11–19.

13 Rothberg, B.E.G., Pena, C.E.A., and Rothberg, J.M. (2005) A systems biology approach th target identification and validation for human chronic disease drug discovery, in *Modern Biopharmaceuticals*, vol. 1 (ed. J. Knaeblein), Wiley-VCH, Weinheim, Germany, pp. 99–125.

14 Jarvis, L.M. (2007) Christopher Lipinski. *Chemical & Engineering News*, November 5, p. 18.

15 Challis, G.L. and Hopwood, D.A. (2007) Chemical Biotechnology: bioactive small molecules – targets and discovery technologies. *Current Opinion in Biotechnology*, **18** (6), 475–477.

16 Zhang, M.-Q. and Wilkinson, B. (2007) Drug design beyond the "rule-of-five". *Current Opinion in Biotechnology*, **18** (6), 478–488.

17 Höfle, G. (2007) personal communication.

18 Yonath, A. (2005) Ribosomal crystallography: peptide bond formation, chaperone assistance and antibiotics activity. *Molecules and Cells*, **20** (1), 1–16.

19 Steitz, T.A. and Moore, P.B. (2003) RNA, the first macromolecular catalyst: the ribosome is a ribozyme. *Trends in Biochemical Sciences*, **28**, 411–418.

20 Frank, R. (2007) lecture, HZI (Helmholz Zentrum für Infektionsforschung, former GBF), 20.9.07; R. Frank, Die chemische Pipeline. In: Forschungsbericht/Annual report 2006/2007, pp. 32–43), HZI – Helmholtz Zentrum für Infektionsforschung, Braunschweig, Germany.

21 Amzel, L.M. (1998) Structure based drug design. *Current Opinion in Biotechnology*, **9** (4), 366–369.

22 Erhardt, P.W. and Proudfoot, J.R. (2007) Drug discovery: Historical perspective, current status, and outlook, in *Comprehensive Medicinal Chemistry II*, vol. I (ed. P.D. Kennewell), Elsevier, Amsterdam, pp. 29–94.

23 Gabius, H.J. (2006) Cell surface glycans: the why and how of their functionality as biochemical signals in lectin-mediated information transfer. *Critical Reviews in Immunology*, **26**, 43–79.

24 Alavi, A. and Axford, J.S. (2008) Sweet and sour: the impact of sugars on disease. *Rheumatology*, **47**, 760–770.

25 Jarvis, L.M. (2007) Biopharmaceuticals. Novartis gets preliminary nod for new flu vacine. *Chemical & Engineering News*, May 7, 15.

26 Jarvis, L.M. (2007) Refueling a biotech growth engine. *Chemical & Engineering News*, June 22, 22–23.

27 Jarvis, L.M. (2007) Roche joins RNAi drug Fray. *Chemical & Engineering News*, July 16, 8.

28 Jarvis, L.M. (2007) Pharma's mixed bag. *Chemical & Engineering News*, August 20, 32–34; Biotechs rely on old products. pp. 34–35.

29 Jarvis, L.M. (2007) *Chemical & Engineering News*, August 20, 9.

30 Jarvis, L.M. (2007) *Chemical & Engineering News*, August 20, 15–20.

31 Jarvis, L.M. (2007) Pharma's tough balancing act. *Chemical & Engineering News*, November 19, 32–33. Biotech challenges, Chemical & Engineering News, November 19, 34–35.

32 Ullrich, A. (2008) Next wave – multitargeted therapies. *European Biotechnology News,* **7** (1–2), 36.

33 Brorson, K., Swann, P.G., Brown, J. *et al.* (2005) Considerations for Developing Biopharmaceuticals: FDA Perspective, in Modern Biopharmaceuticals, Vol. IV (ed. J. Knaeblein), Wiley-VCH, pp. 1637–1668.

34 Wenzel, A.F. and Sonnega, C. (2005) The regulatory environment for biopharmaceuticals in the EU, in *Modern Biopharmaceuticals,* vol. IV (ed. J. Knaeblein), Wiley-VCH Verlag GmbH, Weinheim, Germany, pp. 1669–1710.

35 (2007) *Nature Biotechnology,* **25**, Editorial, 1375.

36 (2007) *Nature Biotechnology,* Biotech drugs cost $ 1,2 billion., **25**, 9.

37 Thayer, A. (2008) Pipeline woes. *Chemical & Engineering News,* May 5, 12.

38 Rader, R.A. (2008) Paucity of biopharma approvals raises alarm. *Genetic Engineer and Biotechnologist News,* 10–14.

39 Lähteenmäki, R. and Lawrence, S. (2005) Public biotechnology 2004 – the numbers. *Nature Biotechnology,* **23**, 663–671.

40 Lähteenmäki, R. and Lawrence, S. (2006) Public biotechnology 2005 – the numbers. *Nature Biotechnology,* **24**, 625–634.

41 Höfle, G. (2009) History of Epothilone discovery and development, in *Progress in the Chemistry of Organic Natural Products (Zechmeister),* vol. 90 (eds A.D. Kinghorn, H., Falk and J. Kobayashi), Springer, Wien (Austria), pp. 5–15.

42 Höfle, G., Bedorf, N., Steinmetz, H., Schomburg, D., Gerth, K., and Reichenbach, H. (1996) Epothilon A und B – neuartige, 16gliedrige Makrolide mit cytotoxischer Wirkung: Isolierung, Struktur im Kristall und Konformation in Lösung. *Angewandte Chemie,* **108**, 1671–1673.

43 Höfle, G. and Reichenbach, H. (2005) Epothilone, a Myxobacterial metabolite with promising antitumor activity, in *Anticancer Agents From Natural Products* (eds G.M. Cragg, D.G.I. Kingston and D.J. Newman), CRC Press, Taylor and Francis, Boca Raton, pp. 413–450.

44 Höfle, G. (2009) Natural Epothilones, in *Progress in the Chemistry of Organic Natural Products (Zechmeister),* vol. 90 (eds A.D.

Kinghorn, H. Falk and J. Kobayashi), Springer, Wien (Austria), pp. 16–28.

45 Gellissen, G. (ed.) (2005) *Production of Recombinant Proteins,* Wiley-VCH Verlag GmbH, Weinheim, Germany.

46 Knaeblein, J. (ed.) (2005) *Modern Biopharmaceuticals,* Wiley-VCH Verlag GmbH, Weinheim, Germany.

47 Minas, W. (2005) Production of Erythromycin with Saccharapolyspora erythrea, in *Microbial Processes and Products* (ed. J-.L. Barredo), Humana Press, Totowa, New Jersey, USA.

48 Krahe, M. (2003) Biochemical engineering, in *Ullmann's Encyclopedia of Industrial Chemistry,* Wiley-VCH Verlag GmbH, Weinheim, New York, www.mrw. interscience.wiley.com.

49 Anderson, A.S. and Diers, I. (2005) Advanced expression of biopharmaceuticals in yeast at industrial scale: The insulin success story, in *Modern Biopharmaceuticals,* vol. IV (ed, J. Knaeblein), Wiley-VCH Verlag GmbH, Weinheim, Germany, pp. 1033–1144.

50 Li, H., Sethuraman, N., Stadheim, T.A. *et al.* (2006) Optimization of humanized IgGs in glycoingeneered *Pichia pastoris.* *Nature Biotechnology,* **24**, 210–215.

51 Harrison, R.L. and Jarvis, D.L. (2006) Protein *n*-glycosylation in the baculovirus– insect cell expression system and engineering of insect cells to produce "mammalianized" recombinant glycoproteins. *Advances in Virus Research,* **68**, 159–191.

52 Sethuraman, N. and Stadheim, T.A. (2006) Challenges in therapeutic glycoprotein production. *Current Opinion in Biotechnology,* **17** (4), 341–346.

53 Wurm, F.M. (2005) Manufacture of recombinant biopharmaceutical proteins by cultivated mammalian cells in bioreactors, in *Modern Biopharmaceuticals,* vol. 3 (ed. J. Knaeblein), Wiley-VCH Verlag GmbH, Weinheim, Germany, pp. 723–759.

54 Behrendt, U. (2007) Protein purification. Dechema-Conference, Frankfurt, Germany.

55 Gnoth, S., Jenzsch, M., Simutis, R., and Lübbert, A. (2008) Control of cultivation processes for recombinant protein

production: a review. *Bioprocess and Biosystems Engineering*, **31**, 21–39.

56 Zhou, T.-C., Zhou, W.-W., Hu, W., Zhong, J.-J. (2010) Bioreactors, Cell Culture. *Encyclopedia of Industrial Biotechnology*, John Wiley & Sons Inc., New York, http://tinyurl.com/wileyEIB

57 Behrendt, U. (2009) personal communication.

58 Wilson, J.S. (2006) A fully disposable monoclonal antibody manufacturing train. *Bioprocess International*, (Suppl), 4 (6), 34–36.

59 Eibl, R. and Eibl, D. (eds.) (2009) *Disposable Bioreactors. Advances in Biochemical Engineering/Biotechnology*. Vol. 115.

60 *CHEManager*, 17/2007, pp. 13–15.

61 Wahl, J. (2007) *CHEManager*, 17/2007, p. 14.

62 Moosmayer, D. (2007) Lecture SFB-colloquium Braunschweig June 2007; Berichtskolloquium TU Braunschweig, p. 131.

63 Moo-Young, M. (1985) *Comprehensive Biotechnology*, vol. **1 and 2**, Pergamon Press, Oxford.

64 Ludwig, F., Heim, E., Menzel, D., and Schilling, M. (2006) Investigation of superparamagnetic Fe_3O_4 nanoparticles with fluxgate magnetorelaxometry for use in magnetic relaxation immunoassays. *Journal of Applied Physiology (Bethesda, Md: 1985)*, **99**, 08P 106-1-3.

65 Meyer, A., Berensmeier, S., and Franzreb, M. (2007) Direct capture of lactoferrin from whey using magnetic micro-ion exchangers in combination with high-gradient magnetic separation. *Reactive and Functional Polymers*, **67**, 1577–1588.

66 FDA (2007) www.fda.gov/cder/OPS/PAT.htm.

67 *CHEManager*, 19/2007, p. 20.

68 Bud, R. (2007) *Penicillin. Triumph and Tragedy*, Oxford University Press, Oxford.

69 Buchholz, K. (1979) Die gezielte Förderung und Entwicklung der Biotechnologie, in *Geplante Forschung* (eds W. van den Daele, W. Krohn and P. Weingart), Suhrkamp Verlag, Frankfurt, pp. 64–116.

70 Buchholz, K. (2007) Science – or not? the status and dynamics of biotechnology. *Journal of Biotechnology*, **2**, 1154–1168.

71 Elander, R.P. (2003) Industrial production of ß-lactam antibiotics. *Applied Microbiology and Biotechnology*, **61**, 385–392.

72 Hubschwerlen, C. (2007) ß-Lactam antibiotics, in *Comprehensive Medicinal Chemistry II*, vol. 7 (eds J.J. Plattner and M.C. Desai), Elsevier, Amsterdam, pp. 497–517.

73 Drahl, C. (2010) Dirt tells resistance tales. *Chemical & Engineering News*, January 4, 10.

74 Young, L.S. (2007) Bacteriology, major pathogens, and diseases, in *Comprehensive Medicinal Chemistry II*, vol. 7 (eds J.J. Plattner and M.C. Desai), Elsevier, Amsterdam, pp. 469–477.

75 Ohno, M., Otsuka, M., Yagisawa, M. *et al.* (2002) Antibiotics, in *Ullmann's Encyclopedia of Industrial Chemistry*, Wiley-VCH Verlag GmbH, Weinheim, New York, www.mrw.interscience.wiley.com/emrw/.

76 Bruggink, A. (2001) *Synthesis of ß-Lactam Antibiotics*, Kluwer Academic Publishers, Dordrecht.

77 Buchholz, K., Kasche, V., and Bornscheuer, U. (2005) *Biocatalysts and Enzyme Technology*, Wiley-VCH Verlag GmbH, Weinheim.

78 Tischer, W. (1990) Umweltschutz durch technische Biokatalysatoren. Symposium Umweltschutz durch Biotechnik, (Boehringer Mannheim GmbH, ed.): Boehringer Mannheim GmbH, ISBN: 3-88630-130-3.

79 Kirst, H.A. and Allen, N.E. (2007) Aminoglycosides antibiotics, in *Comprehensive Medicinal Chemistry II*, vol. 7 (eds J.J. Plattner and M.C. Desai), Elsevier, Amsterdam, pp. 629–652.

80 Nelson, M.L. and Ismail, M.Y. (2007) The antiobiotic and nonantibiotic tetracyclines, in *Comprehensive Medicinal Chemistry II*, vol. 7 (eds J.J. Plattner and M.C. Desai), Elsevier, Amsterdam, pp. 597–627.

81 Kaneko, T., Dougherty, T.J., and Magee, T.V. (2007) Macrolide antibiotics, in *Comprehensive Medicinal Chemistry II* vol. 7 (eds J.J. Plattner and M.C. Desai), Elsevier, Amsterdam, pp. 519–565.

82 Erickson, B. (2009) Containing swine flu. *Chemical & Engineering News*, May 4, p. 8.

83 Knop, D.R., Draths, K.M., Chandran, S.C. *et al.* (2001) *Journal of the American Chemical Society*, **123**, 10173.

84 Schedel, M. (2000) *Biotechnology*, vol. 8a, Wiley-VCH Verlag GmbH, pp. 295–307.

85 Roche (2003) Roche Lexikon Medizin, Urban & Fischer, Munich, 5th edition.

86 Holland, H.L. (1998) hydroxylation and dihydroxylation, in *Biotechnology*, vol. 8a (eds I. Biotransformations and D.R. Kelly), Wiley-VCH Verlag GmbH, Weinheim (Germany), pp. 475–533.

87 Turner, M. (1998) Perspectives in Biotransformation, in *Biotechnology*, vol. 8a, Biotransformations I, (D.R. Kelly ed.) Wiley-VCH Verlag GmbH, Weinheim (Germany), pp. 5–23.

88 Walsh, G. (2006) Biopharmaceutical benchmarks. *Nature Biotechnology*, 24, 769–778.

89 *Chemical & Engineering News* (2009) 7, July 6, 30–46.

90 *Chemical & Engineering News* (2009) 8, August 24, 20.

91 Ainsworth, S.J. (2007) Pharma adepts. *Chemical & Engineering News*, December 3, 17–29.

92 Lawrence, S. (2007) Billion dollar babies – biotech drugs as blockbusters. *Nature Biotechnology*, 25, 380–382.

93 *CHEManager* (2007) 21, 3, 4.

94 Läsker, K.,(1/2008a), Sueddeutsche Zeitung, January, 23, p. 22.

95 Roche (2008) Geschäftsbericht 2008, F. Hoffmann La-Roche AG, Basel (CH).

96 Djerassi, C. (2008) Doping, aber korrekt. Sueddeutsche Zeitung, 31.5.08, p. 2.

97 Buckland, B.C. and Lilly, M. (1993) Fermentation: an overview, in *Biotechnology*, vol. 3 (ed. G. Stephanopoulos), VCH Verlag GmbH, Weinheim, Germany, pp. 7–22.

98 Kamena, F., Tamborrini, M., Liu, X., Kwon, Y.U., Thompson, F., Pluschke, G., and Seeberger, P.H. (2008) Synthetic GPI array to study antitoxic malaria response. *Nature Chemical Biology*, 4, 238–240.

99 Zhu, J., Wan, Q., Ragupathi, G., George, C.M., Livingston, P.O., and Danishefsky, S.J. (2009) *Journal of the American Chemical Society*, 131, 4151–4158.

100 Lalezari, J.P., Eron, J.J., Carlson, M. *et al.* (2003) A phase II clinical study of the long-term safety and antiviral activity of enfuvirtide-based antiretroviral therapy. *AIDS*, 17, 691–698.

101 Robert-Guroff, M. (2007) Replicating and non-replicating viral vectors for vaccine development. *Current Opinion in Biotechnology*, 18 (6), 546–556.

102 biobanks (2008) www.biobanks.eu.

103 Euro Biotech News (2008) Scientists kick-off EU biobank project. 7, No 1–2, p. 5.

104 Moss, D.W., Maire, I., Calam, D.H., Gaines Das, R.E., Lessinger, J.M., Gella, F.J., and Férard, G. (1994) Reference materials in clinical enzymology: preparation, requirements and practical interests. *Annales de Biologie Clinique*, 52, 189–198.

105 Knaeblein, J. (2005) Executive summary, in *Modern Biopharmaceuticals*, vol. 1 (ed. J. Knaeblein), Wiley-VCH Verlag GmbH, Weinheim, Germany, pp. XLI–CXXII.

106 Bergmeyer, H.U. (ed.) (1983–1986) *Methods of Enzymatic Analysis*, vol. 12, VCH, Weinheim, Germany.

107 Wingren, C. and Borrebaeck, C.A.K. (2008) Antibody microarray analysis of directly labelled complex proteomes. *Current Opinion in Biotechnology*, 19 (1), 55–61.

108 Simon, R. (2008) Microarray-based expression profiling and informatics. *Current Opinion in Biotechnology*, 19 (1), 26–29.

109 Wang, D., Liu, S., Trummer, B.J., Deng, Ch., and Wang, A. (2006) Carbohydrate microarrays for the recognition of cross-reactive molecular markers of microbes and host cells. *Nature Biotechnology*, 20, 275–281.

110 Seibel, J., Hellmuth, H., Hofer, B., and Schmalbruch, B. (2006) Identification of new donor specifities of glycosyltransferases R via the aid of substrate microarrays. *Chembiochem*, 7, 310–320.

111 Hanashima, S. and Seeberger, P.H. (2007) Total synthesis of sialylated glycans related to avian and human influenza virus. *Asian Journal of Chemistry*, 2, 1447–1459.

112 Glycominds (2009). http://www. glycominds.com/.

113 Lavé, T. and Funk, C. (2007) *In vivo* absortion, distribution, metabolism, and excretion studies in discovery and development, in *Comprehensive Medicinal Chemistry II*, vol. 5 (eds B. Teste and

H. van de Waterbeemd), Elsevier, Amsterdam, pp. 31–49.

114 Tillement, J.-P. and Trembley, D. (2007) Clinical pharmacokinetic criteria for drug research, in *Comprehensive Medicinal Chemistry II*, vol. 5 (eds B. Teste and H. van de Waterbeemd), Elsevier, Amsterdam, pp. 11–28.

115 Veronese, F.M. and Pasut, G. (2007) Drug-polymer conjugates, in *Comprehensive Medicinal Chemistry II*, vol. 5 (eds B. Teste and H. van de Waterbeemd), Elsevier, Amsterdam, pp. 1043–1067.

116 Borman, S. (2007) Proteins from birth to death. *Chemical & Engineering News*, August 27, 34–38.

117 Tsuchiya, T., Komori, T., Hirano, M. *et al.* (2009) A Polysaccharide-based container transportation system powered by molecular motors. *Angewandte Chemie (International Edition in English)*, **122**, 736–739.

118 *CHEManager* (21/2007b) p. 28 (source: IMS Health).

119 Holt, B. (2008) Aktuelle Ratings für die Pharmabranche. *CHEManager*, 6/2008, p. 27.

120 Demain, A. (2002) Prescription for an ailing pharmaceutical industry. *Nature Biotechnology*, **20**, 331.

121 *Chemical & Engineering News*, 12/2007, December 10, p. 10, 11.

122 *Chemical & Engineering News*, 12/2007, December 24, p.8.

123 Mullin, R. (2007) Agents of change. *Chemical & Engineering News*, June 25, p. 27–30.

124 Jarvis, L.M. (2008) *Chemical & Engineering News.*

125 Jarvis, L.M. (2008) Roche's reversal. *Chemical & Engineering News*, July 28, 13.

126 Thayer, A.M. (2009) Drug firms face down economy. *Chemical & Engineering News*, May 18, p. 24–26.

127 Newton, C.G. (2007) The role of small- or medium-sized enterprizes in drug discovery, in *Comprehensive Medicinal Chemistry II*, vol. 1 (ed. P.D. Kennewell), Elsevier, Amsterdam, pp. 489–524.

128 Ernst and Young (2001) cited in Jahresbericht, p. 24, VCI (Verband der chemischen Industrie), Frankfurt, 2001.

129 Heinemann, A., Bressan, F., and Röhm, T. (2007) Nachhaltigkeit der Biotechnologie in Deutschland. *CHEManager*, 16/2007, p. 1, 4, 11/2007, 27.

130 *CHEManager*, 1/2009, Jan. 29, p. 1.

131 FAZ 3/2009, Frankfurter Allgemeine Zeitung, Frankfurt (Germany), March 10, p. 11.

132 *Chemical & Engineering News*, 11/2009, Nov., p. 22.

133 Lee, D. and Dibner, M. (2005) The rise of venture capital and biotechnology in the US and Europe. *Nature Biotechnology*, **23**, 672–676.

134 Läsker, K. (2008) Sueddeutsche Zeitung, 19 May 2008, p. 20; 27 May 2008, p. 23.

135 Läsker, K. (2008) Milliardenfusion in der Pharmabrache. Sueddeutsche Zeitung, July 19, p. 27.

136 Kaltenbach, T. and Braeß, A. (2008) Entwicklung von Biosimilars gewinnt an Bedeutung. *CHEManager*, 9/2008, p. 11.

137 Schweizer, R.W. (1998) Solving a split personality. *Chemistry & Industry*, 969–975.

138 Hoffritz, J. (2002) Pillenknick der deutschen Art. Die Zeit, Nov. 21.

139 Short, P.L. (2008) Building a towering success. *Chemical & Engineering News*, March 17, p. 26–29.

140 BMS (2009) Company report.

141 Klinge (2002) Chemie- und Pharmamärkte im Wandel. *CHEManager*, 22/2002, p. 21–32.

142 *Chemistry & Industry* (2010) March 1, p. 29., (1999) March 1, p. 166.

143 *Chemistry & Industry* (1999) Nov. 15, p. 869.

144 *Chemistry & Industry* (1999) Nov. 15, p. 863.

145 *Chemistry & Industry* (2000) Feb. 21, p. 125.

146 *Chemistry & Industry* (1998) Feb. 2, p. 71.

147 *Chemistry & Industry* (2000 Jan. 1, p. 43.

148 *Chemistry & Industry* (2001) June 4, p. 327.

149 *Chemical & Engineering News* (2008) April, p. 10.

150 *CHEManager*, 6 (2009) p. 5.

151 SZ 1/2009, Sueddeutsche Zeitung, January 14, p. 21.

152 Klein, J. (2007) Die nächste Welle an Biotech-Wirkstoffen: Oligonucleotide. *Laborwelt*, **8** (5), 32–33.

153 *Chemical & Engineering News* (2007) 10, October 29, p. 6, 17.

154 Tremblay, J.-F. (2007) Lilly works the world. *Chemical & Engineering News*, 39.

155 *Chemical & Engineering News* (2007) 11, November 26, p. 10.

156 Rolletschek, A. and Wobus, A.M. (2009) *Biol. Chem.*, **390**, 845–849.

157 Schulte, M. (2007) Merck AG, Darmstadt, (Germany), Lecture, SFB-Colloquium, Braunschweig, 29.1.07.

158 Chemical & Engineering News (2010), March 22, p. 25.

159 Chemical & Engineering News (2010), June 21, p. 30.

18
Plant Biotechnology

18.1
Introduction

The promise of transgenic plant, or green biotechnology (BT) is to create crops with higher yields, which can grow on less fertile land, to feed a growing, hungry world population. Crops should be resistant to pests and need less chemicals, notably less insecticides, fungicides, herbicides and fertilizers. The use of genome analysis and quantitative trait linkage has not only accelerated plant breeding programs, but allows definition of new cultivars that need standards for patent protection even though no recombinant DNA methodology was directly applied, i.e. the plant concerned is not a GMO. The majority of agricultural scientists are convinced that crops of high yield, high quality, low cost, and low environmental impact can be delivered by the exploitation of the techniques for plant biotechnology, in particular by involving molecular breeding strategies ([1], Foreword). *Absolute Food production* has risen considerably over decades, a 'Green Revolution', notably in developing countries, even though the proportion of the population directly involved in food production continues to fall. However, the increase in per capita food supply has been small, as a result of a sharp growth in the world population [2]. Hence, research in genetically modified (GM) food production is considered a necessary part of the strategies to ensure adequate nutrition. However, debates over the risks of the technology have, from the beginning, evoked conflicts, and created critical, even negative publicity, particularly in Western Europe (see also Section 8.1).

Transgenic plant biotechnology involves 'the introduction of foreign genes into economically important plant species, resulting in crop improvement and allows the production of novel products in plants' ([3], p. 6). Another definition states that it involves technologies enabling the isolation, amplification, insertion and/or activation of genes from unrelated species, thereby changing the molecular makeup of the host organism, commonly understood by the term 'genetically modified' plants [4].

Relevant *aims* of research and application include enhanced yield (starch, sugar etc.) per hectare, the quality of the products (e.g. amylose vs. amylopectin content in starch production; an optimized fatty acid spectrum of plant oil; enhanced vitamin content in cereals; wood quality for industrial purposes) and crop protection. Crop

Concepts in Biotechnology: History, Science and Business. Klaus Buchholz and John Collins
Copyright © 2010 WILEY-VCH Verlag GmbH & Co. KGaA, Weinheim
ISBN: 978-3-527-31766-0

protection has been a general and major tool in order to provide food in all countries. The most successful, in terms of application, has been herbicide resistance, where reduced utilization of herbicides has been claimed. Resistance against diseases (pests) has been one of the priorities in designing GM plants. This also implies a decrease in the dependence on chemical pesticides. Breeding for disease resistance represents a classical means of crop protection, feasible in host–pathogen systems of significant economic importance, but requires long term commitment of resources. Genetic modification (GM) has been shown to provide an additional and less time-consuming tool, and to be a useful means of creating novel resistance, compared to classic breeding that led to, for example, decades of selecting for aphid- (carrier of plant viruses) resistant potatoes, that yielded only bitter varieties. Further aims are to improve the response to, and tolerance of, a number of biotic and abiotic stresses, such as salt, drought and oxidative stress [1, 2, 5].

'*Genes to gasoline*' – such headlines provoke expectations for second-generation biofuels.

Secondary metabolites represent further topics of interest; enhancing the content of, or improving the spectrum of vitamins, amino acids in proteins, and other ingredients relevant for food quality and health have found attention in research on transgenic plants. Future expectations have been evoked with respect to pharmaceuticals of high, and specific (e.g. favourable glycosylation pattern) quality at low price. However, major problems arise due to the complexity of plant genetics in several respects.

Heated controversies and media attention have affected the discussion on GM crops – few novel technologies have provoked so much opposition, and consequently governmental control, as did plant biotechnology (BT). The critical public perception and resistance of consumers, throughout the world, has slowed down, and even prevented applications, notably in Europe. A real sense of outrage has been expressed by environmental, consumer, and political groups (see also Section 8.1).

18.2
Political, Ethical and Biosafety Aspects

18.2.1
Political and Ethical Aspects, Patenting

Plant BT is subjected to major political, economic, social, ethical and environmental scrutiny, much more than other fields of BT. It is impossible to cover all of the many aspects and divergent trends; therefore a few important issues have been selected for summary; a considerable number of aspects remain controversial. The capacity of transformation to create genetic variability has been demonstrated in many crops. Its routine use, however, has been limited to only a few species, in part due to economic reasons, but also restricted in many countries, notably European, by consumer resistance. There is significant public confusion and concern over the safety (with respect to health and the environment) of GM crops, especially foods, even in the case of disease resistance ([1, pp. 319–342 [5, 6]).

Environmental, consumer, and political groups express outrage at the introduction of this technology, citing concerns about it being a new technology and thus with a potential for undetected risks, or disgust at the profit-seeking activities of the owners of the technology. One overriding concern with BT is its rapidly changing economic structure. In the past 20 years governments around the world have extended private property rights for new BT, for genes and gene constructs and (in some countries) for the entire plants transformed by new genes. Today more than half of the total global agrifood research activity is privately funded and directed. Current corporate concentration poses some potential problems. Many development agencies and non-governmental organizations are concerned that these large consortia command significant market shares and can extract significant profits at the expense of farmers and consumers (see Section 18.5). Another reason for unease is that most of the social institutions related to biotechnology suffer from a deficit of public trust, which is partly due to opaque processes and limited opportunity for public involvement in decision-making. This latter point may be seen differently in the EU, where environmental groups and green parties gained considerable influence in such processes. The debate also opened a rift between wealthy industrial consumers, and some consumers and governments from less developed countries that see the application of BT as one of their best opportunities to increase food supplies and reduce malnutrition. Some express concern that this technology may exacerbate many of the trends started by the Green Revolution – that is, destabilizing societies by the infusion of technology and capital into traditional agricultural systems in peasant economies [4].

The International Assessment of Agricultural Science and Technology for Development (IAASTD), initiated by the World Bank in 2005, brings together 400 experts from government, non-governmental organizations, companies and academia to evaluate science and agriculture to address global poverty and hunger. This initiative has been criticized by seed companies as underestimating the valuable role of modern science [7]. However, yield improvement has been limited up to now, except for cotton and maize, and to a lesser degree soy bean. Claims for providing sufficient food by GM crops are not realistic. Significant potential would comprise a combination of classical methods of agriculture and transgenic crops. This would emphasize agriculture with respect to regional aspects and needs [8].

Charges have been made about the degree of profit-motivated 'control' exercised by agrochemical companies over small farmers, who will not be able to afford the new products, and will lose out in the increased competition, particularly in poorer countries. A key question is how trade-offs between risk and benefit are to be evaluated: a value judgement that will involve the weighing of potential risks against potential benefits [4].

In the European Union (EU) a *de facto* moratorium existed on genetically engineered products. This was officially lifted in 2004, when the European Commission cleared Syngenta to market GM maize (genetically modified for protection against corn borers) that is permitted in 13 non-EU countries for food use. It was the first EU approval of a GM product in more than five years [9]. Approval by the EU of GM maize of Pioneer Hi-Bred, as well as BASF Plant Science's GM potato Amflora (producing starch for non-food industrial applications), has been delayed in May 2008 ([10], see also http://www.biosicherheit.de/de/kartoffel/staerke/32.doku.html, 2008).

The possible failure of industry, researchers and public authorities to address concerns has been discussed by Lassen *et al.* [11]. An Asian perspective has been discussed by Teng [12]. GM crops cultivated in significant areas in four Asian countries with government regulatory approval are cotton, corn and canola. The issues that concern regulators and the public are, similar to other countries, biosafety, food safety, ethics and social justice. The International Food Policy Research Institute (ifpri) advocates advancing agriculture in developing countries, where most people are involved in farming, in order to find sustainable solutions for ending hunger and poverty [13].

A major *crisis* occurred in mid 2008, when food prices increased dramatically. Growing cereals and rape seed for biofuels was thought to be partly responsible. Mainly poor people in developing countries suffered and protested. Riots occurred around the world, in two dozen countries, for example in Haiti and Bangladesh [14, 15]. When the global financial crisis hit in late 2008 food prices declined, and agricultural concerns lost their sense of immediacy. However, critics argue that the food crisis is not over [16, 17].

Ethical aspects, a complex matter, raise fundamental concerns from unease that the technology interferes with the workings of nature and creation, that it involves irresponsible risk-taking for commercial profit, or that it exploits and harms vulnerable individuals and communities. An intrinsic moral point of view is that modern BT is trying to 'play God'. Ethics can be a useful tool in examining the basis of such concerns, to probe the justifications that are offered for moral claims, and to encourage judgements to be made on a more rational and considered basis. Extrinsic (non-moral) concerns focus on two main areas – safety and socioeconomic effects. Change challenges many social norms, and many see it as a threat. In determining the ethics of biotechnology, one must consider: the distribution of winners and losers, who owns and controls the technology (see Section 18.5); food safety and consumer acceptance; and the impact on the environment[1] [4].

An emotive and complex issue is that of *patenting*. The moral concern expressed is that it is wrong to think of 'life forms' as objects that can be 'invented' and 'owned'. However, as discussed in more detail in Chapter 14, it is not the natural product in a form found in nature that is the subject of these biotech patents, but the result of human technical intervention resulting in a new product of commercial value. In contrast, plant genetic resources, as defined by the Food and Agriculture Organization (FAO), are based on the original natural distribution of genetic materials before the intervention of man [19]. Another issue raised in the controversy is that patents could result in restriction of the free exchange of genetic material and the flow of information about it. On the other hand it can be argued that there is an obvious ethical justification for patenting in terms of fairness and justice. If a new product is developed at great expense is it fair or just to deny the inventors recompense for their work and investment? New techniques and products, it is claimed, are needed to feed hungry people and to start to redress the balance between the food supplies available to rich and poor countries, for example with 'Golden Rice' (see Section 18.3.4) [4].

---

1) In an early approach, an extended discussion of issues involved in *green BT* was organized by the Science Center Berlin (Wissenschaftszentrum Berlin), including aspects of law, ethics, sociology, a large range of risks, and molecular biology [18].

18.2.2
Risks and Environmental Aspects

The discussion of *risks* in BT in general goes back to the formulation of recombinant DNA (rDNA) guidelines in Asilomar in 1975; the impact of continuing debate that resulted in the guidelines has been analyzed by Bud [20] and, with respect to plant BT, by van den Daele *et al.* [18] in detail[2] (see Section 8.1). Recently, Ammann [8] discussed the risks of agriculture with GM crops, stating that no modern agriculture is without any risk. However, true accidents, or casualties, due to consumption have only occurred with non-GM food up to now. The Cartagena-protocol on biological safety, that is applied by EU countries, unfortunately, investigates only GM crops, whereas in the USA and Canada all novel crops are submitted to investigation. Thus organic agriculture also uses the *Bacillus thuringiensis* toxin. Further, it must be considered that fungal toxins are a real threat. GM insect-resistant Bt corn is 'demonstrably safer' than regular corn, due to the diminution of mycotoxins, which in turn is due to the reduced damage by insects to the GM plant and, hence, reduced entry sites for fungi [21]. Proponents point to the demonstrated benefits of reduced chemical usage. A 2002 review of 40 major studies concluded that the adoption of eight biotechnology cultivars in the USA increased crop yields by 2 million tonnes, lowered annual growers production costs by 1.2 bn $, and reduced pesticide use by 20 000 t [4]. Allergenicity could be considered to be a risk of GM crops. A detailed analysis of allergenic risks is limited since not all tests currently being applied have a sound scientific basis [22].

Driven by public sensivity to possible biodiversity change following pollen (seed)-mediated transgene flow, some research has been performed and there are now data on the natural prevalence rates and also on the diversity and potential fitness impacts of a handful of viruses that infect crop relatives. However, there is still too little general knowledge to justify broad generalizations about potential 'ecological/biodiversity' impacts. Thus perceived risk concerns the possible role that transgenic plants might play in the enhancement of virus diversity as a result of increasing virus resistance by expressing viral genes, for example virus capsid protein in GMOs. Whenever viral sequences are expressed in plants there is the possibility of new genome combinations resulting from recombination between the transgene transcripts and infecting viral RNA [24]. There is concern about the incorporation of antibiotic marker genes in direct gene transfer (DGT) vectors as they will subsequently be present in the genome of the transgenic plants. This has motivated changes in the design of transformation vectors, and recent strategies substitute antibiotic resistance genes for other markers such as D-amino acid oxidase. Other techniques available include site specific recombination and, preferably, cotransformation [25–28]. According to Ammann [8], amongst the

2) The improvement of crop performance by genetic engineering requires that inserted genes are expressed at the desired level, in the intended tissue, at the required stage of development, and/or following exposure to a specific stimulus. In addition, the transgenes should be faithfully inherited accordingly to Mendelian principles [23].

many studies on food safety that were undertaken, no one has shown horizontal gene transfer. In more than 10 000 publications on the influence of GM crop agriculture no negative effects on biodiversity were identified. Risk concerns remain associated with long term effects, where no knowledge exists, transgenic crops having been investigated thoroughly for 10–15 years. In summary, the focus must be on an analysis of the balance of risk and advantage, however difficult this task may be [8].

With regard to *environmental aspects* Aslaksen and Ingeborg Myhr [29] rhetorically ask 'how much is a wildflower worth?', in order to critically discuss the application of cost-benefit analysis in evaluating environmental impacts of GM crops. The analysis should be supplemented with other methods, such as processes for assessing uncertainty, accommodation of scientific disagreements, and integration of stakeholders' interests and perspectives. The perspective is to recognize the multi-dimensional nature of environmental qualities and risks, such as irreplaceability, irreversibilty, uncertainty and complexity.

Agricultural biotechnology, with GM crops, has extended the dominant 'agri-industrial paradigm', counterposed to the alternative 'agrarian-based rural development paradigm', and led to the controversy over the prospect that GM crops would become inadvertently mixed with non-GM crops. In response, the European Commission developed a policy framework for 'coexistence' between GM, conventional and organic crops, in order to ensure that farmers could freely choose amongst different systems, yet this has become another arena for contending agricultural systems, which may not so readily coexist in practice [30]. Munro [31] raises the concern that the spatial impact of GM crops may reduce, or even eliminate, the planting of organic varieties and other crops where consumers have a preference for non-GM crops. In this context it is relevant to distinguish between those GM that find natural crossing partners in their vicinity, such as canola/rape seed and those that do not readily outcross to wild species such as maize [4, 28]. A considerable number of gene containment strategies in transgenic crops have been developed, including maternal inheritance, male sterility, seed sterility, and others. However, no strategy has been broadly applicable to all crop species, and combinations of approaches may prove to be the most effective solution [32].

A review of worldwide experimental field research and commercial cultivation, covering 1997 till 2007, provides no evidence that the cultivation of the presently commercialized GM crops has caused environmental harm. The study comprised data for maize, oilseed rape, and soybean, with herbicide tolerance and insect resistance as the two main traits [33]. In Europe, the EUMon project aims at monitoring possible adverse effects of GM crops on biodiversity by sampling metadata (i.e. condensed data from previous data studies or databanks) [34].

An analysis of benefits of growing herbicide (glyphosate) resistant (Roundup Ready) soybean of 1998/1990 claimed to be of net benefit to the environment, since the total use of herbicides decreased, and the producers of herbicide tolerant crops (with respect to glyphosate and Basta) claim as well that the adoption of their crops will have a positive effect on the environment. Other concerns have, however, been raised, such as that 'genetic pollution' might result from the spread of indestructible weeds, resistant to pests and herbicides (super-weeds). Further fears have been voiced that GM crops might lead to a loss of genetic diversity, including efficient removal of weed species, and a loss of wildlife habitats, although this has been a general side effect of intensive monoculture practices before ([1], pp. 105–132, [4]).

18.2.3
Regulations

Regulations are a highly complicated matter; they have been dealt with in detail by ([1], pp. 331–342) and Halford [35]; they are treated here only briefly. The application of GM crops is determined by the regulatory framework that is applied to their growth and processing. Regulations should, first of all, ensure that the GM crops and products derived from them are safe, both to the environment and to human and animal health. They, however, cover additional aspects and public concerns, such as labelling of GM foods and GM-derived products so that consumers can exercise choice over whether to eat GM food or ingredients. In fact all of the GM foods on the market have been subjected to extensive regulatory review in one or more countries, often involving between 3 and 7 years of testing and evaluation [4].

Regulators in the EU, Australia and New Zealand, in contrast to those in the USA, have expressed concerns about uncertainties in the science, and have postponed most approvals in recent years. In the EU countries, national laws implement various Council Directives and regulations concerning GM crops and products. These include a directive on the deliberate release of GMOs into the environment, the requirement for extensive environmental risk assessment, approval for research trials, approval for marketing releases, and others. One regulation concerns the traceability and labelling of GMOs and food and feed products from GMOs, in order to ensure the safety of biotech plants and foods and to give the consumer freedom of choice. Products should be labelled as containing GM material unless it is less than 0.9% admixture. Due to the *de facto* moratorium on licensing new GM products in the EU, and due to the activity of various pressure groups, the amount of GM crops grown in the EU is very small compared with worldwide production ([1], pp. 331–342, [4]).

The USA, Canada, Mexico and Japan generally make consistent safety rulings based on science and have approved most of the new GM products for production and consumption. Consequently, the regulators in the USA, Canada and Argentina have concluded that mandatory labels will only be required to signal known health and safety aspects, such as nutritional and compositional change, or new allergens. In the USA, regulations pertaining to GM crops are implemented by three federal agencies, amongst them the US Department of Agriculture (USDA), responsible for testing, and the Food and Drug Administration (FDA) that regulates foods and animal feeds derived from transgenic plants. Genetically engineered foods have to meet the same safety standards as all other foods. However, many GM crops do not need pre-market approval from the FDA because they do not contain substances that are significantly different from those already present in other foods ([1], pp. 331–342, [4]).

Consumers and governments from developing countries have a substantially different perspective. With an estimated 800 million malnourished citizens, many in those countries are vigorous supporters of the new technology. They look to its potential to increase domestic food production and, with new products such as β-carotene-enhanced rice, to substantially reduce malnutrition and disease (see Section 18.3.4) [4]. India was one of the first Asian countries to invest in agricultural biotechnology research and to set up a biosafety system to regulate the approval of genetically modified (GM) crops. However, only one GM crop has been approved, Bt (*Bacillus thuringiensis*) cotton, which was planted on 1.3 million hectares by 1 million farmers in 2006. The GM crop, Bt cotton, is considered to have been an overall success for Indian farmers, referring to ifpri [13]. China is developing an independent strategy to the development of GM crops, relying on local solutions to local problems ([1], p. 340).

18.2.4
Consumer Acceptance and Public Perception

Heated controversy and media attention have affected the discussion on GM crops. The top public concern about GM foods is whether they are safe. While the scientific community has concluded that there are no new measurable or anticipated risks to human or animal health from the crops currently grown, consumers in wealthy countries are not convinced. In the absence of any perceptible consumer benefits, consumers focus on the uncertainties of the new technology. They often express great concern about GM ingredients, yet are for the most part incapable of describing what they dislike. Generally, the response is that they want to be able to choose whether to eat GM foods or not. The ultimate challenge is to provide, despite many difficulties, a credible transparent, and accountable system that provides effective choice to consumers between GM and non-GM food [4, 21].

An important reason for the relatively slow uptake of GM *virus resitant crops* is the perception (particularly in Europe) that too few data are available to form a reliable scientific basis for the assessment of risk to the environment. This, with concerns for public health, is the key issue that must be satisfied under regulations that respond to EU directives [24]. Furthermore, there is public concern about the use of antibiotic resistance markers (see before).

A general problem is that views are being formed on a very limited knowledge base. Moreover, it is argued, that views are formed by interest groups that are prejudiced, for example anti-biotechnology, and that there is a lack of trust in information given out by governments and the big corporations that control (part of) agriculture, both of whom are seen as having their own agendas. Major surveys on public perception have been conducted in the USA by a number of different organizations, and in the UK, with the aim to gauge public opinion in 2003 about GM food and crops. The majority, 85% of the public sampled in the UK, thought that we do not know enough about the potential long-term effects of GM food on our health. Results include for example that only 14% of the responders thought GM food to be a good thing, and 40% thought it was a bad thing (for details see www.gmsciencedebate.org.uk). In the USA, most people do not want the technology to be stopped; rather they want robust regulation put in place to ensure safety and have moral and ethical issues covered in the regulations. Most surveys indicated a 50/50 split for and against GM food. For companies it was bad publicity to impose contracts to prevent farmers from planting harvested seeds in subsequent years or to introduce terminator technology that leads to infertile seed, even though such practices had been common with non-GM cultivars and F1-hybrids ([1], pp. 316–342).

Ultimately, public perception and acceptance comes down to a question of trust, notably in farmers, and in legitimate experts [21, 36]. A robust safety assessment, for example by the International Food Biotechnology Committee, might be reasonable, as it has been presented with a case study on Golden Rice 2. Other considerable biosafety research efforts in response to the critical perception of GMOs include ecological studies, risk/cost/benefit analysis, socio-economic impact studies (see e.g. http://www.ebr-journal.org/; http://www.isbr.info/). They could be an opening for participation and expert–lay interplay [37, 38].

18.3
Research and Development

The scientists view is that 'we all need to work together to feed the world with affordable, nutritious food in an environmentally and socially sustainable manner' shared by many [7].

The majority are convinced that crops of high yield and high quality can be delivered by the exploitation of the techniques of plant biotechnology in molecular breeding strategies. Aims include, for example, improved fatty acid spectrum of plant oil, increased vitamin content in cereals, resistance against diseases (pests) and stress, low cost, and low environmental impact ([2, 39], [1], Foreword). Amongst long term goals are genomic strategies expected to provide clues for unravelling the large resources of cellulosic biomass to produce biofuels [40] (see Section 18.3.2).

18.3.1
General Principles, Methods

Plant BT deals largely with genes that encode proteins, and the relevant regions of the genome (transcription units) that serve as templates ([1], pp. 1–36). As a basis, a primary database – a 'plant integrative omics database' – has been established in the frame of the German initiative for Genome Analysis of the Plant Biological System (GABI). Data from genomics, trancriptomics, proteomics and metabolomics, originating from 14 different model or crop species have been incorporated [41]. A survey of available data bases for gene expression analysis has been presented by Hehl and Bulow [88], for omics approaches by Fukushima *et al.* [89].

Tools for cloning foreign DNA, and the traits associated, into plants have been developed since the end of the 1970s, and understanding of plant genetics has progressed significantly [42]. They comprise genomics (gene mapping, sequencing, functional and expression analysis, data banks, evolution of genome), molecular markers for selection, gene transfer, metabolic engineering, protoplast fusion, regeneration of transformants, and cell culture, with quick propagation. Challenges are the analysis and understanding of sequence differences, and the correlations of phenotype and genotype ([1], pp. 37–76, [43] (see also Section 12)).

Plant *cell and tissue culture* techniques involve the culture of cells, protoplasts, tissues, and organs. Plant *in vitro* technologies are used mainly for the regeneration of somatic embryos for propagation, virus elimination, the transformation and generation of transgenic plants, including the ability to control metabolic pathways of *in vitro* cultured cells and tissues, for example for the production of secondary metabolites. Two concepts, plasticity and totipotency, are central to plant BT. The plasticity allows plants to alter their metabolism, growth and development to best suit their environment. Totipotency (and cell competence) enables propagation in tissue culture, that is, the ability of plant cells to re-express their genetic potential, to undergo dedifferentiation and redifferentiation, and to regenerate new plants. The technology has an advantage over traditional plant propagation, due to its potential for rapid, large scale multiplication of new genotypes ([44], [1], pp. 37–53).

General principles of *gene cloning* in plants and of screening of libraries have been summarized by Twyman [45], Halford [35], [39 and ([1], pp. 54–104). A few genome projects have been completed, including *Arabidopsis thaliana* (as a model species), rice (finished in 2005), canola (in 2009, by Bayer CropScience), 'Golden Delicious' apple (in August 2010) and poplar, *Populus trichocarpa*, a fast growing tree. Further crop-plant genome-sequencing projects include maize, barley, and wheat ([1], pp. 1–36, [40, 46]). Structural annotation and the elucidation of gene functions (functional annotation) is undertaken as well, representing the challenge ahead. The bottleneck is the identification and isolation of specific genes from the large background of sequences in genomic DNA or cDNA [41]. For functional analysis, DNA microarrays have been used to identify genes induced by drought and oxidative stress, wounding, insect feeding, and many other processes in several species, particularltly in *Arabidopsis* and rice. Microarrays have also been used for the analysis of the molecular basis of disease resistance mechanisms in *Arabidopsis*. Proteomics has been used to characterize proteins involved in the response to a number of biotic and abiotic stresses. Viral-based gene silencing and RNA interference represent newer tools for systematic gene inactivation, which are more rapid than the production of mutants [45].

Genetic transformation in order to obtain GM plants, involves the insertion of genes, usually in the form of an expression cassette, into plant cells, generally into the nuclear genome. The production of transgenic crops relies on three technologies ([1], pp. 1–34, [47]):

- The isolation and manipulation of small sequences of DNA (gene cloning)
- The introduction of such exogenous DNA sequences into plant tissues (transformation)
- The selection of transformed cells or tissues and their regeneration into adult plants capable of further propagation.

Transgenes have to be expressed. The translation product (the protein) has to be processed properly and targeted to the correct cellular compartment, by appropriate design of the transgene, including promoters and their function.

Prior to application, classical procedures of plant development in field trials (that must be approved) and the approval procedure by governmental authorities for commercial application have to be undertaken. Thus the *time* required for development of a *new breed (cultivar)*, including the steps mentioned, took 8–12 years at a cost of 2–6 mn € ([1], pp. 331–342, [2]).

The breakthrough in plant transformation technology came, in the early 1980s, with the introduction of *Agrobacterium* vector systems (Table 18.1). They deliver DNA into plant cells at reasonably high efficiency and facilitate the incorporation of the exogenous DNA into the genome of the host plant cell. *Agrobacterium*, a Gram-negative soil bacterium, is a pathogen of dicotyledons (broad-leafed plant species). The ability to cause crown galls (tumourous tissue growth) depends on the ability of *Agrobacterium* spp. to transfer bacterial genes into the plant genome by means of a plasmid known as the Ti (tumour-inducing) plasmid. It is transferred from the bacterium into the plant cell, where it becomes integrated into the genome of the host plant. Commercial GM crop breeding using this technique includes cereals such as rice, corn and barley ([1], pp. 54–76), [47].

Direct gene transfer (DGT) includes four techniques that are used routinely for the production of transgenic plants. These are protoplast transformation, tissue electroporation, silicon carbide fibre vortexing, and particle bombardment (projectile mediated techniques, or biolistics). The development of *particle bombardment technology* became the method most widely used for monocotyledon (cereals, grasses) GM. In particle bombardment, dense microscopical particles, usually of gold, have DNA precipitated onto them and are then accelerated into plant cells. DNA is subsequently released from the particles, allowing transformation to take place ([1], pp. 67–71).

Nowadays, the use of improved strains of *Agrobacterium* and the refinement of DNA delivery via particle bombardment (biolistics) means that virtually all *dicotyledonous* (broad-leafed) species can be transformed with reasonable efficiency. Transgenes are normally inserted as part of a multigene construct that also contains regulatory elements and a selectable marker, often an antibiotic, or herbicide resistance, or D-amino acid oxidase genes [47].

Table 18.1 Selected essential steps in plant biotechnology [42, 48, 49].

1983	First application of the Ti-plasmid as gene vector in plants
	The first transgenic plant, tobacco, created by Schell and Montagu, Gent, Belgium
1985	Plants accepted for patenting in the USA
1986	Generation of virus resistant plants by introduction of virus protein envelope
	First field trials with transgenic plants
1987	Development of particle bombardment for direct gene transfer
1988	Development of the antisense technique
1990	Transformation of cereals as monocotyledons
1993	First patents of insect resistant plants in the USA
1994	FlavrSavr tomato approved and marketed as first GM food in the USA
1996	First large scale commercial planting of transgenic crops

Many of the world's most important crops are *monocotyledonous*, including cereals for human and livestock nutrition. They were much more difficult to culture and regenerate *in vitro* and were not naturally susceptible to infection by *Agrobacterium*. The majority of established transformation techniques target cells in cultured tissues. Generally explants are taken for transformation from immature embryos, immature leaf tissue and isolated shoot meristems. Techniques used preferentially are DGT and *Agrobacterium* transformation, developed and applied more recently in the major cereal species [25]. Plant transformation vectors are used to deliver genes into a recipient genome. The active components of a transformation vector are similar, regardless of the transformation method (i.e. *Agrobacterium* or DGT), and are promoter, structural gene, and terminator sequences. The promoter–gene– terminator unit is cloned, in this order into a plasmid DNA molecule [25]. In addition to transformation techniques metabolomics has been established, as an advanced tool in plant genetics and breeding, as well as standards (e.g. formal data description), further siRNAs and miRNAs as regulatory molecules [50].

Overall, between 1986 and 1997 biotechnology was used to genetically modify 60 crops for 10 different classes of traits. During that period, more than 25 000 field trials were conducted in over 45 countries. As of 2002, 15 crops modified for one or more of 47 phenotypic traits were commercialized, most with attributes related to input and yield performance, including primarily insect resistance and herbicide tolerance (for current application see Section 18.4, Tables 18.2 and 18.3) [4].

18.3.2
Whole Plants and Crop Protection

18.3.2.1 Crop Yield and Quality
The general aims of plant BT envisage improved crop yield and quality. The yield of a crop is ultimately determined by the amount of solar radiation intercepted by the crop canopy, the photosynthetic efficiency, and the harvest index (the fraction of dry matter allocated to the harvest part of the crop). Crop yield is first dependent upon light and its conversion (light harvesting and electron transport) into usable energy (ATP, NADPH) that drives the dark reaction of photosynthesis (carbon dioxide conversion into carbohydrates (Figure 18.1). Losses are due mainly to weeds, fungal and viral infections, each contributing about one third to overall losses of about 37% ([1], pp. 258–266).

A most relevant topic aims at the development of dwarf plants that require less of the plant's resources to be committed to the growth of the stem, allowing more dry matter to be partitioned to the grain rather than the straw. This directly improves yield, and, in addition, the dwarf crops are less prone to damage by wind and rain (a major feature of wheat and maize crops in the Green Revolution, while still based on conventional breeding ([1], pp. 237–266).

Photosynthetis comprises complex processes, suggesting that the task of increasing yield will require whole suites of genes – nuclear and chloroplast – to be enhanced. These would include a series of signal transduction pathways. In experiments with tobacco plants some exhibited reduced stem elongation and with enhanced allocation

Table 18.2 Transgenic crop cultivation. (A) Transgenic crop cultivation by country (main adopters of transgenic crops) ([1], pp. 316–342, [77]).

Country	Area (million ha)		
	1996	2001	2007
USA	1.53	5.7	57.7
Argentina	0.1	11.8	19
Brazil	—	5.7	15
Canada	0.1	3.2	7
India	—	—	6.2
China	—	1.5	3.8
South Africa	—	—	1.8

(B) Transgenic crop cultivation by crop

Crop	Area (million ha)	Proportion of total area (%)
	2006	
Soybean	58.6	68
Maize	25.2	18
Cotton	13.4	39
Canola	4.8	21

(C) Transgenic crop cultivation by trait

Trait	Area (million ha) (2006)
Herbicide resistance	69.9
Insect resistance (Bt)	19
Insect resistance + herbicide resistance	13.1

Table 18.3 Crops grown with herbicide resistance, compound (herbicide), and companies ([1], pp. 104–132).

Crops	Herbicide	Companies
Alfalfa, cotton, oilseed rape, soybean, wheat, maize	Glyphosate (Roundup)	Monsanto
Sugar beet	Glyphosate	Monsanto/Syngenta
Maize, rice, wheat, cotton, oilseed rape, potato	Phosphinothricin (Basta), (Liberty)	Bayer CropScience
Tomato, sugar beet	Phosphinothricin	Syngenta
Rape, rice, flax, tomato, sugar beet, maize	Chlorsulfuron (Glean)	Dupont-Pioneer Hi-Bred
Canola, maize, rice, sunflower, wheat	(Lightening, Beyond)	BASF
Soybean	Atrazine (Lasso)	DuPont
Cotton, oilseed rape, potato, tomato	Bromoxynil (Buctril)	Calgene/Bayer CropScience

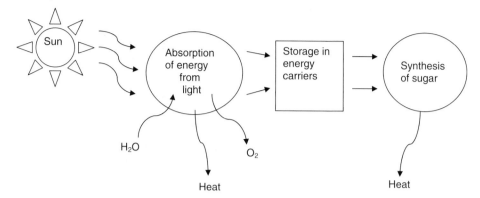

Figure 18.1 Principle of photosynthesis.

of assimilates to leaves increases in the harvest index approached 20%. Dwarfing effects were found in transgenic potato plants with overexpression of the phytochrome, an overall increase in photosynthesis per leaf area and significantly increased tuber yields. However, the enhancement of photosynthetic pathways will require much more fundamental work in order to improve photosynthetic efficiency and increase yield levels, and is expected to be realized only beyond 2015 ([1], pp. 258–266, [49]).

Crop quality with reference to food and nutritional value comprises several traits and properties, including fruit ripening, seed protein quality, and oil composition. *Fruit ripening* is a complex process that brings about the softening of cell walls, sweetening and the production of compounds that impart colour, flavour and aroma. The process is induced by the production of a plant hormone, ethylene. Down-regulating the expression of an endogenous gene using antisense and co-suppression techniques has been applied. The delay of ripening and improved post-harvest shelf life was achieved with tomato (however, with poor consumer acceptance, see Section 18.4), and also with papaya, mango, pineapple and other fruits, although there are no commercially available varieties yet (for improved and modified oil content see Section 18.3.3). Antisense ethylene technology is applicable to all climacteric fruits, and other systems triggered by ethylene, for example spoilage of fruits and vegetables. Reduced spoilage results in an increased storage capacity and thus a tremendous potential for reduced losses during storage and transport, notably in developing countries ([39], [1], pp. 237–266).

The *overall yield* for second generation fuels or energy may include the whole plant. Taking the example of biogas from corn, the aim of breeding is seen to shift from maximal starch yield towards maximal yield of biomass dm (dry matter). The current yield of $180 \, dt \, ha^{-1}$ gives around $6000 \, m^3$ of methane; the aim for 10 years breeding is a biomass yield to give $10\,000 \, m^3$ of methane (Figure 18.2). Optimization of overall production must also consider crop rotation, including corn, rye and sorghum [2]. Bayer CropScience is working on a Jatropha-project that aims at utilizing the Jatropha-plant for biofuels production. The plant produces a non-edible oil and grows on soil where plants for food production do not grow, mainly in South America,

Figure 18.2 The figures are from conventional breeds (Foto: KWS; with kind permission of KWS SAAT AG, Einbeck, Germany); (a) The figure illustrates the aim to increase, within 8 years, the energy yield by 100%, from 5000 up to 10 000 m³/ha (up to 300 dt/ha, right); (b) Variance in energy maize lines; (c) Differences in dry stress tolerance.

Figure 18.2 (*Continued*)

Africa and Asia [51] (See also the considerable efforts invested at http://www.energybiosciencesinstitute.org/index.php).

Osmotic stress, another complex topic, represents a target in order to grow plants on soils otherwise not used for production. An approach, using an artificial neural network, was developed to identify genes genome-wide responding to various osmotic stresses in *Arabidopsis thaliana*. It revealed that gene expression is largely controlled at the transcriptional level. Available gene expression profiling data were used for cross-validation. The results showed that transcript levels of 27 out of 41 top-ranked genes altered under various osmotic stress treatments [52].

18.3.2.2 Herbicide-Tolerance
Crop protection represents a major tool in order to provide food in all countries. It includes all measures of crop maintenance, from sowing to harvest and even to processing. The selection of crops for rotation is crucial. Crop protection includes the prevention of weeds, fungal and viral infections, and quarantine methods against pests or pathogenic organisms. Chemical agents are used on a very large scale, with estimated expenditures of roughly 14 billion (bn) $, about 6 bn $ for herbicides, 4 bn $ for insecticides, and 2.5 bn for fungicides in 1985. These few data (even if some 20 years old) should illustrate the general frame of crop protection and the perspectives for GM crops [53]. Weeds have a significant effect on the yield and quality of crops, as a result of competition for light and nutrients. The reduction in world crops caused by weeds is of the order of 10%.

Herbicide-tolerant GM plants allow more efficient herbicide application, and it has been claimed that the amount of herbicides applied can be reduced significantly. Actually, herbicide-tolerant crops released make up more than three-quarters of the total commercial GM crops released (see Section 18.4). Scientific reasons why herbicide-tolerant crops have been developed so rapidly compared with other GM traits are that genes for resistance were available from bacteria and tolerant plants (resistant crops and weeds), and single-gene mechanisms to obtain tolerance were relatively simple to devise. The major commercial impetus for their development lay in the advantage to the agrochemical/seeds industries of producing crops tolerant to specific herbicides manufactured by the same company. A herbicide-tolerant crop allowed it to be treated for the optimal time span to combat weeds, hence, greatly extending the effectiveness of that herbicide (see also Section 18.5) ([1], pp. 105–132).

By definition herbicides affect plant-specific biological processes, for example acting frequently on plant specific pathways located in the plastids, often found to result from the specific inhibition of a single enzyme/protein. There are four distinct strategies for engineering herbicide tolerance. An important example is the detoxification of the herbicide, using a single gene from a foreign source. Glyphosate, as a prominent example, is a broad-spectrum herbicide that is effective against 76 of the world's worst 78 weeds, and is marketed as Roundup by Monsanto (USA). It is a simple glycine derivative acting as a competitive inhibitor of a key enzyme (EPSPS: 5-enolpyruvylshikimate -3-phosphate synthase) in the biosynthetic pathways of three aromatic amino acids. This pathway supplies the aromatic precursors for a range of phenolic compounds essential for plant growth. One out of three strategies for glyphosate tolerance is based on (mutated, optimized) genes of EPSPS, producing an enzyme that is resistant to glyphosate. This gene is incorporated into the current range of Monsanto's major Roundup Ready crops (soybean, cotton, and oilseed rape). In other strategies glyphosate is detoxified by degradation using an oxidoreductase, or by acetylation using a microbial N-acetyltransferase, the acetylated product not being an effective inhibitor of EPSPS (Figure 18.3) ([1], pp. 105–132). A new technology of DowAgroSciences claims to provide tolerance to multiple herbicide classes in many different crop species. The rapid adoption of glyphosate-tolerant traits has spurred resistance to that herbicide in key weeds [54].

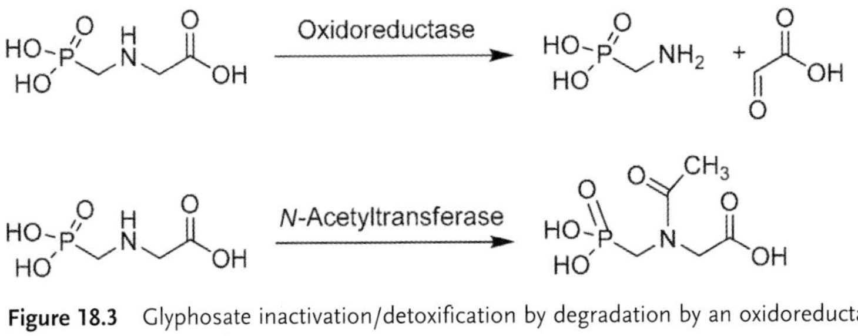

Figure 18.3 Glyphosate inactivation/detoxification by degradation by an oxidoreductase, or acetylation, respectively ([1], p. 112).

18.3.2.3 **Disease Resistance**

Considerable progress in understanding *plant pathogen defence* has been made recently [55]. *Plant diseases* are of global importance, such as rust diseases of wheat and uredo-spores that can be wind-dispersed over long distances. Rust epidemics (fungal diseases) in various parts of the world have caused huge losses. It has been estimated that crop losses in the USA alone cost some 33 bn $ per year. These losses are partly due to the use of monocultures, which encourage epidemic pests and diseases. In Australia breeding for resistance to rust diseases of wheat is estimated to save the industry in excess of 300 mn Aus$ annually. On the other hand they may be of more regional or localized significance. Late blight of potato and brown spot of rice caused famines in Ireland (1845–6) and Bengal (1942), respectively. Rice blast epidemics have brought famine to parts of Japan, and corn blight devastated the corn industry in the USA in 1970. These examples may illustrate the importance of breeding for disease resistance, as long established, where genetic modification now represents an additional useful means of creating resistance ([5], [39], [1], pp. 156–183).

Genetic engineering is likely to aid the labour intensive and time-consuming classical breeding process that can take decades to incorporate resistance, using wild relatives of the cultivated crop as sources of resistance genes ([1], pp. 156–183, [6]). There are different types of resistance based on the mode of inheritance and/or how resistance is expressed. One type is often simply inherited and controlled by single or few genes with a large effect, whereas the alternative being an additive effect of many genes each making a small contribution. Simply inherited genes are likely to be overcome either by viruses or by new pathotypes. However, some simply inherited resistances have remained effective after long term deployment [5].

Plants recognize *microorganisms* on a first level by means of their immune system and activate immune response. Recent insights into the plant immune system suggest that plant cells assemble a pathogen-inducible machinery at the cell surface. On a secondary level several cultivars can recognize microbial species specifically [55–58]. Most *resistance* (R) genes, that occur in naturally disease-resistant plants, mediate recognition and signalling to the plant that it is under attack from a pathogen. A series of defenses are then activated to prevent the pathogen from invading, for example the accumulation of antimicrobial toxins, such as phytoalexins, cell wall modifications, and production of disease resistance proteins [6].

Durability of resistance cannot be assured. Therefore genetic variability must be deployed to buffer against widespread losses in the event of pathotypic change. A good strategy is to make use of genetic diversity for resistance. Cultivars from different geographic regions are the most attractive sources. The main utility of GM is the ability to create plants with multiple disease resistances. Gene pyramiding, or stacking, of resistance traits for a particular pathogen is especially desirable [5, 6].

The 'BASF potato' *Solanum tuberosum* ('Amflora') has a range of dominant genes that code for resistance to late blight. Over many years, resistance genes have been introduced into the edible potato from wild-type varieties, but the effect is soon overcome by the pathogen. BASF has succeeded in introducing two R genes from a wild Mexican species which is resistant to late blight. However, its application for field trials of the GM potato has met much opposition in the EU countries with the main concern being that the encoded proteins have not been tested for allergenic response ([1], pp. 156–183). Approval came finally, after 13 years, in the EU for industrial use in 2010 [90].

Insect and *viral* resistance are relatively simple in nature and can be achieved with monogenetic insertions, for example *Bacillus thuringiensis* (*Bt*) toxin and viral gene-antisense constructs ([1], pp. 133–155, [6]). Symptoms of virus infection are growth retardation, growth distortion, mosaic patterning of leaves, yellowing and wilting. Some plant viruses can be a major problem to agriculture, with severe epidemics periodically causing losses adding up to billions of dollars. The main, antiviral transgenic approach used has been that of '*pathogen-derived resistance*' (PDR), whereby an infectible organism could be transformed with genes from one of its parasites to induce resistance. The presence of a pathogen sequence, for example for genome copying (replicase) enzymes, or specific proteases, was shown to act protectively, directly interfering with the replication of the pathogen or inducing some host defence mechanism. Tobamoviruses (i.e. tobacco mosaic virus) were amongst the first characterized in molecular detail. For PDR the 'gene' coding for the principle structural protein (the capsid, or coat, protein) was available, and it was shown that transgenic tobacco plants, expressing the virus-derived capsid protein, were tolerant to infection. Examples of capsid or coat protein (CP) mediated resistance, out of a considerable number, are potato with potato virus X (PVX), rice with rice yellow mottle virus (RYMV) and wheat with barley yellow dwarf virus (BYDV). Another successful approach uses ribosome-inactivating glycosidases that stop protein synthesis/elongation ([1], pp. 184–211, [24]).

Gene silencing represents a very active area of research and many examples of RNA-mediated resistance to viruses have been described. Silencing has been shown to be one of the mechanisms by which plants naturally defend themselves against viruses. Examples of three transgenic approaches comprise: capsid-mediated resistance, with for example potato, rice and tobacco viruses, replicase (enzymes for virus nucleic acid replication) -mediated, with African cassava, pea and potato viruses, and RNA-mediated resistance, with e.g. potato and tobacco viruses. Chimeric transgenes derived from two distinct viruses have been shown to confer multivirus resistance through gene silencing. Field release trials have been authorized in several countries. Commercial use of squash containing two potyvirus-derived capsid coding sequences was authorized in the USA in 1994. Transgenic virus resistant papaya was the second crop with obtained virus resistance to be commercialized in the USA. Papaya ringspot virus, ubiquitous in plantations, is responsible for 70–90% yield reduction in papaya [24].

Fungal warfare – that is generating resistance against *fungal diseases* by GM – has lagged behind GM for insect and viral resistance since fungal resistance is usually complex, involving multiple genes that are more difficult to manipulate. New promising approaches, however, might be expected from recent insights concerning plant immune responses [59]. The first commercially released GM crop with resistance to fungal disease occurred in the mid 1990s; it was a side benefit of controlling insect damage in GM corn (maize) using the *Bt* gene, which carried also resistance to ear and stalk disease caused by a number of different fungal pathogens, for example *Fusarium* and *Aspergillus* species. Interestingly, the control is due to the reduction of feeding damage by insects, which limits infection by fungi via wounds. Other transgenes encode proteins (introduced from other plants, fungi or bacteria) that fall into two major categories: proteins with a direct activity against the fungal invader, and proteins that increase the plant's own natural defense. Amongst the first group of antifungal proteins are the pathogenesis related (PR) proteins. These include enzymes, for example chitinases and glucanases, that attack the major components of fungal cell walls. Often the addition of two or more transgenes, for example chitinases and glucanases together, act synergistically. Chitinases have been used to create a variety of disease resistant GM crop species, including rice, apple, tobacco, oilseed rape/canola, tomato and peanut ([1], pp. 156–183, [6]).

Induced defense response represents a challenge for creating resistant plants. *Phytoalexins* are non-protein natural antimicrobial toxins that require a series of enzymes for their production. They are part of the induced defense response of plants and their production is increased upon pathogen challenge. GM plants with transgenes for enzymes that increase the quantity and type of phytoalexin have been produced and have increased resistance to fungal infection. Most plants produce the precursors of phytoalexin but require the additon of the grape enzyme stilbene synthase to produce the active, antimicrobial compound. Addition of this enzyme has been shown to increase disease resistance in GM tobacco and tomato [6]. *Active oxygen species*, for example hydrogen peroxide, are generated by several enzymes and aid in fungal disease resistance. The expression of a transgenic enzyme, such as glucose oxidase, which produces hydrogen peroxide, confers resistance in potato to several plant pathogens. *Ribosome-inactivating proteins* (RIP), produced by several plants, can deactivate the protein manufacturing mechanism of ribosomes in fungi. The action of these transgenic proteins is enhanced by the addition of transgenic chitinases, which are thought to increase the access of the RIP to the fungal ribosomes by degrading the fungal cell wall. Transgenic expression of defensins – small peptides, thionins, lectins, and lipid transfer proteins – can confer resistance against fungal pathogens [6]. Recently, a glucosinolate metabolic pathway has been identified in plant cells, and has been recruited for broad-spectrum antifungal defence responses [60].

Insect pests, including caterpillars of butterflies and moths, mosquitoes and adult grasshoppers and beetles, cause a major part of losses due to pests in general, that is estimated at 13% of the world potential crop yield. Modern agricultural practice – growing monocultures – has exacerbated the problem, with current elite cultivars having less natural resistance to pests than their predecessors. Reliance on chemical pesticides caused cycles of resistance to the previous pesticide. Amongst the large range of insect pests affecting crops (given in brackets) a few shall be mentioned: cotton bollworm (cotton, maize), European corn borer (maize, also outside Europe), Colorado beetle (potato), confused flour beetle (cereal flours), and brown plant hopper (rice). Whereas some adult insects feed off plants, most of the problems are caused by insect larvae. Two approaches have been used, bacterial insecticidal genes, and endogenous plant-protection mechanisms (the 'Copy Nature' approach) ([1], pp. 133–155, [6]).

The most widely used bacterial insecticidal genes are the *cry* endotoxin genes from *Bacillus thuringiensis* to produce *Bt* crops (e.g. *Bt* cotton, etc.). The corresponding proteins (Cry proteins) are extremely toxic, at relatively low concentration, for insect larvae. The sequences of *cry* genes had to be modified to optimize expression in plants.[3] First *Bt* crops gained approval for commercial planting in the USA in the mid 1990s, and the success of the approach led to the development of a range of *Bt* crops by several major biotechnology companies involved in crop protection (see Section 18.4). The specificity of Cry proteins allows the targeting of specific pests by particular transgenes, and different crops may have different *cry* genes inserted. Resistance against more than one insect has been achieved. In addition to introducing several transgenes, the build-up of resistance in the insect population must be managed by good agricultural practice, including crop rotation ([1], pp. 133–155).

The *Copy Nature approach*, involving a rational approach to the development of pest-resistant crops, includes several steps, thus the identification of leads (natural plants that show resistance to pests), screening, isolation and purification of insecticidal proteins, mammalian toxicity testing, genetic engineering, followed by selection and testing of the transgenic plants. One successful example is the development of the cowpea trypsin inhibitor as an anti-insect, pest-control mechanism, originating from strains of cowpea growing in Africa. It has been transformed into a range of crop plants including rice, potato, wheat and cotton ([1], pp. 133–155).

18.3.3
Products of Primary Metabolism

A remarkable boost in *sugar content* by 100%, could provide a new perspective for enhanced food and biofuel production. A gene expressing sucrose isomerase that converts sucrose into the isomer isomaltulose (see Section 16.5) has been successfully introduced in sugar cane lines. Isomaltulose was accumulated in storage tissues without any decrease in stored sucrose concentration, resulting in up to double the total sugar concentration in harvested juice [61].

The major *starch* producing crops – cereals and potatoes – are widely grown to produce starch for both food (about 30%) and industrial (about 70%) purposes (see Section 16.2.1). Conventional starch is essentially made up of 25–30% amylose with mainly linear chains of glucose molecules, made up of α-1,4 bound glucopyranosyl units with a molar mass of up to one million Da. The major part, 70–85%, is

---

3) Preparations of *B. thuringiensis* spores or isolated protein crystals have been used as an 'organic' pesticide for half a century. Optimization of bacterial *cry* gene expression in plants had to take into account the copy number and stability of integration of the transgene, as well as the stability and level of transgene expression. In order to tackle the problem of resistant pests, pyramiding, in which transgenes are successively stacked by conventional crosses between different transgenic lines, had to be applied ([1], pp. 133–155).

amylopectin, with main chains as in amylose, but with side chains linked to the main chain by α-D-(1 → 6)-glycosidic bonds, and a molar mass ranging from 16 to 160 million Da. Modifying the content of both compounds has been of significant interest in order to design starch products for different applications: for food processing and other industrial purposes (e.g. paper manufacturing). Since there is a branch in the biosynthetic pathway leading either to amylose, or amylopectin, this pathway could be manipulated at three points in order to enhance or decrease either of the two varieties, resulting in different physicochemical and application properties. First, increasing the overall level of starch biosynthesis was addressed. Second, the inhibition of starch branching enzymes results in 'high-amylose' starch with favourable gel forming properties. Mainly maize and potato have successfully been engineered to overproduce starch varieties, for example special types containing up to 80% amylose. Such starches can be used for confectionary and other industrial purposes. Third, waxy (high amylopectin) starch varieties have been produced in potato, with no amylose, by antisense inhibition of different starch synthase enzymes for starch products gelatinizing easily, giving clear pastes that are ideal for thickening, and with high freeze–thaw stability food products ([1], pp. 267–315, see also [62]).

The *seed quality for food and feed* has attracted some attention, since the protein quality of the major seed crops is determined by the storage proteins. Legume seeds are *deficient* in cysteine and methionine, and most cereals, but not oats and rice, in lysine and methionine and, in maize, also tryptophan. Hence, these are usually combined when considering protein nutritional quality. Feed material has been supplemented with lysine and methionine produced by fermentation for several decades. Approaches to improve the amino acid composition in GM seeds revealed the existence of complex regulatory networks in amino acid biosynthesis. Thus, to increase the lysine level in tobacco seeds, the introduction of an enzyme in the synthetic pathway and the removal of two enzymes in a degradation pathway were necessary. The combination of the two traits gave an 80-fold increase in free lysine, compared with the wild type seed. It is important in this context to ensure that the modifications do not result in unwanted allergenic properties, and that bioavailability studies are undertaken. The first commercial high lysine corn was introduced by Monsanto in 2006 [39, 63].

Functional foods (including nutraceuticals) are defined as 'any modified food or food ingredient that may provide a health benefit. However, 'once touted as the next wave in the food industries, they suffered some major setbacks and disappointing flops,. . . a meagre number of products have ripened while the rest turned sour' [64]. Examples are plants that contain polyunsaturated fatty acids, for example omega-3- and omega-6-fatty acid rich plants which are considered as essential components of the human diet. Amongst different beneficial health traits they are expected to improve brain functions, like attentiveness and concentration, to have anti-inflammatory properties, to be efficacious for the health of the heart and circulatory systems, and to provide positive effects on dysfunction, like dementia (Alzheimers disease) [39, 65]. Examples are the development of oil from soybeans containing high oleic acid (enriched in stearidonic acid, SDA) and reduced saturated fat (Monsanto, USA), and transgene soybean for high oleic acid for increased stability in frying (DuPont and Bunge, USA), that is going through regulatory approval [64]. Monsanto hopes to put SDA-enriched soybean oil in items such as salad dressings and yoghurt smoothies. Researchers of Dow Agrosciences (USA) have inserted the algal genes required to make decosahexanoic acid (DHA) for the human diet in canola seeds [65].

'Genes to gasoline' – such headlines provoke expectations for second generation biofuels based on lignocellulosic materials. However, the recalcitrance of the plant material – the inability to quickly and cost-efficiently break down lignocelluloses that has been evolved to protect and stabilize the plant – stands in the way of utilizing this large resource. A major goal of DOE (the US Department of Energy) is to identify and characterize genes involved in cell-wall biosynthesis and to establish which genes can have an impact on making lignocellulose more easily accessible to breakdown into sugars. Plans are to come up with plants that are more easily digested and might just need a hot-water pretreatment without chemicals. One important step to solve the problem has been sequencing the genomes of *Populus trichocarpa*, a fast growing poplar tree, and further those of switchgrass and miscanthus grass [40, 66]. To improve *forest productivity*, in the context of wood, fibre (cellulose) and paper quality, as well as renewable resources production, *in vitro* cloning (by infection with *Agrobacterium* or particle bombardment) and clonal propagation of transgenic forest trees will play a role [67, 68]).

18.3.4
Products of Secondary Metabolism

A major project, notably for third world countries, has been 'Golden Rice', that entered the scene around 2000. Rice, being the most important food crop worldwide, is eaten by some 3.8 billion people. In some regions of the world where rice forms a staple component of the diet, vitamin A deficiency is a major nutritional problem, which can cause blindness, and other health problems, for example diarrhoea. It has been estimated that around 124 million children are vitamin A deficient, causing about 500 000 children to go blind each year. It is even estimated that improved vitamin A nutrition could prevent 1–2 million childhood deaths per year. One solution proposed for this problem is to engineer rice to produce provitamin A (Figure 18.4). Since the successfully transformed rice grains exhibit a characteristic yellow/orange colour this variety has been termed Golden Rice. A number of technical improvements and further work by Syngenta (CH) made β-carotene (provitamin A) synthesis in a rice variety effective, such that it could provide 50% of the recommended daily intake of vitamin A. Golden Rice appears to meet all the objections to GM crops that have been presented by anti-GM groups:

It benefits primarily the poor and disadvantaged
It will be given free of charge or restrictions to subsistence farmers
It can be re-sown every year from the saved harvest
It was not developed by and for the biotechnology industry, and industry does not benefit from it

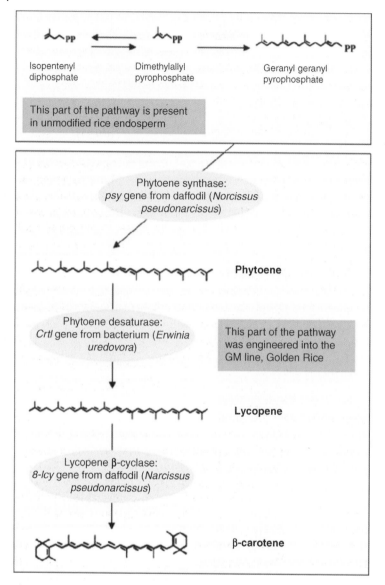

Figure 18.4 Biosynthetic pathway for the production of the vitamin A precursor, β-carotene, in Golden Rice [39].

It is a sustainable, cost-free solution to vitamin A deficiency, and it underwent a robust safety assessment by the International Food Biotechnology Committee (see Section 18.2.2).

Nevertheless Golden Rice has been attacked by groups such as Greenpeace and Friends of the Earth, mainly on the grounds that it could be a 'Trojan horse' for other

GM crops. The nature of this opposition derives from a fundamental belief that all GM crops are undesirable – owing to mere ideology ([1], pp. 250–256, [37, 69]).

Potent and specific pharmacological effects in humans have made *alkaloids* from plants an important source for medicals. Known traditionally for centuries as mixtures, for example opiates, a range of alkaloids has been isolated from plant extracts and produced commercially since the nineteenth century ([70], pp. 223–254). (For medicinal plants in China, and the Ayurvedic system in India, see Kinghorn and Chávez [71]). Due to their complexity it is, in most cases, impossible to produce such metabolites by total chemical synthesis under economic conditions. Since plants produce complex mixtures at low levels, specific alkaloids are very expensive [72]. Pharmacological use includes for example caffeine, cocaine, codeine, morphine, nicotine, quinine and, more recently, taxol (see below). As genetic manipulation of plants becomes more sophisticated, research has focussed on the engineering of alkaloid biosynthesis to generate transgenic plants that overproduce specific alkaloids. Other alkaloids of interest are terpenes, such as vitamins (A, D, and E), flavors and fragrances, that may be commercialized as 'natural' when produced in plants, with better consumer acceptance and at higher prices, in contrast to synthetic substitutes [72]. The drug *morphine* (Figure 18.5a) has probably served as a 'chemical template' to a greater extent than any other plant product. Thus morphine as a lead compound has been modified, while maintaining its potent analgesic activity, to curtail its side effects, and to afford commercially viable drugs. An example of an anticancer agent is paclitaxel (originally known as taxol), a diterpenoid (Figure 18.5b), that is currently obtained from extracts of *Taxus* sp. The spectrum of use of this drug and semisynthetic derivatives has gradually increased to include ovarian cancer resistant to other chemotherapeutics and breast cancer. A challenge is to generate transgenic varieties that overproduce taxol [71]. *Steroids* are amongst the many plant drugs used traditionally, in particular steroidal glycosides, where glycosidation may play specific roles, both for activity and solubility. An important example is digitoxin used as a reliable therapy to treat congestive heart failure (Figure 18.5c).

The production of *biodegradable plastics* (polymers) in plants is one of the most imaginative examples of molecular farming. Polyhydroxybutyrate (PHB), a polyester of hydroxybutyric acid, has been produced commercially by bacteria. The pathway of PHB synthesis is shown in Figure 18.6. It is a relatively simple three-stage pathway starting from acetyl-CoA. The genes for the three enzymes involved have been cloned from *Alcaligenes eutrophus* and transformed into plants, first in Arabidopsis (targeted to the chloroplast), then into oilseed rape, in order to obtain higher yields, which have reached 7.7% of fresh seed weight ([1], pp. 267–315). Another example of the production of biodegradable polymers in transgenic plants has been published by Hühns *et al.* [73] (see also [91]).

The ability to control metabolic pathways of *in vitro* cultured cells and tissues represents a challenge. One aim is the production of functional proteins, such as antibodies and vaccines, in plants, which does not suggest too many difficulties ([74], [1], pp. 267–315, [44, 75]). The production of a vaccine for animals (hemeagglutinin-antigen of paramoxy virus, an atypical variant of bird pest) has been approved by USDA in 2006. It is produced in tobacco cell fermentation [76].

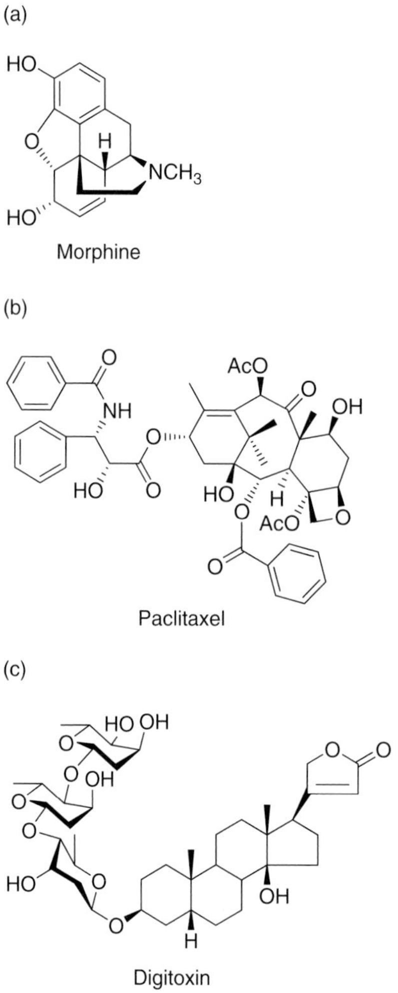

(a)

Morphine

(b)

Paclitaxel

(c)

Digitoxin

Figure 18.5 (a) Morphine; (b) paclitaxel (taxol), (c) digitoxin.

18.4
Application of Modified Plants and Products

'It's Red, Plump, Juicy and out of a Lab. Grocery Shelves Are Awaiting First Genetically altered Food… a tomato that stays firmer and ripens longer on the vine'. That's how the International Herald Tribune announced Calgene's GM tomato on Friday, May 20, 1994. It was not on the shelves for long. It was amongst the first GM products to reach the market: Calgene's (USA) FlavrSavr™ fresh tomatoes, and also processed tomato products containing delayed-softening tomato developed by Zeneca (UK) and the University of Nottingham. In order to improve handling and to

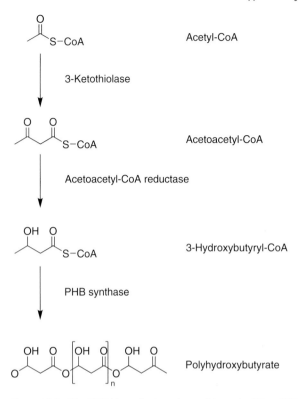

Acetyl-CoA

3-Ketothiolase

Acetoacetyl-CoA

Acetoacetyl-CoA reductase

3-Hydroxybutyryl-CoA

PHB synthase

Polyhydroxybutyrate

Figure 18.6 The PHB biosynthetic pathway of bacteria ([1], p. 283).

develop the full taste of tomato (sub-optimal due to premature harvest), the fruit was engineered to express low levels of two enzymes involved in ripening, polygalacturonase and pectin methylesterase. Despite their initial success, both the Zeneca and the Calgene product were withdrawn from the market, in part following anti-GM backlash ([1], pp. 237–266, [4]).

Applications include first *food*, *feed* and *renewable resources*, with over 15 crops modified for one or more of some 47 phenotypic traits commercialized since 2000, most with attributes related to input and yield performance. In 2006, 102 million (mn) ha were cultivated with GM crops, and in 2007 there were 114 mn ha in 22 countries. Seven countries dominate the global market, USA, Argentina, Brazil, Canada, India, China and South Africa. Amongst the products are soy bean, maize, cotton, rape, cucurbit (squash), tomato, papaya, alfalfa, poplar, petunias and paprika (Table 18.2). The growth of the area cultivated since 1996 over 11 years is obvious. Application was mainly with the combined traits of insecticide resistance and herbicide tolerance. Four crops account for nearly 100% of the commercially grown transgenic crops: soy bean (64%), cotton (43%), maize (24%) and rape (20%). About 9% (11.2 mn ha) of GM maize, soy bean and rape accounted for fuel production.

Remarkably, biotech crops were grown by approximately 10 million farmers in 22 countries in 2006, and some 90% were small resource-poor farmers from developing countries ([4, 8, 39], [1], pp. 316–342, [77]).

Herbicide resistance is the trait found in by far the largest area of transgenic crops, Table 18.3 shows crops grown with herbicide resistance, and companies involved. The most widespread herbicides (and corresponding resistant crops) in application are glyphosate (Roundup) and phosphinothricin (PPT) (Basta, Liberty). Together with insect resistance, these two traits account for effectively all the area of transgenic crops (Table 18.2C) ([1], pp. 316–342). Risks were identified with glyphosate resistant weed species [92].

Crop resistance mainly comprises *resistance* to fungi, virus, and herbicide tolerance. With respect to *virus resistance*, commercial use of squash containing two potyvirus-derived capsid coding sequences was authorized in the USA in 1994. A subsidiary of Monsanto markets several varieties of squash and zucchini that are resistant to three viral diseases.

Conventional resistance routes have been ineffective. Transgenic virus resistant papaya was the second crop with obtained virus resistance to be commercialized in the USA. The rapid timescale of its production and commercialization, in 1998, was due to a very serious outbreak of papaya ringspot virus (PRSV) in 1992 that reduced Hawaii's papaya production by over 40% within 5 years. Other virus-resistant plants are available but have been withdrawn from sale due to a low take-up of these lines ([1], pp. 204–211, [24]).

Bt technology represents one of the most successful transgenic crop types that use genes of *Bacillus thuringiensis* that code for pesticides (see Section 18.3.2.3). A range of important transgenic *Bt* crops have been commercialized (Table 18.4). An important example of *Bt* technology concerns the European corn borer, a major pest of maize

Table 18.4 Examples of commercialized *Bt* crops ([1], pp. 133–155).

Crop	Trade name	Insect pests	Company
Potato	NewLeaf	Colorado beetle	Monsanto (USA)
Cotton	Bollgard II	Tobacco budworm, cotton bollworm, pink bollworm	Monsanto (USA)
Cotton	WideStrike	Tobacco budworm, cotton bollworm, pink bollworm	Dow Chemical (USA)
Maize	Agrisure CB	European corn borer	Syngenta (CH)
Maize	Agrisure RW	Corn rootworm	Syngenta (CH)
Maize	Herculex 1	European corn borer	Mycogen (Dow) Pioneer (DuPont) (all USA)
Maize	Herculex Xtra	European corn borer + corn rootworm	Dow/Pioneer (USA)
Maize	YieldGard Plus	European corn borer + corn rootworm	Monsanto (USA)

that causes considerable damage worldwide, the larvae tunnelling into the central pith of the stalks and ears. *Bt* corn accounted for 14% of global maize production in 2004. The cotton crop in the (USA) is prone to damage by lepidopteran larvae, in particular cotton bollworm, pink bollworm, and tobacco bollworm. Consequently the development of insect-resistant cotton has been one of the major targets for plant biotechnology (in the USA), with success (Table 18.4) ([1], pp. 133–155). The build-up of resistance in the insect population must be managed by good agricultural practices. Thus integrated pest management considers the wider context of pest-crop interaction (natural predator, adjacent plant species, crop rotation, rotating *Bt crops with non-Bt* crops, and chemical sprays) in order to prevent the build-up of resistance in the insect population. Resistance management plans are required by the EPA when a GM variety is approved for commercial growing ([1], pp. 133–155). The first commercially released GM crop with *resistance to fungal disease* occurred in the mid 1990s; it was a side benefit of controlling insect damage in GM corn (maize) using the *Bt* gene (see Section 18.3.2.3) [6].

With a *product of primary metabolism* Monsanto introduced, in 2006, the first commercial high lysine corn, combining the improved feed quality with resistances to herbicide and European corn borer [39].

18.5
Economic Aspects

Sales of some big companies were, Monsanto (USA) 11.4 billion (bn) $ in 2007/08, Syngenta (CH) 9.2 bn $ in 2007, and Bayer CropScience (Germany) 382 million (mn) € in 2007 [78, 79]. Syngenta emerged from the combination of the agrobusiness of Novartis and AstraZeneca [80]. DuPont (USA) earnt 2.5 bn $ in its agriculture and nutrition business in the second quarter of 2008 (giving an estimated total of 10 bn $ for the year, an increase of 23% in one year) [81, 82]. The leading seed producing companies strategically aligned themselves with marketing, processing, and retail companies. Monsanto had started selling recombinant soy seeds in 1996; until 1999 Monsanto bought several seeds companies. DuPont acquired Pioneer Hi-Bred, the leading seeds company for corn, and started to cooperate with Dow Chemical. Aventis, sold its crop science division to Bayer [83]. The innovators are located almost exclusively in the USA and the EU. This may suggest a highly concentrated industry that is extracting monopoly profits, which are in general in the range of over 20%. Thus Bayer crop science made a profit of 22.7% in 2007. An alternative view is that the unexpectedly low and uncertain returns realized by most investors in agricultural BT has discouraged new entrants. The resistance of consumers led to a major crisis in this field, followed by reoriented strategies at Monsanto, including the mergers mentioned [4]. Continuing these tendencies, BASF (Germany) and Monsanto (USA) have cooperated since early 2007 in GM crops. They will invest 1.5 bn $ in common research and development. As the first product from this cooperation a GM corn should be marketed, probably in the period from 2013 to 2015. It will exhibit 6–10% higher yields. The main aims are, besides increased yield (up to 20% in the long term), an enhanced stress tolerance. Crops under investigation are maize, soy bean, cotton and rape. The market value of GM seeds rose from 1 mn $ in 1995 to an estimated 3 bn $ in 2001, and to 6.9 bn $ in 2007, up 11.5% over 2006 [4, 8, 82, 84, 85].

Market share and potential for pay back on investment are key elements in each crop/pathogen commercial decision. In the current climate of public uncertainty and nonacceptance, very few transgenic crops justify the risk and the investment while waiting for regulatory approvals. Producers may also lose due to lower market prices caused by increased productivity and supply [4, 24]. Thus, returns to the producer at the product level are considered to have been mixed. Studies on *Bt* corn, *Bt* cotton, Roundup Ready™ soybeans, and herbicide-tolerant canola show that gross returns can be variable, depending critically on weed or insect infestation rates (if the infestation rates are low, returns will be low). Resistant GM crops are assumed to lower prices of competing pesticides (on conventional varieties), and cause an increase in surplus for farmers who adopt the GM varieties. The studies of yield-enhancing inventions suggest that a significant portion of the benefits appear to go to consumers ([1], pp. 133–155, [4, 86]). Yield improvement has been limited up to now, except for cotton and maize, and to a lesser degree for soy bean [8]. A 2002 review concluded that crop yields of eight GM cultivars increased by 2 million tonnes and lowered annual production costs for growers by 1.2 bn $ [4]. An analysis of the benefits of herbicide (glyphosate) resistant (Roundup Ready) soybean of 1997/1998 resulted in a significant reduction in production costs of the crop derived principally from the reduced use of herbicides, and most of the economic benefits accrue to the farmer. It was also claimed to be of net benefit to the environment. As an example, Table 18.5 shows the distribution of benefits of herbicide resistant soybean ([1], pp. 105–132).

However, the distribution of benefits is highly controversial. In contrast to the data of Table 18.5 it has been estimated that the innovating companies are in many cases the big winners. As both the major investors in the technology and the major beneficiaries of the returns, they have a significant influence on the distribution of the benefits. It is important to include innovator's monopolistic or oligopolistic profits in the total calculation of the returns on the technology. A number of studies have estimated that innovators capture between 37 and 60% of the gross benefits generated by yield-enhancing technologies. In contrast, in the case of herbicide-tolerant canola it has been estimated that although the gross annual returns were

Table 18.5 Distribution of benefits of growing herbicide resistant soybean (Data from [93], cited from [1], pp. 105–132).

Beneficiaries	Estimated benefits (Million US $)	Distribution of benefits (%)
Seed companies	32	3
US consumer	42	4
Technology inventor	74	7
US farmer	769	76
Total benefits	1061	100

large, the gross investment by the industry was only recovered in the seventh year of adoption and, even then, industry had not recouped the value of pre-existing chemical markets that were cannibalized by the new technology [4]. Diverging domestic trade rules are causing the lead GM adopters, especially the USA and Canada, to abandon or lose markets and divert trade to new markets. USA and Canadian exports of corn and soybeans to the EU dropped from the end of the 1990s by some 50 to over 90% [4].

18.6
Summary and Outlook

Significant scientific progress has been made during recent decades. Much has been translated into the development of new GM crops with traits of resistance to pests, improved food quality, and even pharmaceuticals. Global business has been established on a large scale, with large companies operating worldwide, and a continuing trend towards mergers, acquisitions and cooperation.

Prospects for future development include both technological advances relevant for desired plant traits for application, and less critical aspects that cause resistance by opponents. Thus, as ([1], p. 316) ask, what are the concerns of the public? Of a fundamental nature are new approaches that make use of data from genome-sequencing projects together with increased understanding of the fundamental biology underlying agronomically important traits, and that rely on plant genes (and less on bacterial, or fungal, that raise criticisms) to produce the desired effects. This includes avoiding the use of antibiotic marker genes. Another concern about the environmental impact of GM crops is the potential for spread of transgenes in pollen. Chloroplast engineering offers an alternative approach to the delivery of transgenes with a number of advantages while reducing this environmental impact. In the majority of cases chloroplasts are not found in pollen, thus gene transfer to weedy relatives is reduced. Chloroplasts are also ideal targets for multigene engineering.

Enhanced food production might be of long range and major practical importance, improving world wide food supply, both with respect to quantity and quality. It would involve complex agronomics traits. One strategy addresses the acceleration of rice flowering to allow two crops per year, with only a small reduction in yield per crop. Another strategy for improving yields are dwarf cereal strains that are generally higher yielding as less of the photoassimilate goes into growth and more goes into the grain. GM technology is now being used to manipulate the mechanism responsible for stem growth, for example for designing dwarf rice plants. Dwarf strains can take a considerable time to produce by conventional breeding but may be relatively quick using GM technology ([1], pp. 343–366).

Much hope relies on *second generation biofuels* that do not compete with food production, based on trees (e.g. poplar), or other plants with modified lignocelluloses. This would also allow a more environmentally benign paper-making process. Amongst beneficial GM plants are also plants engineered for various *phytoremidiation* purposes ([1], pp. 343–366).

A new paradigm, *molecular breeding*, may create a strategy, utilizing a set of approaches that integrate genetic/genomic tools with conventional breeding. The elucidation of pathways and systems that underpin complex traits will be important, not only to identify targets for genetic manipulation, but also for understanding the consequences of modifying complex systems. Amongst the aims of molecular mapping is to produce a sufficiently fine-scale map to pinpoint the location of genes that play a role in determining relevant agronomic traits, often influenced by multiple genetic and environmental factors. Genomics comprises functional genomics, including understanding the gene function and transcriptomics, using microarrays and gene chips. Further tools include proteomics, metabolomics, and their integration into systems biology (see Sections 13.6 and 15.6) ([1], pp. 343–366).

Interestingly in the USA strategies that were successful in the past (e.g. penicillin production, see Section 4.3.4) are again being applied to solve highly complex problems seen as a national priority: DOE's BioEnergy Science Centre (BESC) is employing considerable effort – a task force including 300 scientists in different institutions, including academia and industry – aimed at designing a strategy to solve the problem of using lignocellulosic biomass for biofuel production.

In contrast to optimistic views, the concerns of the public lead to scepticism, in part due to the lack of trust in information given out by governments and the big corporations that control agriculture, as a consequence of examples of big business imposing its wishes on farmers and consumers. However, the main issues over biotech food are trust and safety, and the surveys of public feeling highlight the central issue as being the influence of biotech crops on health ([1], pp. 316–340).

In conclusion, contradicting views and expectations, either optimistic or sceptic, continue to feed heavy controversies. A continued debate and dialogue on all these aspects and concerns seems necessary and mandatory – the great potential of GM crops depends on acceptance.

References

1 Slater, A., Scott, N.W. and Fowler, M.R. (2008) *Plant Biotechnology: The Genetic Manipulation of Plants*, Oxford University Press, Oxford.

2 Harling, H.(KWS) (2008) Logistik für Agrarrohstoffe an den Beispielen Stroh, Zuckerrüben und Weizen. Lecture, Dechema conference Industrielle Nutzung nachwachsender Rohstoffe, February, 18/19, 2008, Frankfurt, Germany.

3 Becker, T., Breithaupt, D., Doelle, H.W. et al. (2007) Biotechnology, in *Ullmann's Encyclopedia of Industrial Chemistry*, Wiley-VCH Verlag GmbH, Weinheim.

4 Phillips, P.W.B. (2003) Development and commercialization of genetically modified plants in: Thomas et al., 2003 [87], pp. 273–279.

5 Bariana, H.S. (2003) Breeding for disease resistance. in: Thomas et al., 2003 [87], pp. 244–253.

6 Plummer, K.M. (2003) Genetic modification of disease resistance, fungal pathogens. in: Thomas et al., 2003 [87], pp. 253–257.

7 Aldrige, S. (2008) Plant biotechs defect. *Nature Biotechnology*, 26, 255.

8 Ammann, K. (2008) Balance zwischen Risiko und Nutzen (Interview). *CHEManager* 20, 2008, 1, 2.

9 Chem. Market Reporter (2004) European Union approves GM maize. *Chemical Market Reporter*, **265** (21), 6.

10 Anonymus (2008) EU lays tracks for future GMO policy. *European Biotechnology Science & Industry News*, **7** (5–6), 5.

11 Lassen, J., Madsen, K.H. and Sandoe, P. (2002) Ethics and Genetic Engineering-lessons to be learned from GM foods. *Bioprocess and Biosystems Engineering*, **24**, 263–271.

12 Teng, P.P. (2008) An Asian perspective on GMO and biotechnology issues. *Asia Pacific Journal Clinical Nutrition*, **17** (Suppl 1), 237–240.

13 ifpri (2008) International Food Policy Research Institute - www.ifpri.org/ Economic Considerations of Biosafety and Biotechnology Regulations in India. Proceedings of a Conference in New Delhi, India, 24–25 August 2006.

14 Süddeutsche Zeitung (2008) May 30, p. 19.

15 The New York Times (2008) June 23, pp. 1, 4.

16 von Braun, J. (2008) The food crisis isn't over. *Nature*, **456**, 701.

17 Modi, A. (2008) Global grain prices skid, local prices hold firm. *Business Standard*, December 15.

18 van den Daele, W., Pühler, A. and Sukopp, H. (1996) *Grüne Gentechnik im Widerstreit*, VCH, Weinheim (Germany).

19 Phillips, P.W.B. (2003) Ownership of plant genetic resources. in: Thomas *et al.*, [87], pp. 289–295.

20 Bud, R. (1994) In the engine of industry: Regulators of biotechnology, 1970–1986, in *Resistance to New Technology* (ed. M. Bauer), Cambridge University Press, Cambridge, pp. 293–309.

21 McHughen, A. (2007) Public perceptions of biotechnology. *Biotechnology Journal*, **2**, 1105–1111.

22 Goodman, R.E., Vieths, S., Sampson, H.A. *et al.* (2008) Allergenicity assessment of genetically modified crops – what makes sense? *Nature Biotechnology*, **26**, 73–81.

23 Jones, H.D. and Cannell, M. (2003) Transgene stability and inheritance. in: Thomas *et al.*, [87], pp. 402–405.

24 Cooper, J.I. and Walsh, J.A. (2003) Genetic modification of disease resiastance, viral pathogens. in: Thomas *et al.*, [87], pp. 257–262.

25 Lazzeri, P.A. and Barcelo, P. (2003) Transformation in monocotyledons. in: Thomas *et al.*, [87], pp. 383–392.

26 Scheid, O.M. (2004) Either/or selection markers for plant transformation. *Nature Biotechnology*, **4**, 398–399.

27 Bradford, K.J., Van Deynze, A., Gutterson, N., Parrott, W., and Strauss, S.H. (2005) Regulating transgenic crops sensible: lessons from plant breeding, biotechnology and genomics. *Nature Biotechnology*, **23**, 439–444.

28 Hehl, R. (2010) personal communication.

29 Aslaksen, I. and Ingeborg Myhr, A. (2007) "The worth of a wildflower": Precautionary perspectives on the environmental risk of GMOs. *Ecological Economics*, **60**, 489–497.

30 Levidow, L. and Boschert, K. (2008) Coexistence or contradicion? GM cops versus alternative agriultures in Europe. *Geoforum*, **39**, 174–190.

31 Munro, A. (2008) the spatial impact of genetically modified crops. *Ecological Economics*, **67** (4), 658–666.

32 Daniell, H. (2002) Molecular strategies for gene containment in transgenic cops. *Nature Biotechnology*, **20**, 581–584.

33 Sanvido, O., Romeis, J. and Bigler, F. (2007) Ecological impacts of genetically modified crops: Ten years of field research and commercial cultivation. *Advances in Biochemical Engineering/Biotechnology*, **107**, 235–278.

34 Schmeller, D.S. and Henle, K. (2008) Cultivation of genetically modified organisms: resource needs for monitoring adverse effects on biodiversity. *Biodiversity and Conservation*, **17**, 3551–3558.

35 Halford, N. (2006) *Plant Biotechnology*, John Wiley & Sons, New York, USA.

36 Stewart, P.A. and Mclean, W.P. (2008) Public perceptions of benefits from and worries over plant-made industrial products and plant-made pharmaceuticals: The influence of institutional trust. *Review of Policy Research*, **25**, 333–348.

37 Glenn, K.C. (2008) Nutritional and safety assessment of foods and feeds nutritionally improved through biotechnology – case studies by the International Food biotechnology Committee of ILSI. *Asia Pacific Journal Clinical Nutrition*, **17** (Suppl 1), 229–232.

38 Delgado, A. (2008) Opening up for participation in agro-biodiversity Conservation: the expert-lay interplay in a Brasilian social movement. *Journal of*

Agricultural and Environmental Ethics, **21**, 559–577.

39 Shewry, P.R., Jones, H.D. and Halford, N.G. (2008) Plant biotechnology: transgenic crops. *Advances in Biochemical Engineering/Biotechnology*, **111**, 149–186.

40 Ritter, S.K. (2008) Genes to gasoline. *Chemical and Engineering News*, December 8, 10–17.

41 Riano-Pachón, D.M., Nagel, A., Neigenfind, J. *et al.* (2008) GABIPD: the GABI primary database - plant integrative "omics database". *Nucleic Acids Research* 37 (Issue Suppl. 1, D954–D959).

42 Vasil, I.K. (2008) A short history of plant biotechnology. *Phytochemistry Reviews*, **7**, 387–394.

43 Bauer, E. and Wenzel, G. (2008) Mit Biotechnologie zur besseren Pflanze für die Rohstoffnutzung. Lecture, Dechema conference Industrielle Nutzung nachwachsender Rohstoffe, February, 18/ 19, 2008, Frankfurt, Germany.

44 Ziv, M. and Altman, A. (2003) Tissue culture. General principles. In: Thomas *et al.*, [87], pp. 1341–1353.

45 Twyman, R.M. (2003) Gene cloning, general principles. In: Thomas *et al.*, [87], pp. 359–369.

46 Schwarz, H. (2009) Eine Frage der Akzeptanz. Bayer entschlüsselt Raps-Genom, Süddeutsche Zeitung, 10/ 11.Oktober, p. 27.

47 Murphy, D.J. (2003) Transformation in dicotyledons. In: Thomas *et al.*, [87], pp. 382–3.

48 Kunz, M. (1995) personal communication.

49 Vasil, I.K. (2003) The science and politics of plant biotechnology. *Nature Biotechnology*, **21**, 849–851.

50 Meyers, B.C., Green, P.J. and Lu, C. (2008) miRNAs in the plant genome: All things great and small. *Genome Dynamics*, **4**, 1008–1118.

51 Wenning, W. (2008) Science for a better Life. *CHEManager*, **6**, 21.

52 Kant, P., Gordon, M., Kant, S., Zolla, G. Davidov, O., Heimer, Y., Chalifa-Casp, V., Shaked, R., and Barak, S. (2008) Functional-genomics-based identification of genes that regulate Arabidopsis responses to multiple abiotic stresses. *Plant, Cell & Environment*, **31**, 697–714.

53 Cramer, H.-H. (2002) Crop protection, in *Ullmann's Encyclopedia of Industrial Chemistry*, Wiley-VCH Verlag GmbH, Weinheim, New York, www.mrw .interScience.wiley.com/emrw/.

54 Short, P. (2007) An agrochem rebound. *Chemical and Engineering News*, October 1, 23–25.

55 Kwon, C., Panstruga, R. and Schulze-Lefert, E. (2008) *Trends in Immunology*, **29**, 159–166.

56 Bittel, P. and Robatzek, S. (2007) Microbe associated molecular patterns (MAMPs) probe plant immunity. *Current Opinion in Plant Biology*, **10**, 335–341.

57 Jones, J.D. and Dangl, J.L. (2006) The plant immune system. *Nature*, **444**, 323–329.

58 Zipfel, C., Robatzek, S., Navarro, L. *et al.* (2004) Bacterial diease resistance in *Arabidopsis* through flagellin reception. *Nature*, **428**, 764–767.

59 Everts, S. (2008) Fungal warfare. *Chemical and Engineering News*, December 22, 8.

60 Bednarek, P., Pislewska-Bednarek, M., Svatos, A. *et al.* (2009) A glucosinolate metabolic pathway in living plant cells mediates broad-spectrum antifungal defence. *Science*, **323**, 101–106.

61 Wu, L. and Birch, R.G. (2007) Doubled sugar content in sugar cane plants modified to produce a sucrose isomer. *Plant Biotechnology Journal*, **5**, 109–117.

62 Chibbar, R.N. and Baga, M. (2003) Carbohydrates. In: Thomas *et al.*, [87], pp. 449–459.

63 Hirschi, K. (2008) Nutritional improvements in plants. *Trends in Plant Science*, **13**, 459–463.

64 Powell, K. (2007) Functional foods from biotech-an unappatizing prospect? *Nature Biotechnology*, **25**, 525–531.

65 Arnold, C. (2008) Fish out of water. *Chemical and Engineering News*, August 11, p. 39–41.

66 Yuan, J.S., Tiller, K.H., Al-Ahmad, H., Stewart, N.R. and Stewart, C.N. Jr. (2008) Plants to power: Bioenergy to fuel the future. *Trends in Plant Science*, **13**, 421–429.

67 Bonga, J.M. and Park, Y.S. (2003) Clonal propagation, forest trees. In: Thomas *et al.*, [87], pp. 1395–1402.

68 Boudet, A.-M. (2003) Wood quality. In: Thomas *et al.*, [87], pp. 504–512.

69 *Nature Biotechnology* (2005) Editorial: Reburnishing Golden Rice, **23**, 395.

70 Ullmann (1915) *Enzyklopädie der Technischen Chemie*, vol. **1**, Urban & Schwarzenberg, München, Berlin, pp. 223–254.

71 Kinghorn, A.D. and Chávez, D. (2003) Pharmaceuticals, plant drugs. In: Thomas *et al.*, [87], pp. 1145–1152.

72 Twyman, R.M., Verporte, R., Memelink, J. and Christou, P. (2003) Genetic modification of secondary metabolism. Alkaloids. In: Thomas *et al.*, [87], pp. 493–500.

73 Hühns, M., Neumann, K., Hausmann, T. *et al.* (2009) Tuber specific *cph*A Expression to enhance cyanophycin production in potatoes. *Plant Biotechnology Journal*, **7**, 883–898.

74 Cox, K.M., Sterling, J.D., Regan, J.T. *et al.* (2006) Glycan optimization of a human monoclonal antibody in the aquatic plant *Lemna minor. Nature Biotechnology*, **24**, 1591–1597.

75 Wang, Y. (2008) Needs for new plant-derived pharmaceuticals in the post-genome era. An industrial view in drug research and development. *Phytochem Reviews*, **7**, 395–406.

76 transcript (2006) 12, No. 3, p. 31.

77 transcript (5 2008), 14, Nr.3, p. 38.

78 *CHEManager*, (3/2008) p. 3.

79 FAS (4 2009) Frankfurter Allgemeine Sonntagszeitung, April 19, p. 38.

80 Short, P.L. (2008) Building a towering success. *Chemical and Engineering News*, March 17, 26–29.

81 Voith, M. (2003) Agriculture saves chemical earnings. *Chemical and Engineering News*, August 18, 17–19.

82 Voith, M. (2008) BASF is betting on the farm. *Chemical and Engineering News*, May 26, 24–25.

83 LeMonde (2001) 7 September, p. 22.

84 Bryner, M. (2008) Agricultural biotech. Chemical Week, 170, issue 17, June 2.

85 *CHEManager*, (19/2008) p. 6.

86 Huso, S.R. and Wilson, W.W. (2006) Producer surplus distributions in GM crops: the ignored impacts of Roundup Ready wheat. *Journal of Agricultural and Resource Economics*, **31** (2), 339–354.

87 Thomas, B., Murphy, D.J., and Murray B.G. (eds) (2003) *Encyclopedia of Applied Plant Sciences*, vol. 3, Elsevier, Oxford.

88 Hehl, R. and Bülow, L. (2008) Internet resources for gene-expression analysis in *Arabidopsis thaliana. Curr. Geneomics*, **9**, 375–380.

89 Fukushima, A., Kusano, M., Redestig, H., Arita, M., and Saito, K. (2009) Integrated omics approaches in plant systems biology. *Current Opinion in Chemical Biology*, **13**, 532–538.

90 Chemical & Engineering News (2010), March 8, p.18.

91 Voith, M. (2010) Growing chemicals. Chemical & Engineering News, June 7, pp. 25–28.

92 Voith, M. (2010) Chemical & Engineering News, April 4, p. 9.

93 Falck-Zepeda, J.B., Traxler, G., and Nelson, R.G. (2000) Recent creation and distribution from biotechnology innovations. *Agribusiness*, **16**, 21–32.

Index

Concepts in Biotechnology: History, Science and Business. Klaus Buchholz and John Collins
Copyright © 2010 WILEY-VCH Verlag GmbH & Co. KGaA, Weinheim
ISBN: 978-3-527-31766-0